Genetics and Physiology of *ASPERGILLUS*

THE BRITISH MYCOLOGICAL SOCIETY
SYMPOSIUM SERIES NO. 1

Genetics and Physiology of *ASPERGILLUS*

edited by

JOHN E. SMITH
Department of Applied Microbiology,
University of Strathclyde, Glasgow

and

JOHN A. PATEMAN
Institute of Genetics,
University of Glasgow, Glasgow

1977

ACADEMIC PRESS
London: New York: San Francisco
A Subsidiary of Harcourt Brace Jovanovich, Publishers

ACADEMIC PRESS INC. (LONDON) LTD.
24/28 Oval Road,
London NW1

United States Edition published by
ACADEMIC PRESS INC.
111 Fifth Avenue
New York, New York 10003

Library of Congress Catalog Card Number: 77–71836
ISBN: 0–12–650960–3

Printed in Great Britain by Galliard (Printers) Ltd, Great Yarmouth, Norfolk

CONTRIBUTORS

AL OBAIDI, Z., Department of Applied Microbiology, University of Strathclyde, Glasgow, G1 1XW.

AL ZUBAIDY, T.S., Department of Biology, Science College, Mosul University, Iraq.

ANDERSON, J.G., Department of Applied Microbiology, University of Strathclyde, Glasgow, G1 1XW.

ARST, H.N., Department of Genetics, University of Cambridge, Cambridge, CB2 3EH.

BAILEY, C.R., Department of Genetics, University of Cambridge, Cambridge, CB2 3EH.

BARBESGAARD, P., Novo Industri A/S, DK-2880 Bagsvaard, Denmark.

BARTNIK, E., Department of Genetics, University of Warsaw, Warsaw, Poland.

BERRY, D.R., Department of Applied Microbiology, University of Strathclyde, Glasgow, G1 1XW.

BIRKETT, J.A., Department of Genetics, University of Sheffield, Sheffield.

CHMIEL, A., Technical University, Lodz, Poland.

CLUTTERBUCK, A.J., Department of Genetics, University of Glasgow.

COCKER, R., Department of Applied Microbiology, University of Strathclyde, Glasgow, G1 1XW.

COHEN, B.L., Department of Genetics, University of Glasgow, Glasgow.

COVE, D.J., Department of Genetics, University of Cambridge, Cambridge, CB2 3EH.

CROFT, J.H., Department of Genetics, University of Birmingham, Birmingham.

DAVIS, B., Department of Biology, University of Sheffield, Sheffield.

DEANS, S.G., Department of Applied Microbiology, University of Strathclyde, Glasgow, G1 1XW.

FELDEN, R.A., Department of Pharmacology, C.M.D.N.J. - Rutgers Medical School, Piscataway, New Jersey 08854, U.S.A.

FENNELL, D.I., Northern Regional Research Center, A.R.S. U.S. Department of Agriculture, Peoria, Illinois 61604, U.S.A.

GEALT, M.A., Department of Pharmacology, C.M.D.N.J. - Rutgers Medical School, Piscataway, New Jersey 08854, U.S.A.

GORTON, D., Department of Biology, University of Essex, Colchester, Essex.

GREENSHIELDS, R.N., Department of Biological Sciences, University of Aston, Birmingham.

GUZEWSKA, J., Department of Genetics, University of Warsaw, Warsaw, Poland.

EDWARDS, J.H., MRC Pneumoconiosis Unit, Llandough Hospital, Penarth, Glamorgan.

JAKUBOWSKA, J., Institute of Fermentation Technology and Microbiology, Technical University, Lodz, Poland.
JINKS, J.L., Department of Genetics, University of Birmingham, Birmingham.
KINGHORN, J.R., Department of Genetics, University of Glasgow, Glasgow.
KLIMCZUK, J., Department of Genetics, University of Warsaw, Warsaw, Poland.
McCULLOUGH, W., Department of Genetics, University of Leicester, Leicester, LE1 7RH.
MORRIS, N.R., Department of Pharmacology, C.M.D.N.J. - Rutgers Medical School, Piscataway, New Jersey 08854, U.S.A.
MOSS, M.O., Department of Microbiology, University of Surrey, Guildford, Surrey.
NARDI, R.V., Department of Pharmacology, C.M.D.N.J. - Rutgers Medical School, Piscataway, New Jersey 08854, U.S.A.
PATEMAN, J.A., Department of Genetics, University of Glasgow, Glasgow.
PAYTON, M.A., Department of Genetics, University of Leicester, Leicester, LE1 7RH.
PIOTROWSKA, M., Department of Genetics, University of Warsaw, Warsaw, Poland.
ROBERTS, C.F., Department of Genetics, University of Leicester, Leicester, LE1 7RH.
ROPER, J.A., Department of Genetics, University of Sheffield, Sheffield.
ROWLANDS, R.T., Department of Bacteriology, University of Bristol, Bristol.
SANDERS, M.M., Department of Pharmacology, C.M.D.N.J. - Rutgers Medical School, Piscataway, New Jersey, 08854, U.S.A.
SCAZZOCCHIO, C., Department of Biology, University of Essex, Colchester, Essex.
SHEIR-NEISS, G., Department of Pharmacology, C.M.D.N.J. - Rutgers Medical School, Piscataway, New Jersey 08854, U.S.A.
SMITH, J.E., Department of Applied Microbiology, University of Strathclyde, Glasgow, G1 1XW.
THOMAS, A., Department of Applied Microbiology, University of Strathclyde, Glasgow, G1 1XW.
TURNER, G., Department of Bacteriology, University of Bristol, Bristol.
WEGLENSKI, J., Department of Genetics, University of Warsaw, Warsaw, Poland.
WOOD, B.J.B., Department of Applied Microbiology, University of Strathclyde, Glasgow, G1 1XW.
ZONNEVELD, B.J.M., Genetisch Laboratorium der Rijksuniversiteit, Kairerstraat 63, Leiden, The Netherlands.

EDITORS' PREFACE

The Symposium held in Birmingham in September 1976 under the auspices of the British Mycological Society brought together for the first time workers representative of most of the main areas of current Aspergillus research. A feature of the meeting was the realisation by many of us of just how wide was the range of research areas to which Aspergillus workers were making a contribution. This diversity is reflected in the articles collected here and emphasises the value of a volume containing accounts which would normally be widely scattered in specialist journals.

The aspergilli have a worldwide distribution and are undoubtedly one of the most ubiquitous of all groups of fungi. As filamentous eukaryotic organisms they represent an important link between the truly unicellular state and the multicellular condition of higher organisms. It must be anticipated that a fuller understanding of the factors which regulate growth and development in the aspergilli could well be of major importance and relevance not only to fundamental studies on differentiation in eukaryotes but also to specific industrial processes utilizing these organisms.

The study of *Aspergillus* genetics effectively began with the classical work of Pontecorvo and his collaborators on *A. nidulans* in the 1950's. Subsequently a number of small and widely dispersed groups have made considerable progress in the interrelated areas of physiology, biochemistry and genetics. The eight articles on various aspects of metabolism and genetic regulation in *A. nidulans* provide a unique coverage both comprehensive and up to date of the important recent advances in these areas. The particular facility of *A. nidulans* for studies of the regulation of gene expression in a simple eukaryote is a key theme in these contributions. Other articles such as those on mitochondrial genetics and population genetics give an authoritative and exciting foretaste of advances to come in hitherto rather neglected areas of *Aspergillus* genetics.

The aspergilli are possibly the greatest contaminants of man-made and natural organic products and some species can cause infections in man and animals. Certain aspergilli can form mycotoxins during growth as contaminants of cereal and other food products.

The aspergilli are not entirely destructive in nature and several species have had their metabolic powers harnessed for commercial exploitation.

We thank our contributors for producing manuscripts which were such a pleasure for us to read and to Mrs. Evelyn Smith and Mrs. Margaret Provan we are indebted for typing and arranging the manuscripts with intelligence and precision.

Department of Applied Microbiology JOHN E. SMITH
University of Strathclyde

Department of Genetics JOHN A. PATEMAN
University of Glasgow

CONTENTS

Aspergillus Taxonomy

D. I. FENNELL

Northern Regional Research Center, Agricultural Research Service, U.S. Department of Agriculture, Peoria, Illinois, U.S.A.

Taxonomy, defined by Webster as classification according to natural relationships, has been for the most part a descriptive and nonexperimental science made up of generalizations based on judgments of observed habits of nature as Rogers philosophized in 1958. Taxonomy of *Aspergillus* has been no exception. Delineation and classification of species within the genus have been based on differences in cultural and morphological characters (Raper & Fennell, 1965) that admittedly are the visible expressions of only a fraction of their genomes. With the advent of molecular biology, mycological taxonomy may be at the threshold of major advances.

BIOCHEMISTRY

For many years, Japanese investigators have been using various physiological characteristics in combination with morphological differences to distinguish between species of *Aspergillus* (Ohara, 1951, 1952, 1953; Nehira & Nomi, 1956; Nehira, 1957). Because of the widespread use of *A. oryzae* and related taxa in oriental fermented foods, their interest has been directed primarily to the members of the *A. flavus* group. More recently they have been understandably anxious to show that the strains used commercially in their food production processes do not produce aflatoxins. On the basis of computer analysis of a large number of components (including morphological characteristics, ability to grow on anisaldehyde medium, change of conidial colour on this medium, kojic acid production, browning of rice koji and aflatoxin production) Murakami (1970a,b,1971a,b,c,1972a,b,c) and his colleagues (Murakami *et al.*, 1970; Murakami & Nasuno,1972) believe they can separate the species within the group. Hesseltine, Sorenson & Smith (1970), however, found it impossible to separate species of the *A.flavus* group on the basis of: utilization of various carbon sources, production of kojic acid, production of aflatoxins, production of amylase, temperature response, and sclerotium production at various concentrations of $NaNo_3$ and sucrose.

Cell Wall Composition

Few of the increasingly numerous reports of application of molecular biological approaches to fungal phylogenetics deal directly with *Aspergillus* taxonomy. In Bartnicki-Garcia's review (1968) of fungal cell wall chemistry as it relates to morphogenesis and taxonomy, Aspergilli are classified in the chitin-glucan cell wall category, a large and diverse group that includes the Chytridiomycetes, Ascomycetes, Basidiomycetes,* and Deuteromycetes.* To my knowledge, no general survey of *Aspergillus* species has been made, and inclusion of *Aspergillus* in this great conglomerate of fungi has resulted from study of only a limited number of species *A.oryzae* (Horikoshi & Iida, 1964), *A.phoenicis* (Bloomfield & Alexander, 1967), *A.parasiticus* (Distler & Roseman, 1960), *A. niger* (Johnston, 1965), and *Aspergillus* sp. (Ruiz-Herrera, 1967). Investigators generally agree (Bartnicki-Garcia, 1968) that N-acetylglucosamine, glucose, mannose, and galactose are present in *Aspergillus* cell walls. Galactosamine has not always been reported. Mycelium of *A.parasiticus* was found to have the highest content of galactosamine glycan of the fungi tested by Distler & Roseman (1960). It is doubtful that cell wall components vary sufficiently to be taxonomically significant at the species level but sugar ratios of alkali-soluble fractions of cell walls may be useful at the generic level (O'Brien & Ralph, 1966; Jones, Farmer, Bacon & Wilson, 1972).

Proteins

When Hall (1969) reviewed molecular approaches to taxonomy of the fungi, he believed that the value of gel electrophoretic protein patterns in delimiting species of fungi had been abundantly confirmed *Aspergillus* was referenced only once. Nealson & Garber (1967) had

*Except certain families of yeasts.

by then concluded that esterase and phosphatase zymograms obtained by this technique proved a biochemical means of distinguishing between 15 species of *Aspergillus* represented by 32 strains. Invariant enzyme sites were found for four of the five species represented by more than one strain. Of eight strains of *A. rugulosus* one yielded esterase and phosphatase zymograms strikingly different from the other seven and was considered to be misidentified. Morphological comparisons supported this conclusion. Garber & Rippon (1968) and Clare, Flentje & Atkinson (1968) reviewed the literature on the use of protein profiles and zymograms in fungal taxonomy and concluded that these electrophoretic patterns offered promise.

Garber (1973,1974) and his students at the University of Chicago have since applied electrophoretic techniques in obtaining enzyme profiles of a number of *Aspergillus* species and enthusiastically endorse the value of this method to taxonomy of the genus.

Beebe's(1970) report that amylolytic patterns of strains assignable to *Aspergillus carbonarius*, *A.foetidus* and *A.tubingensis* were comparable to those of typical strains of *A.niger* and the apparent corroboration of this finding by induction of heterokaryons or vegetative diploids in all possible combinations of the four species(Hayhome, 1970) are difficult to interpret in view of the significant morphological differences between the four species.

Kurzeja & Garber (1973), by acrylamide gel electrophoresis of extracellular amylolytic enzymes, detected 6-9 sites of amylolytic activity accommodated by 9 patterns among 90 strains assigned to *Aspergillus nidulans* and four of its varieties. Thirty-six of the strains showed one pattern and 45 a second pattern. Only one or two strains fell into each of the remaining seven patterns. The three varieties, *latus*, *echinulatus*, and *acristatus* each displayed a unique pattern; variety *dentatus* fell in pattern 2. Production of fertile cleistothecia in crosses between strains showing the two predominant patterns, and the similarity of their esterase and phosphatase zymograms, led these investigators to conclude that only these strains were *A.nidulans*. The remaining strains were considered sibling or variant species.

Nasuno (1971,1972a,b,1974) has used polyacrylamide gel disc electrophoresis to demonstrate significant differences in mobilities of the alkaline and acid proteinases, cellulases, and pectinlyases of *Aspergillus oryzae* and *A.sojae*. However, electrophoretic patterns of amylase, neutral proteinase, phosphatase, α and β glucosidase, α and β galactosidase, and leucine aminopeptidase were similar in both species and of no value for their differentiation.

Fractionation of soluble proteins by polyacrylamide gel electrophoresis was used by Kulik & Brooks (1970) in attempts to define seven species of *Aspergillus:* three strains each of *fumigatus*, *flavus*, *oryzae*, *parasiticus*, *ochraceus*, *tamarii*, and *leporis*.

Although they found two soluble protein fractions common to all seven species, they hesitated to accord generic significance to their presence on the basis of this small sample. Two other protein fractions, found in all strains of the five species of the *flavus* group, supported the phylogenetic soundness of the group concept. However, a comparison of homologous protein fractions found in composites of strains of each species appeared to indicate that *A.parasiticus* is as closely related to *fumigatus* as it is to *oryzae*.

Much more data will be required before electrophoretic mobility patterns of proteins and specific enzymes can be confidently applied to generic and specific determinations. Changes in enzyme activities that have been correlated with stages of differentiation in *Aspergillus* species (Smith & Anderson, 1973), and the fact that extraction procedures disrupt organelles and permit normally compartmentalized enzymes and metabolites to meet and react to cause gross artifacts, suggest the need for caution in interpretation of results. Credibility of electrophoretic studies also hinges on the accuracy of initial identifications of strains and on the inclusion of a reasonably large number of representatives of each taxon. Standardized techniques will be essential to acceptance of this tool by taxonomists.

Pyrolysis Products

Pyrolysis-gas-liquid chromatography (PGLC) has been used to characterize bacteria (Reiner, 1965,1967) and fungi (Oyama & Carle, 1967). Vincent & Kulik (1970, 1973) have examined the potential application of computer comparison of pyrolytic elution patterns, which reflect actual differences in chemical composition of the micro-organisms, to species and strain differentiation of members of the *Aspergillus flavus* group. Differences between strains of both *A.flavus* and *A.oryzae* appeared to be as great as differences between the two species. Composited pyrochromatograms for all strains of each of the four species (*flavus, oryzae, parasiticus, tamarii*) shared 23 peaks. *A.flavus* and *A.tamarii* showed a 58% similarity value, closest of all. Comparison of similarity values again showed *A.parasiticus* to be least similar to the other species, as had electrophoresis of protein fractions. *A.parasiticus* is also morphologically unique; it is the one species in the *A.flavus* group with strictly uniseriate phialides.

Vincent & Kulik (1973) also applied numerical analysis to the interpretation of PBLC data in characterizing species variation among three strains each of four species of the *glaucus* group. Their results correlated closely with conventional taxonomic placement of these species. However, strains of *Aspergillus ruber* showed less intraspecific similarity than interspecific similarity; this suggested to the authors "that some strains of *A. ruber* might be more closely related to other *A. glaucus* group members or to other groups of fungi" than they are to one another. The intergroup similarity between these 12 strains of the *A. glaucus* group was compared to *A. flavus* group speces. Again, *A. parasiticus* was unique, showing characters in common with strains of the *A. glaucus* group.

Strangely, in intergeneric comparisons of composite pyrolysis profiles of conidia from nine *Penicillium* species of the Asymmetrica-Fasciculata with those from members of the *A. flavus* group, *A. parasiticus* showed a somehwat higher similarity value to the nine *Penicillium* spp. than to the other three species in the *flavus* group (Kulik & Vincent, 1973).

Although reproducibility of pyrolysis profiles has been demonstrated for the bacteria (Reiner, 1967), nonreproducibility of pyrograms of fungi has plagued investigators (Carmichael, Sekhon & Sigler, 1973; Sekhon & Carmichael, 1971, 1972; Burns, Stretton & Jayatilake, 1976). Variation between replicates is often as great as between species. Apparently homogeneity is more readily achieved in a bacterial population than in fungi where physiological age of different portions of the mycelium varies.

It is regrettable that difficulties encountered and time consumed in application of PGLC to fungal taxonomy so limits the number of strains included in any one study that results of the investigation are of questionable value. If reproducibility is so elusive with single strains of a single species, within a single lab and using the same apparatus, is there hope for comparing results between laboratories? Will there ever be a vast body of information collated from all investigations that will yield information useful for automated taxonomy?

Could it be that the term 'fingerprinting by PGLC' is more apt than intended -- that every strain of every species of every genus will be as unique as in *Homo sapiens?*

Nucleic Acids

Molecular analysis of the genetic code can be approximated by analysis of nuclear DNA. Determination of the relative proportions of the four bases: adenine, thymine, guanine, and cytosine, and their order in the DNA strand are of potential taxonomic importance. DNA homology analysis has been applied with impressive success to taxonomic problems in bacteria (DeLey, 1968), but until very recently little application has been made to taxonomy of the fungi.

Storck, with various colleagues, has been most actively engaged in the study of RNA and the various DNAs of fungi as they relate to taxonomy and phylogeny. In 1965, he reported guanine and cystine

(GC) contents of RNA (expressed as moles per cent) ranging from 44.1-49.3 for the Zygomycetes, 47.4-54.4 for the Ascomycetes, 48.2-54.5 for the Deuteromycetes, and 50.4 to 60.5 for the Basidiomycetes.

Comparison of the nucleotide composition of total, soluble, and ribosomal RNA, the subject of Storck's earlier investigations, had shown that soluble RNA of *Neurospora crassa* differed from both total RNA and ribosomal RNA in having more guanylic and cytidylic acids and in containing pseudourylic acid. In *N.crassa* the ratio of cytidylic (36%) to guanylic (25%) acids is reversed in comparison with other organisms, including yeasts, and may be a unique and significant characteristic (Henney & Storck, 1963a,b). In *Aspergillus niger*, ribosomal RNA, like total RNA, was found to contain 53% guanylic plus cytidylic acids while soluble RNA contains 58%, with the increase being due to cytidylic acid (Moyer & Storck, 1964).

Molecular weights of the two components of ribosomal RNA of a limited number of fungi representing all major classes were determined (Lovett & Haselby, 1971) by electrophoresis in acrylamide gel. The molecular weight of the 18s RNA was found to be similar for all fungi (0.71-0.75 million); whereas, the 25s RNA molecular weights ranged from 1.45 million in the Myxomycetes to 1.30-1.31 million for the Ascomycetes and Basidiomycetes.

Storck (1966) reported GC contents of DNA to be 38-48% for 9 species of Zygomycetes; 38-54% for 14 Ascomycetes; 47-62% for 9 Deuteromycetes, and 44 to 63% for the 9 Basidiomycetes.

Villa & Storck (1968) found by CsCl density gradient equilibrium centrifugation that the per cent GC of nuclear (nu-DNA) and mitochondrial DNA (mit-DNA) was the same for three Mucorales (two genera; three species), whereas average differences of 15 and 29% were detected for nine species of Ascomycetes and two species of Basidiomycetes, respectively, with mit-DNA being lower than nu-DNA. These findings were interpreted as indicating that the base composition of mitochondrial DNA, as well as that of nuclear DNA, might have phylogenetic significance.

Following a survey of 492 isolates covering 322 species of Oömycetes, Zygomycetes, Ascomycetes, and Deuteromycetes, including 41 species of *Aspergillus*, Storck & Alexopoulos (1970) concluded that, on the basis of its population of DNA molecules, a genus is characterized by a standard deviation value of less than 10% and, on the average, close to 3%. For the species of *Aspergillus* included, GC contents ranged from 48 to 61%. Except for *A.conicus* with 61% GC, all would have fallen within 10 percentage points of one another.

The G +C content of DNA has been found to range from 50.0-51.0% for the Hymenochaetaceae and from 51.0-59.5% for the Polyporaceae. The authors (Storck,Nobles & Alexopoulos, 1971) concluded that, although their results suggested the possible taxonomic significance of the GC content of DNA at the species, generic, family, and higher levels in the Homobasidiomycetes, their sample of three species in the Hymenochaetaceae and 29 in the Polyporaceae was too small to provide a basis for firm conclusions.

It is of interest to note that the GC range from 50-60 is not unique for the Homobasidiomycetes. Storck (1966) and Storck, Alexopoulos & Phaff (1969) had found that for all species studied of the yeast genera *Cryptococcus, Rhodotorula,* and *Sporobolomyces* per cent GC of DNA averaged 58.1 (none under 50) as compared with 35-48 for ascosporogenous and anascosporogenus yeasts. However, all three yeast genera are now known to have basidiomycetous perfect states.

Amino Acid Biosynthesis

In a study of 22 species of fungi, Hütter & DeMoss (1967), identified five different association patterns of the specific enzymes involved in the synthesis of tryptophan. The Eurotiales, as represented by one strain each of *Aspergillus nidulans* and *Byssochlamys nivea,* shared the Type I pattern with the Chytridiales, as represented by *Rhizophlyctis rosea,* and all higher Ascomycetes and Holobasidiomycetes tested. On the basis of these findings, the authors postulated a more direct relationship between the Chytridiales and Eurotiales than had previously been suggested, with the Zygomycetes and Endomycetales being eliminated as intermediates.

The Aspergilli, or at least the Plectascales, have been shown (Vogel, 1965) to share with most other fungi the α-aminoadipic pathway for lysine biosynthesis, as contrasted with the diaminopimelic acid pathway present in bacteria, blue-green and green algae, vascular plants, and the fungal orders Saprolegniales, Leptomitales, Lagenidales, Peronosporales (all Oömycetes) and the Hyphochytriales. This biosynthetic difference suggests evolution along two lines for the two groups of fungi.

Hydrocarbon Metabolism

The literature on aliphatic hydrocarbons of fungi has been reviewed by Weete (1972). First evidence of the presence of paraffinic hydrocarbons in fungi was reported by Baker & Strobel (1965) when they identified an homologous series of n-alkanes extracted from uredospores of *Puccinia striiformis.* The following year, Oro, Laseter & Weber (1966) found that the alkane components of two species of *Ustilago* and one of *Sphacelotheca,* determined by gas chromatography, differed sufficiently to suggest the possibility that hydrocarbon patterns might offer a method of taxonomic differentiation of plant pathogens.

Investigations undertaken by Jones (1969) to determine the role microbial alkanes play in the hydrocarbon spectrum of the soil showed average fungal lipids to be higher and hydrocarbon patterns less variable than those in bacteria. Major hydrocarbon peaks were observed on the C_{29} region, with a minor component in the C_{17-20} range. The C_{odd} to C_{even} ratio of the paraffins in *Aspergillus* and *Penicillium* tended to unity, while those of *Trichoderma viride* contained an odd to even ratio of less than one, and algal paraffin showed a strong C_{odd} predominance.

Both alkane synthesis and utilization by *Aspergillus nidulans* have been verified by Bird, Lynch & Pirt (1974). An homologous series of n-alkanes from C_{15}-C_{35} and squalene were detected by GLC of a chloroform:alcohol extract of the mycelium from which polar materials had been removed by chromatography on alumina.

The major saturated fatty acids found by Dart (1975), in conidia of six distinctive species of *Penicillium*, were palmitic and stearic; the major unsaturated were three C_{18} fatty acids that co-chromatographed with oleic, linoleic, and α-linolenic acids. In addition, all species showed traces of n-C_{14}, n-C_{15}, and C_{17}, except *P. chrysogenum* which contained 0.4% of the C_{17}. No unusual fatty acids were found, apart from trace amounts of odd-numbered saturated and unsaturated fatty acids.

Changes induced by temperature in the lipid composition of *Cephalosporium kiliense*, a pathogenic fungus (Sawicki & Pisano, 1976), suggests caution against the use of this character as a taxonomic criterion.

Inorganic Elements

Through use of X-ray microanalysis applied in wavelength dispersive spectroscopy, Thibaut & Ansel (1973) detected the presence of sodium and chloride in *Aspergillus niger* in addition to calcium, potassium, sulfur, phosphorus, and magnesium which were detected in both *A. fumigatus* and *A. niger*. The authors suggest that closely related species within the genus may be differentiated by the presence or absence of one rather unusual element and that relationships between species may be discovered through their chemical affinities.

Although Dalby & Gray (1974) attributed no taxonomic significance to their results, the distinct differences in the concentration of Zn in the mycelia and the Cu content of spores could be interpreted as showing generic correlations in their limited study of *Linderina pennispora*, *Aspergillus niger*, *Spicaria elegans*, *Penicillium claviforme*, and *P. roqueforti*.

MORPHOLOGY

Until a much larger body of biochemical information is available, identification of *Aspergillus* will continue, as in the past, to be based on their morphological, developmental, and cultural characterization. Whether or not one agrees with Thom & Church (1926), Thom & Raper (1945), and Raper & Fennell (1965) that classification in the *A.nidulans* group is just as meaningful as segregation in *Emericella* --and much more useful to nonmycologists --the morphology of the asexual states will continue as a primary characteristic for determination of placement within the genus or genera.

Raper & Fennell (1965) recognized 132 species and 18 varieties in the genus. These they classified in 18 groups of one to several species having certain basic characters in common. It should be emphasized that these 'groups' are not accredited taxa and serve only as convenient slots for fast characterization of strains.

The characters utilized for delineation of groups include:

Head colour - black,white,yellow,brown, and various shades of green.
Head shape - globose,radiate,loosely or compactly columnar.
Phialides in a single or double series - or,by presently more generally approved terminology - phialides developed directly from the vesicle or from metulae.
Vesicle shape - clavate,globose,subglobose,subclavate, turbinate,hemispherical, or terminally flattened and with or without a constriction in the conidiophore immediately below the vesicle.
Conidiophores - smooth or variously roughened,coloured or uncoloured.
Hülle cells - presence or absence and shape.

These major separating characters are usually clearly evident, and with this information alone, a group identification of an unknown, can be made with nearly complete confidence. There are exceptions,of course, and an occasional species shows characters that appear to align it with more than one group. In these instances, species are assigned where the preponderance of their characters indicate. All of the species described since 1965, except *Aspergillus bisporus* (Kwon-Chung & Fennell,1971) can be integrated in this group format. *A.bisporus*, with radiate heads, uniseriate phialides, black conidia, and globose hülle cells fits no place at the moment. Except for the presence of Hulle cells, *A.bisporus* suggests the *A.cervinus* group both morphologically and physiologically.

A perfect state is now known for at least some species in 7 of the 18 subdivisions and, in general, differences in ascocarps correlate with differences in the imperfect states. This statement is not as true as it once was. As the perfect states of more and more species of *Aspergillus* are discovered, the situation becomes more complex. Until the *Emericella* perfect state of *A.niveus* was described in 1973, no one perfect state had been found to be associated with more than one of Raper and Fennel's 'groups'. Species representative of the *A.glaucus* group have two different perfect states, while species classified in the *A.ornatus* group have four. These departures proved to Subramanian (1972) "that the classification of a form genus into groups, though extremely useful and suggestive of relationships sometimes, may not always be so" - and he may be right. These departures may also prove that the species are not properly grouped. *A.athecius* should perhaps have been segregated in a separate group with the nonascosporic and osmophilic species, *A.flaschentraegeri* (Stolk, 1964); this relationship was suggested by Raper & Fennell (1965). Their *A. ornatus* group was admittedly heterogeneous and it will be surprising if it is not further fragmented in the future. *A.niveus*, removed from the Thom & Raper (1965) *A.terreus* group and placed in the *A. flavipes* group by Raper & Fennell (1965), has recently been found to have a perfect state more closely related to that of the *A. nidulans* group than to that described for *A.flavipes* (Wiley & Fennell, 1973; Wiley & Simmons, 1973). The perfect state of *A. niveus* differs from that of *A.nidulans* in having hyaline to yellow rather than red-orange to violet-blue ascospores. Perhaps the ascospores of both *A.terreus* and *A.carneus* --when they are found--also will be hyaline to yellow.

Benjamin (1955), after comparison of the perfect stages of *Aspergillus* separated these forms into three sharply defined series which correlated with the 'groupings' of Thom & Raper (1945). As had others before him, Benjamin considered each of his "series" to be of generic rank and selected, from previously published names, *Eurotium* (Link, 1809) for some members of the *A.glaucus* group *Emericella* (Berkeley, 1857) for the *A.nidulans* Group, and *Sartorya* (Vuillemin, 1927) for the *A.fischeri* series of the *A.fumigatus* Group. The description of the ascosporic stage of *A.citrisporus* and *A.ornatus* (Raper,Fennell & Tresner, 1953) had apparently eluded him; no generic name was proposed and the species were not assigned to a perfect state genus. Raper (1957) objected to Benjamin's treatment. In 1959, Fennell and Warcup described the unique ascosporic state of *Aspergillus alliaceus*. Strangely enough, no one leaped into the breach with a new generic name until after Raper & Fennell (1965) had committed numerous similar public sins by describing (with Kwon) a number of new species of *Aspergillus* with three different ascosporic states. Even then, the reaction was slow --but, ultimately all-inclusive.

Sarbhoy & Elphick (1968) described the perfect state of *Aspergillus paradoxus* as *Hemicarpenteles paradoxa.* Subramanian (1972), accepting *Eurotium, Emericella, Sartorya* and *Hemicarpenteles* (the genus proposed by Sarbhoy & Elphick (1968) for the perfect state of *A.paradoxus)*, selected the name *Edyuillia* for *A.athecius* of the *A.glaucus* group, *Chaetosartorya* for the ascosporic species of the *A.cremeus* group, *Warcupiella* for *A.spinulosus*, *Sclerocleista* for *A.ornatus* and *A.citrisporus*, and *Syncleistostroma* for *A. alliaceus.*

Almost simultaneously, Malloc & Cain (1972a) accepted *Eurotium, Emericella,* and *Hemicarpenteles* but, recognizing the inadequacies of Vuillemin's *Sartorya,* proposed and described *Neosartorya* for the members of the *A.fischeri* series and assigned *A.ornatus* and *A.citrisporus* to this genus. Malloch & Cain (1972a) recorded an imperfect state in the *A.ornatus* group for the perfect state genus *Dichlaena (D. lentisci).* They further proposed *Gymnoeurotium* (for *A.athecius*), *Harpezomyces* (for *A.cremeus* group), *Sporophormis* (for *A.spinulosus*), and *Petromyces* (for *A.alliaceus*). Three of their four new generic names were obligate synonyms when published. Protected by Article 59 of the International Code of Botanical Nomenclature, *Petromyces* becomes the valid name for the ascocarpic stage of *A.alliaceus* replacing Subramanian's *Syncleistostroma.* If one subscribes to this fragmentation of *Aspergillus,* Malloch and Cain's *Neosartorya* should replace *Sartorya;* Subramanian's segregation of *A.ornatus* and *A.citrisporus* from *Neosartorya* was justified, and his genus *Sclerocleista* should stand.

Wiley & Simmons (1973) proposed and described the genus *Fennellia* for the perfect state of *A.flavipes* with a unique ascocarp initial. At the same time, the perfect state of *A.niveus* was assigned to *Emericella* as the first species of the genus to produce hyaline ascospores.

As a means of orienting the reader in this monumental maze, two keys are provided. The first is based on the relationship between Raper & Fennell's 'groups' and these genera, and the second on the characteristics of the perfect states.

Based on Imperfect State

Imperfect state: *A.glaucus* group
- Asci produced in the ascogonial coil; no ascocarp ---*Edyuillia*
- White to yellow thin-walled cleistothecia without investing hyphae --------------------------------*Eurotium*

Imperfect state: *A.ornatus* group
Ascocarps prosenchymatous; asci in nonperidiate locules *Warcupiella**
Asci produced in one nonperidiate locule within a pseudoparenchymatous stroma; ascospores with two equatorial crests *Hemicarpenteles*
Ascocarps peridiate, thin-walled, produced at 1-3 locules within a sclerotioid stroma *Dichlaena*
Ascocarp with pseudoparenchymatous peridium, produced in the dark, purple when mature *Sclerocleista*
Imperfect state: *A.fumigatus* group *Neosartorya*
Imperfect state: *A.ochraceus* group *Petromyces*
Imperfect state: *A.cremeus* group *Chaetosartorya*
Imperfect state: *A.nidulans* group and *A.niveus* *Emericella*
Imperfect state: *A.flavipes* *Fennellia*

Based on Perfect State

1. Phialides uniseriate 2
1^{1} Phialides biseriate, strictly or predominantly 8
2. Osmophilic 3
2^{1} Not osmophilic 6
3. Ascocarp lacking; asci produced within the ascogonial coil *Edyuillia*
3^{1} Ascocarps present 4
4. Ascocarps white to yellow, thin walled, without investing hyphae *Eurotium*
4^{1} Ascocarps within a stroma 5
5. Stroma sclerotioid; asci produced in 1-3 peridiate locules *Dichlaena*
5^{1} Stroma pseudoparenchymatous; asci in 1 nonperidiate locule *Hemicarpenteles*
6. Ascocarps hyphal, nonperidiate *Warcupiella*
6^{1} Ascocarps with pseudoparenchymatous peridium 7
7. Mature ascocarps white to yellow; ascospores with two well-separated equatorial crests *Neosartorya*
7^{1} Mature ascocarps purple, produced only in the dark, ascospores with single equatorial crest *Sclerocleista*

*The imperfect state of *Warcupiella* has recently been transferred to a new genus, *Raperia*, by Subramanian & Rajendran (1975).

8. Ascocarp embedded in, or at least associated with, hülle cells ... 9
8^1 Ascocarps not associated with hülle cells ... 10

9. Hülle cells globose to subglobose or elliptical: ascocarp initial a conspicuous coil; ascospores hyaline to yellow, red-orange, red-violet, or blue ... *Emericella*
9^1 Hülle cells elongate to helical, ascocarp initial a subglobose ascogonium; ascospores hyaline to pale yellow ... *Fennellia*

10. Asci produced in one to several peridiate locules within conspicuous upright sclerotioid stroma; initials unknown ... *Petromyces*
10^1 Ascocarps tomentose; peridium one to several cell layers thick; ascocarp initial a coiled ascogonium; osmophilic ... *Chaetosartorya*

Cleistothecial Ascomycetes with *Aspergillus*, *Penicillium* and *Paecilomyces* imperfect states have been classified in either the Eurotiaceae or the Aspergillaceae by most authors (En gler & Prantl, 1897; Clements & Shear, 1931; Bessey, 1950; Gaumann, 1952; von Arx & Müller, 1954; Alexopoulos, 1962; Raper & Fennell, 1965; Ainsworth, 1971; Fennell, 1973). However, Malloch & Cain (1972a), after forcibly inserting *Trichocoma paradoxa* -- a fungus without a cleistothecial perfect state but with a *Penicillium* imperfect-- into the family, selected Trichocomataceae (Fischer, 1897), because it predates Eurotiaceae (Clements & Shear, 1931), as the appropriate family name for these fungi. Malloch & Cain (1972b) later corrected this name to Trichocomaceae. According to these authors, the family name Aspergillaceae applies only to Fungi Imperfecti and cannot be used legitimately for Ascomycetes. Batra (1975) has followed Malloch & Cain (1972a) in classifying all perfect state genera with *Aspergillus* imperfect states in the family Trichocomaceae Fischer.

Using the morphological approach to taxonomy of *Aspergillus* and all of those associated perfect state genera, mycologists have made considerable progress in unravelling the mysteries of the genus since it was described by Micheli in 1729. More progress will be made as the results of transmission and scanning electron microscopic studies (Locci, 1969, 1972; Locci & Quaroni, 1970, 1971, 1972a,b; Fennell, St.Julian, Bulla & Baker, 1974) are incorporated in the present body of developmental and morphological data.

Traditional mycologists are reluctant to abandon their conventional approach to taxonomy, primarily, I think, because, with a microscope, a few books, and a familiar terminology one can make acceptable determinations with a fair degree of confidence.

We taxonomists feel some impatience with the molecular biologists who seem to be offering tantalizing bits of new information, generally confirmatory of classifications based on morphology but not yet contributing significantly to a solution of recognized classification problems. The new methods are laborious and time consuming, and investigations to date have been restricted to analysis of a few authenticated strains of a limited number of well-known species. Difficulties encountered in replication of experiments emphasize the need for adoption of standardized procedures for growth, extraction, and analysis. Problems of interpretation remain to be solved. However, even the most conservativ taxonomists welcome the new dimensions that are being added to the knowledge of the fungi by this new information. Before we can incorporate these techniques into classification studies, procedure must be simplified and standardized - perhaps even mechanized - and the flood of data that will be generated must eventually be stored in a computerized central clearing house. Despite the present limitations, it appears possible that, in far fewer than the 247 years that *Aspergillus* has been known, chemotaxonomic identification of Aspergilli will be a reality and that through a combination of conventional and molecular approaches, the phylogenetic relationships between fungal families will have been mapped.

REFERENCES

AINSWORTH,G.C. (1971). *Ainsworth & Bisby's dictionary of the fungi* 6th ed., Commonwealth Mycological Inst., Kew, Surrey,England.

ALEXOPOULOS,C.G. (1962). *Introductory mycology*, 2nd ed., Wiley, New York.

ARX,J.A.von & Müller,E. (1954). Die Gattungen der amerosporen Pyrenomyceten. *Beitr. Krypt. Fl. Schweiz*, 2, 1-434.

BAKER,K. & STROBEL,G.A. (1965). Lipids of the cell wall of *Puccinia striiformis* uredospores. *Monterey Academy of Science Proceedings* 25, 83-6.

BARTNICKI-GARCIA,S. (1968). Cell wall chemistry, morphogenesis, and taxonomy of fungi. *Annual Review of Microbiology* 22, 87-108.

BATRA,L.R. (1975). Ascomycetes of Pakistan: Plectomycetes. *Biologia* 21(1), 1-37.

BEEBE,J.D. (1970). Amylases of selected species of *Aspergillus*. Ph.D. Thesis, University of Chicago.

BENJAMIN,C.R. (1955). Ascocarps of *Aspergillus* and *Penicillium* *Mycologia* 47, 669-87.

BERKELEY,M.J. (1857). *Introduction to cryptogamic botany* Balliere: London.
BESSEY,E.A. (1950). *Morphology and taxonomy of the fungi.* McGraw-Hill (Blakiston): New York.
BIRD,C.W.,LYNCH,J.M. & PIRT,S.J. (1974). Concurrent synthesis and metabolism of alkanes by *Aspergillus nidulans*. *Chemistry & Industry* 15, 617-8.
BLOOMFIELD,B.J. & ALEXANDER,M. (1967). Melanins and resistance of fungi to lysis. *Journal of Bacteriology* 93, 1276-80.
BURNS,D.T., STREETON,R.J. & JAYATILAKE,S.D.A.K. (1976). Pyrolysis gas chromatography as an aid to the identification of *Penicillium* species. *Journal of Chromatography* 116, 107-15.
CARMICHAEL,J.W., SEKHON,A.S. & SIGLER,L. (1973). Classification of some dermatophytes by pyrolysis-gas- liquid chromatography. *Canadian Journal of Microbiology* 19, 403-7.
CLARE,B.G.,FLENTJE,N.T. & ATKINSON, M.R.(1968). Electrophoretic patterns of oxidoreductases and other proteins as criteria in fungal taxonomy. *Australian Journal of Biological Science* 21, 275-95.
CLEMENTS, F.E. & SHEAR,C.L. (1931). *The genera of fungi*, Wilson: New York.
DALBY,D.K. & GRAY,W.D. (1974). Inorganic ion content of fungal mycelia and spores. *Canadian Journal of Microbiology* 20, 935-6.
DART,R.K. (1975). Long-chain fatty acids in spores of *Penicillium*. *Transactions of the British Mycological Society* 65, 312-5.
DELEY, J.(1968). DNA base composition and hybridization in the taxonomy of phytopathogenic bacteria. *Annual Review of Phytopathology* 6, 63-90.
DISTLER,J.J. & ROSEMAN,S. (1960). Galactosamine polymers produced by *Aspergillus parasiticus*. *Journal of Biological Chemistry* 235, 2538-41.
ENGLER,A. & PRANTL,K. (1897). Die Naturlichen Pflanzenfamilien,etc. Teil I, Abt. 1, pp.301-304; Abt. 1**, pp.415-516 (1900). W.Englemann, Leipzig.
FENNELL,D.I. (1973). Plectomycetes: Eurotiales. In *The fungi* Vol.IVA Edited by Ainsworth, G.C., Sparrow,F.K., and Sussman,A.S. Academic Press: New York.
FENNELL,D.I., ST.JULIAN, G., BULLA,L.A., & BAKER,F.L. (1974). Scanning electron microscopy of conidiophore ornamentations in *Aspergillus* species. SEM/1974 (Part II). Proceedings of Workshop on SEM and Plant Sciences, IIT Research Inst.
FENNELL,D.I. & WARCUP,J. (1959). The ascocarps of *Aspergillus alliaceus*. *Mycologia* 51, 411-3.
FISCHER,E. (1897). Plectascineae. In Engler and Prantl. qv.
GARBER,E.D.(1973). Electrophoresis as a taxonomic and genetic tool for fungi. *Bulletin Torrey Botanical Club* 100, 308-12.

GARBER,E.D. (1974). Enzymes as taxonomic and gentic tools in *Phaseolus* and *Aspergillus*. *Israel Journal of Medical Science* 10, 268-77.

GARBER,E.D. & RIPPON,J.W. (1968). Proteins and enzymes as taxonomic tools. *Advances in Applied Microbiology* 10, 137-54.

GAUMANN,E.A. (1952). The fungi, (translated by F.L.Wynd). Hafner: New York.

HALL,R. (1969). Molecular approaches to taxonomy of fungi. *Botanical Review* 35, 285-304.

HAYHOME,B.A. (1970). Genetic studies on the regulation of the amylolytic enzymes of *Aspergillus niger*. Ph.D.Thesis, University of Chicago.

HENNEY,H. & STORCK,R. (1963a). Nucleotide composition of ribonucleic acid from *Neurospora crassa*. *Journal of Bacteriology* 85, 822-6.

HENNEY, H. & STORCK,R. (1963b). Ribosomes and ribonucleic acid in three morphological states of *Neurospora*. Science 142,1675-7.

HESSELTINE,C.W., SORENSON,W.G. & SMITH,M. (1970). Taxonomic studies of the aflatoxin-producing strains in the *Aspergillus flavus* group. *Mycologia* 62, 123-32.

HORIKOSHI,K. & Iida, S. (1964). Studies of the spore coats of fungi. I. Isolation and composition of the spore coats of *Aspergillus oryzae*. *Biochimica et Biophysica Acta* 83, 197-203.

HÜTTER,R. & DEMOSS,J.A. (1967). Organization of the tryptophan pathway: a phylogenetic study of the fungi. *Journal of Bacteriology* 94, 1896-907.

JOHNSTON,I.R. (1965). The composition of the cell walls of *Aspergillus niger*. *Biochemical Journal* 96, 651-8.

JONES,D., FARMER,V.C., BACON,J.S.D., & WILSON,M.J. (1972). Comparison of ultrastructure and chemical components of cell walls of certain plant pathogenic fungi. *Transaction of the British Mycological Society* 59, 11-23.

JONES, J.G. (1969). Studies on lipids of soil microorganisms with particular reference to hydrocarbons. *Journal of General Microbiology* 59, 145-52.

KULIK,M.M. & BROOKS,A.G. (1970). Electrophoretic studies of soluble proteins from *Aspergillus* spp. *Mycologia* 62, 365-376.

KULIK,M.M. & VINCENT,P.G. (1973). Pyrolysis-gas-liquid chromatography of fungi: observations on variability among nine *Penicillium* species of the section Asymmetrica, subsection Fasciculata. *Mycopathologica Mycologica Applicata* 51, 1-18.

KUREZA,K.C. & GARBER,E.D.(1973). A genetic study of electrophoretically variant extracellular amylolytic enzymes of wild-type strains of *Aspergillus nidulans*. *Canadian Journal of Genetics and Cytology* 15, 275-87.

KWON-CHUNG, K.J. & FENNELL,D.I. (1971). New pathogenic species of *Aspergillus*. *Mycologia* 63, 478-89.

LINK,H.F. (1809). Observationes in ordines plantarum naturales. *Gesellschaft Naturforschender Freunde zu Berlin, Magazin* 3, 1-42.

LOCCI,R. (1969). Investigations on fungal features by scanning electron microscopy. I. The life cycle of *Aspergillus amstelodami*. *Rivista di Pathologia Vegetale*,Ser.IV,5, 223-32.

LOCCI,R. (1972). Scanning electron microscopy of ascosporic Aspergilli. Supplement to *Rivista di Pathologia Vegetale*,IV.8,1-172.

LOCCI, R. & QUARONI,S. (1970). Investigations on fungal features by scanning electron microscopy. II. Morphological characteristics of the perfect stage of *Aspergillus* species. *Rivista di Pathologia Vegetale, Ser.VI 3*, 141-92.

LOCCI,R. & QUARONI,S. (1971). Investigations on fungal features by scanning electron microscopy. III. Observations on additional cleistothecial species of *Aspergillus*. *Rivista di Pathologia Vegetale*,Ser.IV, 7,141-60.

LOCCI,R. & QUARONI,S. (1972a). Investigations on fungal features by scanning electron microscopy. IV. Perfect stage structures additional *Aspergillus* species. *Rivista di Pathologia Vegetale* Ser.IV,8, 103-25.

LOCCI, R. & QUARONI,S. (1972b). Investigations on fungal features by scanning electron microscopy. V. Further studies on ascosporic *Aspergillus* species. *Rivista di Pathologia Vegetale* Ser.IV,8, 253-320.

LOVETT,J. S. & HASELBY,J.A.(1971). Molecular weights of the ribosomal ribonucleic acid of fungi.*Archiv für Mikrobiologie* 80,191-204.

MALLOCH,D. & CAIN,R.F. (1972a). The Trichocomataceae: Ascomycetes with *Aspergillus, Paecilomyces*, and *Penicillium* imperfect states. *Canadian Journal of Botany* 50, 2613-28.

MALLOCH,D. & CAIN,R.F. (1972b). The Trichocomaceae (Ascomycetes): synonyms in recent publications. *Canadian Journal of Botany* 51, 1647-8.

MICHELI,P.A. (1729). Nova plantarum genera juxta tournefortii methodum disposita, 234 pp., Florence, Italy.

MOYER, R.C. & STORCK,R. (1964). Properties of ribosomes and RNA from *Aspergillus niger*. *Archives of Biochemistry and Biophysics* 104, 193-201.

MURAKAMI,H. (1970a). Taxonomic studies on Japanese industrial strains of *Aspergillus*. (Part 13). Distribution of various mycological characters in all of the strains and some important characters. *Report of Research Institute of Brewing, Japan* 142, 15-22.

MURAKAMI,H. (1970b). Taxonomic studies on Japanese industrial strains of *Aspergillus*. (Part 15). Component analysis of every 20 mycological character of 406 strains. *Report of Research Institute of Brewing, Japan* 143, 5-14.

MURAKAMI,H. (1971a). Taxonomic studies on the Japanese industrial strains of *Aspergillus*. (Part 17). Classification (*Aspergillus oryzae* group). *Report of Research Institute of Brewing, Japan* 143, 5-14.

MURAKAMI, H. (1971b). Taxonomic studies on the Japanese industrial strains of *Aspergillus* (Part 18). Molds other than the koji mold and mycological characters of all strains. *Report of Research Institute of Brewing, Japan* 143, 16-29.

MURAKAMI,H. (1971c). Classification of the kojic mold. *Journal of General & Applied Microbiology* 17, 281-309.

MURAKAMI, H. (1972a). Taxonomic studies on the Japanese industrial strains of *Aspergillus*. (Part 21). Relation between the component analysis and the classification of key characters. *Report of Research Institute of Brewing, Japan* 144, 7-12.

MURAKAMI,H. (1972b). Taxonomic studies on the Japanese industrial strains of *Aspergillus*. (Part 22). Clusters of *A.oryzae* group and *A.flavus* group including 32 additional strains. *Report of Research Institute of Brewing, Japan* 144, 13-19.

MURAKAMI, H. (1972c). Taxonomic studies on the Japanese industrial strains of *Aspergillus*. (Part 23). Summary of the corrections and a supplementary system of the classification. *Report of Research Institute of Brewing, Japan* 144, 20-25.

MURAKAMI, H. & NASUNO, S. (1972). Taxonomic studies on the Japanese industrial strains of *Aspergillus*. (Part 20). Exclusion of three strains from the group of Murakami. *Report of Research Institute of Brewing, Japan* 144, 3-6.

MURAKAMI,H., HARA,S., TAKAHASHI,T., RYU,G., OHBA,T., SUITO,K. & TAKANO,F. (1970). Taxonomic studies on the Japanese industrial strains of *Aspergillus*. (Part 12). Pink colouration of conidia by anisic acid and the related compounds, re-examination on the production of kojic acid and aflatoxin, and the determination of decomposition of hydrogen peroxide. *Report of Research Institute of Brewing, Japan* 142, 1-12.

NASUNO, S. (1971). Polyacrylamide gel disc electrophoresis of alkaline proteinases from Aspergillus species. *Agricultural and Biological Chemistry* 35, 1147-50.

NASUNO,S. (1972a). Differentiation of *Aspergillus sojae* from *Aspergillus oryzae* by polyacrylamide gel disc electrophoresis. *Journal of General Microbiology* 71, 28-33.

NASUNO,S. (1972b). Electrophoretic studies of alkaline proteinases from strains of *Aspergillus flavus* group. *Agricultural and Biological Chemistry* 36, 684-9.

NASUNO,S. (1974). Further evidence on differentiation of *Aspergillus sojae* from *Aspergillus oryzae* by electrophoretic patterns of cellulase, pectin-lyase, and acid proteinase. *Canadian Journal of Microbiology* 20, 413-6.

NEALSON, K.H. & GARBER, E.D. (1967). An electrophoretic survey of esterases, phosphatases, and leucine amino-peptidases in mycelial extracts of species of *Aspergillus*. *Mycologia* 59, 330-6.

NEHIRA, T. (1957). Taxonomic studies of the genus *Aspergillus*. *IX* The *Aspergillus flavus-oryzae* group and related species in the Orient. *Journal of Fermentation Technology* 35, 50-7.

NEHIRA, T. & TOMI, R. (1956). Taxonomic studies on the genus *Aspergillus*. VIII. The relation between the morphological characteristics and the amylolytic and proteolytic properties in the *Aspergillus* I-III. *Journal of Fermentation Technology* 34, 391-9; 423-8; 457-63.

O'BRIEN, R.W. & RALPH, B.J. (1966). The cell composition and taxonomy of some Basidiomycetes and Ascomycetes. *Annals of Botany, N.S.* (London) 30, 831-43.

O'HARA, I. (1951). Classification of the *Aspergillus tamarii-oryzae* group. I. The production of kojic acid from various compounds as an aid in identification. *Research Bulletin of Faculty of Agriculture, Gifu University*, No.1, 71-85.

O'HARA, I. (1952). Classification of *Aspergillus tamarii-oryzae* Group. Part 2. The production of kojic acid from various compounds as an aid for identification (Suppl. to Part I). *Journal of Agricultural Chemistry Society, Japan* 26(10), 547-51.

O'HARA, I. (1953). Classification of *Aspergillus tamarii-oryzae* Group. Part 3. Diagnosis of the series, species, and subspecies. *Researcg Bulletin Gifu Imperial College of Agriculture* 28(3), 75-85.

ORO, J., LASETER, J.L., & WEBER, D. (1966). Alkanes in fungal spores. *Science* 154, 399-400.

OYAMA, V.I. & CARLE, G.C. (1967). Pyrolysis-gas-chromatography application to gas detection and chemotaxonomy. *Journal of Gas Chromatography* 5, 151-4.

RAPER, K.B. (1957). Nomenclature in *Aspergillus* and *Penicillium*. *Mycologia* 49, 644-62.

RAPER, K.B. & FENNELL, D.I. (1965). *The genus Aspergillus*. Williams and Wilkins Co.: Baltimore, Md.

RAPER, K.B., FENNELL, D.I. & TRESNER, H.D. (1953). The ascosporic stage of *Aspergillus citrisporus* and related forms. *Mycologia* 45, 671-92.

REINER, E. (1965). Identification of bacterial strains by pyrolysis-gas-liquid chromatography. *Nature* 206, 1272-4.

REINER, E. (1967). Studies on differentiation of microorganisms by pyrolysis-gas-liquid chromatography. *Journal of Gas Chromatography* 5, 65-7.

ROGERS, D.P. (1958). Philosophy of taxonomy. *Mycologia* 50, 326-32.

RUIZ-HERRERA, J. (1967). Chemical components of the cell wall of *Aspergillus* sp. *Archives of Biochemistry and Biophysics* 122, 118-25.

SARBHOY,A.K. & ELPHICK,J.J. (1968). *Hemicarpenteles paradoxus* gen. & sp. nov.: The perfect state of *Aspergillus paradoxus*. *Transactions of the British Mycological Society* 51, 155-7.
SAWICKI,E.H. & PISANO,M.A. (1976). Temperature induced changes in the lipid composition of a pathogenic fungus. *Abstract Annual Meeting of American Society Microbiology* 76, F32.
SEKHON,A.S. & CARMICHAEL,J.W. (1971). Classification of some Gymnoascaceae by pyrolysis-gas-liquid chromatography using added marker compounds. *Sabouraudia* 13, 83-8.
SEKHON,A.S. & CARMICHAEL,J.W. (1972). Pyrolysis-gas-liquid chromatography of some dermatophytes. *Canadian Journal of Microbiology* 18, 1593-601.
SMITH,J.E. & ANDERSON, J.G. (1973). Differentiation in the Aspergilli. *Symposium Society General Microbiology* 23, 295-337.
STOLK,A.C. (1964). *Aspergillus flaschentraegeri* sp. nov. *Transactions of the British Mycological Society* 47, 123-6.
STORCK,R. (1965). Nucleotide composition of nucleic acids of fungi. I. Ribonucleic acids. *Journal of Bacteriology* 90, 1260-4.
STORCK,R. (1966). Nucleotide composition of nucleic acids of fungi. II. Deoxyribonucleic acids. *Journal of Bacteriology* 91, 227-30
STORCK,R. & ALEXOPOULOS,C.J. (1970). Deoxyribonucleic acid of fungi. *Bacteriological Reviews* 34(2), 126-54.
STORCK,R., ALEXOPOULOS,C.J. & PHAFF, H.J. (1969). Nucleotide composition of deoxyribonucleic acid of some species of *Cryptococcus, Rhodotorula, and Sporobolomyces*. *Journal of Bacteriology* 98, 1069-72.
STORCK,R., NOBLES, M.K. & ALEXOPOULOS,C.J. (1971). The nucleotide composition of deoxyribonucleic acid of some species of Hymenochaetaceae and Polyporaceae. *Mycologia* 63, 38-49.
SUBRAMANIAN,C.V. (1972). The perfect states of *Aspergillus*. *Current Science* 41, 755-61.
SUBRAMANIAN,C.V. & RAJENDRAN,C. (1975). *Raperia,* a new genus of Hyphomycetes. *Kavaka* 3, 129-33.
THIBAUT,M. & ANSEL, M. (1973). Mycological applications of X-ray microanalysis. *Journal of Bacteriology* 116, 1181-4.
THOM,C. & CHURCH,M. (1926). *The Aspergilli,* Williams & Wilkins: Baltimore, Md.
THOM,C. & RAPER,K.B. (1945). *A manual of the Aspergilli,* Williams & Wilkins: Baltimore, Md.
VILLA,V.D. & STORCK,R. (1968). Nucleotide composition of nuclear and mitochondrial deoxyribonucleic acid of fungi. *Journal of Bacteriology* 96, 184-90.
VINCENT,P.G. & KULIK,M.M. (1970). Pyrolysis-gas-liquid chromatography of fungi: differentiation of species and strains of several members of the *Aspergillus flavus* group. *Applied Microbiology* 20, 957-63.

VINCENT,P.G. & KULIK,M.M. (1973). Pyrolysis-gas-liquid chromatography of fungi: numerical characterization of species variation among members of the *Aspergillus glaucus* group. *Mycopathologica et Mycologica Applicata* 51, 251-65.

VOGEL,H.J. (1965). Lysine biosynthesis and evolution. *In Evolving genes and proteins.* Edited by Bryson,W. and Vogel, H.J. Academic Press, Inc.: New York.

VUILLEMIN,P. (1927). *Sartorya,* nouveau genre de Plectascinees angiocarpes. *Comptes rendre* 184, 136-7.

WEETE,J.D. (1972). Aliphatic hydrocarbons of the fungi. *Phytochemistry* 11, 1201-5.

WILEY,B.J. & FENNELL,D.I. (1973). Ascocarps of *Aspergillus stromatoides, A.niveus,* and *A.flavipes. Mycologia* 65, 752-60.

WILEY,B.J. & SIMMONS,E.G. (1973). New species and a new genus of Plectomycetes with *Aspergillus* states. *Mycologia* 65, 934-8.

Asexual Development in *Aspergillus*

J. E. SMITH, J. G. ANDERSON, S. G. DEANS and B. DAVIS*

Department of Applied Microbiology, Strathclyde University, Glasgow, U.K.

INTRODUCTION

The complex process of fungal sporulation must be viewed as the integration of numerous biological events. The unique ability to change from a vegetative filamentous growth state to a reproductive and generally spherical spore form must be viewed as the end result of a genetic competence reacting to particular physico-chemical environmental changes which in turn promote the internal stimulus to regulate cellular activity. The end result of this interplay is the creation of a new developmental pattern with its characteristic and unique metabolism and morphology.

Fungal sporulation has been extensively reviewed in recent years and most reviews contain numerous references to the aspergilli (Smith & Galbraith, 1971; Smith & Anderson, 1973; Lovett, 1975; Turian, 1974,1975,1976; Smith, 1977). The complexity and variation of the conidiophore apparatus in the aspergilli and its role in classification of the genus has been authoratively considered by Raper & Fennell (1965).

ULTRASTRUCTURAL CHANGES DURING CONIDIATION

Conidiophore development

The first stage in the development of the *Aspergillus*

* *Present Address: Department of Biology, Sheffield Polytechnic, Sheffield, U.K.*

conidiophore involves the transformation of a hyphal cell into the characteristic foot cell from which the conidiophore will emerge. Presumably, because of the difficulty of identifying cells which are destined to become foot cells, there is no information available concerning the ultrastructural changes which take place during this first but important aspect of conidiophore development. Developing and mature conidiophores are, however, easy to distinguish from mycelium because of their greater diameter and thickened walls.

The earliest ultrastructural study on the conidiophore appears to be that on *Aspergillus niger* by Tanaka & Yanagita (1963). In this study it was observed that the conidiophore foot had a very thick cell wall which was composed of up to eight layers of different densities. With *A.nidulans*, however, Oliver (1972) observed that the foot had a two-layered wall. The outer layer (primary wall) was continuous with the wall of the basal mycelium and the inner layer (secondary wall) lined the sides and ends of the foot. The conidiophore stalk also possessed a two-layered wall except at growing apices where the two layers could not be distinguished. On the external surface of the wall was a superficial dense lamina which invested the entire conidiophore.

The conidiophore wall of *Aspergillus flavus* is also two-layered and in this species the surface of the outer layer is ornamented with spines (Bojovic-Cvetic & Vujicic, 1974). Spines are a characteristic feature of the conidiophore wall of *A.flavus* and of some other *Aspergillus* species and thus provide useful taxonomic characteristics (Raper & Fennell, 1965). A fibrillar layer between or over the spines was also present on the surface of *A.flavus* conidiophores. The fibrillar layer was not continuous but was present as fragments and at late stages of conidiophore development was almost absent. Bojovic-Cvetic & Vujicic (1974) consider that this external fibrillar layer in *A.flavus* corresponds to the superficial lamina which is found in *A.nidulans* although for the latter species it forms a complete layer. Oliver (1972) has suggested that the lamina in *A.nidulans* is formed by the transmission of materials through the secondary wall and also that the development of the brown pigmentation of the conidiophore is associated with lamina deposition. The conidiophore develops pigmentation before the basal mycelium and Oliver (1972) suggests that this is because excretion products accumulate more rapidly at the conidiophore surface than at the basal mycelium where they are dispersed into and cause melanisation of the medium prior to colouration of the hyphal walls.

Studies on the intracytoplasmic structure of *Aspergillus* conidiophores have given some indication of the subcellular changes which take place during development. The cytoplasm of young conidiophores appears dense with an abundance of ribosomes and other organelles which show a distribution pattern similar to that seen in growing hyphae. Thus with *A.nidulans* Oliver (1972) observed that the conidiophore apex contained groups of cytoplasmic vesicles similar to those observed in hyphal apices (McClure, Park & Robinson, 1968; Grove & Bracker, 1970). Between the apex and the nearest nuclei was a densely packed zone of ribosomes and mitochondria. Vacuolation developed in the older regions of the stalk in association with bodies morphologically similar to autophagic vacuoles. Plasmalemmasomes and lomasomes were sparsely and non-specifically distributed. With both *A.nidulans* (Oliver, 1972) and *A.flavus* (Bojovic-Cvetic & Vujicic, 1974) it has been observed that organelles, particularly nuclei and mitochondria are elongate and run parallel to the long axis of the conidiophore. With *A.flavus* microtubules occurred both in the cytoplasm and immediately below the plasma membrane where they ran parallel to the conidiophore wall. With *A.nidulans*, however, Oliver (1972) commented that microtubules were sparsely distributed and that filaments of around 10nm in thickness and up to 150nm in length were more commonly found. These filaments were aligned parallel to or at low angle to microtubules and elongated nuclei, mitochondria and cell surfaces. In both studies the elongated nature of organelles and the presence and orientation of the microtubules and filaments were attributed to the process of cytoplasmic streaming which can be observed in living conidiophores.

Ultrastructural changes in the conidiophore during ageing and degeneration have been described with *Aspergillus niger* (Tanaka & Yanagita, 1963) and *A.flavus* (Bojovic-Cvetic & Vujicic, 1974). Older conidiophores showed an increase in vacuolation and marked changes in the appearance of organelles particularly mitochondria and nuclei. In old degenerating conidiophores many of the nuclei and mitochondria disappeared from the cytoplasm.

Vesicle development

When the conidiophore reaches its full length there is a gradual enlargement of the apex to form the vesicle (Fig. 1). This swelling is accompanied by extensive thickening of the vesicle wall. Studies with *Aspergillus niger* (Tsukahara, 1970) and *A.nidulans* (Oliver, 1972) using transmission electron microscope (TEM) indicated that the vesicle wall, like that of the conidiophore stalk, is two layered. With *A.clavatus* Hanlin (1976) observed that the number of apparent layers in the vesicle wall was influenced by fixation technique and up to eight layers were seen in permanganate fixed material. However, observations on broken vesicle walls made by scanning electron microscopy (SEM) (Tokunaga, Tokunaga & Harada, 1973) clearly indicate that the vesicle wall is composed of two major structurally different zones (Fig. 2).

Details of the subcellular structure of the vesicle during development have been described for several species. With *Aspergillus nidulans* Oliver (1972) observed that during the

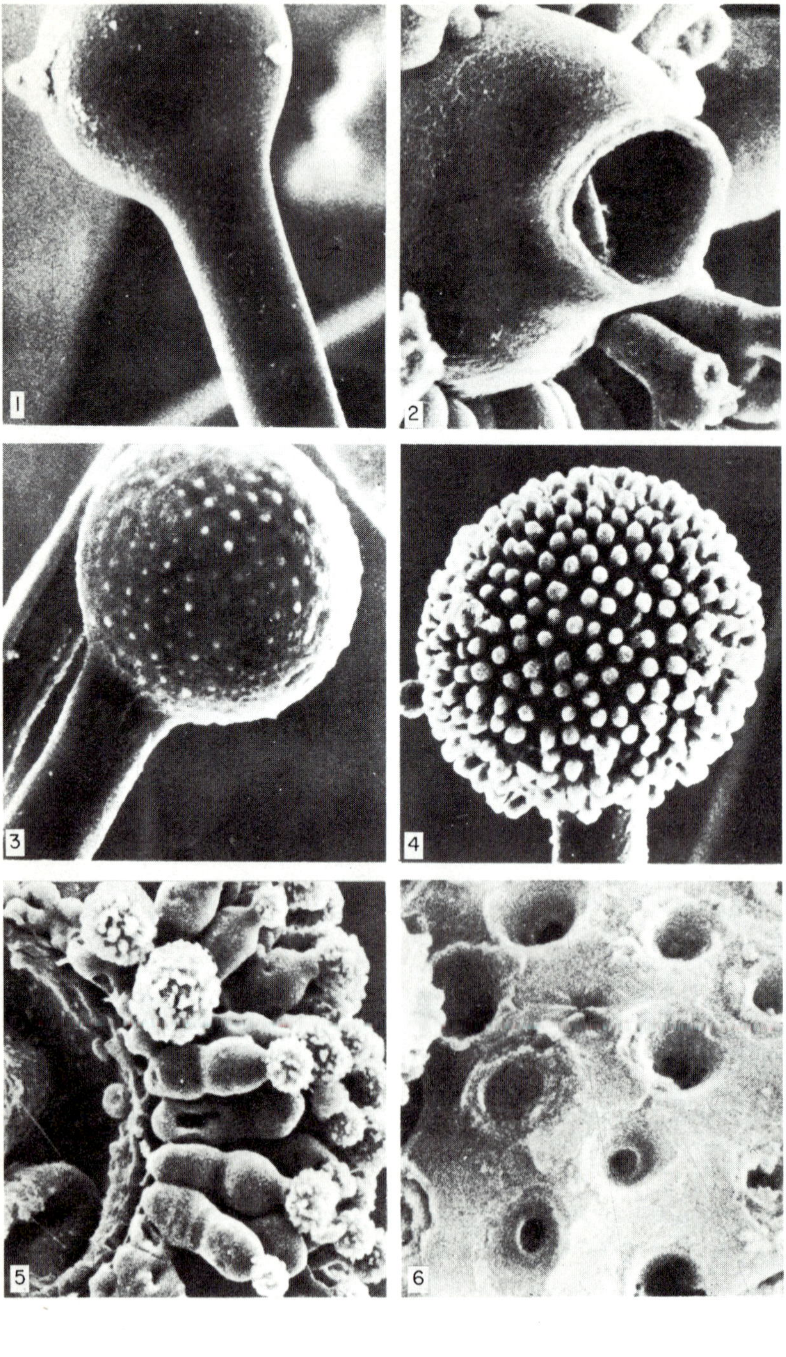
1
2
3
4
5
6

formation of the vesicle the groups of cytoplasmic vesicles which were previously concentrated in the pointed conidiophore apex became scattered as swelling progressed. Also with *A.flavus* Bojovic-Cvetic & Vujicic (1974) observed that organelles were dispersed uniformly throughout the vesicle and that they were no longer elongated presumably because the circulation of the cytoplasm was not strictly orientated as in the conidiophore stalk. Another distinctive feature of the *A.flavus* vesicle was an increase in the number of microbodies and their arrangement in groups. With *A.clavatus* Hanlin (1976) observed that lomasomes were common along the vesicle wall and these appeared to be of the plasmalemmasome type described by Heath & Greenwood (1970).

Legends to Fig. 1 - 6.

Fig. 1. SEM showing formation of the vesicle at the conidiophore tip. From Tokunaga *et al.*(1973).

Fig. 2. SEM of a portion of the conidial head of *Aspergillus niger* showing branched secondary sterigmata and a broken portion of the vesicle wall. From Tokunaga *et al.*(1973).

Fig. 3. SEM of an early stage of sterigmata formation; the vesicle wall shows the synchronous protrusion of young sterigmata. From Tokunaga *et al.* (1973).

Fig. 4. SEM showing continued development of the sterigmata from the vesicle. From Tokunaga *et al.* (1973).

Fig. 5. SEM of the under-side of the conidial head of *A.oryzae*, showing a single series of sterigmata which are producing conidiospores. From Tokunaga *et al.*(1973).

Fig. 6. SEM of a portion of an exposed vesicle wall, showing concavities (with central pore) which represents the points of attachment of the sterigmata. From Tokunaga *et al.* (1973).

Studies by Trinci,Peat & Banbury (1968) have shown an interesting feature of the vesicle of *Aspergillus giganteus*. This species does not show the uniform dispersal of organelles in the vesicle as was observed with *A.flavus* (Bojovic-Cvetic & Vujicic, 1974). Instead there are two distinct zones of protoplasm comprising a peripheral layer containing large numbers of mitochondria and nuclei and a central area which is almost completely devoid of these organelles, containing instead densely packed granules resembling glycogen granules. This arrangement was interpreted as representing a peripheral area of high metabolic activity and a central region in which food reserves are accumulated. It is possible that this feature of the *A.giganteus* vesicle is related to the comparatively large size of the sporulating structure in this species.

Ageing and degeneration changes in the vesicle appear to be essentially similar to those described for the conidiophore stalk region, namely an increase in vacuolation and the disorganisation and eventual disappearance of many organelles. Bojovic-Cvetic & Vujicic (1974) have pointed out that at the time of advanced degeneration of the vesicle and conidiophore stalk of *Aspergillus flavus* the metulae and phialides are still actively producing conidia. It appears therefore that the cytoplasm of the vesicle and stalk matures and senesces synchronously but almost independently of the processes in the sterigmata and phialides.

Sterigmata development

The production of numerous sterigmata from the vesicle represents the final stage in the development of the spore bearing apparatus. With all species of *Aspergillus* examined by electron microscopy it has been observed that the spherical protrusions which represent the sterigmata primordia arise more or less synchronously from the entire surface of the vesicle. About 400-500 protrusions are produced from the vesicle of *A.niger* and the synchronous nature of the process in this species has been attractively demonstrated by Tokunaga *et al.* (1973) using scanning electron microscopy (Fig. 3,4).

In some species of *Aspergillus* only primary sterigmata (direct phialide formation) are formed whereas with others the sterigmata are composed of both primary and secondary sterigmata (metulae and phialides). When both metulae and phialides are produced the more complex sterigmata take on an appearance which is characteristic of the species. Visualisation by scanning electron microscopy demonstrates that the sterigmata of *A.niger* form complicatedly branched secondary sterigmata at their distal ends (Fig. 2) whereas with *A.oryzae* (Fig. 5) the sterigmata are not branched but are simply extended rods which are constricted in the middle (Tokunaga *et al.* 1973). When sterigmata are removed from the vesicle wedge-shaped depressions which possess a central pore are left in the vesicle wall (Fig. 6).

The changes in wall structure during the outgrowth and development of the sterigmata have been described in particular detail for *Aspergillus nidulans* (Oliver, 1972) and *A. clavatus* (Hanlin, 1976). An early indication of sterigmata formation in both species was the appearance of numerous depressions or conical pits in the inner surface of the vesicle wall probably by dissolution of wall material (Fig. 7). With *A.clavatus* depressions also formed on the outer wall directly opposite those on the inner wall. Dissolution of the inner wall continued until the protoplast pushed out to form small spherical buds invested by a thin cell wall (Fig. 8).

The walls of sterigmata appear to remain thin and flexible for some time during their development. With *Aspergillus giganteus* (Trinci *et al.*,1968) and several other *Aspergillus* spp. (Tokunaga *et al.*,1973) it has been observed that as the sterigmata increase in size they come in contact with one another and thus, presumably by pressure effect, they lose their spherical shape and become more cylindrical (Fig. 9). With species which produce only phialides and no metulae these cylindrical projections become narrowed at the apex. Also the wall at the apex becomes much thicker due to the formation of a secondary wall layer inside the outer primary wall as was observed with *A.clavatus* (Hanlin, 1976). These changes create the 'bottle-necked' extension which is a characteristic feature of many phialides.

When both metulae and phialides are produced these are separated by a septal wall which is an inward growth of the inner wall layer. The septum possesses a central pore with a diameter of about 0.07μ m (Tokunaga *et al.*,1973). The sterigmata are also cut off from the vesicle by a basal septum which forms in the middle of the narrow channel through the vesicle wall. A pore of similar diameter is also present in these septa and in a detailed study on *A.giganteus* these septal pores were always observed to be unplugged (Trinci *et al.*,1968).

Considerable attention has been paid to the appearance and localisation of organelles during sterigmata development. With both *Aspergillus nidulans* (Oliver, 1972) and *A.clavatus* (Hanlin,1976) cytoplasmic vesicles were observed to occur in concentrated clusters adjacent to the depressions in the inner surface of the vesicle dome where the sterigmata primordia would emerge (Fig. 7). These cytoplasmic vesicles persisted in the tips of sterigmata throughout their development (Fig. 8,10). It is generally accepted that these vesicles are involved in fungal cell wall growth and their observed localisation at areas of wall dissolution and growth during sterigmata development is consistent with the lytic and synthetic properties which are thought to be associated with these structures.

With *Aspergillus nidulans* Oliver (1972) has made interesting observations regarding a possible spatial relationship in the sterigmata between the cytoplasmic vesicles, plasmalemmasomes and lomasomes The plasmalemmasomes, lomasomes and a variety of intermediary forms were observed to occur flattened against the cell wall of the sterigmata. By careful observation at various stages of sterigmata development it was observed that cytoplasmic vesicles were strictly localised in apices whereas plasmalemmasomes showed subapical and basal localisation. This and the later association of plasmalemmasomes with lomasomes led Oliver (1972) to conclude that this presented strong support for the hypothesis of Heath & Greenwood (1970) that plasmalemmasomes arise from the incorporation of apical cytoplasmic vesicle membranes into the cell surface in excess of the

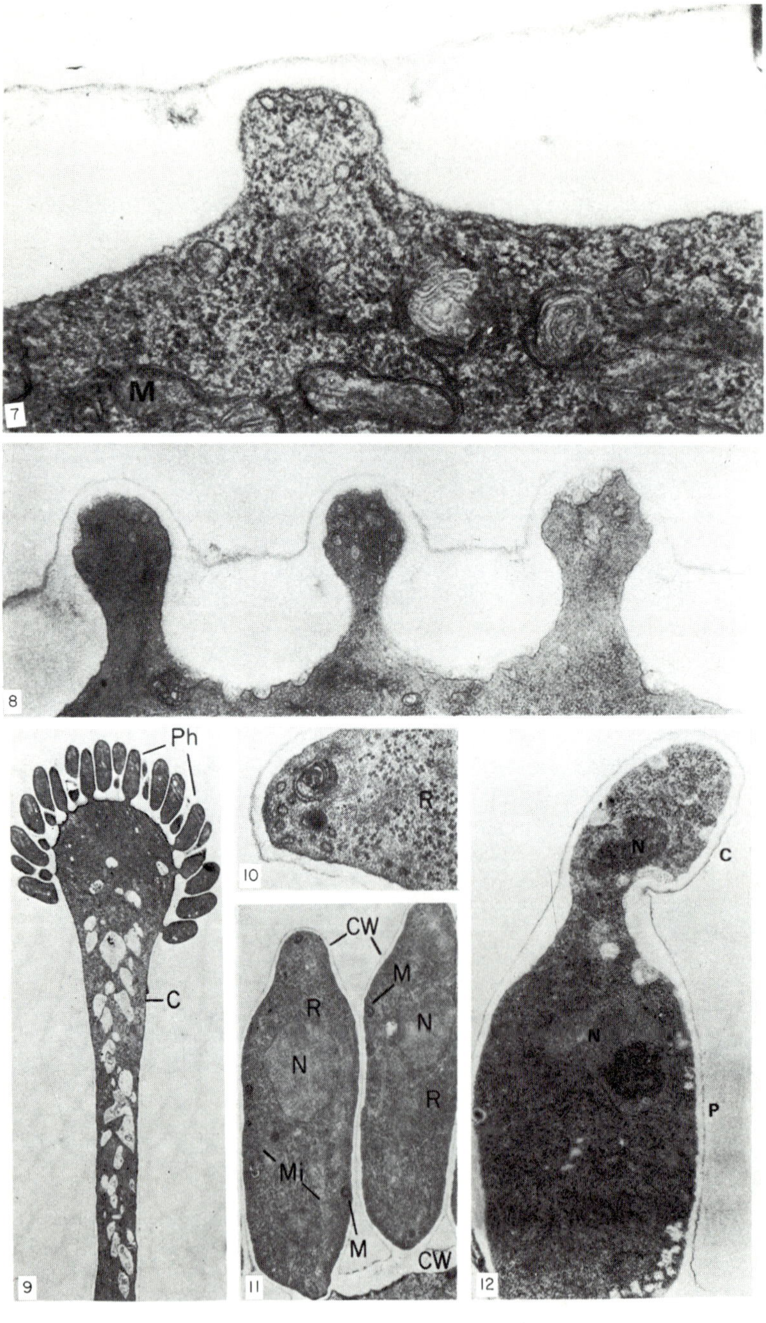
M
7
8
Ph
C
9
R
10
CW
M
R
N
N
R
Mi
M
CW
11
N
C
N
P
12

Legends to Fig. 7 - 12.

Fig. 7. Section showing an early stage of phialide development in *Aspergillus clavatus*. A thin area of the vesicle wall is just beginning to protrude beyond the periphery of the wall. From Hanlin (1976).

Fig. 8. Later stage in the synchronous development of the phialides from the vesicle of *A. clavatus*. From Hanlin (1976).

Fig. 9. Section through a conidiophore (C) of *Aspergillus fumigatus* bearing young phialides (Ph), prior to conidia production. From Ghiorse & Edwards (1973).

Fig. 10. Enlargement of young phialide apex of *A. fumigatus* showing vesicles and ribosomes (R). From Ghiorse & Edwards (1973).

Fig. 11. Young phialides of *A. fumigatus* with nuclei (N), mesosome-like structures (M), ribosomes (R) and slightly visible mitochondria (Mi). From Ghiorse & Edwards (1973).

Fig. 12. Section through phialide (P) of *A. clavatus*. The phialide is forming the first conidium (C) and the position of nuclei (N) and the smooth connection between the conidium and phialide are shown. From Hanlin (1976).

---

increase in wall surface area. Thus they may be a collection of exce membrane which can be sequestered later. In the sterigmata of *A. nidulans* plasmalemmasomes were frequently observed in association with the septal pore and thus they may play a role in septum formation (Oliver, 1972).

A good deal of interest has also been shown in the appearance and localisation of nuclei during sterigmata development and the subsequent production of conidia. From the combined observations made of *Aspergillus giganteus* (Trinci *et al.*, 1968), *A. nidulans* (Oliver, 1972) and *A. clavatus* (Hanlin, 1976) some idea can be gained of nuclear behaviour during sterigmata development. Nuclear division does not apparently take place until the buds (sterigmata primordia) are fully enlarged. A single nucleus and other organelles then migrate into each sterigmata and this is soon followed by cross-septum formation. The event of nuclear migration appears to be difficult to detect in most of the species examined, however, some sections were obtained with *A. giganteus* which showed nuclei apparently in the process of migration from the vesicle into the phialide primordia. Comparison of events observed with *A. nidulans* (Oliver, 1972) and *A. giganteus* (Trinci *et al.*, 1968) indicates that a greater degree of synchronised nuclear division and movement occurs in the former species. With *A.giganteus* although excellent localised synchrony occurred the development of the vesicle as a whole was not perfectly coordinated. On the same vesicle areas were found where the phialide primordia contained nuclei, others where the primordia did not contain nuclei and still others where nuclei were apparently in the process of migration into the primordia. Not enough work has been done with other species to know whether this lack of synchronisation is a feature peculiar to the *A.giganteus* vesicle perhaps as a consequence of its comparatively large size.

Fully developed phialides of the various species examined show essentially similar ultrastructural features (Fig. 11, 12). A single nucleus is present, usually in a central position, along with mitochondria, glycogen granules, sparse endoplasmic reticulum and occasionally vacuoles which become particularly prominent in older speciments. Dark crystalline structures of unknown composition were often observed in vacuoles in the phialides of *Aspergillus clavatus* (Hanlin, 1976). A consistent feature of the primary sterigmata (metulae) of *A.niger* was the presence of a large vacuole with a spongy structure (Tsukahara, 1970). Electron-dense, membrane-bound bodies, probably homologous with Woronin bodies were typically associated with the septal pores of sterigmata of *A.giganteus* (Trinci *et al.*, 1968) and *A.clavatus* (Hanlin, 1976). Cytoplasmic vesicles, plasmalemmasomes and other membranous structures are also an important ultrastructural feature of mature phialides and these will be discussed more fully in relation to conidia production.

Conidia development

An understanding of the precise mechanisms involved in the formation of the primary and subsequent conidia from the phialides of *Aspergillus* species is important for both development and taxonomic considerations. The fine structural changes associated with primary conidium production have been observed in only a few studies notably those on *A.giganteus* (Trinci *et al.*, 1968) and *A.clavatus* (Hanlin, 1976) using TEM and on several other *Aspergillus* spp. using SEM (Tokunaga, Harada & Tokunaga, 1971; Tokunaga *et al.*, 1973).

The primary conidia arise as spherical swellings from the apex of the phialide. Both the inner and outer walls of the phialide apex are continuous around the young conidium (Fig. 12). This newly synthesised wall material which also distends to surround the emerging conidium in *Penicillium* spp. (Zachariah & Fitz-James, 1967; Cole & Kendrick, 1969a) has been called the "apical plug" by Fletcher (1971). Once the primary conidium has enlarged the nucleus of the phialide divides and one of the daughter nuclei migrates into the developing spore (Fig. 12) whilst the other remains in the phialide (Trinci *et al.*,1968; Oliver, 1972; Hanlin, 1976).

The changes in wall structure which accompany the formation of the first and second conidia have been described in particular detail for *Aspergillus clavatus* by Hanlin (1976). The wall of the enlarging primary conidium appeared thin, smooth and homogeneous except for a dark outer wall layer. When the conidium had attained full size a septum with a central pore formed at the base, just inside the phialide apex, by invagination of the plasmalemma (Fig. 13,14). Following septum formation the phialide protoplast pushed out to form the second conidium. As this occurred there was a clearly defined break on the outside of the phialide which demarcated the phialide apex. The wall of the second conidium was continuous with the inner wall of the apex and the outer wall of the phialide remained as a small collarette around the mouth of the phialide (Fig. 14). These observations by Hanlin (1976) are in agreement with other recent studies on the mechanism of production of the first and second conidia in phialogenous species. The character of the first formed conidium has been considered an important feature for distinguishing between phialidic and annellidic modes of conidium development (Cole & Kendrick, 1969a,b). However, as more ultrastructural information becomes available the distinction between the mechanism of primary conidium production in the two types is less clear than originally thought and consequently this would appear to reduce the value of this aspect of conidiation for taxonomic purposes (Hammill, 1974; Hanlin, 1976).

Whereas the first formed conidia of *Aspergillus* spp. arise as spherical protrusions from the phialides the second and subsequent conidia have a cylindrical shape at first and they only assume the typical spherical shape of the mature spore at a later stage in their development (Trinci *et al.*, 1968; Tokunaga *et al.*, 1973). As the second conidium enlarges it is cut off from the phialide by a septum and at the same time the septum separating it from the first conidium thickens and forms a connective between the two conidia. Subsequently formed conidia are also held in the chain by similar connections. Observations by Hanlin (1976) on *A.clavatus* indicate that a narrow channel remains open between adjacent conidia even when the septum thickens greatly (Fig.14). Later as the conidial chain increases in length the connective begins to break down (Fig. 16) and eventually the conidia separate.

The interstitial (basipetal) budding pattern of sporogenesis which occurs at the phialide apex must involve rapid synthesis of conidial wall material and it is of interest to note observations which have been made on the localisation of cytoplasmic vesicles and other membranous structures during this process. Observations on *Aspergillus nidulans* (Oliver, 1972), *A.fumigatus* (Ghiorse & Edwards, 1973) and *A.clavatus* (Hanlin, 1976) clearly demonstrate that cytoplasmic vesicles regularly occur in the peripheral cytoplasm of phialides engaged in conidia production. With *A.nidulans* Oliver

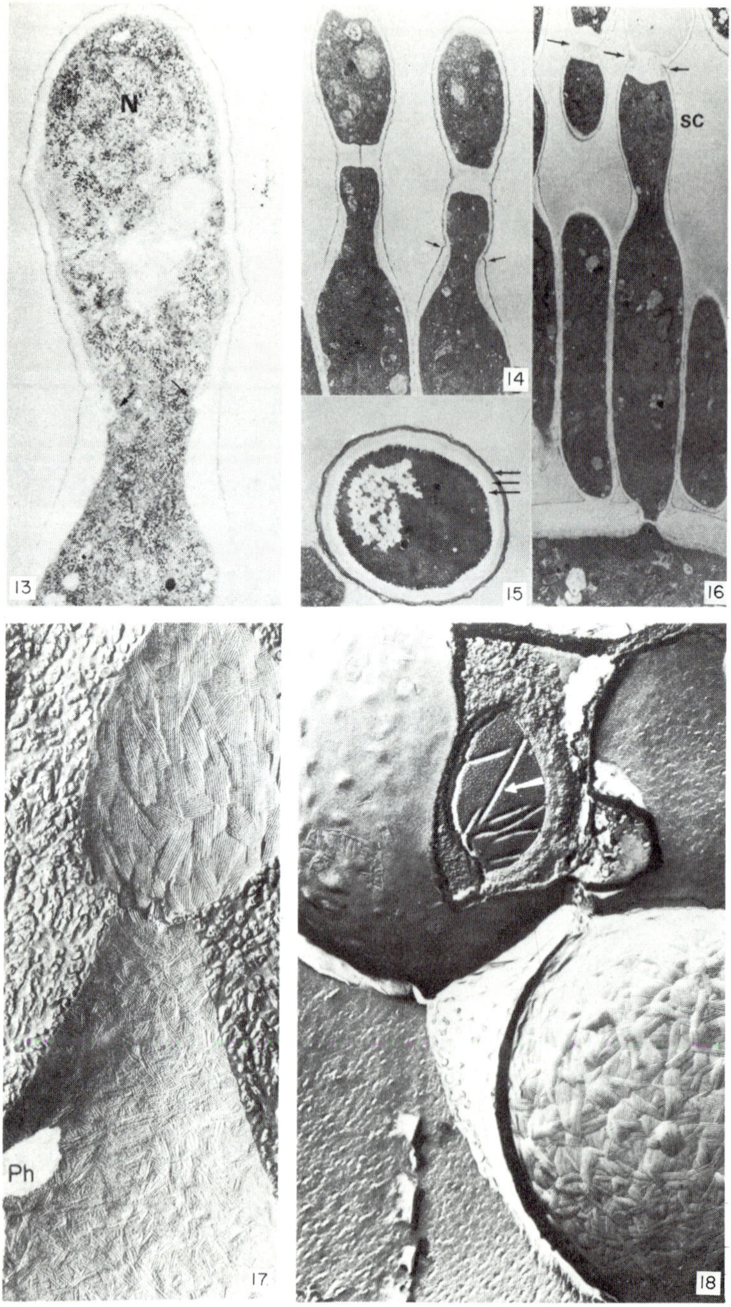
N
13
14
15
SC
16
Ph
17
18

Legends to Fig. 13 - 18

Fig. 13. Phialide and young conidium of *Aspergillus clavatus* at the time of initiation of septum development (arrows). From Hanlin (1976).

Fig. 14. Two phialides of *A.clavatus* showing the formation of subsequent conidia. A small collarette (arrows) at the phialide apex and a pore through the thickened septum at the base of the distal conidium are shown. From Hanlin (1976).

Fig. 15. Section through a mature conidium of *A.clavatus* showing three wall layers (arrows) and an electron transparent area. From Hanlin (1976).

Fig. 16. Section through phialide of *A.clavatus* bearing two conidia. Dissolution of the connective (arrows) between terminal and second conidium (SC) has begun. From Hanlin (1976).

Fig. 17. Replica of a frozen-etched preparation of *A.fumigatus* showing a phialide (Ph) bearing a young conidium. Rodlets are present on the surface of the phialide and conidium but show differences in distribution. From Ghiorse & Edwards (1973).

Fig. 18. Replicas of frozen-etched mature conidia of *A.fumigatus*. Lower conidium shows typical surface rodlets and upper conidium shows the rodlets covered by an external layer. The upper conidum also shows grooves (arrow) on the exposed surface of the plasma membrane. From Ghiorse & Edwards (1973).

(1972) also observed that plasmalemmasomes occurred in the phialide close to the collarette especially near the end of each budding cycle. Membranous structures resembling mesosomes were seen by Ghiorse & Edwards (1973) in the peripheral cytoplasm of *A.fumigatus* phialides. It was suggested that these structures might be functionally related to the cytoplasmic vesicles and that they participated in the synthesis of new wall material. Similar mesosome-like structures have previously been found in phialides of *Penicillium claviforme* where they are also thought to be involved in conidial wall synthesis (Zachariah & Fitz-James, 1967). Interestingly, few if any cytoplasmic vesicles have been observed within the developing conidia although with *A.fumigatus* Ghiorse & Edwards (1973) again detected the presence of the mesosome-like structures which might contribute to wall development in enlarging conidia.

The ageing or maturation changes which take place after the conidia have been delimited from the phialide apex represent an important aspect of conidia development. Comparison of the newly formed conidia with their mature counterparts clearly illustrates that substantial differentiation of the cell occurs during the maturation period. The cytoplasm of mature conidia of *Aspergillus fumigatus* appeared denser and much more compact than newly formed conidia (Ghiorse & Edwards, 1973). Cytoplasmic changes observed during the maturation of *A.nidulans* conidia led Oliver (1972) to suggest that autophagic processes reduced the total number of organelles without eliminating any one type. Indeed during the early stages of spore development some mitochondria were observed in this study lying within autophagic vacuoles.

The most striking of the changes in subcellular morphology which take place during maturation are those involving cell wall structure. A considerable number of studies have been made, with both TEM and SEM, on the conidial wall of various *Aspergillus* spp. Most of those studies have been concerned with mature rather than developing conidia and a comprehensive list of references to this work has been presented by Mangenot & Reisinger (1976). There has been considerable variation in the interpretation of wall subunits and the situation can be further confused by the use of different terminologies. For mature conidia of *Aspergillus* spp.the wall has been reported as containing either two or three layers. The most recent studies however on *A.aureolatus* (Vujicic & Muntanjola-Cvetkovic, 1973), *A.clavatus* (Hanlin, 1976), *A.nidulans* (Border & Trinci, 1970; Weisberg & Turian, 1971; Florance, Denison & Allen, 1972; Oliver, 1972), *A.tamarii* (Roquebert, 1973) and several other *Aspergillus* spp. (Tokunaga *et al.*,1973) do show that the mature conidia possess a three-layered wall (Fig.15).

A detailed account of the development of the three-layered conidial wall during maturation has been provided with *Aspergillus clavatus* (Ghiorse & Edwards, 1973). The outermost layer of young conidia was a thin 'cuticle' or electron-dense zone (10-20nm in width) which is characteristic of mature conidia. Concomitantly, an innermost wall layer, also of relatively high electron density became more apparent. This layer not only overlayed the plasma membrane, but also appeared to be intimately associated with it. This layer can apparently be destroyed by fixation with $KMnO_4$ (Ghiorse, 1969) and according to Ghiorse & Edwards (1973) this may account for the failure of some studies to detect this layer in *Aspergillus* conidia. Between the 'cuticle' and the innermost layer was a zone of medium electron density which did not change significantly as the conidia matured.

Pigmentation is another feature which develops during conidial maturation. Oliver (1972) observed with *Aspergillus nidulans* that the newest (lowest) five or so spores in each chain were white and that spores above these developed the green pigmentation. Development of the green pigmentation was apparently restricted to an exterior site under the superficial lamina following cytoplasmic maturation.

Studies on fine sectioned material using the TEM technique have proved invaluable for the elucidation of the changes in intracellular morphology during conidium development. Likewise the use of the SEM technique has proved extremely effective for the visualisation of the surface characteristics of developing and mature conidia. The maturation of *Aspergillus* conidia involves the elaboration of surface ornamentations and knowledge of the distribution and forms of these ornamentations is of considerable developmental and taxonomic interest. Locci & Quaroni (1972) have delimited the species of *Aspergillus* into five different categories according to the topography of conidial surfaces as visualised by SEM. Development of the surface ornamentations during the maturation of *A.niger* and *A.oryzae* conidia has been followed by Tokunaga *et al.*(1973).

Some of the observed changes in wall structure and pigmentation which take place during maturation clearly result from exposure to the aerial environment. When conidia are formed in liquid media or within agar they usually lack the surface ornamentations which are produced in subaerial culture (Mangenot & Reisinger, 1976). Conidia of *Aspergillus niger*, for example, when produced in submerged culture do not develop the thick outer cuticle or the black pigmentation which is characteristic of the species (Anderson & Smith, 1971b).

Conidia which form in an aerial environment possess an external film which is irregularly raised from the underlying coat and it is in this space, between the film and coat, that ornamentation products accumulate (Mangenot & Reisinger, 1976). In liquid media this external film is replaced by a mucilagenous zone. Thus Mangenot & Reisinger (1976) have concluded that while the appearance of ornamentations depends on the genetic constitution of the species the morphological expression of this ornamentation is closely dependent on hydrometric conditions which allow the formation of the external film.

Freeze-etching and replica techniques have also provided important information regarding the surface characteristics of developing and mature conidia. Freeze-etched preparations of mature conidia of *Aspergillus* and *Penicillium* have shown that the wall surfaces are covered with rodlets arranged in intricate textile-like patterns characteristic of the species (Hess, Sassen & Remsen, 1968; Hess & Stocks, 1969). More recently Ghiorse & Edwards (1973) observed the rodlet pattern of *A.fumigatus* conidia as they were formed at the phialide apex (Fig.17). Remnants of a thin external layer, covering the rodlet zone, were observed in the freeze-etched preparations (Fig. 18). Ghiorse & Edwards (1973) interpreted this to be remnants of a polyphosphate film as was suggested previously for a similar film found with *Penicillium* conidia (Fisher & Richmond, 1970). Another interesting feature of *A.fumigatus* conidia which was revealed by freeze-etching was the appearance of long deep groves

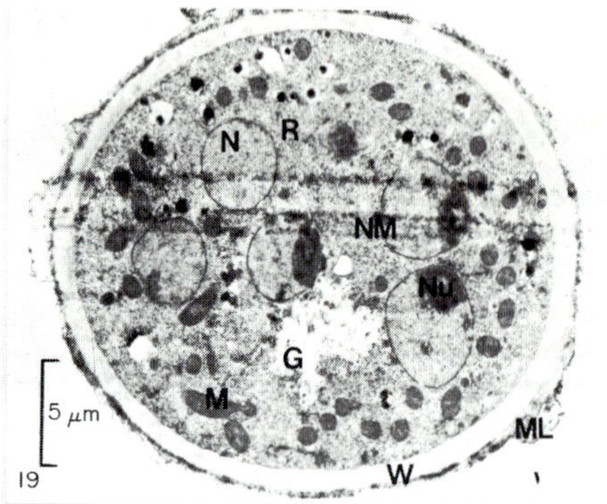
N
R
NM
Nu
G
M
5 μm
ML
W
19

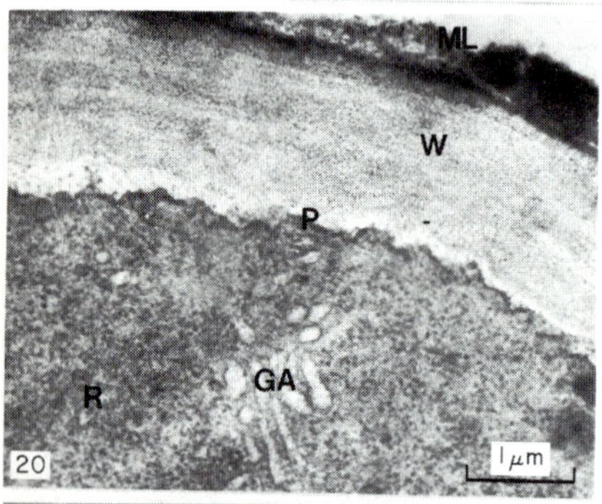
ML
W
P
GA
R
1 μm
20

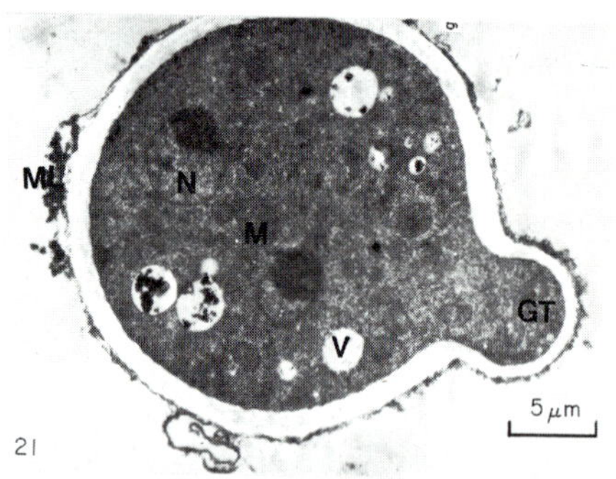
ML
N
M
GT
V
5 μm
21

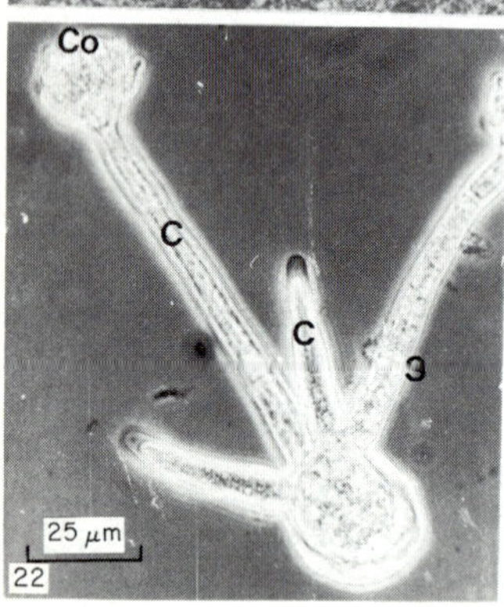
Co
C
C
25 μm
22

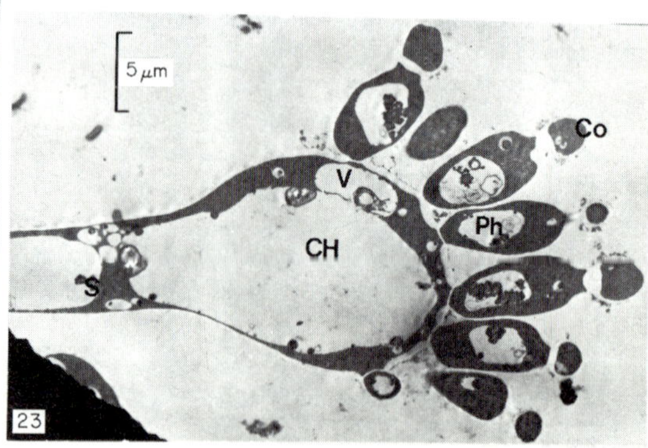
5 μm
Co
V
Ph
CH
S
23

Legends to Figs. 19 - 23.
Fig. 19. 24h giant cell showing general cytoplasmic contents, multinuclear arrangement, mitochondria, thick cell walls. Wall striations are apparent.

Fig. 20. Cell wall of giant cell showing Golgi apparatus, plasmalemma and wall striations.

Fig. 21. Early conidiophore emergence 4 h after dropping temperature to 30°C showing new wall formation from inside layers.

Fig. 22. Giant cell emitting two mature conidiophores and two immature conidiophores.

Fig. 23. Vesicle head illustrating phialides and conidia. There are no metulae present. Septa can be seen between vesicle and phialides. There is no melanin present on the secondary conidia.

Scale bar = 5μm except Fig. 20 bar = 1μm
Abbreviations: N, nucleus; M, mitochondrion; W, cell wall; P, plasmalemma; G, glycogen particles; R, ribosomes; GT, germ tube; V, vacuole; NM, nuclear membrane; ML, melanin layer; NU, nucleolus; GA, Golgi apparatus; C, conidiophore; CH, conidiophore head; Ph, phialides; Co, conidium; S, septum.

---

in the plasma membrane (Fig. 18).

While a number of studies have demonstrated the presence of rodlets on the surface of dormant conidia Ghiorse & Edwards (1973) also observed that rodlets occurred on the wall surfaces of the phialides and conidiophore of *Aspergillus fumigatus*. The rodlets of the phialides and conidia were of the same average diameter but they differed in number and distribution (Fig. 17). It appears therefore that these structures are not limited solely to the conidial wall surface and studies with other species will no doubt confirm this widespread occurrence.

Microcycle conidiation

Microcycle conidiation is the recapitulation of conidiation following spore germination without an intervening phase of mycelial growth. Microcycle conidiation or sporulation is now well documented in fungi (Anderson, 1977; Smith, 1977).

A feature common to most forms of microcycle conidiation is the increase in the size of the spore by spherical growth and is normally initiated by elevated temperature. During normal spore germination the first stage following activation usually involves the formation of a slightly enlarged cell which results from wall changes involving the uniform isotropic deposition of cell wall (Smith, Gull, Anderson & Deans, 1976). When the spore has achieved maximum or near maximum size the pattern of wall synthesis becomes polarised in that wall polymer deposition becomes limited to normally one or a few areas within the spore and these will subsequently become the site(s) of incipient germ tube emergence (Bartnicki-Garcia, 1973). Studies originally with *Aspergillus niger* (but now extended to other aspergilli) have shown that the incubation of conidia at 44°C or at a slightly lower temperature in the presence of added CO_2 (Anderson & Smith, 1972; Kuboye, Anderson & Smith 1976) caused massive conidial enlargement from *c* 3.5 μm to *c* 25 μm while inhibiting germ tube formation. Ultrastructural changes observed during the formation of these large spherical cells (Fig. 19) included an increase in the number of nuclei and the production of an excessively thick (2 μm) multi-layered wall (Fig. 20). This clearly indicated that isotropic wall deposition continued during the prolonged spherical growth which occurred at the elevated temperature. The organizational relationships of wall synthesis and spherical cell growth have been extensively discussed by Anderson (1977).

When these giant cells of *Aspergillus niger*, which have been maintained at 44°C for 24h, are then incubated at 30°C there is a direct outgrowth or production of one or more large conidiophores (Fig.21) (Anderson & Smith, 1971b). Such conidiophores will produce abundant normal conidia (Fig.22) capable of typical growth at 30°C but also able to recapitulate the microcycle if incubated at 44°C. The conidiophore apparatus differs from the normal subaerial conidiophore by being somewhat smaller, shows occasional septa in the conidiophore and produces the phialides directly from the swollen vesicle (Fig.23).

THE CONTROL OF CONIDIATION

The process of conidiogenesis in the aspergilli is under genetic control (see Clutterbuck, this volume). The genes involved can be considered as: (1) those controlling inductive competence and (2) those controlling morphological competence. When the genes controlling inductive competence are altered by mutation the fungus cannot respond to an inductive stimulus. Similarly, when genes controlling morphological competence are altered by mutation abortive development occurs blocked at specific stages of conidiophore or conidia formation.

In genetically competent aspergilli conidiation normally occurs following a period of vegetative growth. It is often believed that vegetative growth and conidiation are incompatible and that the transition to conidiation is usually preceeded by some environmental parameter that limits the forward vegetative growth process. It is probably more correct to consider vegetative growth and conidiation as cellular processes which compete for limiting metabolic intermediates rather than as mutually exclusive phenomena. The point at which a vegetative mycelium changes to a conidiating system may well be the consequence of nutrient exhaustion or imbalance or to some other environmentally controlled limiting process. There are numerous examples in the literature where conidiation in the aspergilli can be associated with or follows from some degree of limitation of vegetative expression (for ref. see Smith & Anderson, 1973).

Induction of conidiation

Growth and differentiation under static subaerial conditions is the 'natural' manner of development of the aspergilli. Few of the aspergilli require vitamins or other growth factors and can grow well on simple media (Raper & Fennell, 1968). Nevertheless, the nature and composition of the medium can produce marked variation in growth characteristics and in the colouration and morphology of the conidial heads. The heterogeneous nature of surface growth and the consequent problems that this creates in analytical biochemical studies (Smith & Anderson, 1973) has prompted the use of submerged culture to study conidiation in the aspergilli. Favourable conidiation can now be achieved under flask culture and batch and continuous fermenter conditions (Galbraith & Smith, 1969a; Anderson & Smith, 1971a; Ng, Smith & McIntosh, 1973a). Submerged cultivation can lead to more homogeneous mycelium particularly if growth is in the filamentous rather than pellet form. Aspergilli growing in submerged culture will normally remain vegetative in growth form; however, induction of conidiation can usually be achieved by the reduction or exhaustion of assimilable nitrogen while carbohydrate is still available (Galbraith & Smith, 1969a; Vezina & Singh, 1975). Before considering some noticeable examples of conidiation induction studies with *Aspergillus niger* it is pertinent to examine the interesting studies of Axelrod on the induction of conidiation in *A.nidulans*.

Before a decision or commitment to conidiate is achieved there must be a specific amount of biomass or colony growth. With *Aspergillus nidulans* this point is reached by colonies grown from single conidia on surface cultures and also by mycelium in liquid cultures in about 20h at 30°C (Axelrod, 1972). Up to this time no exposure to a normal conidiation stimulus will lead to conidiation whereas after this point the mycelium becomes receptive to stimuli which in turn will lead to conidiation about 7-8h later. Mature colonies of filamentous fungi growing on agar surfaces normally have a zone of vegetative mycelium bordering the central area of conidiation. Since this purely vegetative area is approximately the distance travelled by the hyphal tips in 7-8h it would appear that there is no latent period before competence is achieved by hyphae in a mature colony. It is now believed that this period of vegetative growth is related to the known gene control of conidiophore induction (Axelrod, Gealt & Pastushok, 1973; Gealt & Axelrod, 1974) and indicates the time required to complete the endogenous gene-controlled activity (Pastushok & Axelrod, 1976).

Studies with the microcycle system of *Aspergillus niger* have further confirmed these observations. Giant cells must achieve a minimum diameter or volume before they are competent to support conidiation. Any limitation to their normal size (15-25μ m) either by early removal from the giant cell medium or by using a wide range of metabolic inhibitors leads either to reversion to a filamentous vegetative growth form or remaining as giant cells incapable of producing a conidiophore (Deans & Smith, 1977a).

Conidiogenesis in filamentous fungi is subject to a multitude of internal and external regulatory controls. Early studies aimed to unravel the physiological and biochemical complexities of conidiogenesis in the aspergilli made use of simple flask culture techniques (Galbraith & Smith, 1969a,b; Valenzuela-Perez & Smith, 1971; Smith & Valenzuela-Perez, 1971; Smith, Valenzuela-Perez & Ng,1971; Lloyd, Morris & Smith, 1970). Although such cultivation methods allowed adequate all-round contact between mycelium and medium (a serious defect in static surface cultures) they did possess other less advantageous aspects such as poor aeration, variation between samples and inoculum size and difficulty in regulating pH, etc. Thus although these studies disclosed some important aspects of conidiogenesis the methodology left many environmental parameters unexplored.

Using modern fermentation techniques (Bull & Bushell, 1976) it has become possible to examine conidiogenesis in the aspergilli under more exacting reproducible conditions including aeration, temperature, mechanical agitation, pH, incremental feeding and replacement culture. By such means the environmental regulatory factors can be critically analysed under experimental conditions not possible in the 'natural' surface or simple submerged cultivation. Yet it must always be recognised that this submerged conidiation is not the natural developmental form of the fungus and caution must prevail when extrapolating information derived from fermenter studies to traditional surface culture. Notwithstanding, the advantages of the fermenter system greatly outweigh the disadvantages.

An obvious disadvantage of submerged fermentation systems can be the heterogeneous nature of mycelial form and subsequent biochemical analysis will represent only a summation of many physiological states. To ensure a more worthwhile analysis it must be possible to achieve some degree of synchronous development of morphological events. Moderately synchronous conidiogenesis has now been achieved with *Aspergillus niger* using either medium replacement culture

(Anderson & Smith, 1971a) or by microcycle conditions (Kuboye *et al.*, 1976; Deans & Smith, 1977a).

Using the medium replacement technique (Anderson & Smith, 1971a) involving several litres of relatively synchronous mycelial culture the various developmental phases of conidiogenesis have been studied morphologically and biochemically (Smith & Ng, 1972; Ng, Smith & Anderson, 1972; Lloyd, Anderson, Smith & Morris, 1972; Ng, Smith & McIntosh, 1973b). The first morphological event of conidiogenesis, foot cell formation (stage 1), is induced by growth in a medium in which nitrogen is the limiting nutrient (LN medium). Foot cells cannot be induced in a high nitrogen medium in which both the nitrogen and carbon sources become exhausted simultaneously or in a medium of low oxygen tension. Conidiophore elongation (stage 2) occurs in LN medium after the exhaustion of exogenous nitrogen although the continued presence of the carbon source (glucose) is essential. If the mycelium is left in this medium no further morphological development will occur other than the continued growth of the immature pointed conidiophore. In this way it resembles the *bristle* mutant of *A.nidulans* which forms immature conidiophores of indefinite length (Martinelli & Clutterbuck, 1971; see also Clutterbuck this volume). Replacement of the culture into a new medium containing a nitrogen source and a TCA cycle intermediate (citrate) as the carbon source will induce vesicle and phialide formation (stage 3). At this stage conidia will not develop from the phialides (a condition not unlike the *abacus* mutant of *A.nidulans*) and indeed prolonged exposure to this medium can often result in dedifferentiation with the formation of conidiophores from the phialides. Multiple dedifferentiations are not uncommon. Conidia formation (stage 4) can most effectively be induced by transferring the culture into a medium with glucose as the carbon source and nitrate as the nitrogen source. Thus by means of nutrient replacement in the controlled environment of the fermenter it has been possible to follow the synchronous maturation of conidiophores. The biochemical relevance of these studies will be discussed later.

One of the most interesting and novel developments in fungal sporulation investigations has been the discovery of the microcycle system in several fungi including the aspergilli (for ref. see Smith, 1977). Subsequently it has been possible to develop the full microcycle of *Aspergillus niger* in fermenter culture affording rigorous control on the developmental phases but also producing large volumes of biomass for biochemical analysis. This system undoubtedly offers many advantages over conventional mycelial conidiation because of the limited amount of vegetative growth present before induction of the conidiophore. Also the entire conidiating apparatus arises from a single cell - the giant cell, which will facilitate a more accurate measurement of relevant changes. The success of the fermenter system depended greatly on the synchronous activation and development of the giant cell. To increase the synchrony index of the inoculated conidia a variety of methods encompassing both induction synchrony and selection synchrony were considered. However, washing the conidia repeatedly in sterile distilled water was found to yield the greatest number of conidia in a condition that permitted a high degree of synchrony to be achieved in the microcycle process. It has been found that establishing a synchronous giant cell phase greatly facilitates subsequent synchrony in later stages of conidiogenesis. The relationship of the development pattern to time is shown in Fig.24.

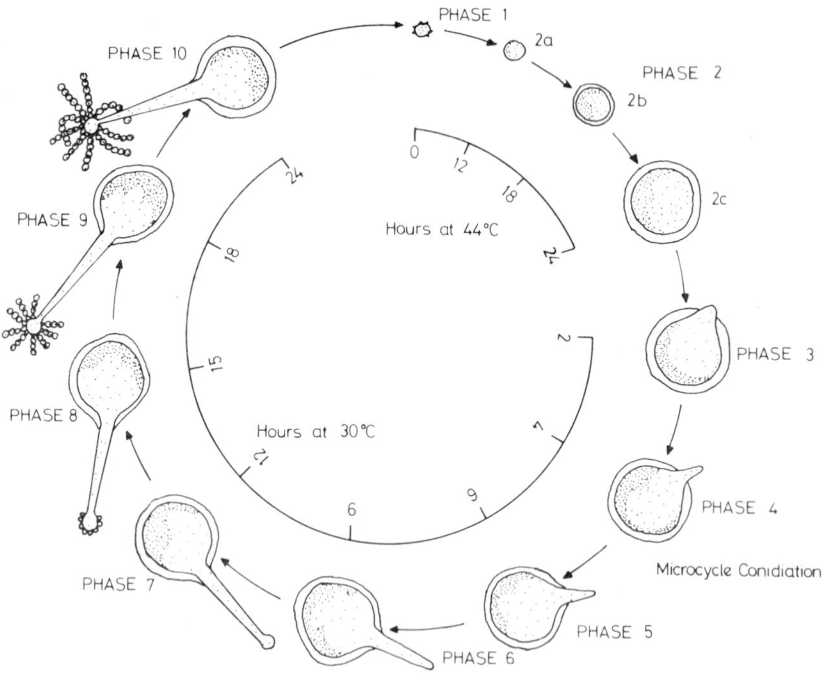

Fig. 24. The microcycle clock of *Aspergillus niger*.

In batch-type fermentation systems the chemical nature of the medium environment is by nature transient because of nutrient uptake and waste product release into a closed system. For this reason it becomes impossible to deduce from a batch culture system whether conidiation results from nutrient exhaustion or limitation or from a limitation of growth rate imposed by these conditions. Attempts to separate these phenomena have been sought by using continuous culture techniques which will allow a study of an organism at various growth rates and under various metabolic steady states.

Sporulation has been studied in several fungi by means of continuous culture techniques (for ref. see Smith, 1977) but only the studies with *Aspergillus niger*, in particular, will be examined (Ng, Smith & McIntosh, 1973a).

Before examining the factors regulating conidiogenesis it is pertinent first to consider the concepts of balanced and unbalanced growth (Bu'lock, 1975). When fungal growth is controlled by the intrinsic characteristics of the fungus rather than the environment i.e. the rate of nutrient uptake into the mycelium and rate of utilisation is constant, the fungus is exhibiting balanced growth. The rate at which growth rate declines is genetically specific and environmentally dependent and should be a reflection of the degree of imbalance in the various metabolic functions within the organism. Resulting from growth rate changes there will be qualitative and quantitative changes in metabolic intermediates with concomitant changes in metabolic pathways and resultant morphological expression. Growth becomes unbalanced when the growth rate is limited by the reduction in availability of an essential nutrient factor or by the influence of other environmental factors which can influence cellular activity viz. temperature, pH, O_2 or CO_2 tensions etc. Symptoms of unbalanced growth include excretion of intermediary metabolites, alternative metabolic pathways, accumulation of storage polymers, and morphological differentiation.

It has long been considered that there is a relationship between growth limitation and sporulation in fungi. Numerous studies have shown that sporulation in fungi often follows on from nutrient exhaustion (for ref. see Hawker, 1966; Smith & Galbraith, 1971; Smith & Berry, 1974). Although most of these batch studies showed that sporulation occurred after the fungus had been exposed to growth limiting conditions it is not always the case that unbalanced growth will lead to sporulation type differentiation. For instance, carbon exhaustion can often result in autolysis. Undoubtedly, the main requirement for sporulation is a movement away from balanced growth but the manner of achieving the unbalanced state is the determining feature. Continuous cultivation studies represent the only experimental means of elucidating this particular aspect of conidiogenesis.

Ng *et al.* (1973a) have examined the effect of growth rate and the nature of several substrate limitations on conidiation of *Aspergillus niger*. When citrate was the limiting factor conidiation intensity varied inversely with dilution rate. Under glucose limitation conidiation was difficult to achieve; at low dilution rates when limited conidiation occurred steady state could not be maintained because of the onset of autolysis. When a steady state was achieved at higher dilution rates there was no conidiation while nitrate limitation with citrate gave conidiophore initiation at low dilution rates but also early wash-out. However, with glucose limited nitrate medium good conidiation can be achieved.

The results of these experiments are very illuminating. Although the experiments in which dilution rate was varied under citrate limitation show that conidiation can be controlled by growth rate, similar studies with glucose limitation further indicate that medium composition must also be considered. Thus conidiogenesis in *Aspergillus niger* is determined by the concentration and nature of the medium components in the culture medium. Limitation of growth brought about by the control of growth rate/dilution rate implies that the intensity of the limitation promotes conidiation. Furthermore, the type and character of this limitation is also critical. Consequently, it can be seen that conidiation in *A.niger* results from specific types of limitation to the normal unlimited pattern of vegetative growth. When growth is balanced a vegetative growth pattern will prevail without conidiation. Although nutrient limitation is one of the more obvious ways of limiting fungal growth other environmental factors such as light, temperature, pH etc., which have been shown to influence conidiation could well act indirectly by influencing essential growth limiting reactions. Thus growth rate must certainly be the central factor determining the nature and extent of conidiation in fungi.

Finally, it should be noted that sporulation of filamentous fungi in continuous culture may often show considerable reduction in the complexity of the conidiophore (Righelato, Trinci, Pirt & Peat, 1968; Ng *et al.*,1973a). With *Aspergillus niger* conidiophores were often characterised by possessing small vesicles and few phialides and sometimes conidia developed from modified hyphal tips. This may well indicate that the complex conidiophore apparatus typical of many filamentous fungi may not necessarily need to precede conidia formation in liquid culture.

BIOCHEMICAL ASPECTS OF CONIDIATION

The complex ultrastructural and enzymological composition of living cells is now well documented. The nature of this organization, in particular that of cellular compartmentalization, presents bewildering problems in the analysis and interpretation of the myriad of biochemical reactions that occur during growth and subsequent development of an organism. As a consequence most meaningful biochemical analyses have been concerned with unicellular organisms although limited studies are now being made with filamentous multicellular fungi (for ref. see Killik & Wright, 1974; Smith & Berry, 1974; Smith, 1977).

It is now generally accepted that differential gene activation is the main factor controlling cellular differentiation and that such activity will be expressed in quantitative and qualitative changes in the role of enzyme synthesis. It is further recognised that enzyme activity can subsequently be controlled by stabilization, activation, modulation and allosteric factors, while the availability of low molecular weight precursors is of critical importance (Smith, 1977). The current accepted practice of crude cell smashing and concomitant analysis of the enzyme homogenate must surely be of limited use and such results can only have restricted value, particularly when complex integrated biochemical pathways are analysed. *In vivo* analysis of cellular activity must be technically advanced because only in this direction can a true realisation of the metabolic activity of a cell be appreciated and documented.

Yet another important factor that must be recorded is the use in such studies of cell populations where the performance of a cell or organism is equated to the average performance of the culture population. Thus from a biochemical point of view all studies carried out with asynchronous populations in batch and continuous culture are truly misleading. Only with the development of synchronous populations will it be possible to approach the recognition of the biochemical activities of the components of a microbial population. Again the filamentous nature of fungi such as the aspergilli in which there is a proven range of biochemical activity along a filament is further evidence of the complexity of these organisms (Smith & Anderson, 1973).

Undoubtedly the microcycle system represents a major experimental discovery that will permit, under synchronous conditions, the examination of fungal development and in particular the factors regulating conidiogenesis.

This section will not attempt to exhaustively examine the full range of biochemical studies that have been carried out during conidiation in the aspergilli but will arbitrarily select a few examples. A fuller appreciation can be obtained from selected reviews (Smith & Anderson, 1973; Smith & Berry, 1974; Smith, 1977).

Cell wall formation

The fungal cell wall is a complex assembly of polymers in which the chemical nature of the polymers and their spatial relationships to each other can change during growth and development (Bartnicki-Garcia, 1968, 1973; Rosenberger, 1976). However, there are few documented analyses of the chemical and physical changes in wall structure associated with sporulation in filamentous fungi. A relationship between the α-1, 3 glucan/α-1, 3 glucanase system in the regulation of cleistothecium and conidiophore formation in *Aspergillus nidulans* has been demonstrated by Zonneveld (1974, and this volume). Cleistothecia were formed when α-1, 3 glucan and α-1, 3 glucanase were present in adequate levels; the more α-1, 3 glucan available the more cleistothecia were formed. When high quantities of glucose were present in the medium there was a concomitant increase in the level of α -1, 3 glucan and a parallel decrease in conidiogenesis. Zonneveld (1974) considered that the synthesis of α-1, 3 glucan prevented conidiophore initials from breaking through the cell wall which had thickened considerably.

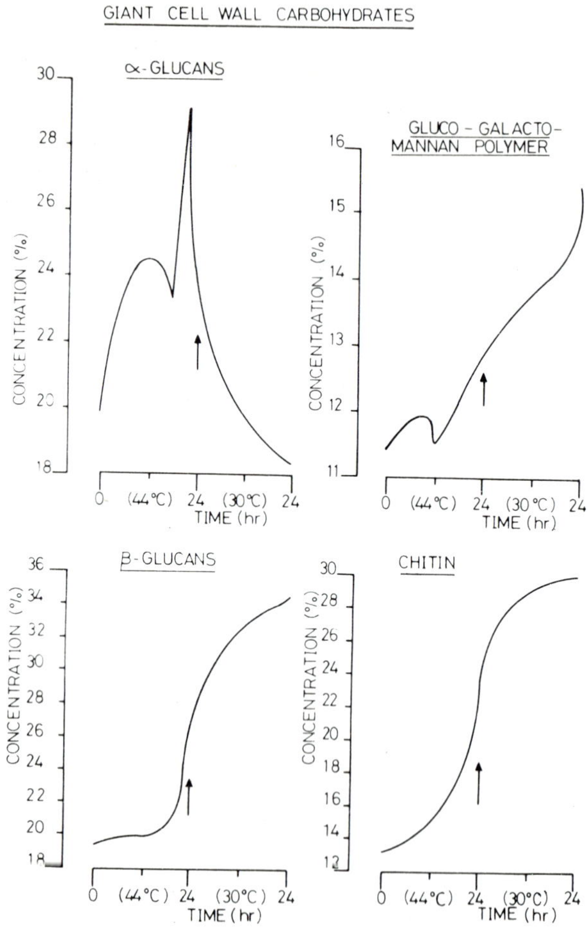

Fig. 25. Changes in cell wall components during microcycle conidiation of *Aspergillus niger*.

Deans & Smith (1977b) have recently analysed the chemical changes in wall composition during synchronous microcycle conidiation in *Aspergillus niger*. During the first 12h of giant cell formation there is a build up of the α -glucan fraction in the cell wall (Fig. 25). Between 12-16h a drop in level occurs but then increases again to a maximum prior to the shift-down in temperature to facilitate conidiophore outgrowth. During conidiophore development there is a continuous decrease in α-glucan level whereas the other wall components continue to increase. This may indicate that the α-glucans function as storage compounds during the vegetative giant cell formation and are subsequently utilised during later conidiogenesis. In contrast, the β-glucans, equal quantitatively to the α-glucans during vegetative growth and which continue to be formed during conidiation, are almost certainly occuping a structural role in wall organisation.

Chitin levels continue to increase throughout the entire microcycle and the synthetic enzyme, chitin synthase exhibits peak activity around 11-13h at 44°C and about 12h at 30°C. The heterogeneous glucose-galactose-mannan fraction does not appear to contribute greatly to the wall fabric, levels not rising above 15%, although the steady overall increase during giant cell formation infers incorporation within the wall matrix.

Cell wall protein levels increased during the first 16h at 44°C coinciding with maximum rate of wall formation. After this there was a steady decrease in wall protein but increasing again during conidiogenesis at 30°C. Cell wall lipid dropped sharply during the first 16h at 44°C possibly being utilised during giant cell formation. Later when the giant cell was nearing its maximum size lipid formation again increased, but dropped again during outgrowth at 30°C. Lipid synthesis again increased when conidia were being formed.

Carbon metabolism

The catabolism of carbon compounds has been extensively studied during conidiogenesis in the aspergilli. These studies have sought to elucidate the relative contributions of the two main pathways viz. the Embden-Meyerhof-Parnas and pentose phosphate, by means of selective inhibitors, measurement of specific enzymes, isotope labelling techniques, and measurement of intermediates (Carter & Bull, 1969; Smith & Valenzuela-Perez, 1971; Valenzuela-Perez & Smith, 1971; Smith *et al.*,1971; Ng *et al.*,1973b; Ng, Smith & McIntosh, 1974).

Vegetative mycelium generally exhibits high EMP activity whereas there is a progressive increase in HMP activity as the conidiophores are initiated and develop. Thus conidiation requires the direct oxidation of glucose through the HMP pathway generating reducing potential to be used in the biosynthetic reactions of conidiogenesis. The manner in which conidiation is initiated can be an important determining factor in enzyme patterns. Thus although an increase in the PP pathway generally accompanies induction of conidiation in most media, such an increase does not occur when glutamate is present in the medium (Smith *et al.*,1971). The presence of glutamate obviates the necessity of generating reducing potential. Clearly, there are many biochemical routes to achieving conidiation in *Aspergillus*.

Oxidative metabolism appears to be an essential function for conidiogenesis in the aspergilli and environmental conditions which favour glycolysis will prevent conidial initiation and development. However, the relative roles of the TCA cycle and the related glyoxylate cycle in conidiogenesis is still not clearly understood. High activity of key enzymes of the TCA cycle during conidiophore development in replacement cultures of *Aspergillus niger* (Ng *et al.*, 1972) as well as the stimulation of conidiogenesis by TCA cycle acids (Galbraith & Smith, 1969a) give weight to the hypothesis that oxidative metabolism is an essential feature of *Aspergillus* conidiation.

As a prelude to a comprehensive examination of the energetics of microcycle conidiation and in particular the role of the mitochondria several methods have been examined for extracting intact organelles from the giant cell system. A cell smash method utilising the Braun MSK rotary homogeniser has been successfully used to break through the exceptionally thick cell wall and release intact mitochondria (Davis & Smith, 1977). Mitochondrial integrity was checked by using a specific marker enzyme, fumarase (Fig. 26). Wall digestion studies have also permitted the formation of protoplasts from giant cells and present studies are aimed at producing protoplasts in quantities suitable for biochemical analysis (Davis, d'Avillez-Paixao, Deans & Smith, 1977).

Isocitrate lyase but not malate synthase has been shown to increase in activity during conidiogenesis in *Aspergillus niger* in flask culture (Galbraith & Smith, 1969b), replacement culture (Ng *et al.*,1972) and in continuous culture (Ng *et al.*,1974). It has been considered that the glyoxylate formed by isocitrate activity is transaminated with alanine to form glycine, a precursor of RNA synthesis. Glyoxysome particles have not yet been successfully isolated from conidiating systems.

Finally, limited studies with *Aspergillus niger* have shown that major fluctuations occur in the levels of glycolytic intermediates during conidiation (Smith & Valenzuela-Perez, 1971; Smith & Ng, 1972). The interpretation of such results is difficult but must surely represent a more accurate and meaningful approach to understanding *in vivo* metabolic activity.

Macromolecular synthesis

There have been few studies on the nature of nuclear expression during sporulation in filamentous fungi (for ref. see Smith,

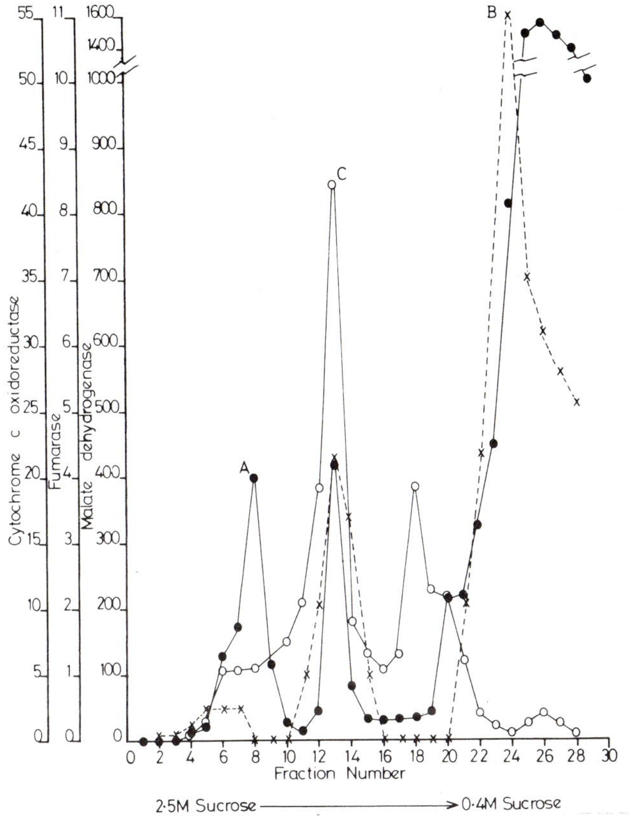

Fig. 26. Distribution of marker enzymes after linear sucrose density gradient centrifugation of cell extracts of *Aspergillus niger* grown under microcycle conidiation conditions.

1 ml fractions were collected from the bottom of the centrifuge tube and were numbered in order of collection. A, malate dehydrogenase (●—●), B, fumarase (X—X) and C, cytochrome c oxidoreductase (o—o) (Davis & Smith, 1977).

1977) and in particular the aspergilli do not appear to have been studied. However, in this laboratory, studies have now commenced on the analysis of the nature of nuclear expression in microcycle conidiation. Preliminary studies using a wide range of selective antimetabolites have set out to determine the relationship between cellular inhibition of specific metabolic functions and expression of this inhibition in morphogenesis (Deans & Smith, 1977a; D.Duncan, D.Berry & J.E. Smith unpublished results).

Table 1. Effects on cell diameter and microcycle development of addition of inhibitors at various stages of giant cell formation of *Aspergillus niger* (Deans & Smith 1977a)

Time of Addition (h at 44°C)	Cell diameter after 24h at 44°C (μ m)			Phase reached in microcycle after 24h at 30°C (*see Fig.24)		
	Actinomycin D (50 μ m ml^{-1})	Chloramphenicol (60 μ m ml^{-1})	Cyclohex-imide (20 μ m ml^{-1})	Actinomycin D (50 μ m ml^{-1})	Chloramphen-icol (60μ m ml^{-1})	Cyclo-heximide (20μm ml^{-1})
Control	23	23	23	10	10	10
Zero	6	6	6	2a	2a	2a
2	6:6*	6:6	6:6	2a:2a	2a:2a	2a:2a
4	6:6	14:6	6:6	2a:2a	5:2a	2a:2a
6	6:6	14:6	6:6	2a:2a	6:2a	2a:2a
8	10:11	18:12	11:11	2b:2b	9:2b	2b:2b
10	21:14	20:15	14:13	9:2b	10:2b	2b:2b
12	20:14	20:21	14:14	9:2b	10:4	2c:2b
14	21:16	21:21	18:20	10:2b	10:8	6:4
16	23:18	20:23	23:19	10:2c	10:9	7:6
18	23:23	23:23	23:21	10:9	10:10	10:7
20	23:23	23:23	23:23	10:9	10:10	10:8
22	23:23	23:23	23:23	10:10	10:10	10:8
24	23:23	23:23	23:23	10:10	10:10	10:8

* Figures before colon refer to inhibitor being removed after 2h contact. Figures after colon refer to no inhibitor removal values.

The inhibitors studied included actinomycin D, mitomycin C, 5-fluorouracil, 8-azaguanine, rifampicin, puromycin, cycloheximide, emetine, tetracycline, 2,4-dinitrophenol and sodium azide. All inhibitors gave some degree of inhibition when added at time of inoculation and in all cases conidiation was more susceptible to inhibition than giant cell formation. Inhibitor concentrations that reduced spherical cell size below *c* 18 μm (control 23 μm) caused total inhibition of subsequent conidiophore outgrowth. Since this effect could be brought about by all inhibitors irrespective of their mode of action these results show a remarkable similarity to the studies of Axelrod (discussed earlier) which showed that a critical cell volume or biomass was necessary before conidiation could occur in *Aspergillus nidulans*. Thus there must be a critical giant cell size before conidiogenesis can occur. Any decrease below a specific cell volume prevents the realization of conidiogenesis even in ideal initiating conditions.

In a second series of experiments actinomycin D, cycloheximide or chloramphenicol were added for limited periods of time to the giant cells and then removed by centrifugation. In this way it was possible to show that the critical period of inhibition of actinomycin D and chloramphenicol was approximately 10h of giant cell formation whereas cycloheximide retained potent inhibition up to 14-16h of giant cell formation (Table 1). Further elucidation of these results is being sought at the molecular level.

REFERENCES

ANDERSON,J.G. (1977). Temperature induction of fungal development. In *The filamentous fungi*. Vol.3. Edited by J.E.Smith & D.R.Berry. London: Edward Arnold in press.

ANDERSON,J.G. & SMITH,J.E. (1971a). Synchronous initiation and maturation of *Aspergillus niger* conidiophores. *Transactions of the British Mycological Society* 56, 8-29.

ANDERSON,J.G. & SMITH,J.E. (1971b). The production of conidiophores and conidia by newly germinated conidia of *Aspergillus niger* (microcycle conidiation). *Journal of General Microbiology* 69,185-97.

ANDERSON,J.G. & SMITH,J.E. (1972). The effects of elevated temperatures on spore swelling and germination in *Aspergillus niger*. *Canadian Journal of Microbiology* 18, 289-97.

AXELROD,D.E. (1972). Kinetics of differentiation of conidiophores and conidia by colonies of *Aspergillus nidulans*. *Journal of General Microbiology* 73, 181-4.

AXELROD,D.E., GEALT,M. & PASTUSHOK,M. (1973). Gene control of developmental competence in *Aspergillus nidulans*. *Developmental Biology* 34, 9-15.

BARTNICKI-GARCIA,S. (1968). Cell wall chemistry, morphogenesis and taxonomy in fungi. *Annual Review of Microbiology* 22, 87-108.

BARTNICKI-GARCIA,S. (1973). Fundamental aspects of hyphal morphogenesis. *Symposium Society of General Microbiology* 23, 245-67.

BOJOVIC-CVETIC,D. & VUJICIC,R. (1974). Ultrastructure of conidiophores in *Aspergillus flavus*. *Transactions of the British Mycological Society* 63, 131-5.

BORDER,D.J. & TRINCI,A.P.J. (1970). Fine structure of the germination of *Aspergillus nidulans* conidia. *Transactions of the British Mycological Society* 54, 143-52.

BULL,A.T. & BUSHELL,M.E. (1976). Environmental control of fungal growth. In *The filamentous fungi*, Vol.2,pp.1-31 Edited by J.E. Smith & D.R. Berry, London: Edward Arnold.

BU'LOCK,J.D. (1975). Secondary metabolism in fungi and its relationship to growth and development. In *The filamentous fungi*, Vol. 1, pp.33-58. Edited by J.E. Smith & D.R. Berry London: Edward Arnold.

CARTER,B.L.A. & BULL,A.T. (1969). Studies of fungal growth and intermediary metabolism under steady and non-steady state conditions. *Biotechnology and Bioengineering* 11, 785-804.

COLE,G.T. & KENDRICK,W.B. (1969a). Conidium ontogeny in hyphomycetes. The phialides of *Phialophora*, *Penicillium* and *Ceratocystis*. *Canadian Journal of Botany* 47, 779-89.

COLE,G.T. & KENDRICK,W.B. (1969b). Conidium ontogeny in hyphomycetes. The annellophores of *Scopulariopsis brevicaulis*. *Canadian Journal of Botany* 47, 925-9.

DAVIS,B. & SMITH,J.E. (1977). Organelle biogenesis in *Aspergillus niger*. *FEMS Letters* 1, 51-3.

DAVIS,B.,d'AVILLEZ-PAIXAO,T., DEANS, S.G. & SMITH,J.E. (1977). Protoplast formation from giant cells of *Aspergillus niger*. *Transactions of the British Mycological Society* (in press).

DEANS,S.G. & SMITH,J.E. (1977a). The influence of metabolic inhibitors on microcycle conidiation in *Aspergillus niger*. Manuscript in preparation.

DEANS,S.G. & SMITH,J.E. (1977b). Wall biogenesis in microcycle conidiation of *Aspergillus niger*. Manuscript in preparation.

FISHER,D.J. & RICHMOND,D.V. (1970). The electrophoretic properties of some surface components of *Penicillium* conidia. *Journal of General Microbiology* 64, 205-14.

FLETCHER,J. (1971). Conidium ontogeny in *Penicillium*. *Journal of General Microbiology* 67, 207-14.

FLORANCE,E.R., DENISON,W.C. & ALLEN,J.T.C. (1972). Ultrastructure of dormant and germinating conidia of *Aspergillus nidulans*. *Mycologia* 64, 115-23.

GALBRAITH,J.C. & SMITH,J.E. (1969a). Sporulation of *Aspergillus niger* in submerged liquid cultures. *Journal of General Microbiology* 59, 31-45.

GALBRAITH,J.C. & SMITH,J.E. (1969b). Changes in activity of certain enzymes of the tricarboxylic acid cycle and the glyoxylate cycle during the initiation of conidiation of *Aspergillus niger*. *Canadian Journal of Microbiology* 15, 1207-12.

GEALT,M. & AXELROD, D.E. (1974). Co-ordinate regulation of enzyme inducibility and developmental competence in *Aspergillus nidulans*. *Developmental Biology* 41, 224-32.

GHIORSE,W.C. (1969). Morphological and cytological changes in germinating *Aspergillus fumigatus* conidia. M.S.Thesis. Rensselaer Polytechnic Institute, Troy, New York.

GHIORSE,W.C. & EDWARDS,M.R. (1973). Ultrastructure of *Aspergillus fumigatus* conidia. Development and maturation. *Protoplasma* 76, 49-59.

GROVE,S.N. & BRACKER,C.E. (1970). Protoplasmic organisation in hyphal tips among fungi: vesicles and Spitzenkörper. *Journal of Bacteriology* 104, 989-1009.

HAMMILL,T.M. (1974). Electron microscopy of phialides and conidiogenesis in *Trichoderma saturnisporum*. *American Journal of Botany* 61, 15-24.

HANLIN,R.T. (1976). Phialide and conidium development in *Aspergillus clavatus*. *American Journal of Botany* 63, 144-55.

HAWKER,L.E. (1966). Environmental influences on reproduction. In *The fungi* Vol. 2. pp.435-469. Edited by G.C.Ainsworth and A.S. Sussman. New York, London: Academic Press.

HEATH,I.B. & GREENWOOD,A.D. (1970). The structure and formation of lomasomes. *Journal of General Microbiology* 62, 129-37.

HESS,W.M. & STOCKS,D.L. (1969). Surface characteristics of *Aspergillus* conidia. *Mycologia* 61, 560-71.

HESS,W.M., SASSEN, M.M.A. & REMSEN, C.C. (1968). Surface characteristics of *Penicillium* conidia. *Mycologia* 60, 290-303.

KILLICK,K.A. & WRIGHT,B.E. (1974). Regulation of enzyme activity during differentiation in *Dictyostelium discoideum*. *Annual Review of Microbiology* 28, 139-66.

KUBOYE, A.O., ANDERSON,J.G. & SMITH,J.E. (1976). Control of autolysis of a spherical cell form of *Aspergillus niger*. *Transactions of the British Mycological Society* 67, 27-31.

LLOYD,G.T., MORRIS,E.O. & SMITH, J.E. (1970). A study of the esterases and their function in *Candida lipolytica, Aspergillus niger* and a yeast-like fungus. *Journal of General Microbiology* 63, 141-50.

LLOYD,G., ANDERSON,J.G., SMITH, J.E. & MORRIS,E.O. (1972). The effect of medium nitrogen and of conidiation on esterase synthesis in *Aspergillus niger*. *Transactions of the British Mycological Society* 59, 63-70.

LOCCI,R. & QUARONI, S. (1972). Investigations on fungal features by scanning electron microscopy. V. Further studies on ascosporic *Aspergillus* species. *Rivista di patologia vegetale* 4, 253-320.

LOVETT,J.S. (1975). Growth and differentiation in the water mold *Blastocladiella emersonii:* cytodifferentiation and the role of ribonucleic acid and protein synthesis. *Bacteriological Review* 39, 345-404.

MANGENOT,F. & REISINGER,O. (1976). Form and function of conidia as related to their development. In *The fungal spore: form and function* pp.789-847. Edited by D.J. Weber and W.M.Hess: New York: John Wiley and Sons.

MARTINELLI,S.D. & CLUTTERBUCK,A.J. (1971). A quantitative survey of conidiation mutants in *Aspergillus nidulans*. *Journal of General Microbiology* 69, 261-8.

McCLURE,W.K., PARK,D. & ROBINSON,P.M. (1968). Apical organisation in the somatic hyphae of fungi. *Journal of General Microbiology* 50, 177-82.

NG,A.M.L., SMITH,J.E. & Mc INTOSH,A.F. (1973a). Conidiation of *Aspergillus niger* in continuous culture. *Archiv für Mikrobiologie* 88, 119-26.

NG,A.M.L., SMITH,J.E. & McINTOSH,A.F. (1973b). Changes in activity of tricarboxylic acid cycle and glyoxylate cycle enzymes during synchronous development of *Aspergillus niger*. *Transactions of the British Mycological Society* 61, 13-20.

NG,A.M.L., SMITH,J.E. & McINTOSH,A.F. (1974). Influence of dilution rate on enzyme synthesis in *Aspergillus niger* in continuous culture. *Journal of General Microbiology* 81, 425-34.

NG,W.S., SMITH,J.E. & ANDERSON,J.G. (1972). Changes in carbon catabolic pathways during synchronous development of *Aspergillus niger*. *Journal of General Microbiology* 71, 495-504.

OLIVER,P.T.P. (1972). Conidiophore and spore development in *Aspergillus nidulans*. *Journal of General Microbiology* 73, 45-54.

PASTUSHOK,M. & AXELROD,D.E. (1976). Effect of glucose, ammonium and media maintenance on the time of conidiophore initiation by surface colonies of *Aspergillus nidulans*. *Journal of General Microbiology* 94, 221-4.

RAPER,K.B. & FENNELL, D.I. (1965). *The genus Aspergillus* : Baltimore, Williams and Wilkins.

RIGHELATO,R.C., TRINCI,A.P.J., PIRT,S.J. & PEAT, A. (1968). The influence of maintenance energy and growth rate on the metabolic activity, morphology and conidiation of *Penicillium chrysogenum*. *Journal of General Microbiology* 50, 399-412.

ROQUEBERT,M.F. (1973). Modalites de la sporogenese chez deux phialides *Aspergillus tamarii* kita et *Stilbothamnium nudipes* Haum. *Revue de mycologie* 38, 3-8.

ROSENBERGER,R.F. (1976). The cell wall. In *The filamentous fungi*. Vol.2, pp.328-344. Edited by J.E.Smith and D.R.Berry. London: Edward Arnold.

SMITH,J.E. (1977). Asexual sporulation in filamentous fungi. In *The filamentous fungi* Vol.3. Edited by J.E.Smith and D.R.Berry. London: Edward Arnold (in press).

SMITH,J.E. & ANDERSON,J.G. (1973). Differentiation in the Aspergilli. *Symposium Society of General Microbiology* 23, 295-337.

SMITH,J.E. & BERRY,D.R. (1974). *An introduction to biochemistry of fungal development.* London: Academic Press.

SMITH,J.E. & GALBRAITH,J.C. (1971). Biochemical and physiological aspects of differentiation in the fungi. *Advances in Microbial Physiology* 5, 45-134.

SMITH,J.E. & NG,W.S. (1972). Fluorometric determination of glycolytic intermediates and adenylates during sequential changes in replacement culture of *Aspergillus niger*. *Canadian Journal of Microbiology* 18, 1657-64.

SMITH,J.E. & VALENZUELA-PEREZ,J. (1971). Changes in intracellular concentrations of glycolytic intermediates and adenosine phosphates during the growth cycle of *Aspergillus niger*. *Transactions of the British Mycological Society* 57, 103-10.

SMITH,J.E.,VALENZUELA-PEREZ,J. & NG, W.S. (1971). Changes in activities of the Embden-Meyerhof-Parnas and pentose phosphate pathways during the growth cycle of *Aspergillus niger*. *Transactions of the British Mycological Society* 57, 93-101.

SMITH,J.E., GULL,K., ANDERSON,J.G. & DEANS,S.G. (1976). Organelle changes during fungal spore germination. In *The fungal spore: form and function*. pp. 301-354. Edited by D.J.Weber & W.M.Hess, New York: John Wiley & Sons.

TANAKA,K. & YANAGITA,T.(1963). Electron microscopy on ultrathin sections of *Aspergillus niger*. II. Fine structure of conidia-bearing apparatus. *Journal of General and Applied Microbiology* 9, 189-203.

TOKUNAGA,J., HARADA,K. & TOKUNAGA, M. (1971). A scanning electron microscopy of the spore formation process in genus *Aspergillus*. *Journal of Electron Microscopy* 20, 224-5.

TOKUNAGA,M., TOKUNAGA,J. & HARADA,K. (1973). Scanning and transmission electron microscopy of sterigma and conidiospore formation in *Aspergillus* group. *Journal of Electron Microscopy* 22, 27-38.

TRINCI,A.P.J., PEAT,A. & BANBURY,G.H. (1968). Fine structure of phialide and conidiospore development in *Aspergillus giganteus* 'Wehmer'. *Annals of Botany* 32, 241-9.

TSUKAHARA,T. (1970). Electron microscopy of conidiospore formation in *Aspergillus niger*. *Sabouraudia* 8, 93-7.

TURIAN,G. (1974). Sporogenesis in fungi. *Annual Review of Phytopathology* 12, 129-37.

TURIAN,G. (1975). Differentiation in *Allomyces* and *Neurospora*. *Transactions of the British Mycological Society* 64, 367-80.

TURIAN,G. (1976). Spores in Ascomycetes, their controlled differentiation. In *The fungal spore: form and function*, pp.715-88. Edited by D.J.Weber and W.M. Hess. New York: John Wiley & Sons.

VALENZUELA-PEREZ,J. & SMITH,J.E. (1971). Role of glycolysis in in sporulation of *Aspergillus niger* in submerged cultures. *Transactions of the British Mycological Society* 57, 111-9.

VEZINA,C. & SINGH,K. (1975). Transformation of organic compounds by fungal spores. In *The filamentous fungi,* Vol.1 pp.158-92. Edited by J.E.Smith & D.R. Berry. London: Edward Arnold.

VUJICIC,R. & MUNTANJOLA-CVETKOVIC,M. (1973). A comparative ultrastructural study on conidium differentiation in *Cladosarium*-like mutant 22B of *Aspergillus aureolatus. Journal of General Microbiology* 79, 45-51.

WEISBERG,S.H. & TURIAN,G. (1971). Ultrastructure of *Aspergillus nidulans* conidia and conidial lomasomes. *Protoplasma* 72, 55-67.

ZACHARIAH,K. & FITZ-JAMES,P.C. (1967). The structure of phialides in *Penicillium claviforme. Canadian Journal of Microbiology* 13, 249-56.

ZONNEVELD,B.J.M. (1974). α-1,3 glucan synthesis correlated with α-1,3 glucanase synthesis, conidiation and fructification in morphogenetic mutants of *Aspergillus nidulans. Journal of General Microbiology* 81, 445-51.

Biochemistry and Ultrastructure of Sexual Development in *Aspergillus*

B. J. M. ZONNEVELD

Genetisch Laboratorium der Rijksuniversiteit, Leiden, The Netherlands

INTRODUCTION

Biochemical studies on differentiation in *Aspergillus* dealing with germination, growth, conidiospore formation and fructification, are largely devoted to the process of transcription and translation. The implicit hypothesis is that differentiation can be defined as selective gene activation. It is clear that gene activity and its product, the enzyme, play an essential role in differentiation. Equally important but much less realised is the influence of substrates and food source on the regulation of differentiation. The organism uses its whole machinery to react to environmental and subsequent intracellular changes in food supply. Enzymes are important and essential but only tools to adapt to the changed substrate levels. If environmental changes are profound, such as the exhaustion of the carbohydrate supply, specific messenger synthesis may take place as a reaction to the changed environment. The carbon source provides elements for the synthesis of cell structures and energy for synthesis and basal metabolism. Exhaustion of e.g. glucose will at the same time affect the synthetic capacity and the energy level of the cell. Also exhaustion of other components of the medium such as nitrate and phosphate will effect synthesis and energy supply via protein and ATP synthesis.

In the present context attention is given to sexual

differentiation in *Aspergillus*. In particular, the composition of the medium is studied in relation to growth and sexual differentiation, knowing how strongly it might influence these processes. Some remarks are made about other environmental factors that might influence differentiation and attention is given to ultrastructure. In *Aspergillus* the main reserve material forms part of the cell wall, therefore the composition of the cell wall is described, and the significance of this reserve material in differentiation is discussed, followed by a description of the degrading enzymes. Synthesis and degradation of reserve material can be influenced by morphological mutation, a glucose analogue and manganese. However, glucan degradation is not the only possibility for *Aspergillus* to provide carbon and energy for differentiation; protein degradation can also play a decisive role. The role of cytoplasmic factors has been related to the above described biochemical parameters. Finally, cyclic adenosine monophosphate (cAMP) is discussed in relation to external circumstances, to changes in the energy level of the cell and to sexual differentiation and the importance of energy supply in differentiation is stressed.

THE ROLE OF ENVIRONMENTAL FACTORS ON SEXUAL DIFFERENTIATION

The composition of the medium

SALTS AND GLUCOSE. The medium generally used for *Aspergillus* contains $NaNO_3$,6g; KH_2PO_4, 1.52 g; $MgSO_4$, 0.52 g; and KCl, 0.52g/l (Pontecorvo, 1953). However, it appears that the salt concentration can be lowered by up to 5 times even if a high (6%) glucose concentration is used. Therefore, KCl can be omitted in the 1% glucose medium without any demonstrable effect allowing the experimenter to make a 50 x concentrated stock solution. *Aspergillus nidulans* grows on a wide range of glucose concentrations. In 9 cm petri dishes with 50 ml of medium, covered with a spore suspension (10 x 10^6 spores) it takes about 32h to consume 1% glucose, 2 days to consume 3% glucose, 3 days to consume 6% glucose and 6 days to consume 12% glucose. The higher glucose concentration leads to a decrease in conidiospore formation and an increase in cleistothecial development. The ratio at which glucose and $NaNO_3$ are jointly exhausted is 10 g glucose to 1.1 g $NaNO_3$/l.

TRACE ELEMENTS. The most commonly used trace element suspension is described by Cove (1966). It contains Zn, Mo, Bo, Mn, Cu and Fe. By adding 10 ml concentrated HCl to a 1000 times concentrated suspension a clear solution is obtained. The main source of contaminating trace elements is the agar, although the omission of

Mo, Bo or Fe does not have any effect on growth with 1% Difco agar and with Oxoid No 3 also Zn and Mn can be omitted (see also Agnihotri, 1967). Moreover, agar can also supply low amounts of carbon, nitrogen and sulphur (Payton, McCullough & Roberts, 1976). The role of manganese in growth and development and its effect on cAMP levels will be described later. By using a Ca specific chelator we could show that Ca found in the agar was necessary for asexual and sexual differentiation under circumstances where dry weight was not much affected. It is estimated that about 100 μM of Ca is needed per litre medium for 1% glucose. Because this element is not included in the above trace element solution it is possible that liquid cultures are hampered by shortage of calcium.

pH. During growth the pH of the medium may change depending mainly on the nitrogen source used. Sodium nitrate will lead to a pH of about 8, due to the no longer neutralized sodium ion. However, an ammonium salt such as NH_4Cl will give an acid medium which ultimately arrests growth due to the remaining chloride. This can be prevented by using a buffer or a neutralizer such as $CaCO_3$ or by using ammonium tartarate because in this case both ions, the ammonium and the tartarate, are used without a change in pH. When NH_4NO_3 is used, first the ammonium and then the nitrate is metabolized resulting in strong acidity. This can be prevented by decreasing the nitrogen amount used (remember that in Pontecorvo's medium with 1% glucose there is a 5 times surplus of nitrogen). Substituting $NaNO_3$ by peptone will result in a pH of about 5. So it is possible to vary the resulting pH of the medium by changing the nitrogen source, although of course the different nitrogen sources are not equivalent concerning their effects on metabolism.

Other factors influencing differentiation

Several other environmental factors have been described as important to differentiation in *Aspergillus* but this has not been sustained by careful biochemically orientated investigation. Factors such as pH, humidity, temperature, oxygen and osmotic pressure seem to be rather indifferent to *Aspergillus* whereas light and CO_2 are of considerable influence. A sterol mutant of *A. fenelliae* has poor asexual reproduction and has lost sexual reproduction (Kim & Kwon-Chung, 1974). This might indicate a role for steroids in differentiation.

LIGHT. Several reports show that light favours conidial production and that cleistothecia are predominantly found in the dark (Curran, 1971; Rai, Tewari & Sinha, 1967; Muntanjola-Cvetkovik &

& Vukic, 1972). Darkness during the first 24h seems important for cleistothecial production in *Aspergillus nidulans* (Zonneveld, unpublished results). It seems useful for *Aspergillus* in nature to make conidia if the mycelium is exposed (light) and cleistothecia if the mycelium is buried in the substrate (darkness).

CO_2. When a plate culture was placed in a closed space containing an alkali solution this trapping of the CO_2 was sufficient to prevent cleistothecial formation (Zonneveld, unpublished results). Biochemical analysis revealed that there was a low amount of reserve material synthesized and low α-1,3 glucanase activity, but pH was not affected. However, CO_2 trapping after the reserve glucan was formed still strongly inhibited α-1,3 glucanase. Both before and after glucose exhaustion, glucose is the main carbon and energy source and is catabolized via C_3 intermediates. The metabolism of the cell is strongly dependent on a C_2 source and CO_2 seems to be essential in that respect (anaplerotic carbon fixation) via pyruvate carboxylase (Skinner & Armitt, 1972).

ULTRASTRUCTURE

Information about the morphology of cleistothecial development is rather scarce.

The initials of *Emericella* (the name for the perfect stage of an *Aspergillus* group to which also *Aspergillus nidulans* belongs) are described by Benjamin (1955) as follows: A loosely looped coil of hyphae without differentiated ascogonia or antheridia arise usually at the base of a conidiophore. The hyphae branch profusely forming a clump of rather wide hyphae. Some of these branches anastomose while others form intercalary or terminal swellings which enlarge and eventually form the Hülle cells covering the ascocarp (cleistothecium). This description is confirmed by our own observations. The Hülle cells are partly (only?) intercalary or subterminal and not only terminal as suggested by Ellis, Reynolds & Alexopoulos (1973) (Fig. 1).

Later stages are difficult to describe. According to Benjamin (1955) the very young ascocarps are composed of wide intertwined hyphae making up 2-3 layers. In older ascocarps these wide hyphae are less clear and surrounded by 6-8 layers of narrow hyphae on which the Hülle cells are born. As the ascocarps ripen, however, these hyphae become more compressed and may ultimately be only one layer thick. Inside the ascocarps the asci each contain 8 wine-red ascospores. They are different for each species as shown by Locci (1972) with the SEM for many *Aspergillus* species and this also applies for the conidiospores (Fig. 2). In some *Aspergillus* species sclerotia are found. These are considered by some people as degenerated ascocarps (Rudolph,

1962; Rai *et al.*, 1967). They seem to develop under the same circumstances that are conducive to ascocarp formation.

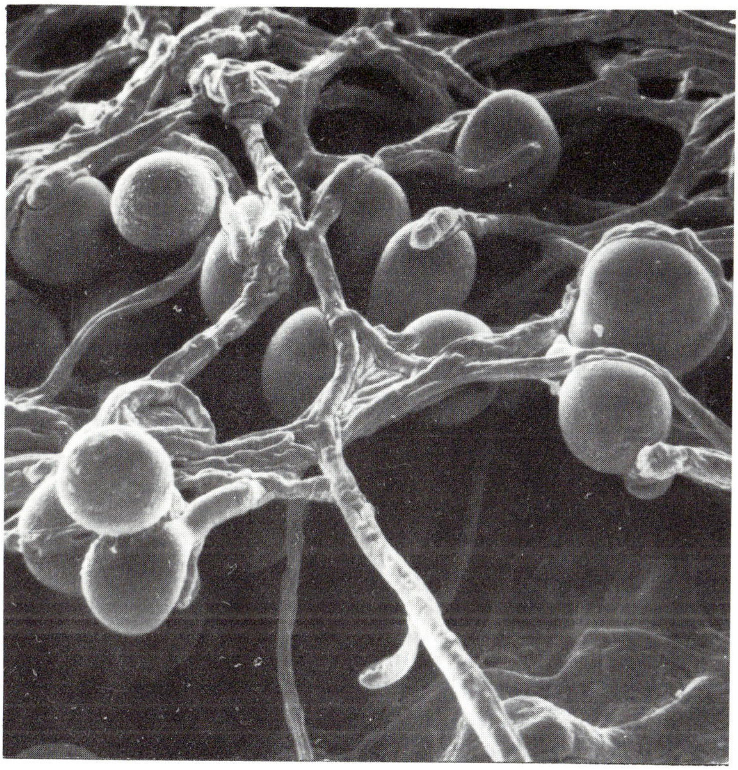

Fig. 1. Hülle cells of *Aspergillus nidulans* seen by scanning electron microscopy.

THE HYPHAL CELL WALL

The hyphal wall fulfills several functions to an organism. It gives the hyphae shape, protects against the environment, maintains the osmotic pressure and sometimes contains reserve material. Studies on part or the whole cell wall of different *Aspergillus* species (*A. niger*, Johnston, 1965a; *Aspergillus* spp. Ruiz-Herrera, 1967; *A. nidulans*, Bull, 1970; Zonneveld, 1971) can be summarised as follows:

The cell wall comprising at least 60% of the total dry weight of the mycelium is composed of several polymers. The most resistant, chitin, gives the cell its shape. A glucosamine mutant isolated by Katz & Rosenberger (1970), that forms swollen hyphae with intermediate concentrations of an osmotic stabilizer, illustrates this function of chitin. Intermingled with chitin is an alkali-insoluble glucan polymer, containing β -1,3 and β -1,6 linkages and according to Johnston (1965a) and Zonneveld (1971) also α-1,4 linkages. It comprizes about 30% of the dry weight of the cell wall and seems to be mixed with small quantities of mannose and galactose polymers. Lipids and proteins can also be found in the cell wall. The amount of lipid varies enormously and may constitute more than 40% of the dry weight in *Aspergillus nidulans* (Naguib, 1959). Melanin can also be an important component in older hyphae making up about 18% of the dry weight (Bull, 1970). In *A. niger* (Johnston, 1965b) and *A. aculeatus* (Gold, Mitzel & Segel, 1973) nigeran is also found. Nigeran has alternating α-1,3 and α-1,4 linkages (Johnston, 1965b is water-soluble, and can be up to 30% of the dry weight. In both (acleistothecial) organisms the accumulated nigeran is not degraded during starvation. The same applies to the alkali-soluble polymer of *A. niger*, α-1,3 glucan, found also in *A. nidulans*. The latter organism, however, is capable of degrading α-1,3 glucan during starvation and because it forms up to 40% of the dry weight it is the main reserve polymer during cleistothecial formation.

The results of Peberdy & Buckley (1973) with a fluorescent brightener on the hyphae of *A. nidulans* seems to indicate that the hyphal tips are devoid of α-1,3 glucan, suggesting that it is only accumulated in older hyphae.

An extracellular galactose derivative was isolated in small amounts from liquid cultures by Gorin & Eveleigh (1970).

Because the degradation of the cell wall plays an important role in differentiation glucanases will be considered.

GLUCANASES AND GLUCOSIDASES

The *Aspergilli* produce a range of glucanases and glucosidase The discussion here is restricted to the cleistothecial species *Aspergillus nidulans*. The following enzymes have been reported: β -galactosidase (Fantes & Roberts, 1973; Paszewski, Chojnacki, Litwinska & Gajewski, 1970); amyalse (7 isoenzymes, Kurzeja & Garber, 1973) and amylase, α-1,3 glucanase, β-1,3 glucanase (Zonneveld, 1972b).

α-1,3 glucanase is an exo-splitting enzyme whose only product is glucose. It is specific for α-1,3 glucan and does not split substrates with alternating α-1,3, α-1,4 linkages (nigeran) (Zonneveld, 1972a). Furthermore, it has been shown t be under strong glucose repression. But glucose is not the only sugar that gives repression. In fact, all sugars and some polysaccharides on which *Aspergillus* gives good growth show this repression (lactose, maltose, sucrose, sorbitol, glycerol, starch, arabinose, but not pustulan, α-methylglucoside, dextran or nigeran). The catabolite repression negative mutation *cre* A1 (Arst & Cove, 1973; Baily & Arst, 1975) does not affect the repression of β-1,3 glucanase.

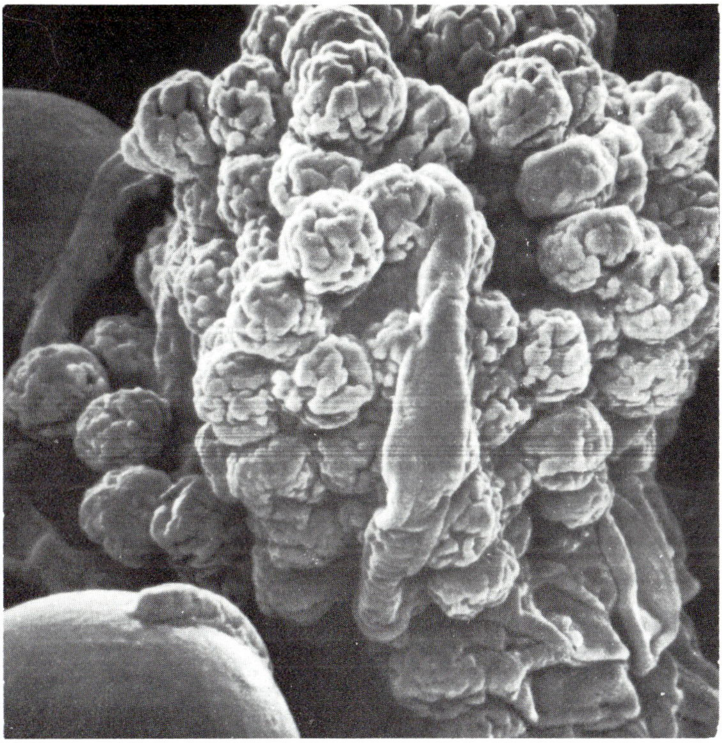

Fig. 2. Conidia of *Aspergillus nidulans* by scanning electron microscopy.

The other glucanases reported here and also invertase, β-glucosidase, isomaltase, maltase and lichenase are produced even when glucose is still present in the medium. Seemingly, no induction is needed, although addition of the substrate usually doubles enzyme activity. Activities toward pustulan (β-1,6 glucan) and trehalose are very low and no activity could be found with dextran or cellulose. The activity of most of the reported enzymes increases strongly after glucose exhaustion. This suggests that *Aspergillus* uses a whole battery of degrading enzymes when glucose is exhausted and it might then depend on which available substrates will function as energy supplier in fructification.

GLUCAN SYNTHESIS AND DEGRADATION DURING FRUCTIFICATION

Bainbridge *et al.* (1971) and Naguib (1959) have described what happens with several of the components of the mycelium in liquid cultures after glucose exhaustion. They reported that not only glucan but also lipid, protein, DNA and RNA were degraded. The function of this autolysis was, however, not indicated except for maintaining the metabolism of the cell.

Zonneveld (1972b) working with plate cultures has shown that α-1,3 glucan degradation is of prime importance during fructification. α-1,3 glucan forms the outer fibrillar layer of the cell wall of *Aspergillus nidulans*. This water-insoluble glucan is found in considerably varying quantities in the cell wall, increasing with the amount of glucose in the medium. This suggests a role as reserve material, substantiated by the finding that it is completely degraded after glucose exhaustion. Moreover, the addition of extra glucose is used mainly to form a α-1,3 glucan.

Low amounts of glucose (0.8%) result in an absence of both α-1,3 glucan and cleistothecial formation and the same is found when trace elements are omitted, mainly due to a manganese deficiency as will be shown later.

The other glucans of the cell wall (β-1,3, β-1,6 and α-1,4) do not take over as carbon and energy source when α-1,3 glucan is absent although a minor contribution cannot be excluded However, except for the fact that it is found in large quantities and that it is degraded by a specific enzyme α-1,3 glucanase, there is nothing exclusive about α-1,3 glucan. It provides a steay flow of carbon and energy which can also be provided by adding starch to the medium. The degradation of α-1,3 glucan is not influenced by a surplus of nitrate. This is because after glucose exhaustion nitrate reductase is quickly degraded (Hynes, 1973), and this stops nitrogen uptake. On the other hand, a surplus of glucose relative to nitrate is mainly directed to a α-1,3 glucan synthesis. In Fig. 3, dry weight, glucan content, α-1,3 glucanase activity and glucose exhaustion are shown for *Aspergillus nidulans*.

To analyse the importance of α-1,3 glucan in differentiation mutants or phenocopies can be used. The latter can be obtained by low glucose content in the medium as described earlier, by the use of the glucose analogue 2-deoxyglucose (Zonneveld, 1973) and by the omission of the trace element manganese. In all these cases α-1,3 glucan synthesis is absent and so also is cleistothecial formation. α-1,3 glucanase synthesis, absent with 2-deoxyglucose (0.1%) is low with 0.8% glucose but normal with manganese deficiency. This suggests that under completely different circumstances there is no direct correlation between

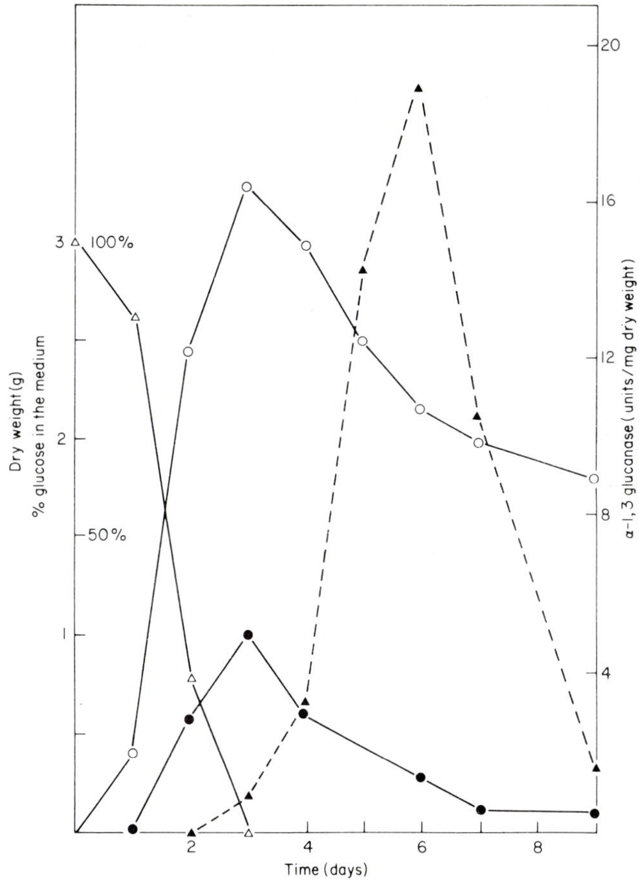

Fig. 3. The dry weight of total mycelium (○), dry weight of the α-1,3 glucan fraction (●), the depletion of glucose in the medium (△) and the activity of the enzyme α-1,3 glucanase (▲) of an *Aspergillus nidulans* plate culture grown with 3% glucose.
α-1,3 glucanase activity is expressed as nM glucose liberated/min (=U)/mg dry weight.

the amount of glucan and glucanase as the results with mutants suggested (Zonneveld, 1974). 2-deoxyglucose used in a ratio of 1 to 30 g glucose/l does not impair growth, although dry weight is lower, due to absence of α-1,3 glucan. If 2-deoxyglucose is added after glucan formation, it still completely prevents cleistothecial formation. If halfway during the growth period a large amount of glucose is added, this is mainly directed to α-1,3 glucan synthesis. However, the simultaneous addition of 2-deoxyglucose directs most of the extra sugar to the synthesis of other glucans. It is not known where exactly 2-deoxyglucose interferes, although it is now thought (see Slayman, 1973) that the main effect of 2-deoxyglucose is to block the initial reactions in glycolysis, resulting in a lowering of the ATP level of the cell, thereby preventing glucan synthesis or, after glucan synthesis, the development of the cleistothecia. This is indicated by only 30% activity of malate dehydrogenase and aldolase and a slight decrease in glucose-6-phosphate dehydrogenase.

The results with manganese cultures show a normal α-1,3 glucanase and β-1,3 glucanase activity but amylase and protease activity were respectively 60 and 75% lower (Zonneveld, 1975a). There was an increase in glucan synthesis with increasing amounts of manganese. If several enzymes of intermediary metabolism were measured the main difference between manganese deficient and sufficient cultures was the low activity of phosphoglucomutase in the deficient culture. This might indicate that α-1,3 glucan is only synthesized when there is a high activity of phosphoglucomutase especially relative to other metabolic enzymes (Zonneveld, 1975b). This is substantiated by a phosphoglucomutase mutant isolated by Elorza & Arst (1971). This strain is indeed low in phosphoglucomutase activity on the first day, but a high activity later on results in a high glucan synthesis and cleistothecial production.

Cleistothecia developed only partially at 0.125 μM manganese although glucan and glucanase were present and this indicates that manganese was also needed for cleistothecial formation directly and addition of manganese at that stage induces a rapid ripening. DNAases of *Aspergillus nidulans* are according to Winder & Campbell (1973) manganese requiring enzymes. Perhaps DNA synthesis itself is affected by the absence of DNA breakdown.

By the use of mutants it was possible to show that absence of α-1,3 glucan resulted in an absence of cleistothecia. The mutants used, isolated by Clutterbuck (1969) and by Zonneveld (1974), did also indicate that large amounts of α-1,3 glucan resulted in large numbers of cleistothecia. Lowering the amount of nitrogen in the medium resulted in higher amounts of α-1,3 glucan so that one strain, acleistothecial on high nitrogen medium, now did form cleistothecia.

PROTEIN DEGRADATION IN FRUCTIFICATION

Results to be published elsewhere show that protein degradation can also play a decisive role in fructification, contrary to the suggestions of Cohen (1973) and Zonneveld (1975a).

On medium with 120 g glucose/l and 6 g sodium nitrate/l (120-6) nitrogen is exhausted in three days and glucose in six days. On the sixth day dark cleistothecia are found showing that development had started at the time of nitrate exhaustion and while glucose concentration was still high. However, only under circumstances where glucan is absent is it possible to demonstrate the significance of protein as carbon and energy source. Manganese deficient cultures are devoid of α-1,3 glucan. Addition of manganese after glucose exhaustion only induces cleistothecial development when the nitrogen is exhausted resulting in a rise of protease activity.

On 10 g glucose/l cleistothecia are found only when the nitrogen is exhausted at the same time as the glucose. Thus a 10-1 culture results in cleistothecia which are absent in a 10-2 culture (Fig. 4). This figure also shows that amylase and α-1,3 glucanase levels are as usual correlated and inversely related to protease levels. Laminarinase activity is more correlated with protease levels.

When both glucose and nitrogen are exhausted and sufficient amounts of α-1,3 glucan have been formed (60-6 culture), the latter is used preferentially and protease levels are low. The reason is that apart from nitrogen and carbon, protease seems also to be regulated by the energy level of the cell. So, in a 30-6 culture with a surplus of nitrate a low protease activity is found after glucose depletion, but a 10-6 culture has a very high activity due to the absence of α-1,3 glucan (Table 1). This high level is still further increased when the nitrogen is also exhausted (10-1; 10-0,5).

OTHER POSSIBLE FACTORS IN CLEISTOTHECIAL DEVELOPMENT

Preliminary experiments have shown that exhaustion of phosphate seems to lead to cleistothecial development but not that of sulphate, potassium, magnesium and of the trace elements.

Under certain circumstances, *Aspergillus nidulans* is able to synthesize up to 46% of its dry weight as lipids (Naguib, 1959). The results of Naguib (1959) and Bainbridge *et al.* (1971) with autolyzing liquid cultures show a decrease in the amount of lipids when glucose is exhausted. The results suggest that lipid degradation may play a role in differentiation on agar plate cultures as carbon and energy source but this has to be investigated.

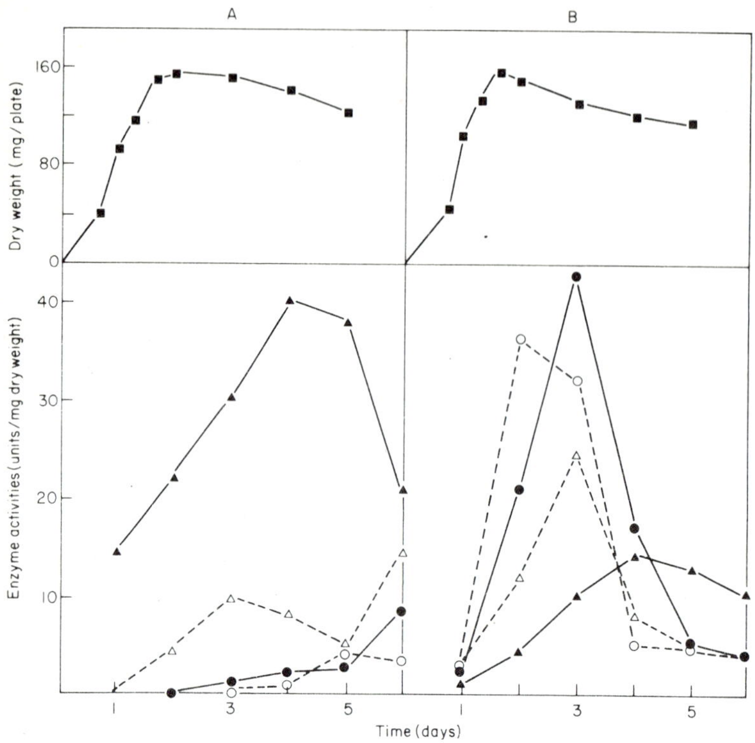

Fig. 4. The dry weight (■) of total mycelium, and activities of the enzymes protease (▲), α-1,3 glucanase (●), amylase (○) and laminarinase (△) of an *Aspergillus nidulans* culture grown with 10 g glucose/l and 1 g (A) or 2 g (B) of $NaNO_3$ per litre. Amylase, laminarinase and α-1,3 glucanase are expressed as nM glucose liberated/min (=U)/mg dry weight. Protease activity is expressed as μg azocasein split/min/mg dry weight.

CYTOPLASMIC FACTORS IN CLEISTOTHECIAL DEVELOPMENT

Mahoney & Wilkie (1962) have studied acleistothecial colonies arising from wild type strains of *Aspergillus nidulans*. These strains showed a continued segregation for this character and it was shown that this was a case of non-Mendelian inheritance controlled by a nuclear gene. Croft (1966) has investigated 175 wild isolates of *A. nidulans* and found half of these unstable with respect to cleistothecial production. In these unstable strains he was unable to select pure breeding cleistothecial or acleistothecial strains. The progeny of ascospores or Hülle cells always gave a fixed percentage of acleistothecial segregates, but the progeny of conidia varied. Arlett (1960) failed to obtain pure breeding acleistothecial strains with *A. nidulans* as did Klasterska (1963) with several other *Aspergillus* species, although they were obtained by Jinks (1954) with *A. glaucus*.

These authors did not reach a conclusion whether the non-Mendelian inheritance was due to an autonomous self-replicating particle or due to a self-perpetuating metabolic system. It might be that the mitochondria are the autonomous particles and that they at the same time are the metabolic system involved. Conidia are found on young actively growing hyphae having high activities of several enzymes of intermediary metabolism (Zonneveld, 1975b), and perhaps highly activate mitochondria.

Ascospores and Hülle cells are formed when the mitochondrial activity might be largely repressed as indicated by the low activity of malate dehydrogenase (Zonneveld, 1975b). This effect may be reinforced because active mitochondria might prevent glucan synthesis and so cleistothecial development. In both cases germination leads again to active mitochondria. However, selection in these unstable strains may lead to a more or less fixed type of mitochondria or always active leading to conidia or alternating active and repressed leading to cleistothecia. Such more or less fixed mitochondria might be of advantage when the environment e.g. light is stable. The *Aspergillus* variants described by Dorn (1970) as invasive because they overgrow neighbouring colonies, may contain continuously repressed mitochondria resulting in undifferentiated or highly cleistothecial colonies. The same internal circumstances may also lead to derepressed synthesis of extracellular enzymes which enables them to be invasive by degrading the hyphae of the invaded colony.

Recently, some cytoplasmic-inherited mitochondrial mutants were obtained. A cold-sensitive mutant was isolated by Waldron & Roberts (1973) and an oligomycin resistant strain was isolated by Rowlands & Turner (1973). The latter showed an altered cytochrome spectrum (80% more cytochrome c). Furthermore, a more chloramphenicol resistant mutant was found by Gunatilleke, Scazocchio & Arst (1975). The commonly used strains of Glasgow origin are of the stable type. However, it might be of interest to replace the mitochondria of the unstable strains via the mutated mitochondria with mitochondria of the stable strains, and to investigate what the relationship is between these unstable strains and the underlying biochemical factors.

Table 1. Maximum protease activities on media containing different amounts of glucose and sodium nitrate.

Medium (g/l)	Day measured	Protease
Gluc-$NaNO_3$		Units/mg dry weight
120 - 6	6	1,6
60 - 6	5	2,1
30 - 6	5	1,2
10 - 6	4	23,1
10 - 3	4	19,4
10 - 2	4	20,0
10 - 1	4	52,9
10 - 0,5	4	62,8

THE ROLE OF cAMP IN GLUCAN SYNTHESIS AND FRUCTIFICATION

The signficance of cAMP in regulation of metabolism, gene expression and differentiation of fungal cells is now substantial (Flawia & Torres, 1973; Scott & Solomon, 1973; Uno, Yamaguchi & Ishikawa, 1974; Wold & Suzuki, 1973). In *Aspergillus nidulans* cAMP levels have been described by Clutterbuck (1975) and Zonneveld (1976) and a phosphodiesterase has been reported by Polya, Brownlee & Hynes (1975). Clutterbuck suggested that high cAMP levels found during early growth, followed by a rapid decrease, is a prerequisite for conidial formation. An explanation for the observed cAMP pattern is derived from several experiments on different media. It is proposed that cAMP levels are in the first place determined by the way the organism grows, and this in turn depends mainly on the composition of the medium. The levels of cAMP seem to depend on the intracellular glucose concentration (or the amount of glucose transport across the membrane). The intracellular glucose concentration in turn depends on the ratio between uptake and metabolism of glucose. The main source of glucose is the medium, and after exhaustion, the degradation product of α-1,3 glucan. The speed of glucose metabolism depends on the particular growth phase. With regard

regard to this, three important biochemical 'switch points' can be noted, i.e.: (1) from dormancy to germination; (2) from logarithmic to steady state growth; and (3) from growth to sexual differentiation. When germinated the organism starts to build up its cell machinery. Glucose metabolism, starting slowly, will soon increase, resulting in a lowering of the intracellular glucose concentration and an increasing cAMP level during logarithmic growth. On a heavily inoculated plate culture the hyphae will touch each other in less than 24h. The resulting relative decrease in growth results in an increased intracellular glucose concentration and a decreasing cAMP level. This switch point also forms the start for conidial formation and the beginning of the synthesis of the reserve material α-1,3 glucan which then functions as a carbon and energy store.

This same pattern of rise and fall of cAMP during growth is also found in several enzymes of intermediary metabolism (Zonneveld, 1975b). This does not indicate that cAMP regulates these enzymes but merely that both react to the same changing circumstances. When glucose is exhausted from the medium α-1,3 glucanase starts to degrade α-1,3 glucan and so prevents a rise in cAMP. The resulting steady flow of glucose may be a prerequisite for cleistothecial development because in this way the levels of cAMP and of α-1,3 glucanase are regulated. In fact, in vitro α-1,3 glucanase is able to degrade all α-1,3 glucan in 3h but in vivo it takes 3 days. This flow of glucose from glucan degradation indeed prevents a sharp increase in cAMP concentration as is examplified by the results with manganese deficient cultures. Here the absence of α-1,3 glucan results in a sharp rise in cAMP, followed by a sharp decrease, probably because the energy sources are exhausted (Zonneveld, 1976). If there is no α-1,3 glucan to be degraded (on 1% glucose), this will result in a rise in cAMP and in a high protease activity.

In Table 2 activities of several enzymes are shown and it seems that laminarinase, invertase, β-glucosidase and protease are high in strains that do not synthesize α-1,3 glucan and presumably have a high cAMP level. On the other hand amylase and α-1,3 glucanase seem to depend on a lower cAMP level, sustained by the degradation of α-1,3 glucan.

CONCLUDING REMARKS

This survey has dealt with cleistothecial initiation and differentiation as far as they are dependent on specific metabolic processes and influenced by the environment. It is clear that a genetical basis for several aspects of sexual differentiation would be welcomed. Still it seems justified to make

Table 2. Maximum enzyme activities and α-1,3 glucan content of several strains on medium containing 30 g glucose and 6 g nitrate/l (except for protease 1,5 where 1,5 g/l of nitrate was used).

	α-1,3 glucan	α-1,3 glucanase	Amylase	Laminarinase	Invertase	β-glucosidase	Protease	Protease 1,5
biA_1	++	8,6	8,3	8,3	8,3	5,3	0,4	2,3
$biA_1 dcl_1$	+++	8,6	8,0	5,0	3,6	3,9	1,1	-
$biA_1 lcl_{14},y$	++	14,3	13,6	4,6	3,3	3,5	0,2	0,1
$biA_1 medA_{15}$	++	7,0	7,3	8,6	7,6	7,7	0,2	0,6
$biA_1 stuA_1$	-	1,3	5,3	19,6	23,2	19,3	0,6	5,9
$biA_1 acl_1$	-	3,3	4,6	15,0	22,6	11,6	0,3	2,5
$biA_1 acl_4,y$	-	0,7	2,6	3,0	3,0	1,9	0,1	0,4
$biA_1 acl_5,y$	-	2,7	4,3	6,3	3,0	4,6	0,2	0,2

some general remarks about differentiation in *Aspergillus* which may also apply to other organisms. Energy supply is proposed to be a decisive factor. This does not mean that other factors are not involved, because it is clear that the organism is a complex system. A fungus is completely dependent on the energy supply in its medium for growth. So it is not surprising in view of the manifold environments to which fungi are adapted that they have developed many ways to obtain and to preserve their energy.

Growth can only be followed by differentiation if sufficient reserve material has accumulated. This can only take place when, during growth, a surplus of energy was available, resulting in the synthesis of reserve material α-1,3 glucan. A few examples may elucidate the dependence on the energy supply. A change from glucose to glycerol is sufficient to prevent any sexual differentiation. On the other hand, if sodium nitrate is replaced by ammonium tartarate, cleistothecia are again formed in the glycerol cultures, probably because no energy is needed to reduce nitrate. It is rather easy to obtain acleistothecial mutants. Many types of mutation will lead to an impaired energy level, resulting in an absence of reserve material and as a consequence cleistothecia are absent. Good examples are the mutants isolated by Houghton (1976, 1971) that seem to have reduced availability of high energy compounds caused by a lesion in their oxidative phosphorylation system and they all are acleistothecial. Some of the conidiophore mutants isolated by Clutterbuck (1969) also did not make cleistothecia. It was found that some of these were devoid of α-1,3 glucan which may indicate an impaired energy level.

As previously discussed, nitrogen starvation can also lead to some cleistothecia and the same can be said for phosphate. The common principle in the effect of carbon, nitrogen or phosphorus depletion on differentiation is perhaps that they all interfere with the energy supply of the mycelium. Carbon depletion may stop the energy supply directly, nitrogen depletion via the terminated enzyme synthesis and phosphorus via a halted ATP synthesis. That for conidiospore development the emergence from the substrate seems the most important factor may be considered as a question of food supply. When the conidiophore is no longer embedded in medium, the footcell can regulate the energy supply at the desired level. After 24h it depends on the glucose concentration in the medium and/or the intracellular energy level whether mainly conidia or mainly α-1,3 glucan are synthesized. The fact that both can be jointly synthesized suggests that the energy levels needed for glucan and conidial synthesis overlap each other. The reason that forced heterokaryans form more cleistothecia than homokaryons may be that the former grow more slowly resulting in a higher energy level, more α-1,3 glucan and so more cleistothecia. These examples may be sufficient to stress the importance of the energy supply in growth, synthesis of reserve material and differentiation.

REFERENCES

AGNIHOTRI, V.P. (1967). Role of trace elements in the growth and morphology of five ascosporic *Aspergillus* species. *Canadian Journal of Botany* 45, 73-9.

ARLETT, C.F. (1960). A system of cytoplasmic variation in *Aspergillus nidulans*, *Heredity*, 15, 377-88.

ARST, H.N. & COVE, D.J. (1973). Nitrogen metabolite repression in *Aspergillus nidulans*. *Molecular & General Genetics* 126, 111-41.

BAILEY, C. & ARST, H.N. (1975). Carbon catabolite repression in *Aspergillus nidulans*. *European Journal of Biochemistry* 51, 573-7.

BAINBRIDGE, B.W., BULL, A.T., PIRT, S.J., ROWLEY, B.I. & TRINCI, P.J. (1971). Biochemical and structural changes in non-growing maintained and autolysing cultures of *Aspergillus nidulans*. *Transactions of the British Mycological Society* 56, 371-85.

BENJAMIN, C.R. (1955). Ascocarps of *Aspergillus* and *Penicillium*. *Mycologia* 47, 669-87.

BULL, A.T. (1970). Chemical composition of wild type and mutant *Aspergillus nidulans* cell walls. The nature of polysaccharide and melanin constituents. *Journal of General Microbiology* 63, 75-94.

CLUTTERBUCK, A.J.(1969). A mutational analysis of conidia development in *Aspergillus nidulans*. *Genetics* 63, 317-27.

CLUTTERBUCK, A.J. (1975). Cyclic AMP levels during growth and conidiation. *Aspergillus Newsletter* 12, 13-15.

COHEN, B.L. (1973). The neutral and alkaline proteases of *Aspergillus nidulans*. *Journal of General Microbiology* 77, 521-8.

COVE, D.J. (1966). The induction and repression of nitrate reductase in the fungus *Aspergillus nidulans*. *Biochimica et Biophysica Acta* 113, 51-6.

CROFT, J.H. (1966). A reciprocal phenotypic instability affecting development in *Aspergillus nidulans*. *Heredity* 21, 565-79.

CURRAN, P.M.T. (1971). Sporulation in some members of the *Aspergillus glaucus* group in response to osmotic pressure, illumination and temperature. *Transactions of the British Mycological Society* 57, 201-11

DORN, G.L. (1970). Genetic and morphological properties of undifferentiated and invasive variants of *Aspergillus nidulans*. *Genetics* 66, 267-79.

ELLIS, T.T., REYNOLDS, D.R.& ALEXOPOULOS, C.J. (1963). Hülle cell development in *Emericella nidulans*. *Mycologia* 65, 1028-35.

ELORZA, M.V. & ARST, H.N. (1971). Sorbose resistant mutants of *Aspergillus nidulans*. *Molecular & General Genetics* 111, 185-93.

FANTES, P.A. & ROBERTS, C.F. (1973). β-Galactosidase activity and lactose utilization in *Aspergillus nidulans*. *Journal of General Microbiology* 77, 471-86.

FLAWIA, M.M. & TORRES, H.N. (1972). Adenylate cyclase activity in *Neurospora crassa*. *Journal of Biological Chemistry* 247 4517-20.

GOLD, M.H., MITZEL, D.L. & SEGEL, I.H. (1973). Regulation of nitrogen accumulation by *Aspergillus aculeatus*. *Journal of Bacteriology* 113, 856-62.

GORIN, P.A.J. & EVELEIGH, D.E. (1970). Extracellular 2-acetamido-2-deoxy-D-galacto-D-galactan from *Aspergillus nidulans*. *Biochemistry* 9, 5023-27.

GUNATILLEKE, I.A.U.N., SCAZZOCCHIO, C. & ARST, H.N. (1975). Cytoplasmic and nuclear mutations to chloramphenicol resistance in *Aspergillus nidulans*. *Molecular and General Genetics* 137, 269-76.

HOUGHTON, J.A. (1970). A new class of slow-growing non-perithecial mutants of *Aspergillus nidulans*. *Genetical Research (Cambridge)* 16, 285-92.

HOUGHTON, J.A. (1971). Biochemical investigation of the slow-growing non-perithecial (*spg*) mutants of *Aspergillus nidulans*. *Genetical Research (Cambridge)* 17, 237-44.

HYNES, M.J. (1973). The effect of lack of a carbon source on nitrate-reductase activity in *Aspergillus nidulans*. *Journal of General Microbiology* 79, 155-7.

JINKS, J.L. (1954). Somatic selection in fungi. *Nature* 174, 410.

JOHNSTON, I.R. (1965a). The composition of the cell wall of *Aspergillus niger*. *Biochemical Journal* 96, 651-8.

JOHNSTON, I.R. (1965b). The partial acid hydrolysis of a highly dextrorotatory fragment of the cell wall of *Aspergillus niger*. *Biochemical Journal* 96, 659-64.

KATZ, D. & ROSENBERGER, R.F. (1970). A mutation in *Aspergillus nidulans* producing hyphal walls which lack chitin. *Biochimica et Biophysica Acta* 208, 452-60.

KIM, S.J. & KWON-CHUNG, K.J. (1974). Polyene-resistant mutants of *Aspergillus fennelliae:* sterol contents and genetics. *Antimicrobial Agents and Chemotherapy* 6, 102-13.

KLASTERSKA, I. (1963). The effect of cytoplasmic factors on the variability of *Aspergillus* strains. *Biologia Plantarum (Praha)* 5, 100-8.

KURZEJA, K.C. & GARBER, E.D. (1973). A genetic study of electrophoretically variant extracellular amylolytic enzymes of wild-type strains of *Aspergillus nidulans*. *Canadian Journal of Genetics and Cytology* 15, 275-87.

LOCCI, R. (1972). Scanning electron microscopy of ascosporic Aspergilli. *Rivista di Pathologie Vegetale* Serie IV, Suppl. Vol. VIII, 135-6.

MAHONEY, M. & WILKIE, D. (1962). Nucleo-cytoplasmic control of perithecial formation in *Aspergillus nidulans*. *Proceedings of the Royal Society Series B* 156, 524-32.

MUNTANJOLA-CVETKOVIC, J. & VIKIC, V.V. (1972). Influence of light on Hülle cell and aleuriospore formation in *Aspergillus*. *Transactions of the British Mycological Society* 58, 67-72.

NAGUIB, K. (1959). Growth and metabolism of *Aspergillus nidulans* Eidam in surface culture. *Canadian Journal of Botany* 37, 353-64.

PAYTON, M., McCULLOUGH, W. & ROBERTS, C.F. (1976). Agar as a carbon source and its effect on the utilization of other carbon sources by acetate non-utilizing (*acu*) mutants of *Aspergillus nidulans*. *Journal of General Microbiology* 94, 228-33.

PASZEWSKI, A., CHOJNACKI, T., LITWINSKA, J. & GAJEWSKI, W. (1970). Regulation of lactose utilization in *Aspergillus nidulans*. *Acta Biochemica Polonica* 17, 385-91.

PEBERDY, J.F. & BUCKLEY, C.E. (1973). Adsorption of fluorescent brightners by regenerating protoplasts of *Aspergillus nidulans*. *Journal of General Microbiology* 74, 281-88.

POLYA, G.M., BROWNLEE, A.G. & HYNES, M.J. (1975). Enzymology and genetic regulation of a cyclic nucleotide-binding phosphodiesterase-phosphomonoesterase from *Aspergillus nidulans*. *Journal of Bacteriology* 124, 693-703.

PONTECORVO, G. (1953). The genetics of *Aspergillus nidulans*. *Advances in Genetics* 5, 142-238.

RAI, J.N., TEWARI, J.P. & SINHA, A.K. (1967). Effect of environmental conditions on sclerotia and cleistothecial production in *Aspergillus*. *Mycopathologia et Mycologia Applicta* 31, 209-24.

ROWLANDS, R.T. & TURNER, G. (1973). Nuclear and extranuclear inheritance of oligomycin resistance in *Aspergillus nidulans*. *Molecular & General Genetics* 126, 201-16.

RUDOLPH, E.O. (1962). The effect of some physiological and environmental factors on sclerotial Aspergilli. *American Journal of Botany* 49, 71-8.

RUIZ-HERRERA, J. (1967). Chemical components of the cell wall of *Aspergillus* species. *Archives of Biochemistry and Biophysics* 122, 118-25.

SCOTT, W.A. & SOLOMON, B. (1973). Cyclic 3',5'-AMP phosphodiesterase of *Neurospora crassa*. *Biochemical and Biophysical Research Communications* 53, 1024-30.

SKINNER, V.M. & ARMITT, S. (1972). Mutants of *Aspergillus nidulans* lacking pyruvate carboxylase. *FEBS Letters* 20, 16-18.

SLAYMAN, C.L. (1973). Adenine nucleotide levels in *Neurospora*, as influenced by conditions of growth and by metabolic inhibitors. *Journal of Bacteriology* 114, 752-66.

UNO, I., YAMAGUCHI, M. & ISHIKAWA, T. (1974). The effect of light on fruiting body formation and adenosine 3',5'-cyclic monophosphate metabolism in *Coprinus macrohizus*. *Proceedings of the National Academy of Sciences USA* 17, 479-83.

WALDRON, C. & ROBERTS, C.F. (1973). Cytoplasmic inheritance of a cold-sensitive mutant in *Aspergillus nidulans*. *Journal of General Microbiology* 78, 379-81.

WINDER, F.G. & CAMPBELL, G.R. (1973). The deoxyribonucleases of *Aspergillus nidulans*. *Heredity* 31, 423.

WOLD, W.S.M. & SUZUKI, I. (1973). Cyclic AMP and citric acid accumulation by *Aspergillus niger*. *Biochemical and Biophysical Research Communications* 50, 237-43.

ZONNEVELD, B.J.M. (1971). Biochemical analysis of the cell wall of *Aspergillus nidulans*. *Biochimica et Biophysica Acta* 249, 506-14.

ZONNEVELD, B.J.M. (1972a). A new type of enzyme, an exo-splitting α-1,3 glucanase from non-induced cultures of *Aspergillus nidulans*. *Biochimica et Biophysica Acta* 258, 541-7.

ZONNEVELD, B.J.M. (1972b). Morphogenesis in *Aspergillus nidulans*. The significance of α-1,3 glucan of the cell wall and α-1,3 glucanase for cleistothecium development. *Biochimica et Biophysica Acta* 273, 174-87.

ZONNEVELD, B.J.M. (1973). Inhibitory effect of 2-deoxyglucose on cell wall α-1,3 glucan synthesis and cleistothecium development in *Aspergillus nidulans*. *Developmental Biology* 34, 1-8.

ZONNEVELD, B.J.M. (1974). α-1,3 glucan synthesis correlated with α-1,3 glucanase synthesis, conidiation and fructification in morphogenetic mutants of *Aspergillus nidulans*. *Journal of General Microbiology* 81, 445-51.

ZONNEVELD, B.J.M. (1975a). Sexual differentiation in *Aspergillus nidulans*. The requirement for manganese and its effect on α-1,3 glucan synthesis and degradation. *Archives of Microbiology* 105, 101-4.

ZONNEVELD, B.J.M. (1975b). Sexual differentiation in *Apergillus nidulans*. The requirement for manganese and the correlation between phospho-glucomutase and the synthesis of reserve material. *Archives of Microbiology* 105, 105-8.

ZONNEVELD, B.J.M. (1976). The effect of glucose and manganese on adenosine-3',5'-monophosphate levels during growth and differentiation of *Aspergillus nidulans*. *Archives of Microbiology* 108, 41-4.

The Genetics of *Aspergillus nidulans*

D. J. COVE

Department of Genetics, University of Cambridge, Cambridge, U.K.

INTRODUCTION

One of my first tasks in preparing this paper, was to read once again the review which appeared in Advances in Genetics in 1953, by Professor Pontecorvo and his co-workers, and with which this paper shares a common title. It still remains today an impressive contribution. It is not even fair to say that in it are laid the foundations for genetic analysis in *Aspergillus nidulans*, for the paper contains much of the superstructure too. There have, of course, been developments and refinements since, but there is no doubt that Pontecorvo and his team had already established *A.nidulans* as a genetic organism twenty three years ago. Where considerable progress has been made more recently is in the establishment of *A. nidulans* as an organism for molecular studies. Starting with the biochemical genetical investigations of the sixties, knowledge of the physiology of *A.nidulans* has been accumulating, such that it now represents one of the best model systems for the study of eukaryote gene expression and its control. Whilst there were those who doubted its authenticity as a eukaryote, even until quite recently, to geneticists, secure in their knowledge that it behaved in breeding experiments as a typical eukaryote, there was no doubt. With the recent finding (Felden, Sanders & Morris, 1976; Morris, these proceedings) that *A.nidulans* has not only nuclei, and chromosomes but histones too, further converts may be expected. In this paper, I will go over again the features of the life cycle of *A. nidulans* which make it so suitable for genetical analysis, and include only a few of the more recent developments, as many of these will be dealt with more competently by the other contributors to this volume.

ASEXUAL REPRODUCTION

Aspergillus nidulans (Eidam) Winter is an ascomycete fungus showing colonial mycelial growth. The mycelium consists of multinucleate branched hyphae, approximately 3 μm in diameter, which develop uninucleate asexual spores, conidia, on vertical branches called conidiophores. Each conidiophore bares as many as a hundred chains of conidia, which in turn may each contain over a hundred spores. Germination and growth of spores is rapid, such that on solid media at 37°, mature conidia are produced within two days, and a single conidium can develop into a colony bearing 10^9 conidia in five days (See Pontecorvo *et al.*, 1953; Cove, 1966; Clutterbuck, 1974 for details of media recipes).

Mutagenesis

Being uninucleate and easily obtainable in large numbers, conidia make ideal material both for studies of mutation *per se*, and for obtaining mutant strains. Techniques involving a wide variety of mutagens have been devised, including X-rays (Pontecorvo *et al.*, 1953), U.V. (Pontecorvo *et al.*, 1953; Alderson & Hartley, 1969), gamma irradiation and a number of chemical mutagens (Alderson & Hartley, 1969). Of the chemical mutagens N-methyl-N'-nitro-N-nitrosoguanidine has probably been used most extensively, but its undoubted high efficacy as a mutagen needs to be balanced against the risks involved in handling it. Ultraviolet light although safer, has the disadvantage of inducing a high frequency of translocations (Käfer, 1965).

Range of mutant types obtainable

The original sample of mutants obtained by Pontecorvo and his co-workers consisted principally of auxotrophs, but also contained a number of visible mutants, including some with altered conidial colour, a mutant type of considerable use in conjunction with many of the techniques of genetic analysis in *A. nidulans*. Pontecorvo *et al.*,(1953) list at least 14 genes in which mutation can lead to auxotrophy, and this number has grown steadily, until today about one hundred such genes are known (Clutterbuck & Cove, 1976). The total number of genes characterised to date, is now in excess of three hundred and fifty (Clutterbuck, 1974; Clutterbuck & Cove, 1976).

Many of these mutate to impair catabolic function, emphasising the suitability of *A.nidulans*, with its ability to utilise a wide range of carbon and nitrogen sources, for the study of this aspect of metabolism. Other genes now characterised, include those coding for products involved in transport of solutes across the cell membrane, and a growing number whose products have roles in the regulation of gene expression both in metabolism and in development. Recently the first genes whose products affect mitochondrial function, have been characterised and at least some of these show a non-Mendelian pattern of inheritance (Rowlands & Turner, 1974a; Gunatilleke, Arst & Scazzocchio, 1975a; Gunatilleke, Scazzocchio & Arst, 1975b; Turner & Rowlands, 1976 and these proceedings).

Mutant selection systems

The non-selective procedure of total isolation still remains the most satisfactory method for isolating novel mutant types, as selective techniques may involve assumptions which cannot be tested until a mutant type is obtained. Total isolation in *A.nidulans* was greatly facilitated by the development of a replica plating technique, in which micro-colonial growth is induced by the inclusion of sodium desoxycholate in the medium of the master plate, and spores are transferred by damp velvet onto replica plates (Mackintosh & Pritchard, 1963). Of selective techniques, the most generally useful are probably those involving starvation either of a biotin auxotroph (Pontecorvo *et al.*,1953) or of a putrescine auxotroph (Herman & Clutterbuck, 1966). Although the techniques involve quite different procedures, each allows the selection of mutations leading to the inability to grow on a particular medium. Neither technique is effective enough to be of much use in the selection of spontaneous mutations. A large number of more specific selection procedures have been described, some of which are very effective in selecting spontaneous mutations (Cove, 1976 describes one such technique).

Growth in bulk for biochemical studies

A.nidulans is an obligate aerobe. It grows well on agar media in petri dishes or vials, providing these do not have an airtight seal. It can also be grown submerged in liquid culture providing aeration is vigorous. To obtain mycelial samples for biochemical studies, it is conveniently cultured on a shaker in 1 litre erlenmeyer flasks containing 200-400 ml of medium, or, if larger quantities are required, in vigorously-aerated carboys (Cove, 1966, 1967).

SEXUAL REPRODUCTION

A.nidulans is homothallic. Although this means that difficulties are sometimes encountered in crossing two strains, the absence of mating types is in general a convenience, enabling as it does any two strains to be crossed together. Cleistothecia are formed usually as a result of a single fertilisation event, and so all the ascospores contained within one cleistothecium will either be of one or other parental type if self fertilisation occurred, or will segregate for the characters of both parents if cross fertilisation occurred instead. By routinely choosing strains with, for example, different conidial colour phenotypes, and growth testing a small sample of ascospores from a number of cleistothecia, a crossed cleistothecium can be identified, and since a single cleistothecium can contain up to half a million ascospores, sufficient progeny for most routine genetic analysis will then be available. Clutterbuck (1974) describes some of the techniques which favour cleistothecium formation. When certain strains are crossed, almost all cleistothecia obtained are themselves of hybrid origin. This finding which was first described by Pontecorvo *et al.* (1953) who termed it "relative heterothalism" is a considerable convenience when particular crosses prove difficult. By outcrossing one or both of the strains involved, and getting the mutations to be investigated into different genetic backgrounds, the cross may subsequently have a much greater chance of success.

Unordered ascus analysis

A.nidulans asci contain eight ascospores, which are not ordered. Intact asci can be isolated from slightly immature cleistothecia, so there is the possibility of performing unordered octad analysis. This is a technique which has not been exploited to any great extent, although Strickland (1958a,b) showed such analysis was possible eighteen years ago.

Segregation and linkage

Genetic analysis in *A.nidulans* usually involves plating, at a low density, ascospores from a cleistothecium, which is known to be hybrid, onto media upon which all types of progeny are able to develop. The phenotype of progeny with respect to biochemical markers segregating, is generally classified by replica plating. In this procedure a multi-wired replicator is employed, the progeny to be tested having first been subcultured in a set pattern on a master plate. This technique is a simple and rapid and allows a large number of replica tests to be made from a single master plate. The construction of linkage maps is greatly facilitated because the technique of haploidisation (see below) allows the rapid allocation of a newly isolated mutation to one of the eight linkage groups. Linkage analysis is usually straightforward, although because of the high rate of recombination found in *A.nidulans*, linkage to mapped genes has not been detected for a number of genes, shown by haploidisation to be located in particular linkage groups. Up-to-date versions of the *A.nidulans* linkage map are to be found in Clutterbuck (1974) and Clutterbuck & Cove (1976).

Little or no clustering of functionally related genes

Sufficient genetic analysis has now been carried out on a number of different metabolic systems, to be able to say that clustering of functionally related genes in *A.nidulans* is far less common than in *Escherichia coli*. The significance of what clustering is found, is still doubtful. It seems likely that in several of the cases known, the genes concerned are coding for products which aggregate together, and the possibility remains that rather than several genes, a single gene coding for a multifunctional protein is involved, as has been shown to be the case for example, for the *his*-3 region in *Neurospora crassa* (Minson & Creaser, 1969). However this is not the explanation in all cases, and at least in some, it is possible that an operon-type structure exists involving the co-ordinate transcription of adjacent genes, which share common regulatory sites (Arst & MacDonald, 1975 and these proceedings). It is difficult to obtain critical evidence for the existance of an operon, unless mutations involving a regulatory recognition site can be identified. Such mutations should only affect the structural genes which lie *cis* to them. Recently a number of such *cis*-acting mutations have been isolated (Hynes, 1975; Arst & MacDonald, 1975; Arst & Scazzocchio, 1975; Lukaszkiewicz & Paszewski, 1976), but only one of these is associated with a gene cluster. The close linkage of the structural genes for nitrate and nitrite reductases is at present being studied, and a brief account of progress to date will be given below. In this case, although the two genes appear to be contiguous, it is unlikely that they comprise an operon. The reason for their close linkage remains uncertain therefore, although I have suggested earlier (Cove, 1970) that this close linkage may have evolved to allow the genetic information for the assimilation of nitrate to be inherited as a unit, thus allowing selection to ensure that the enzymes concerned had individual characteristics compatible with one another.

Intragenic recombination and deletion mapping

Intragenic recombination studies using *A.nidulans* are now fairly numerous. Clutterbuck (1974) lists a number of these. Most were carried out with a view to the study of the recombination process. Recombination frequencies were in general estimated by the selection of prototrophic recombinants, and the ordering of mutant sites within a gene was aided where possible by the inclusion of outside markers in the crosses. Such analyses are time-consuming and it is not surprising therefore that they have not been used to any extent to study the relationship between gene structure and function. Recently Tomsett and Cove (unpublished results) have devised a deletion mapping procedure essentially similar to that used in bacteria and viruses. Since in such procedures, the measurement of recombination is qualitative rather than quantitative, analysis is less tedious. The genes being studied are *nia*D and *nii*A, the structural genes for nitrate and nitrite reductase respectively, and the aim of this work is to extend to eukaryotes, detailed studies which relate the genetic map to the function and structure of the gene products. In particular, the locations of mutations which are known to lead to the production of altered enzymes, are being compared with those of null-mutations, in order to determine how extensive the enzyme-coding regions are and their relationship to one another. The analysis has been aided by the finding that about 5% of spontaneous *nia*D mutants, selected as a result of their chlorate resistance, contain deletions of the *nia*D gene, which extend into the *nii*A gene (Cove, 1976). Since such mutants have the same phenotype as *nii*A *nia*D double point mutants, the two genes must either be contiguous, or at least be separated by genetic material whose function is not essential for growth under normal laboratory conditions. So far, deletions define 6 regions within the *nii*A gene, and 10 regions within the *nia*D gene, but further resolutions should be attainable fairly soon. Fig. 1 shows the current state of this work.

HETEROKARYOSIS AND DIPLOIDY

The hyphae of Glasgow strains of *A.nidulans* and strains derived from them, in general anastomose readily, and heterokaryons can therefore be formed between two strains by inoculating them in close proximity.Heterokaryons can only be maintained by the use of selection. Typically two strains with complementary auxotrophies are maintained in heterokaryotic association by culturing on medium which does not contain the supplements required by either strain. Heterokaryons maintained in this manner have a very characteristic morphology, showing non-uniform growth with more vigorous sectors arising irregularly. Since conidial colour is cell autonomous, a heterokaryon between strains of different conidial-colour genotype will bear conidia of two colours. In such cases, individual heads of conidia are usually all of the same colour, but mixed heads do occur. Conidial chains are however always uniform, reflecting their origin from a single mother nucleus (Pontecorvo *et al.*,1953).

Although the conidia on heterokaryons are almost all haploid, occasionally diploid conidia are produced. The absolute frequency of diploid conidia cannot be estimated, but the frequency of heterozygous diploid conidia borne by a complementary heterokaryon is as

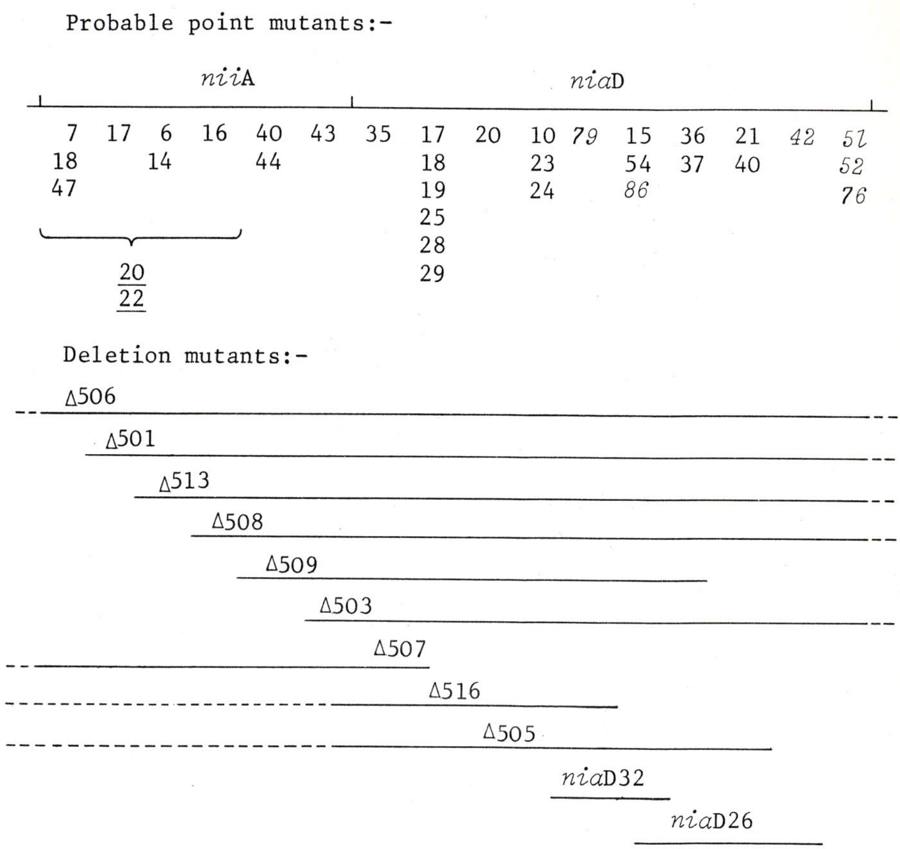

Fig. 1. Fine structure map of the *nii*A *nia*D region of chromosome VIII, constructed by the use of deletion mapping (Tomsett and Cove, unplublished results). *nii*A and *nia*D are the structural genes for nitrate and nitrite reductase. *nii*A *nia*D deletion mutants are prefixed Δ, and are presesented by a bar indicating the extent of the deleted material. Point mutants underlined are temperature sensitive, and therefore probably lie within the coding region of the relevant gene. Point mutations listed vertically may or may not recombine with each other.

as low as 10^{-7} (Roper, 1952). However, it is not uncommon for a heterokaryon to give rise to a sector of diploid mycelium, which is readily identifiable if the heterokaryon is between two complementary conidial-colour mutants. Although diploids can be selected by plating conidiospores on appropriate selective medium, if a large number of different diploids are required, it is often more convenient to maintain the 'parent' heterokaryons by regular subculture, and wait for diploid sectors to arise spontaneously. In this way it has been possible recently to obtain eighty six out of a set of one hundred diploids in ten weeks (Cove and Simpson, unpublished results).

Although almost all strains used for genetic analysis in laboratories form heterokaryons readily, occasionally difficulty is encountered with newly isolated mutant strains. When this occurs, derivative strains obtained by outcrossing can usually be obtained which have the normal ability to form heterokaryons. This heterokaryon incompatibility is therefore presumably the result of the accumulation of an additional mutation during the isolation of the mutant under investigation. Heterokaryon incompatibilities are however widespread among wild isolates of *A.nidulans* (Grindle, 1963a,b).

Complementation tests with heterokaryons and diploids

Both heterokaryons and diploids can be employed to perform complementation tests. Heterokaryons, because of the ease with which they are obtained, tend to be used more often. In general, it is necessary to establish a balanced heterokaryon using additional auxotrophies for selection purposes, and supplementing for the character under investigation. Once established, the heterokaryon can be subcultured onto tester media by the transfer of a small portion of the heterokaryotic mycelium. In some cases it is not necessary to adopt this procedure, and heterokaryon complementation between, for example, mutants unable to utilise nitrate as a nitrogen source, can be tested merely by co-inoculating the strains to be tested on nitrate medium. The ease with which such mutants establish heterokaryons, which is dependent on their characteristic very weak growth on nitrate, is exploited in a rapid test for cytoplasmic inheritance devised by Gunatilleke *et al.*(1975b- see below).

No satisfactory method has been devised for the culture of heterokaryons in liquid medium to obtain mycelium for enzyme studies, although Rever (1965) detected complementation to form nitrate reductase between two mutant strains, by inoculating a mixture of dry spores on the surface of liquid medium, and harvesting the mycelial pad which developed after 28h incubation at 25°. Because of this, enzyme studies have generally employed diploids. Because *A.nidulans* diploids are also somewhat unstable (see below), it is usual to culture them in a liquid medium which continues to select for diploid growth, although otherwise culture procedures are similar to those for haploid strains.

Pontecorvo (1963) pointed out that the occurrence of both a heterokaryotic and a diploid phase in *A.nidulans* made it possible to detect a gene whose product was confined to the nucleus in which that gene was located. In his discussion, he considered in particular, genes whose products had regulatory roles. Differences between the pattern of complementation in heterozygous diploids and in heterokaryons could be accounted for if the product of a regulator gene could only exert its effect on a structural gene whose expression it controlled, when that structural gene was situated within the same nucleus. However, Cassleton & Lewis (1967) pointed out that heterokaryons, with their variable nuclear ratios and patterns of nuclear distribution, were not ideal material for these studies and differences in complementation patterns at the heterokaryon and diploid levels, which are not infrequent, should not be regarded uncritically as providing evidence for nuclear compartmentalisation. For example, if a regulator gene product is made in limiting quantities, the inhomogeneous distribution of nuclei within a heterokaryon may result in the concentration of the regulator product synthesised by the nuclei of one genotype, to be too low in the vicinity of the nuclei of the other genotype to have its normal effect.

The regulation of the *nia*D and *nii*A genes by the *nir*A gene illustrates this situation. Here, complementation for growth on nitrate between an *nir*A$^-$, non-inducible, mutant and mutations affecting either the *nia*D or *nii*A structural genes, is very weak in heterokaryons but strong in diploids. However, when enzyme production is measured, *nir*A$^-$/*nir*A$^+$ diploids are found to produce significantly less nitrate and nitrite reductase than *nir*A$^+$ homozygous diploids (Cove, 1969). The *nir*A$^+$ gene product must therefore be present at a very low concentration, such that one functional *nir*A gene copy cannot produce sufficient product to turn on two copies of the structural genes. Given this, it is plausible to reason that the weak complementation of *nir*A$^-$ mutants with *nia*D or *nii*A mutants in heterokaryons is due to the concentration of *nir*A gene product being too low in the vicinity of the nuclei containing intact *nia*D or *nii*A structural genes to effect full induction. It has recently been shown that the molybdenum-containing cofactor coded for by the *cnx* genes is not under the control of the *nir*A gene (Garrett & Cove, unpublished data), and it is interesting therefore that *cnx*$^-$ mutants complement with *nir*A$^-$ mutants for growth on nitrate, strongly in both heterokaryons and diploids.

Purine catabolism provides two further examples of differences at the heterokaryon and diploid levels in complementation patterns between regulator and structural gene mutants. Non-inducible, *ua*Y$^-$ mutants do not complement in heterokaryons with *hx*B$^-$ mutants, which are unable to make a structural component shared by xanthine dehydrogenases I and II, but do complement in diploids (Scazzocchio & Darlington, 1967). Studies of enzyme levels in diploids, show that, like the *nir*A product, the *ua*Y product is made in near-limiting quantities (Scazzocchio, unpublished data and these proceedings).

A similar situation holds for the *hx*B gene and the *apl*A regulator gene (Scazzocchio, 1973). In both these cases it is probable that the failure of complementation in heterokaryons is a consequence of there being an insufficient concentration of regulator product in the mycelium, rather than that product being confined to the nucleus, although the latter explanation cannot be excluded. Further studies aimed at studying the role of the nuclear membrane, particularly in a regulatory context, would therefore probably be better carried out using a fungus, such as *Coprinus lagopus*, which has a dikaryon phase, as well as heterokaryon and diploid phases, so that intra-cellular maldistribution can be distinguished from nuclear compartmentalisation (Cassleton and Lewis, 1967).

The use of heterokaryons in the study of cytoplasmic inheritance

Since nuclear fusion is rare in *A.nidulans* heterokaryons, and since they produce uninucleate conidia, they are ideal material for the detection of cytoplasmic inheritance. Only cytoplasmically inherited characters can be transmitted in heterokaryons, so as to appear in new combinations with the component nuclear genotypes. Jinks (1963) and more recently Rowlands & Turner (1973, 1974b) and Gunatilleke *et al.* (1975b) have described the detection of cytoplasmic inheritance using heterokaryons, and a contribution by Turner & Rowlands on mitochondrial genetics is in these proceedings.

Haploidisation of diploids as an aid to genetic analysis

Diploids of *A.nidulans* are not completely stable, and break down to give haploid strains (Pontecorvo, Tarr-Gloor & Forbes, 1954). The mechanism is still rather obscure, but it is possible that initially a single chromosome is lost, to give an unstable clone, which then loses further chromosomes, until the haploid state is attained (Käfer, 1961; McCully & Forbes, 1965). The frequency of haploidisation can be greatly enhanced by the use of *p*-fluorophenylalanine (McCully & Forbes, 1965), arsenate (Van Arkel, 1963) or benlate (Hastie, 1970). Haploidisation is itself a powerful genetic tool. Since no recombination event occurs in the formation of haploid segregants from a diploid, whole chromosomes segregate together. If a diploid is constructed between a strain carrying an unmapped mutant gene, and a 'master strain' which has each of the eight linkage groups marked with a mutant allele, the linkage group to which the unmapped gene belongs can be determined. Among the haploid segregants the unmapped gene will be linked (in repulsion) to one only of the mutant alleles of the 'master strain' (McCully & Forbes, 1965). Haploidisation also allows the detection of reciprocal translocations, which occur rather commonly in *A.nidulans* (Käfer, 1962). If one of the component strains of a diploid carries a translocation, the two linkage groups involved will fail to segregate independently upon

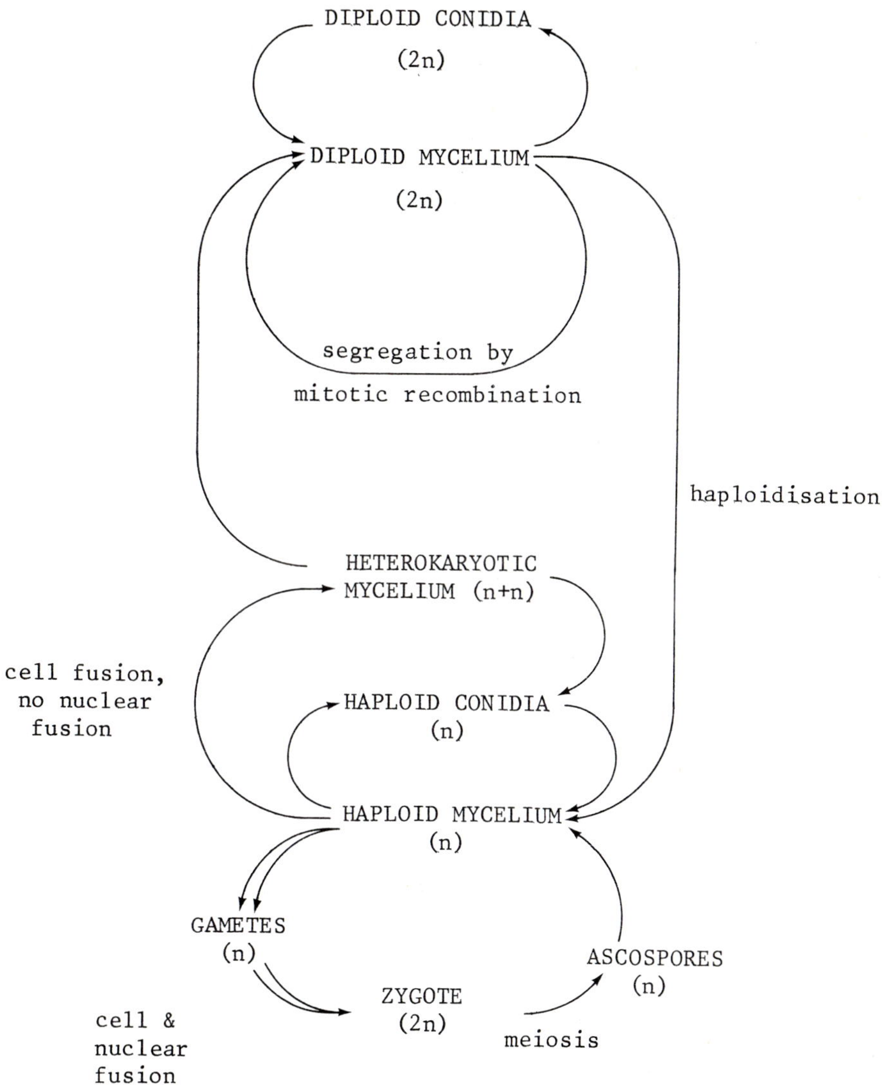

Fig. 2. Life cycle of *Aspergillus nidulans*.

haploidisation, as recombinant segregants would carry either dupli cations or defficiencies and be inviable. Haploidisation is therefore usually the first technique employed in mapping a new mutation, as it not only allows the mutation to be allocated rapid ly to the appropriate linkage group, but also detects whether the mutant strain carries a translocation.

Mitotic recombination

A.nidulans is among the most suitable experimental organisms available for the analysis of mitotic recombination. Mitotic recombination occurs fairly frequently in diploid mycelium (Käfer, 1961) and can be detected readily if diploids of suitable genotype are used. Generally, diploids heterozygous for a recessive visible mutation have been used, where sectors of growth arising as a result of mitotic recombination involving the mutation, can be identified visually. Care must be taken to exclude the possibility, that the sector has not arisen instead as a result of haploid isation. Clutterbuck (1974) lists a number of studies of mitotic recombination in *A.nidulans*, most of which were carried out to stu mitotic recombination *per se*, but which also establish the colinearity of the meiotic and mitotic maps. Mitotic recombination is also of use for genetic mapping, and has been used to locate the centromeres in linkage groups, and more recently, to analyse trans locations (Ma & Käfer, 1974). That mitotic recombination has not been used more widely as a tool for genetic analysis in *A.nidulans* probably reflects the ease with which more conventional meiotic methods can be employed.

CONCLUDING REMARKS

In this paper, I have attempted to give an outline of the li cycle of *A.nidulans* (Figure 2 summarises the basic life cycle) and how this can be exploited for genetic analysis. I have only adde flesh to this skeleton treatment in a few places, as it seemed to me that this was a job better left to the remaining contributors t this volume, who are qualified to add not only flesh, but clothes as well.

REFERENCES

ALDERSON,T. & HARTLEY,M.J. (1969). Specificity for spontaneous ar induced forward mutation at several gene loci in *Aspergillus nidulans*. *Mutation Research* 8,255-64.

ARST,H.N. & MacDONALD,D.W. (1975). A gene cluster in *Aspergillus nidulans* with an internally located *cis* acting regulatory region. *Nature,London* 254, 26-31.

ARST,H.N. & SCAZZOCCHIO,C. (1975). Initiator constitutive mutation with an 'up-promoter' effect in *Aspergillus nidulans*. *Nature, London* 254, 31-4.

CASSLETON, L.A. & LEWIS,D. (1967). Dilution of gene products in the cytoplasm of heterokaryons in *Coprinus lagopus*. *Genetical Research, Cambridge* 9, 63-71.

CLUTTERBUCK,A.J. (1974). *Aspergillus nidulans*. In : *Handbook of genetics*, edited by R.C.King, Vol.1, pp.447-510, New York: Plenum Press.

CLUTTERBUCK,A.J. & COVE,D.J. (1976). The genetic loci of *Aspergillus nidulans*. In: *CRC Handbook of microbiology*, Vol.IV, edited by A.I Laskin and H. Lechevalier, Cleveland, Ohio: Chemical Rubber Publishing Co., (second edition in the press).

COVE,D.J. (1966). The induction and repression of nitrate reductase in the fungus *Aspergillus nidulans*. *Biochimica et Biophysica Acta* 113, 51-6.

COVE,D.J. (1967). Kinetic studies of the induction of nitrate reductase and cytochrome *c* reductase in the fungus *Aspergillus nidulans*. *Biochemical Journal* 104, 1033-9.

COVE,D.J. (1969). Evidence for a near limiting intracellular concentration of a regulator substance. *Nature, London* 224,272-3.

COVE,D.J. (1970). Control of gene action in *Aspergillus nidulans*. *Proceedings of the Royal Society, London, Series B*, 176,267-75.

COVE,D.J. (1976). Chlorate toxicity in *Aspergillus nidulans*: The selection and characterisation of chlorate resistant mutants. *Heredity* 36, 191-203.

FELDEN,R.A., SANDERS,M.M. & MORRIS,N.R. (1976). The presence of histones in *Aspergillus nidulans*. *Journal of Cell Biology* 68, 430-9.

GRINDLE,M. (1963a). Heterokaryon incompatibility of unrelated strains in the *Aspergillus nidulans* group. *Heredity* 18, 191-204.

GRINDLE,M. (1963b). Heterokaryon compatibility of closely related wild isolates of *Aspergillus nidulans*. *Heredity* 18, 397-405.

GUNATILLEKE,I.A.U.N., ARST,H.N. & SCAZZOCCHIO, C. (1975a). Three genes determine the carboxin sensitivity of mitochondrial succinate oxidation in *Aspergillus nidulans*. *Genetical Research, Cambridge* 26, 297-305.

GUNATILLEKE,I.A.U.N., SCAZZOCCHIO,C. & ARST,H.N. (1975b). Cytoplasmic and nuclear mutations to chloramphenicol resistance to *A.nidulans*. *Molecular and General Genetics* 137, 269-76.

HASTIE,A.C. (1970). Benlate-induced instability of *Aspergillus* diploids. *Nature, London* 226, 771.

HERMAN,C. & CLUTTERBUCK,A.J. (1966). A method for selection of auxotrophs by means of 'spidery' growth. *Aspergillus Newsletter* 7, 13-14.

HYNES,M.J. (1975). A *cis*-dominant regulatory mutation affecting enzyme induction in the eukaryote *Aspergillus nidulans*. *Nature,*

London 253, 210-12.

JINKS,J.L. (1963). Cytoplasmic inheritance in fungi. In: *Methodology in basic genetics*, edited by W.J.Burdette, pp.325-5 San Francisco, California: Holden-Day.

KÄFER, E. (1961). The process of spontaneous recombination in vegetative muclei of *Aspergillus nidulans*. *Genetics* 46, 1581-6

KÄFER,E. (1962). Translocations in stock strains of *Aspergillus nidulans*. *Genetica, The Hague*, 33, 59-68.

KÄFER, E.(1965). The origins of translocations in *Aspergillus nidulans*. *Genetics* 52, 217-32.

LUKASZKIEWICZ, Z. and PASZEWSKI,A. (1976). Hyper-repressible operator-type mutant in sulphate permease gene of *Aspergillus nidulans*. *Nature, London* 259, 337-8.

McCULLY,K.S. and FORBES,E. (1965). The use of *p*-fluorophenylalani with 'master strains' of *Aspergillus nidulans* for assigning genes to linkage groups. *Genetical Research, Cambridge* 6, 352-9.

MA,G.C.L. & KÄFER,E. (1974). Genetic analysis of the reciprocal translocation *T2(I;VIII)* of *Aspergillus* using the technique of mitotic mapping in homozygous translocation diploids. *Genetics* 77, 11-33.

MacKINTOSH,M.E. & PRITCHARD,R.H. (1963). The production and repli plating of micro-colonies of *Aspergillus nidulans*. *Genetical Research, Cambridge* 4, 320-2.

MINSON,A.C. & CREASER,E.H. (1969). Purification of a trifunctiona enzyme catalising three steps of the histidine pathway from *Neurospora crassa*. *Biochemical Journal* 114, 49-56.

PONTECORVO,G. (1963). Microbial genetics: retrospect and prospec *Proceedings of the Royal Society, London, Series B* 158, 1-23.

PONTECORVO,G., ROPER,J.A., HEMMONS,L.M., MacDONALD,K.D. & BUFTON, A.W.J. (1953). The genetics of *Aspergillus nidulans*. *Advances in Genetics* 5, 141-238.

PONTECORVO,G., TARR-GLOOR,E. & FORBES, E.(1954). Analysis of mitotic recombination in *Aspergillus nidulans*. *Journal of Genetics* 52, 226-37.

REVER,B.M. (1965). Biochemical and genetical studies of inorganic nitrogen metabolism in *Aspergillus nidulans*. Ph.D.dissertatio University of Cambridge.

ROPER,J.A. (1952). Production of heterozygous diploids in filamentous fungi. *Experientia, Basel* 8, 14-15.

ROWLANDS,R.T. & TURNER,G. (1973). Nuclear and extranuclear inheritance of oligomycin resistance in *Aspergillus nidulans*. *Molecular and General Genetics* 126, 201-16.

ROWLANDS,R.T. & TURNER,G. (1974a). Physiological and biochemica studies of nuclear and extranuclear oligomycin-resistant mutan of *Aspergillus nidulans*. *Molecular and General Genetics* 132, 73-88.

ROWLANDS, R.T. & TURNER,G. (1974b). Recombination between the extranuclear genes conferring oligomycin resistance and cold sensitivity in *Aspergillus nidulans*. *Molecular and General Genetics* 133, 151-61.
SCAZZOCCHIO,C. (1973). The induction of xanthine dehydrogenase II. *European Journal of Biochemistry* 36, 439-45.
SCAZZOCCHIO,C. & DARLINGTON,A.J. (1967). The genetic control of xanthine dehydrogenase and urate oxidase synthesis in *Aspergillus nidulans*. *Bulletin de la Societe de Chimie Biologique* 49,1503-8.
STRICKLAND,W.N. (1958a). Abnormal tetrads in *Aspergillus nidulans*. *Proceedings of the Royal Society, London, Series B* 533-42.
STRICKLAND,W.N. (1958b). An analysis of interference in *Aspergillus nidulans*. *Proceedings of the Royal Society, London, Series B* 149, 82-101.
TURNER,G. & ROWLANDS,R.T. (1976). Cytochrome abnormalities and cyanide-resistant respiration in extranuclear mutants of *Aspergillus nidulans*. *Journal of Bacteriology* 125, 389-97.
VAN ARKEL,G.A. (1963). Sodium arsenate as an inducer of somatic reduction. *Aspergillus Newsletter* 4, 9.

Carbon Metabolism in *Aspergillus nidulans*

W. McCULLOUGH, M. A. PAYTON and C. F. ROBERTS

Department of Genetics, University of Leicester, Leicester, U.K.

INTRODUCTION

Carbon metabolism in filamentous fungi has been reviewed by Cochrane (1958), Niederpruem (1965) and Blumenthal (1965), and was extensively covered by a series of articles in Smith and Berry (1975). A review in bacteria, particularly *Escherichia coli*, describes many genes controlling enzymes of intermediary metabolism (Fraenkel & Vinopal, 1973).

We take a narrow view of the topic and concentrate our attention upon C (carbon) metabolism in *Aspergillus nidulans*. While the major pathways of intermediary metabolism probably occur in *A. nidulans* (Fig. 1), we think it necessary to secure this basic knowledge in the organism especially if it is to be developed for studies in genetic regulation. Moreover *A. nidulans* is a good analogue for many filamentous fungi of economic importance, a point emphasised by the production of penicillin in the species (Holt & Macdonald, 1968; Merrick & Caten, 1975).

The chief attraction of research in *A. nidulans* lies in the powerful techniques available in genetics, both to isolate mutants with specified phenotypes and for their rapid analysis. What can be gained from the use of mutants in C metabolism which could not have been deduced from comparative biochemistry? Apart from knowledge of the mapping of genes controlling C metabolism we think that there are three main reasons for studying

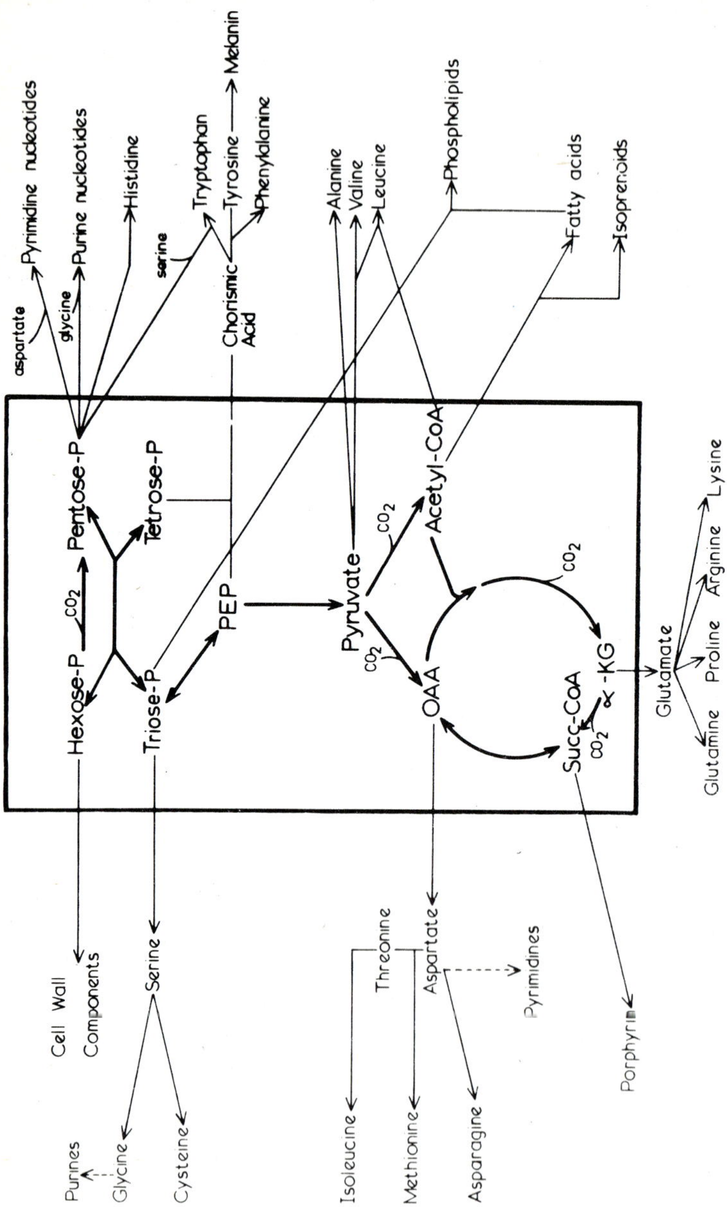

Fig. 1. Biosynthetic pathways in *Aspergillus nidulans*

this aspect of genetics of *A. nidulans*. First, mutant lesions can provide a way of identifying the enzyme of physiological importance among several possible alternative reactions, an argument skillfully expounded and illustrated by Kornberg (1966a). For example, pyruvate carboxylase is the major anaplerotic enzyme in glycolysis (Skinner & Armitt, 1972) and PEP carboxykinase essential for the release of intermediates from the tricarboxylic acid cycle for gluconeogenesis (Armitt, McCullough & Roberts, 1976). Second, mutants can lead to the discovery of previously unsuspected metabolic pathways, for example, in galactose or lactose utilization discussed below. Finally, and perhaps most important, the study of the regulation of metabolic pathways and gene function in eukaryotic cells requires a sound knowledge of those pathways and the availability of structural gene mutations.

COMPOUNDS UTILIZED AS CARBON SOURCE FOR GROWTH

The Aspergilli utilize a wide range of organic compounds as source of carbon and energy for growth (Cochrane, 1958). Most reports on the growth of *A. nidulans* on different C sources use plate tests to estimate visually the amount of mycelium and degree of conidiation relative to growth on glucose (Table 1). Although qualitative, such data are quite useful in the comparative sense for they provide an indication of growth yield. Many C sources produce growth after an obvious lag period indicating their adaptive metabolism. Trinci (1971) has emphasised that comparative specific growth rates cannot be computed from growth on solid media owing to variation in the peripheral growth zone on different C sources.

Carbon source utilization by wild-type strains is little affected by the nature of the N (nitrogen) source, though certain mutants in carbohydrate metabolism differ in growth on nitrate or ammonium (Hankinson, 1974). Moreover the utilization as *sole C source* of compounds supplying both C and N is generally not affected by the presence of alternate N sources in the medium either in wild-type or in mutants with lesions in N metabolism (Arst & Cove, 1973; Hynes, 1975a) or in C metabolism (Payton, McCullough & Roberts, 1976). On the other hand N regulation is dependent upon the nature of the C source (Pateman & Kinghorn, this volume).

Agar as a weak C source

A. nidulans makes appreciable growth on minimal medium plates to which no C source has been added other than agar, without liquifying the gel (Hankinson, 1974; Fiddy & Trinci, 1975; Payton *et al.*, 1976). There is some evidence suggesting that

Table 1. Carbon sources for *Aspergillus nidulans*

Good	Moderate	Poor	No Growth
Sucrose	Cellobiose. Lactose* Maltose. Melibiose Raffinose. Trehalose		
D Glucose	D Galactose*	L Rhamnose	L Sorbose. D Fucose
D Fructose	D Mannose		
	D Mannitol* D Sorbitol*		
	D Glucuronate D Glucuronate-δ-lactone	D Gluconate D Gluconate δ-lactone	D Galacturonate
L Arabinose D Lyxose D Xylose	D & L Arabitol D Ribitol Xylitol	D Arabinose D Ribose L Xylose	
		D Erythrose Erythritol	
	Glycerol* Acetate* Ethanol*	DL Lactate Succinate. Tartrate	Citrate
	Butyrate*	Oleate. Palmitate	

Generally the growth yield on a defined mineral salts agar medium with C source at 1% (w/v) has been scored visually relative to growth on glucose or sucrose over a period of up to 3 days incubation at 37°C. The four headings roughly correspond to categories used by various authors: Good, +++, 6 or 5; Moderate, ++, 4 or 3; Poor, +(+), 2 or 1; No growth, -, 0.

Certain amino acids serve as moderate or poor C sources for growth in addition to providing a source of N (Kinghorn & Pateman this volume). Utilization of amides as C source is described by Hynes (1975b).

Table 1 cont'd....

* Carbon sources utilized adaptively. The lag period can be from 12h (acetate, galactose) to several days (butyrate). Many other C sources are probably also utilized adaptively but this has not been reported.

---

degradation products are utilized through C_2 metabolism rather than glycolysis but the mechanism is obscure. The presence of agar may generate both false positive or false negative growth responses compared with liquid medium which are at variance with the expectations from known enzyme lesions, and we urge caution in the interpretation of growth tests especially for utilization of organic or amino acids done only on agar plates.

Growth in liquid culture

There is little quantitative information upon the effect of varying the nature or concentration of C source upon growth of *A. nidulans* in liquid culture. The maximum growth rates obtained in glucose limited continuous culture (mass doubling time, Td = 8.9h) was much less than that from batch culture (3.5h) in the same apparatus (Carter & Bull, 1969) but the reason for this discrepancy remains obscure (Carter *et al.*, 1971) though it also occurs in *A. niger* (Ng, Smith & McIntosh, 1974). Moreover *A. nidulans* produced melanin in continuous culture, but only upon glucose exhaustion in batch culture (Carter & Bull, 1969).

Data on growth rates in batch culture suggest that these do not differ greatly with C source, though the types of compounds reported are similar (Table 2). It remains possible that the major differences in growth on different C sources are in the lag period and growth yields rather than the (maximum) logarithmic rate, as shown in the dimorphic fungus *Geotrichum candidum* (Caldwell & Trinci, 1973).

Growth on mixed C sources

There is growing interest in the response of *A. nidulans* to mixtures of C sources with respect to both inhibition of uptake and to catabolite repression. As might be expected inhibition and repression may occur between any two potential C sources and has been demonstrated by the use of certain types of mutant (Bailey & Arst, 1975; Arst, this volume). Certain mutants in glycolysis are predicted to require a mixture of C sources for growth (see below).

Table 2. Rates of growth of *Aspergillus nidulans* in batch culture with different carbon sources

Strain	N Source	C Source	Mass Doubling Time (TD,h)	Reference
BWB224	NO_3^-	D Glucose (56 mM)	1.96	Trinci (1971)
(*yA*)	NO_3^-	D Glucose (1 mM)	2.43	
Wild Type	NO_3^-	D Glucose (56 mM)	1.4	Rosenberger & Kessel (1967)
Glasgow 012	NO_3^-	D Glucose (56 mM)	1.8	
(*adF*15, *biA*1)	NO_3^-	L Arabinose (33 mM)	3.7	
	NO_3^-	L Xylose (33 mM)	9.0	
R21	NO_3^-	Sucrose (40 mM)	2.7	Roberts (unpublished)
(*pabaA*1; *yA*)	NO_3^-	Sucrose (20 mM)	2.9,2.4, 3.2	
	NO_3^-	Acetate (200 mM)	2.8	
	NO_3^-	Acetate (100 mM)	3.4	
	NO_3^-	Acetate (100 mM)*	2.7	
	NH_4^+	Acetate (100 mM)*	2.6	
	NO_3^-	Lactose (20 mM)	3.4	Fantes (1972)

Table 2. cont'd.....

Strain	N Source	C Source	Mass Doubling Time (TD,h)	Reference
	NO_3^-	D Galactose (20 mM)	4.1,5.6	
	NO_3^-	Glycerol (40 mM)	3.9	
Glasgow 051	NH_4^+	Glucose (56mM	1.25	Cohen (1973)
(*biA*1)	(tartrate)	Glucose (56 mM)	1.25	

Data are from batch cultures (25 ml minimal medium in 250 ml flasks growing at 37°C with vigorous shaking. The lower growth rates reported by our laboratory are probably due to the volume of medium used (400 ml in 2 litre flasks).

* Medium buffered at pH 6.5 by 200 mM phosphate.

INTERMEDIARY METABOLISM

The occurrence of the Embden-Meyerhoff-Parnas (EMP) sequence and the pentose phosphate pathway (PPP) in filamentous fungi, including species of *Aspergillus*, has been documented by Blumenthal (1965), and the presence of enzymesof the tricarboxylic acid (TCA) cycle by Niederpruem (1965). Many of the enzymes have been demonstrated in *A. nidulans* (Figure 2 and 3) though few attemps made to investigate their intra-cellular location or the fine control of their activity. The enzymes of the EMP and PPP together with those for the metabolism of specific carbohydrates are likely to be cytosolic while the glyoxylate cycle enzymes may occur in a membrane bound organelle (glyoxysome) separate from the mitochondria.

Glucose metabolism

Among the enzymes of glycolysis hexokinase, glucose-6-P (phosphate) dehydrogenase and aldolase were reported by Carter & Bull (1969), phosphohexose isomerase by Elorza & Arst (1971) and

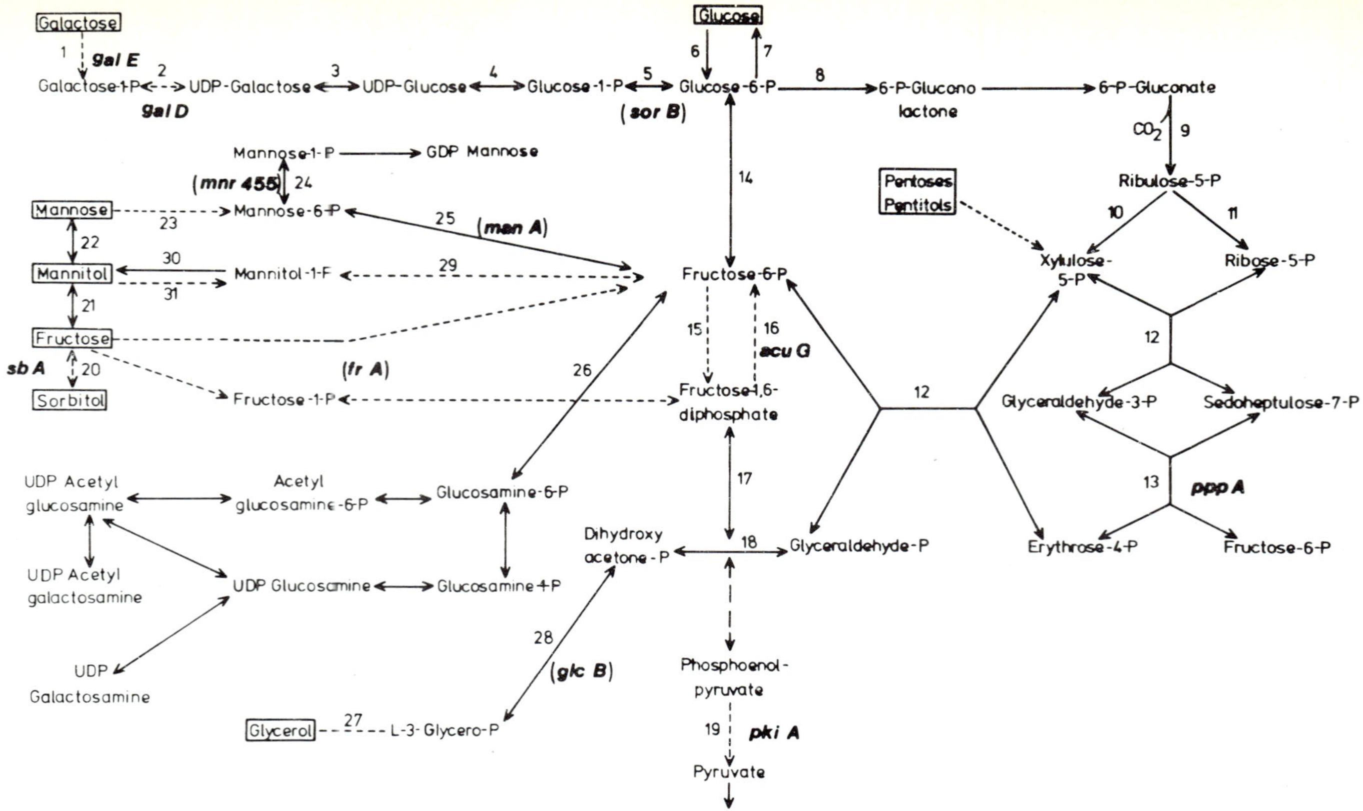

Fig. 2 Carbohydrate metabolism in *Aspergillus nidulans*. Constitutive enzymes are shown by solid lines, and enzymes whose synthesis is regulated by broken lines. Structural genes are in italics, *galE*, and those in parenthesis are probably structural gene loci.

Fig. 2 The enzymes indicated are: 1, galactokinase; 2, UDP-galactose pyrophosphorylase; 3, UDP-glucose epimerase; 4, UDP-glucose pyrophosphorylase; 5, phosphoglucomutase; 6, hexokinase; 7, glucose-6-phosphatase; 8, glucose-6-phosphate dehydrogenase; 9, 6-phosphogluconate dehydrogenase; 10, phosphoketopentoepimerase; 11, phosphoribiosomerase; 12, transketolase; 13, transaldolase; 14, phosphoglucoisomerase; 15, phosphofructokinase; 16, fructose-1,6-diphosphatase; 17, aldolase; 18, trios phosphate isomerase; 19, pyruvate kinase; 20, sorbitol dehydrogenase; 21, mannitol dehydrogenase; 22, mannose dehydrogenase; 23, mannokinase; 24, phosphomannose mutase; 25, phosphomannose isomerase; 26, L-glutamine: D fructose-6-P amidotransferase; 27, glycerol kinase; 28, glycerol phosphate dehydrogenase; 29, mannitol-1-P dehydrogenase; 30, mannitol-1-P phosphatase; 31, mannitol kinase.

---

pyruvate kinase by McCullough & Roberts (1974). In a study of key enzymes in both the EMP and PPP Hankinson & Cove (1974) reported in addition, phosphofructokinase in the EMP and 6-phosphogluconate dehydrogenase, ribose phosphate isomerase, transketolase and transaldolase in the PPP.

Both the EMP and PPP provide intermediates for biosynthetic pathways. They differ in that the EMP provides NADH which is oxidised to generate ATP while the oxidative branch of the PPP (glucose-6 P to ribulose-5-P) is one source of NADPH used to drive biosynthetic reactions. The relative flux through the two paths was investigated in glucose limited continuous culture by Carter & Bull (1969) who concluded, on the basis of the respirometry of ^{14}C1- or ^{14}C6-glucose, sensitivity to inhibitors of the EMP and enzyme assays that 80% of glucose is metabolised through the EMP at low glucose concentrations yielding slow growth rates (Td 28.8 to 12.6h). At a maintenance ratio of glucose (0.029 g glucose/g dry weight/h) or upon carbon source starvation the EMP accounts for 100% of carbon flow (Bainbridge *et al.*, 1971). Several physiological parameters indicate a metabolic shift occurring at growth rates above Td = 9.2h, including an increased flux to 40% of glucose metabolised through the PPP. Lowered oxygen tension ($<$ 30 mm mercury) at an intermediate growth rate produced a similar increase in flux through the PPP (Carter & Bull, 1969). In all these studies nitrate was the N source.

The general conclusions (Carter *et al.*, 1971) are that with increasing rate of glucose supply higher growth rate is paralleled by greater capacity (activity/g dry weight) in enzymes of the EMP and PPP. The qualitative shift at Td $\leq$ 9.2 h) is

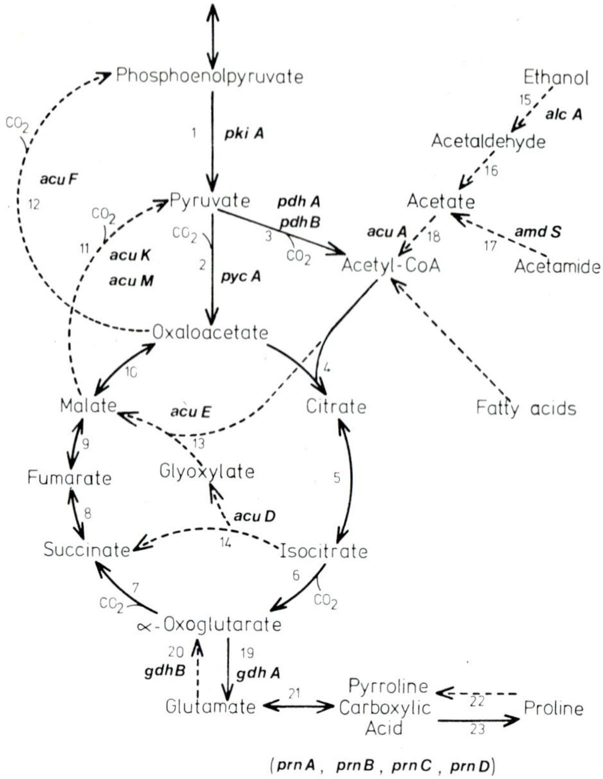

Fig. 3. The tricarboxylic acid cycle and some related pathways in *Aspergillus nidulans*.

Constitutive enzymes are shown by solid lines and enzymes whose syrthesis is regulated by broken lines. Structural genes are in italics, *pkiA*.

The enzymes indicated are: 1, pyruvate kinase; 2, pyruvate carboxylase; 3, pyruvate dehydrogenase; 4, citrate synthase; 5, aconitase; 6, isocitrate dehydrogenase; 7, oxoglutarate dehydrogenase; 8, succinate dehydrogenase; 9, fumarase; 10, malate dehydrogenase; 11, malic enzyme; 12, PEP carboxykinase; 13, malate synthase; 14, isocitrate lyase; 15, alcohol dehydrogenase; 16, aldehyde dehydrogenase; 17, acetamidase; 18, acetyl-CoA synthase; 19, NADP-glutamate dehydrogenase; 20, NAD-glutamate dehydrogenase; 21, pyrroline-5-carboxylate dehydrogenase; 22, proline oxidase; 23, pyrroline-5-carboxylate reductase.

Fig. 3 cont'd......

Although the glyoxylate cycle reactions are shown as part of the TCA cycle it is probable that they are spatially separate in distinct organelles (glyoxysomes) which have been reported from other fungi (Kobr, Vanderhaeghe & Combepine, 1969; O'Sullivan & Casselton, 1973).

---

not understood though in part comprising a disproportionate increase in the PPP enzymes. While the levels of enzyme of central metabolism vary only over a 2 to 5-fold range with glucose limitation, and similarly with glucose or citrate limitation in *A. niger* (Ng *et al.*, 1974), Carter *et al.* (1971) did note a 40-fold change in biosynthetic NADP-linked glutamic dehydrogenase. The possibility that changes in growth rate may *per se* produce disproportionate changes in enzyme levels is often overlooked. It is unfortunate that technical problems restricted these studies to conditions not generally employed with *A. nidulans*, particularly growth at 30°C and Td $\geq$ 9 h.

The effect of nitrate on production of enzymes of the EMP or PPP has been studied in glucose grown batch cultures by Hankinson & Cove (1974). The levels of three enzymes of the EMP (phosphofructokinase, aldolase and pyruvate kinase) remain unchanged in growth on urea with or without nitrate, but hexose phosphate isomerase increased 3-fold with nitrate and four enzymes of the PPP (both dehydrogenases, transketolase, transaldolase) 2-fold. The increased capacity of the PPP is consistent with an increased demand for NADPH for nitrate reduction.

Two unlinked genes affecting enzymes in the PPP have been identified by mutants defective in growth on pentoses, pentitols and glucoronate (Hankinson, 1974). Mutants in *pppA* contain less than 10% of normal transaldolase activity. Mutants in *pppB* are leaky in growth and contain 25-50% normal activities of the four regulated PPP enzymes, phosphohexose isomerase and also mannitol-1-P dehydrogenase. The same enzymes, except transaldolase, are increased 2-fold in the *pppA* mutants, perhaps due to glucose-6-P accumulation. The *ppp* mutants are also defective in utilization of nitrate and nitrite. Moreover, their accumulation of sedoheptulose-7-P, which is enhanced by nitrate, is abolished if the strains also contain mutant lesions in nitrate utilization. Thus the metabolism of nitrate stimulates flux through the PPP. Since NADP linked isocitrate dehydrogenase does not change in response to nitrate (Hankinson & Cove, 1974) and malic enzymes is low in hexose grown mycelium (McCullough & Roberts, 1974) it appears likely that the *ppp* is the major source of NADPH for reduction of nitrate in growth on hexose.

The requirement for pyruvate kinase in glycolysis but not in gluconeogenesis was suggested by its high activity in growth on glycolytic C sources and low activity on gluconeogenic C sources, such as acetate (McCullough & Roberts, 1974). The phenotype of mutants (*pkiA)* lacking the enzyme (Payton & Roberts, 1976) confirm this interpretation. Moreover growth of the *pkiA* mutants on acetate is extremely sensitive to inhibition by hexose. Selection by escape from inhibition should yield mutants defective in the uptake of specific hexoses or with second enzyme lesions before pyruvate kinase preventing accumulation of PEP.

THE TRICARBOXYLIC ACID CYCLE AND TERMINAL RESPIRATION

The Tricarboxylic Acid (TCA) cycle in obligate aerobes is the final respiratory pathway in which intermediates produced by peripheral pathways are oxidized to carbon dioxide and water. In the process NAD is reduced to NADH which serves as electron donor for the electron transport pathway and oxidative phosphorylation in which energy is conserved as ATP. The TCA cycle also functions to supply biosynthetic intermediates such as oxoglutarate and oxaloacetate. In eukaryotes these reactions are largely located within the mitochondrion as integral components of the membranes or in the matrix (Ernster & Kuylenstierna, 1969). However, the TCA cycle has been largely neglected in comparison with peripheral pathways of C metabolism in *A. nidulans* despite its important central position in metabolism (Fig. 3).

The enzymes of glycolysis and the pentose phosphate pathway degrade carbohydrates to pyruvate. These reactions are cytoplasmic and the pyruvate formed is transported into the mitochondria where it is oxidized in the TCA cycle, the initial reaction of which is the condensation of acetyl-CoA with oxaloacetate catalysed by citrate synthase. The acetyl-CoA is formed by the oxidation of pyruvate catalysed by the pyruvate dehydrogenase complex of enzymes, which in *Escherichia coli* consists of three separate components and their co-factors (Koike, Reed & Carroll, 1963). Mutants of *E. coli* lacking each of the components have been isolated (Henning & Hertz, 1964; Guest, 1974) and their tight linkage strongly suggests an operon. A mutant of *A. nidulans* lacking overall pyruvate dehydrogenase activity was isolated by Romano & Kornberg (1968). Further mutants have been isolated (Guest, Payton, McCullough & Roberts, unpublished) and these fall into two groups between which complementation is found both *in vivo* and *in vitro* by mixing cell-free extracts. The mutants define two unlinked genes: *pdhA* (chromosome 1) is defective in dihydrolipoyl transacetylase (E2) and *pdhB* (chromosome V) which lacks pyruvate decarboxylase (E1).

Although the formation of pyruvate dehydrogenase is regulated in *E. coli* (Henning, Dennert, Hertel & Shipp, 1966) it is formed constitutively by *A. nidulans* (Guest *et al.*, unpublished). This means that the enzyme complex is present during growth on acetate, for instance, although the properties of mutants show that it is not required under these conditions. Clearly, pyruvate dehydrogenase activity must be subject to feedback inhibition from the TCA cycle, probably by NADH and acetyl-CoA as in *E. coli* (Hansen & Henning, 1966) and *Neurospora* (Harding, Caroline & Wagner, 1970).

Since one of the functions of the TCA cycle is to generate intermediates for biosynthetic pathways, an anaplerotic reaction (Kornberg, 1966a) is necessary to 'top up' the level of oxaloacetate to maintain the cycle. In most bacteria this reaction is catalysed by PEP carboxylase but in fungi pyruvate carboxylase fulfills this role. Attempts to confirm the presence of PEP carboxylase in *A. nidulans* (Bushell & Bull, 1974) have been unsuccessful (McCullough, unpublished). Single gene mutants (*pycA*) of *A. nidulans* lacking pyruvate carboxylase have been isolated (Skinner & Armitt, 1972), and as expected they fail to grow on glucose unless a TCA cycle intermediate is also supplied. These strains grow poorly on glucose media supplemented with glutamate because of the catabolite repression of NAD-glutamate dehydrogenase (Kinghorn & Pateman, 1974). Surprisingly *pycA* mutants grow (though poorly) on glycerol as sole carbon source which suggests an alternate anaplerotic pathway is operating under these conditions. Further evidence for an alternative pathway to pyruvate carboxylase comes from a study of revertants of *pycA* to growth on sucrose as sole carbon source (McCullough & Roberts, unpublished). Most of the revertants did not contain pyruvate carboxylase activity but segregate *pyc* suppressor mutations in crosses to wild-type. The nature of the alternative pathway is not known.

Pyruvate carboxylase is formed constitutively by *A. nidulans* (Skinner & Armitt, 1972) as in *S. cerevisiae* (Haarasilte & Oura, 1975) although the properties of mutants lacking the enzyme show that it is only required during growth on glycolytic carbon source. In *Saccharomyces* (Miller & Atkinson, 1972) and *Penicillium* (Stan & Schormuller, 1968) the enzyme is controlled by the adenylate energy charge (Atkinson, 1968) and aspartate. The enzyme from yeast is activated by acetyl-CoA (Cooper & Benedict, 1966), but in *A. nidulans* as in *A. niger* (Bloom & Johnson, 1962) there is apparently no requirement for activation by acetyl-CoA.

The occurrence of enzymes of the TCA cycle in fungi was reviewed by Niederpruem (1965). More recent data is available for *Neurospora*(Flavell & Fincham, 1968), *A. niger* (Ng, Smith & McIntosh, 1973), *Saccharomyces* (Gosling & Duggen, 1971) and *Blastocladiella* (Khouw & McCurdy, 1969). However, comparable data for *A. nidulans* has not been published and is given in Table 3.

In general the enzymes of the TCA cycle are elevated in acetate grown mycelium compared with sucrose as was also observed in *Neurospora* (Flavell & Fincham, 1968). Thus citrate synthase, fumarase and malate dehydrogenase are increased 2- or 3-fold as in *Neurosposa* and form a constant proportion group (Pette, 1966). Presumably the increased enzyme activities during growth on acetate reflects the increased flux through the cycle when acetate is the sole carbon source. The co-ordinate derepression of TCA cycle enzymes in *Neurospora* was discussed by Flavell & Woodward (1970).

The presence of NADP linked isocitrate dehydrogenase and the absence of an NAD dependent enzyme (McCullough, unpublished) confirms the result of Hankinson & Cove (1974). This is in marked contrast with other fungi (Sanwal & Stachow, 1965; Kornberg & Pricer, 1951; Ramakrishnan & Martin, 1955) where both nucleotide dependant activities were identified. Isocitrate dehydrogenase was derepressed more than the other TCA cycle enzymes on acetate medium. Another interesting feature of these results is that failure to detect oxoglutarate dehydrogenase activity in sucrose grown mycelium although some activity was apparently present in mycelium grown on acetate. The failure to detect this enzyme in *A. niger* (Ng, Smith & McIntosh, 1973), *Penicillium* (Attwood & Guest, personal communication) and *Blastocladiella* (Khouw & McCurdy, 1969) has been noted previously although the enzyme is known to be present in *Saccharomyces* (Polakis & Bartley, 1965). Flavell & Fincham (1968) detected enzyme activity in *Neurospora* using a manometric method, but the products of the reaction were not characterized. It would be interesting to know if the enzyme is absent from some fungi (and if so, how oxoglutarate is oxidised) or whether the results can be attributed to a particularly labile enzyme.

To our knowledge no mutants of *A. nidulans* are known which have lesions in the TCA cycle although resistance to growth inhibition by the fungicide carboxin has been correlated with the presence of an altered succinate dehydrogenase less sensitive to carboxin (Gunatillake, Arst & Scazzocchio, 1976). In yeast mutants lacking citrate synthase (Burand, Drillien & Bhattacharjee, 1975), aconitase (Ogur, Coker & Ogur, 1964) and oxoglutarate dehydrogenase (Subik, Kolarov & Lachowicz, 1972) have been identified. Citrate synthase and aconitase mutants have a requirement for glutamate. These mutants share the property of respiratory deficiency and inability to grow on nonfermentable

Table 3. Enzymes of the tricarboxylic acid cycle in *Aspergillus nidulans*

Enzyme	Carbon Source for Growth	
	Sucrose	Acetate
Citrate synthase	39	119
Aconitase	4.4	23
NADP isocitrate dehydrogenase[1]	9	102
Oxoglutarate dehydrogenase[2]	none detected	33.0
Succinate dehydrogenase[3]	1.0	1.6
Fumarase	39.3	81.7
Malate dehydrogenase	594	1554

TCA cycle enzyme activities in mycelium of *A. nidulans* grown with either sucrose (20 mM) or acetate (100 mM) as sole carbon source (McCullough, unpublished). All enzyme activities were measured at 37°C (except oxoglutarate dehydrogenase which was 27°C and are expressed as μmoles product formed, h^{-1}, mg^{-1} soluble protein. The following assay procedures were used: citrate synthase (Cooper & Beevers, 1969); aconitase (Khouw & McCurdy, 1969); NADP dependent isocitrate dehydrogenase (Kornberg, 1955); oxoglutarate dehydrogenase (Amarasingham & Davis, 1965); succinate dehydrogenase (Ellis, 1959); fumarase (Hill & Bradshaw, 1969); and malate dehydrogenase (Khouw & McCurdy, 1969).

[1]See also Carter & Bull (1969) and Hankinson & Cove,(1974)

[2]Guest, Payton, McCullough & Roberts, unpublished.

[3]See also Gunatilleke, Arst & Scazzocchio (1976).

nonfermentable carbon sources with strains lacking cytochromes (Subik, Kuzela, Kolarov, Kovac & Lachowicz, 1970); Morita & Mifuchi, 1970) or with defects in oxidative phosphorylation (Beck, Matoon, Hawthorne & Sherman, 1968). It is to be expected therefore that mutants of this class could only be isolated as conditional mutations in *A. nidulans* which requires an intact terminal respiratory pathway.

Although normal cytochromes are present in *Aspergilli* (Rowlands & Turner, 1974; Watson & Smith, 1967), there is also a cyanide insensitive, antimycin insensitive pathway of electron transport to a terminal oxidase in *A. nidulans* (Gunatilleke, Arst & Scazzocchiio, 1976; Turner, this volume). This type of respiration has been identified in mitochondria from some higher plants (Bendall & Bonner, 1971), various fungi (Sherald & Sisler, 1970; Kawakita, 1970; Lambowitz & Slayman, 1971), *Paramecium* (Sainsard, Claisse & Balmefrezol, 1974) and *Euglena* (Sharpless & Buton, 1970). The alternate pathway branches from the normal pathway before Cytochrome B probably at the level of Coenzyme Q (von Jagow, Weiss & Klingenberg, 1973) and is largely wasteful since it may include only phosphorylation site I. Although the function of the cyanide resistant pathway is unknown it has been suggested that it might act to reduce the NADH/NAD ratio when increased glycolytic and TCA cycle activity would be beneficial (Lambowitz, Smith & Slayman, 1972). This is consistent with the fact that the nitrogen source and oxygen tension in the culture medium can modulate activity of the cyanide resistant pathway in *Neurospora* (Sauer, Colvin & Munkres, 1973a,b). It is conceivable that the increased oxidative capacity noted by Carter & Bull (1971) at low oxygen tension could include greater flux through the cyanide insensitive pathway. It would also be interesting to know if the level of cyanide resistant respiration in *A. nidulans* varied with the source of nitrogen in the culture medium.

CARBOHYDRATE METABOLISM

Uptake of carbohydrates

The differences in growth of *A. nidulans* on different carbohydrates could be due to regulation of key catabolic enzymes or of specific uptake systems (Romano & Kornberg, 1968; Romano, 1973). In *Escherichia coli* both types of control occur and it has been shown that transport of lactose and galactose are the rate-limiting steps in their catabolism (for review see Kornberg, 1973). However, most studies in *A. nidulans* so far have been biochemical due to the few uptake mutants available. A constitutive active uptake system for galactose was demonstrated by Roberts (1970). Gajewski *et al.* (1972) demonstrated a gene *lacA* specifying a lactose permease which is readily induced by lactose but poorly induced by galactose. Thus *A. nidulans* possesses both constitutive and inducible uptake systems. However, Mark & Romano (1971) demonstrated that constitutivity may be the general phenomenon, since they showed that glucose and fructose, like galactose, had constitutive uptake. Elorza & Arst (1971) showed a specific uptake system for sorbose, by use of a sorbose resistant mutant (*sorA*) which did not appreciably affect growth on a range of similar sugars. Together, the *lacA* and *sorA* are the only documented uptake mutants in *A. nidulans*. A number of sugar uptake mutants have been isolated in *Coprinus* (Moore, 1973) and *Neurospora* (Klingmuller, Huh & Kaudewitz, 1970).

Brown & Romano (1969) showed that the uptake of hexose by *A. nidulans* does not require their phosphorylation. Thus the organism conforms to the general rule that obligate aerobes lack a phospho-transferase system (Romano, Eberhard, Dingle & McDowell 1970). The activity of the hexose uptake systems can be altered and Romano & Kornberg (1969) demonstrated the regulation of uptake of hexoses by a metabolic derivative of acetate, probably acetyl-CoA. This exclusion by acetate did not equally affect sucrose uptake, providing a strong argument for separate uptake systems for hexoses and sucrose.

Using purely biochemical techniques it is difficult to assess how many systems operate on a single sugar, or the specificity of any single uptake system. Mark & Romano (1971) demonstrated a highly specific transport system for fructose, but overlapping specificaties in the systems for glucose and galactose transport. However as emphasised by Lin (1970) mutant analysis is advisable before definite conclusions are made. Since *A. nidulans* may possess multiple uptake systems for a single sugar as demonstrated for glucose uptake in *E. coli* (for review see Kornberg, 1976) and *Neurospora*(Scarborough, 1970), it would be desirable to isolate further sugar uptake mutants. We are engaged in the isolation of such mutants by selecting revertants insensitive to hexose inhibition in strains normally inhibited by hexose, namely *fr A* (Roberts, 1963a) and *pkiA* (Payton & Roberts, 1976).

Catabolism of certain carbohydrates

D. GALACTOSE: The Leloir pathway was demonstrated in *A. nidulans* by Roberts (1970). The three enzymes galactokinase, galactose-1-P uridyl transferase and UDP-galactose-4-epimerase (Fig. 2) are present at significant constitutive levels and are induced 3 to 5-fold by galactose. Mutants defective in galactose utilization identified six unlinked *gal* genes, including lesions in galactokinase (*galE*), transferase (*galD*) or uninducible for kinase and transferase (*galA*). Mutants in the other three genes grow poorly on galactose; they are not defective in the Leloir pathway nor in the accumulation of ^{14}C-galactose (Roberts, 1970), and all contain normal levels of phosphoglucomutase (Rosenberger, personal communication).

The Leloir pathway enzymes are required for growth on galactose at acid pH but several results indicate a second metabolic route operating at alkaline pH (Roberts, 1970). Katz & Rosenberger (1970a) observed massive incoporation of label from ^{14}C-galactose into the walls of a transferase mutant. Fantes & Roberts (1973) found that galactose induces more β-galactosidase at pH 7.5, and in a galactokinase mutant at this pH. Galactose also induces β-galactosidase in a transferase mutant (Chojnacki *et al.*, 1969). Elorza & Arst (1971) found that the *sorB* mutant deficient in phosphoglucomutase grows on galactose and also when the *sorB* strain contains a *ppp* mutation preventing growth on glucuronate. The nature of the alternate pathway is not known but oxidation to gal actouronate is unlikely since wild-type does not utilize this compound (Hankinson, 1974).

The *galC* locus is implicated in molybdate metabolism and certain other mutants (*gam*) resistant to molybdate toxicity are defective in galactose utilization (Arst & Cove, 1970). A possible relationship between galactose metabolism and morphology is indicated by a further gene (*galG*) which is a suppressor of *brlA*, an aconidial mutant (Clutterbuck, 1974).

LACTOSE. One β-galactosidase with an alkali pH optimum (Paszewski *et al.*, 1970) is found in two molecular forms in *A. nidulans* (Fantes & Roberts, 1973). The enzyme hydrolyses lactose but is induced more effectively by galactose than lactose. Fantes & Roberts (1973) showed that lactose utilization and -galactosidase are separable in *A. nidulans*. A histochemical procedure detecting enzyme activity in colonies showed that strain *lac*105 isolated by poor growth on lactose (*lac*$^-$) and absence of β-galactosidase activity (β*ga*$^-$) was a double mutant. The characters *lac*$^-$ and β*ga*$^-$ segregated independently in a cross to wild-type; one recombinant class (*lac*$^-$; β*ga*$^+$) grew poorly on lactose but formed enzyme on galactose or lactose (indicating lactose uptake); the second (*lac*$^+$; β*ga*$^-$) grew on lactose agare plates but formed no β-galactosidase. Conidia of this second class fail to germinate with lactosein liquid medium but grow when pregerminated with glucose, or if given 1 mM glucose with lactose, indicating a physiological role for β-galactosidase in germination.

Many mutants in *A. nidulans* isolated by poor growth on lactose also grew poorly on other disparate C source. However, Gajewski *et al.*, (1972) showed that mutants only defective on lactose and in which β-galactosidase was not induced by galactose or lactose identify four genes affecting formation of the enzyme. Three similar genes were described by Fantes & Roberts (1973) and one β*gaA* shown to be a structural gene for β-galactosidase by a temperature sensitive allele, and also to contain the original *βga*¯ mutation in the strain *lac*105. Attempts to isolate β-galactosidase constitutive mutants were unsuccessful, though the histochemical stain provides a powerful screening method not fully tested. The alternate mechanism for lactose metabolism remains unexplored but may involve phosphorylation. The alternate metabolic routes for both galactose and lactose and its effectiveness of both sugars in the induction of lactose permease and β-galactosidase presents a complex situation. However, the mutants and techniques available provide a basis for biochemical investigation of the alternate pathways which may have relationships with enzymes involved in cell wall metabolism and perhaps the degradation of agar galactans (Payton *et al.*, 1976).

FRUCTOSE, SORBITOL, MANNITOL, MANNOSE. Hexose and hexitol may be interconverted directly or after phosphorylation, though the specificity of the enzyme systems in fungi is largely unknown (Lewis & Smith, 1967; Speth & Niederpruem, 1976). A mannitol acetyl-P phosphotransferase has been described in *Aspergillus spp* (Lee, 1967). Metabolism of the hexitols and also mannos through fructose in *A. nidulans* is implied by failure of *frA* to utilise any of the sugars, and their inhibition of the growth of *frA* on glucose (Roberts, 1963a; Payton, unpublished). Mutants (*sbA*) failing to grow only on sorbitol (Roberts, 1963a; Elorza & Arst, 1971) or upon mannose (*manA*) have been reported (Valentine & Bainbridge, 1975).

PENTOSE, PENTITOLS. Fungi generally utilize aldo-pentoses by reduction to pentitols which are intercoverted through keto-pentoses and other pentitols to D xylulose, the only compound phosphorylated to enter the *ppp* as D xylulose-5-P (Lewis & Smith, 1967). Failure of the *ppp* mutants to grow on pentose or pentitol (Hankinson, 1974) suggests such a system in *A. nidulans*.

GLYCEROL. A series of recessive mutants defective in glycerol utilization but growing normally on sucrose or acetate have been isolated (Payton, unpublished) and identify two genes for glycerol metabolism. The *glcA* (chromosome V) mutant grows poorly and is partly inhibited by glycerol in growth on other C sources; *glcB* (chromosome I) mutants fail to grow and are strongly inhibited by glycerol. Glycerol inhibition often results from the accumulation of α-glycerol-3-P, due to deficiency in the dehydrogenase. The *glcA* and *glcB* mutants complement poorly in the growth of heterokaryons on glycerol but well in diploids.

L. SORBOSE. While sorbose is accumulated by *A. nidulans* there is no evidence for further metabolism to a toxic product. The basis for sorbose toxicity remains obscure as does the resistance conferred by partial (80%) loss in phospho-glucomutase activity in the *sorB* mutant (Elorza & Arst, 1971).

Cell wall biosynthesis

The major components in the cell walls of Aspergilli are chitin, a β-1 → 4n acetyl-glucosamine polymer, and glucans in which β1 → 3 linked D glucose residues predominate (Bartnicki-Garcia, 1968). Studies by Bull (1970a), Katz & Rosenberger (1970b) and Zonneveld (1971) are substantially in agreement on the composition of the hyphal walls of *A. nidulans*, although the culture conditions used by Bull produced a considerable amount of melanin (15-20% dry weight of cell wall). Neutral polysaccharides account for 60% of cell wall mass, amino polysaccharides 20%, proteins 10% and lipids 10%. Neutral polysaccharides contain mainly D glucose, with lesser quantities of D galactose and D mannose. Katz & Rosenberger (1970a, 1970b) noted a small amount of arabinose; Johnston (1965) recorded low concentrations of D arabinose in walls of *A. niger*. The amino sugars are D glucosamine and D galactosamine, partly acetylated on the amino groups and in distinct polymers (Bull, 1970a). Bull also noted a small proportion of glucuronate. Both Bull (1970a) and Zonneveld (1971) described an alkali and hot water soluble α1 → 3 glucan, while in addition to chitin in the alkali insoluble fraction Bull described β1 → 3, β1 → 6 glucans and Zonneveld and α1 → 4, β1 → 3 glucan associated with a galactomannan. Peptido-galactofuranose mannans are known as extra-cellular allergens from several Aspergilli; and glycoproteins, some with known enzymatic functions, are associated with the cell wall (Gander, 1974). Melanin is regarded principally as a protective material gainst enzymatic digestion (Bull 1970b), and the α1 → 3 glucan as a carbohydrate reserve (Zonneveld, 1972).

The biosynthesis of monomers incorporated into cell walls probably occurs by generally known mechanisms (Fig. 2) though little direct biochemistry has been done in *Aspergillus*. Glucosamine presumably arises by amination of fructose-6-P as shown in *Neurospora* (Russel & Srb, 1974), with subsequent amino-acetylation, epimerization to galactose and incorporation of both amino sugars from the uridine diphosphates. The central roles of glucose-1-P, glucose-6-P and fructose-6-P in cell wall biosynthesis are evident (Fig. 2).

Of particular interest are two studies of mutants defective in cell wall biosynthesis which have turned up among temperature sensitive mutants failing to grow on glucose. Katz & Rosenberger (1970b) describe a mutant (*ts*6) which was osmotically fragile at

41°C, produced distorted hyphae and was deficient in chitin. Cell walls were depleted in both glucosamine and galactosamine, indicating synthesis of the latter from glucosamine. The mutant phenotype was reversed at 41°C by glucosamine and by n-acetyl glucosamine, suggesting a lesion soon after fructose-6-P. Valentine & Bainbridge (1975) describe a mutant which is repaired at 43°C by D mannose. This mutant *mnrA*445 (mannose requiring) is deficient in cell wall mannose. Growth on mannose (0.1%) at 43°C is inhibited by glucose (0.9%). One strain selected for growth on the two sugars and which required *both* sugars proved to contain a second mutation (*manA*) which alone totally prevents growth on mannose. The *mnrA*445 mutant contains a thermolabile phosphomannose mutase and the *manA* mutant is deficient in a phosphomannose isomerase distinct from phosphoglucose isomerase (Valentine, 1975). The mutant lesions in cell wall synthesis permit demonstration that isotopically labelled glucosamine (Katz & Rosenberger, 1971a) or mannose (Valentine & Bainbridge, 1975) are incorporated at the hyphal tip. Katz & Rosenberger (1970a) similarly showed incorporation of ^{14}C galactose at the hyphal tip in a galactose-1-P uridyl transferase mutant. In an elegant series of studies Katz & Rosenberger (1971a;b) show that factors disrupting normal hyphal growth can lead to precursor incorporation at many sites along the hypha and the formation of many new branch points. These studies on mutants defective in the biosynthesis of cell wall components present a refreshing view of carbohydrate metabolism in *A. nidulans* and strongly suggest that new mutants *requiring* specific carbohydrates for growth could be selected by filtration enrichment in glucose and selective plating on specific sugars.

Storage compounds

Glycogen is a major high molecular weight storage material in fungi with trehalose (α1 ⟶ 3 D glucosyl-D glucose) and D mannitol as soluble reserves (Cochrane, 1958). However the nature and relative amounts of such reserve materials may vary with the C or N sources for growth and also between strains (Lewis & Smith, 1967).

A. nidulans can accumulate very high concentrations of neutral fats (Singh & Walker, 1956), but this requires excess hexose and prolonged incubation and there is no accumulation in primary growth (Gregory & Woodbine, 1953). Bainbridge *et al.* (1971) found that although the lipid content of mycelium falls in continuous cultures fed a maintenance level of glucose it rises upon C starvation when other cell constituents are degraded. Enzymes of the glyoxylate cycle are present in freshly harvested conidia (Roberts, unpublished) suggesting possible use of lipid reserves in germination. Zonneveld (1972) has presented evidence that an α1 ⟶ 3 glucan located in the hyphal cell wall is a major C reserve laid down in vegetative growth and essential for further development (Zonneveld, this volume).

Trehalose occurs in conidia of *A. oryzae* and competitive inhibition of trehalase activity by D mannitol (Horikosh, Iida & Ikeda, 1965) suggests preferential utilization of the latter. The formation of mannitol-1-P by NADH linked reduction of fructose-6-P has been shown by Hankinson & Cove (1975) and mannitol probably accumulates by the action of a specific phosphatase shown in other Aspergilli (Lee, 1970; Strandberg, 1969).

High rates of endogenous metabolic activity are found in *A. nidulans* as in many other fungi (Niederpruem, 1965) and about 80% of hexose taken up may be assimilated rather than respired (Roberts, 1963b). It may be possible to circumvent this problem by using suspensions of washed conidia for metabolic studies.

THE METABOLISM OF ACETATE

Two requirements must be met by cells growing on C_2 compounds. The first is to replenish TCA cycle intermediates as they are removed for biosynthesis. This is achieved by the glyoxylate cycle (Kornberg, 1966b) which by-passes the two decarboxylation reactions of the TCA cycle (Fig. 3). The second requirement is for glycolytic intermediates which are provided through gluconeogenesis achieved by the appearance of new enzymes by-passing physiologically irreversible steps in glycolysis (Fig. 2).

Analysis of acetate non-utilizing (*acu*) mutants in *A. nidulans* was undertaken as a probe for the genetic control of intermediary metabolism (Armitt, McCullough & Roberts, 1976), following earlier work establishing the role of the glyoxylate cycle in this organism (Armitt, Roberts & Kornberg, 1970; 1971), and in *Neurospora* (Flavell & Fincham, 1968). These mutants grow normally on hexoses but not on C_2 compounds and some are unable to grow on other carbon sources metabolized directly through TCA cycle intermediates (Fig. 3).

facA, *facB*, *facC*: Apirion (1965) described these three genes which confer resistance to fluoroacetate and failure to grow on acetate. The *facA* (= *acuA*) locus is probably the structural gene for acetyl-CoA synthase (Romano & Kornberg, 1969; Armitt *et al.*, 1976). The *facB* and *facC* loci may affect the intracellular distribution of acetyl-CoA; these mutants are not defective in acetate uptake (Armitt *et al.*, 1976).

acuD, *acuE*: The key enzymes of the glyoxylate cycle are specified by two unlinked genes: *acuD* is identified as the structural gene for isocitrate lyase by temperature sensitive alleles and *acuE* is strongly indicated as the structural gene for malate synthase (Armitt *et al.*, 1976). Mutants at either locus fail to grow on carbon sources metabolized directly to acetyl-CoA. The expression of the two genes is strongly regulated in a co-ordinate manner probably through induction by acetyl-CoA. There is no evidence for sequential induction of malate synthase nor indeed of other enzymes of the acetate pathway. Two unlinked loci have been found at which recessive (*icl^{c}A*) or partially dominant (*icl^{c}B*) mutations produce low constitutive levels of isocitrate lyase, without detectably affecting malate synthase, but this may result from endogenous induction (McCullough & Roberts, 1975).

acuF, *acuG*: The locus *acuG* determines gluconeogenic fructose-1,6-diphosphatase, by-passing phosphofructokinase (Fig. 2). Mutants grow normally on hexose but poorly on triose or TCA cycle precursors. The *acuG* mutant is highly derepressed for PEP carboxykinase (*acuF*) after transfer from sucrose to acetate, suggesting that this enzyme is normally repressed by hexose phosphate. The *acuF* mutants do not grow on any carbon source metabolized through the TCA cycle, confirming the essential role of the enzyme in gluconeogenesis.

acuK, *acuM*: Mutation at either locus leads to loss of malic enzyme activity and failure to grow well on certain TCA cycle precursors. Since pyruvate kinase activity is low during growth on gluconeogenic carbon sources, we argued that the role of malic enzyme was to provide pyruvate under these conditions (McCullough & Roberts, 1974). We predicted that pyruvate kinase constitutive strains might be recovered as revertants of malic enzyme mutants selected for growth on acetate. However, although *acuM* mutants reverted readily, *acuK* and the *acuK*; *acuM* double mutant reverted at a very low rate and all revertants tested provided to have regained malic enzyme activity (Payton, unpublished). This suggests a second role for malic enzyme which may well be generation of NADPH. Malic enzyme is possibly induced by a TCA cycle intermediate (McCullogh & Roberts, 1974).

The functions of the remaining six *acu* loci (including *facB* and *facC*) have not been identified. One possibility lies in structural components of glyoxysomes which have been identified in some fungi (Kobr, Vanderhaeghe & Combépine, 1969; O'Sullivan & Casselton, 1973) or perhaps in duplicate enzymes of the TCA cycle which might be unique to its function.

ACKNOWLEDGEMENTS

We are grateful to colleagues for information provided for this article. Our work is supported by the Science Research Council (Research Grants 5493 and 1667-9).

REFERENCES

APIRION, D. (1965). The two-way selection of mutants and revertants in respect of acetate utilization and resistance to fluoro-acetate in *Aspergillus nidulans*. *Genetical Research, Cambridge* 6, 317-29.

AMARANISINGHAM, C.F. & DAVIS, B.D. (1965). Regulation of α-ketoglutarate dehydrogenase formation in *Escherichia coli*. *Journal of Biological Chemistry* 240, 3664-8.

ARMITT, S., McCULLOUGH, W. & ROBERTS, C.F. (1976). Analysis of acetate non-utilizing (*acu*) mutants in *Aspergillus nidulans*. *Journal of General Microbiology* 92, 263-82.

ARMITT, S., ROBERTS, C.F.& KORNBERG, H.L. (1970). The role of isocitrate lyase in *Aspergillus nidulans*. *Federation of European Biochemical Societies Letters* 7, 231-4.

ARMITT, S., ROBERTS, C.F.& KORNBERG, H.L. (1971). Mutants of *Aspergillus nidulans* lacking malate synthase. *Federation of European Biochemical Societies Letters* 12, 276-8.

ARST, H.N. & COVE, D.J. (1970). Molybdate metabolism in *Aspergillus nidulans*. II. Mutations affecting phosphatase activity and galactose utilization. *Molecular and General Genetics* 108, 146-53.

ARST, H.N. & COVE, D.J. (1973). Nitrogen metabolite repression in *Aspergillus nidulans*. *Molecular and General Genetics* 126, 111-41.

ATKINSON, D.E. (1968). In *The metabolic roles of citrate*, ed. T.W. Goodwin 23-34. New York and London: Academic Press.

BAILEY, C. & ARST, H.N. (1975). Carbon catabolite repression in *Aspergillus nidulans*. *European Journal of Biochemistry* 51, 573-7.

BAINBRIDGE, B.W., BULL, A.T., PIRT, S.J., ROWLEY, B.I. & TRINCI, A.P.J. (1971). Biochemical and structural changes in non-growing maintained and autolysing cultures of *Aspergillus nidulans*. *Transactions of the British Mycological Society* 56, 371-85.

BARTNICKI-GARCIA, S. (1968). Cell wall chemistry, morphogenesis and taxonomy of fungi. *Annual Review of Microbiology* 22 87-108.

BECK, J.C., MATOON, J.R., HAWTHORNE, D.C. & SHERMAN, F. (1968). Genetic modification of energy conserving systems in yeast mitochondria. *Proceedings of the National Academy of Science of the United States of America* 60, 186-93.

BENDALL, D.S. & BONNER, W.D. (1971). Cyanide insensitive respiration in plant mitochondria. *Plant Physiology* 47, 237-45.

BLOOM, S.J. & JOHNSTON, M.J. (1962). The pyruvate carboxylase of *Aspergillus niger*. *Journal of Biological Chemistry* 237, 2718-20.

BLUMENTHAL, H.J. (1965). Glycolysis. In *The fungi*, ed. G.C. Ainsworth & A.S. Sussman, Vol. 1, 229-268. New York and London: Academic Press.

BROWN, C.E. & ROMANO, A.H. (1969). Evidence against necessary phosphorylation during hexose transport in *Aspergillus nidulans*. *Journal of Bacteriology* 100, 1198-203.

BULL, A.T. (1970a). Chemical composition of wild-type and mutant *Aspergillus nidulans* cell walls. The nature of polysaccharide and melanin constituents. *Journal of General Microbiology* 63, 75-94.

BULL, A.T. (1970b). Inhibition of polysaccharases by malanin: Enzyme inhibition in relation of mycolysis. *Archives of Biochemistry and Biophysics* 137, 345-56.

BURAND, J.P., DRILLIEN, R. & BHATTACHARJEE, J.K. (1975). Citrate synthetaseless glutamic acid auxotroph of *Saccharomyces cerevisiae*. *Molecular and General Genetics* 139, 303-9.

BUSHELL, M.E. & BULL, A.T. (1974). Anaplerotic carbon dioxide fixation in steady state and non-steady state fungal cultures. *Proceedings of the Society for General Microbiology* I(3) 69.

CALDWELL, I.Y. & TRINCI, A.P.J. (1973). Kinetic aspects of growth of *Geotrichum candidum* on various carbon sources. *Transactions of the British Mycological Society* 61, 411-6.

CARTER, B.L.A. & BULL, A.T. (1969). Studies on fungal growth and intermediary carbon metabolism under steady and non-steady state conditions. *Biotechnology and Bioengineering* XI, 785-804.

CARTER, B.L.A.& BULL, A.T. (1971). The effect of oxygen tension in the medium on the morphology and growth kinetics of *Aspergillus nidulans*. *Journal of General Microbiology* 65 265-73.

CARTER, B.L.A., BULL, A.T., PIRT, S.J. & ROWLEY, B.I. (1971). Relationship between energy substrate utilization and specific growth rate in *Aspergillus nidulans*. *Journal of Bacteriology* 108, 309-13.

CHOJNACKI, T., PASZEWSKI, A. & SAWICKA, T. (1969). The formation of UDP glucose and UDP galactose in wild-type and mutants of *Aspergillus nidualns*. *Acta Biochimica Polonica* 16, 185-91.

CLUTTERBUCK, A.J. (1974). *Aspergillus nidulans*. In *Handbook of genetics*, ed. R.C. King. Vol. 1 Bacteria, bacteriophages and fungi. 447-510. New York & London: Plenum Press.

COCHRANE, V.W. (1958). *Physiology of fungi*. New York: Wiley.

COHEN, B.L. (1973). Growth of *Aspergillus nidulans* in thin liquid layer. *Journal of General Microbiology* 76, 277-83.

COOPER, T.G. & BEEVERS, H. (1969). Mitochondria and glyoxysomes from Castor Bean endosperm. *Journal of Biological Chemistry* 244, 3507-13.

COOPER, T.G.& BENEDICT, C.R.(1966). The participation of acetyl-CoA in pyruvate carboxylase. *Biochemical and Biophysical Research Communications* 22, 285-90.

ELLIS, H.A. (1959). A colourimetric method for the assay of soluble succinic dehydrogenase and pyridine nucleotide linked dehydrogenases. *Archives of Biochemistry and Biophysics* 85, 561-2.

ELORZA, M.V. & ARST, H.N. (1971). Sorbose resistant mutants of *Aspergillus nidulans*. *Molecular and General Genetics* 111, 185-93.

ERNTSER, L. & KWYLENSTIERNA, B. (1969). Structure, composition and function of mitochondrial membranes. *Federation of European Biochemical Societies Symposium* 17, 5-31.

FANTES, P.A. (1972). Studies on the β-galactosidase system in *Aspergillus nidulans*. Ph.D. Thesis, University of Leicester.

FANTES, P.A. & ROBERTS, C.F. (1973). β-galactosidase activity and lactose utilization in *Aspergillus nidulans*. *Journal of General Microbiology* 77, 471-86.

FIDDY, C. & TRINCI, A.P.J. (1975). Kinetics and morphology of glucose-limited cultures of moulds grown in a chemostat and on solid media. *Archives of Microbiology* 103, 191-7.

FLAVELL, R.B. & FINCHAM, J.R.S. (1968). Acetate non-utilizing mutants of *Neurospora crassa*. II. Biochemical deficiencies and the role of certain enzymes. *Journal of Bacteriology* 95, 1063-8.

FLAVELL, R.B. & WOODWARD, D.D. (1970). The concurrent regulation of metabolically related enzymes. The Krebs cycle and glyoxylate enzymes in *Neurospora*. *European Journal of Biochemistry* 17, 284-91.

FRAENKEL, D.G. & VINOPLA, R.T. (1973). Carbohydrate metabolism in bacteria. *Annual Review of Microbiology* 27, 69-100.

GAJEWSKI, W., LITWINSKA, J., PASZEWSKI, A. & CHOJNACKI, T. (1972). Isolation and characterization of lactose non-utilizing mutants of *Aspergillus nidulans*. *Molecular and General Genetics* 116, 99-106.

GANDER, J.E. (1974). Fungal cell wall glyco-proteins and peptido-polysaccharides. *Annuel Review of Microbiology* 28, 103-19.

GOSLING, J.P. & DUGGAN, P.F. (1971). Activities of tricarboxylic acid cycle enzymes, glyoxylate cycle enzymes, and fructose diphosphatase in Baker's yeast during adaptation to acetate oxidation. *Journal of Bacteriology* 106, 908-14.

GREGORY, M.E. & WOODBINE, M. (1953). The effects of varying pH, temperature, nitrogen, incubation period and sugar concentration on fat producing moulds. *Journal of Experimental Botany* 4, 314-8.

GUEST, J.R. (1974). Gene-protein relationships of the α-keto acid dehydrogenase complexes of *Escherichia coli* K12: chromosome location of the lipoamide dehydrogenase gene. *Journal of General Microbiology* 80, 523-32.

GUNATILLEKE, I.A.U.N., ARST, H.N. & SCAZZOCCHIO, C. (1976). Three genes determine the carboxin sensitivity of mitochondrial succinate oxidation in *Aspergillus nidulans*. *Genetical Research, Cambridge* 26, 297-305.

HAARASILTA, S. & OURA, E. (1975). On the activity and regulation of anaplerotic and gluconeogenic enzymes during growth processes of Baker's yeast. *European Journal of Biochemistry* 52, 1-7.

HANKINSON, O. (1974). Mutants of the pentose phosphate pathway in *Aspergillus nidulans*. *Journal of Bacteriology* 117, 1121-30.

HANKINSON, O. & COVE, D.J. (1974). Regulation of the pentose phosphate pathway in the fungus *Aspergillus nidulans*. *Journal of Biological Chemistry* 249, 2344-53.

HANKINSON, O. & COVE, D.J. (1975). Regulation of mannitol-1-phosphate dehydrogenase in *Aspergillus nidulans*. *Canadian Journal of Microbiology* 21, 99-101.

HANSEN, R.G. & HENNING, U. (1966). Regulation of pyruvate dehydrogenase activity in *Escherichia coli* K12. *Biochimica et Biophysica Acta* 122, 355-8.

HARDING, R.W., CAROLINE, D.F. & WAGNER, R.P. (1970). The pyruvate dehydrogenase complex from the mitochondrial fraction of *Neurospora crassa*. *Archives of Biochemistry and Biophysica* 138, 653-61.

HENNING, U., DENNERT, G., HERTEL, R. & SHIPP, W.S. (1966). Translation of the structural genes of *Escherichia coli* pyruvate dehydrogenase complex. *Cold Spring Harbor Symposia for Quantitative Biology* 31, 227-34.

HENNING, U. & HERTZ, C. (1964). Ein Strukturgen-Komplex für den Pyruvat-Dehydrogenase-Komplex von *Escherichia coli* K12. *Zeitschrift für Vererbungslehr* 95, 260-75.

HILL, R.L. & BRADSHAW, R.A. (1969). In *Methods in enzymology XIII*, 91-99.

HOLT, G. & MACDONALD, K.D. (1968). Penicillin production and its mode of inheritance in *Aspergillus nidulans*. *Antonie von Leeuwenhoek* 34, 409-16.

HORIKOSHI, K., IIDA, S. & IKEDA, Y. (1965). Mannitol and mannitol dehydrogenase in conidia of *Aspergillus oryzae*. *Journal of Bacteriology* 89, 326-30.

HYNES, M.J. (1975a). Studies on the role of the *areA* gene in the regulation of nitrogen catabolism in *Aspergillus nidulans*. *Australian Journal of Biological Science* 28, 301-13.

HYNES, M.J. (1975b). Amide utilization in *Aspergillus nidulans*. Evidence for a third amidase enzyme. *Journal of General Microbiology* 91 99-109.

JAGOW von, G., WEISS, H. & KLINGENBERG, M. (1973). Comparison of the respiratory chain of *Neurospora crassa* wild-type and *mi*-mutants, *mi*-1 and *mi*-3. *European Journal of Biochemistry* 33, 140-57.

JOHNSTONE, I.R. (1965). The composition of the cell wall of *Aspergillus niger*. *Biochemical Journal* 96, 651-7.

KAWAKITA, M. (1970). Studies on the respiration system of *Aspergillus oryzae*: I. Development of respiratory activity during germination in the presence and absence of antimycin A. *Plant Cell Physiology* 11, 377-84.

KATZ, D. & ROSENBERGER, R.F.(1970a). The utilization of galactose by an *Aspergillus nidulans* mutant lacking galactose phosphate-UDP glucose transferase and its relation to cell wall synthesis. *Archiv für Mikrobiologie* 74, 41-51.

KATZ, D. & ROSENGERGER, R.F. (1970b). A mutatation in *Aspergillus nidulans* producing hyphal walls which lack chitin. *Biochimica et Biophysica Acta* 208, 452-60.

KATZ, D. & ROSENBERGER, R.F. (1971a). Lysis of an *Aspergillus nidulans* mutant blocked in chitin synthesis and its relation to wall assembly and wall metabolism. *Archiv für Mikrobiologie* 80, 284-92.

KATZ, D. & ROSENBERGER, R.F. (1971b). Hyphal wall synthesis in *Aspergillus nidulans*: effect of protein synthesis inhibition and osmotic shock on chitin insertion and morphogenesis. *Journal of Bacteriology* 108, 184-90.

KHOUW, B.T. & McCURDY, H.D. (1969). Tricarboxylic acid cycle enzymes and morphogenesis in *Blastocladiella amersonii*. *Journal of Bacteriology* 99, 197-205.

KINGHORN, J.R. & PATEMAN, J.A. (1974). The regulation of NAD L-glutamate dehydrogenase in *Aspergillus nidulans*. *Genetical Research, Cambridge* 23, 119-24.

KLINGMULLER, W., HUH, H. & KAUDEWITZ, F. (1970). Regulierende Wirkung des Gens C der Sorbose-Resistenz von *Neurospora crassa* auf den Zuckertransport. *Molecular and General Genetics* 109 27-41.

KOBR, M.J., VANDERHAEGHE, F. & COMBEPINE, G. (1969). Particulate enzymes of the glyoxylate cycle in *Neurospora crassa*. *Biochemical and Biophysical Research Communications* 37, 640-5.

KOIKE, M., REED, L.J. & CARROLL, W.F. (1963). Resolution and reconstitution of pyruvate dehydrogenase from individual components. *Journal of Biological Chemistry* 238, 30-9.

KORNBERG, A. (1955). Isocitrate dehydrogenase of yeast. In *Methods in enzymology* 1, 705-707.

KORNBERG, A. & PRICER, W.E. (1951). Di- and tri-phosphopyridine nucleotide isocitrate dehydrogenase in yeast. *Journal of Biological Chemistry* 189, 123-36.

KORNBERG, H.L. (1966a). Anaplerotic sequences and their role in metabolism. *Essays in Biochemistry* 2, 1-31.

KORNBERG, H.L. (1966b). The role and control of the glyoxylate cycle in *Escherichia coli*. *Biochemical Journal* 99, 1-11.

KORNBERG, H.L. (1973). Carbohydrate transport by microorganisms. The Leeuwenhoek Lecture, 1972. *Proceedings of the Royal Society, London B* 183, 105-23.

KORNBERG, H.L. (1976). The nature and control of carbohydrate uptake by *Escherichia coli*. *Federation of European Biochemical Societies Letters* 63, 3-9.

LAMBOWITZ, A.M. & SLAYMAN, C.W. (1971). Cyanide resistant respiration in *Neurospora crassa*. *Journal of Bacteriology* 108, 1087-96.

LAMBOWITZ, A.M., SMITH, E.W. & SLAYMAN, C.W. (1972). Oxidative phosphorylation in Neurospora mitochondria. Studies on wild-type, poky and chloramphenicol induced wild-type. *Journal of Biological Chemistry* 247, 4859-65.

LEE, W.H. (1967). Carbon balance of a mannitol fermentation and the biosynthetic pathway. *Applied Microbiology* 15, 1206-10.

LEE, W.H. (1970). Enzymatic and ^{14}C-mannitol studies of the *Aspergillus* mannitol metabolism. *Canadian Journal of Microbiology* 16, 363-7.

LEWIS, D.H. & SMITH, D.C. (1967). Sugar alcohols (polyols) in fungi and green plants. 1. Distribution, physiology and metabolism. *New Phytologist* 66, 143-84.

LIN, E.C.C. (1970). The genetics of bacterial transport systems. *Annual Review of Genetics* 4, 225-62.

MARK, C.G. & ROMANO, A.M. (1971). Properties of the hexose transport systems of *Aspergillus nidulans*. *Biochimica et Biophysica Acta* 249, 216-26.

McCULLOUGH, W. & ROBERTS, C.F. (1974). The role of malic enzyme in *Aspergillus nidulans*. *Federation of European Biochemical Societies Letters* 41, 238-42.

McCULLOUGH, W. & ROBERTS, C.F. (1975). Isocitrate lyase consitutive mutants in *Aspergillus nidulans*. *Heredity* 36, 291.

MERRICK, M.J. & CATEN, C.E. (1975). The inheritance of penicillin titre in wild-type isolates of *Aspergillus nidulans*. *Journal of General Microbiology* 86, 283-93.

MILLER, A.L. & ATKINSON, D.E. (1972). Response of yeast pyruvate carboxylase to the adenylate energy charge and other regulatory parameters. *Archives of Biochemistry and Biophysics* 152, 531-8.

MOORE, D. (1973). Mutants of *Coprinus* selected for resistance to D-glucosamine and L-Sorbose. *Genetical Research, Cambridge* 22, 205-9.

MORITA, T. & MIFUCHI, I. (1970). Characterisation of respiration deficient mutants of *Saccharomyces cerevisiae* lacking cytochrome *a* and *c*. *Biochemical and Biophysical Research Communications* 38, 191-6.

NG, A.M.L., SMITH, J.E. & McINTOSH, A.F. (1973). Changes in activity of tricarboxylic acid cycle and glyoxylate cycle enzymes during synchronous development of *Aspergillus niger*. *Transactions of the British Mycological Society* 61, 13-20.

NG, A.M.L., SMITH, J.E. & McINTOSH, A.F. (1974). Influence of dilution rate on enzyme synthesis in *Aspergillus niger* in continuous culture. *Journal of General Microbiology* 81, 425-34.

NIEDERPRUEM, D.J. (1965). Tricarboxylic acid cycle. In *The Fungi* ed. G.C. Ainsworth & A.S. Sussman, Vol. 1, 269-300. New York & London: Academic Press.

OGUR, M., COKER, L. & OGUR, S. (1964). Glutamate auxotrophs in Saccharomyces. I. The biochemical lesion in the *glt*1 mutants. *Biochemical & Biophysical Research Communications* 14, 193-7.

O'SULLIVAN, J. & CASSELTON, P.J. (1973). The subcellular localization of glyoxylate cycle enzymes in *Coprinus lagopus* (sensu Buller). *Journal of General Microbiology* 75, 333-7.

PASZEWSKI, A., CHOJNACKI, T., LITWINSKA, J. & GAJEWSKI, W. (1970). Regulation of lactose utilization in *Aspergillus nidulans*. *Acta Biochimica Polonica* 17, 385-91.

PAYTON, M., McCULLOUGH, W. & ROBERTS, C.F. (1976). Agar as a carbon source and its effect on the utilization of other carbon sources by acetate non-utilizing (*acu*) mutants of *Aspergillus nidulans*. *Journal of General Microbiology* 94, 228-33.

PAYTON, M. & ROBERTS, C.F. (1976). Mutants of *Aspergillus nidulans* lacking pyruvate kinase. *Federation of European Biochemical Societies Letters* 66, 73-6.

PETTE, D. (1966). Mitochondrial enzyme activities. In *Regulation of metabolic processes in mitochondria*. ed. J.M. Tager, S. Papa, E. Quagliariello & E.C. Slater, 28-50. Amsterdam: Elsevier.

POLAKIS, E.S. & BARTLEY, W. (1965). Changes in the enzyme activities of *Saccharomyces cerevisiae* during anaerobic growth on different carbon sources. *Biochemical Journal* 97, 284-97.

RAMAKRISHNAN, C.V. & MARTIN, S.M. (1955). Isocitrate dehydrogenase in *Aspergillus niger*. *Archives of Biochemistry and Biophysics* 55, 403-7.

ROBERTS, C.F. (1963a). The genetic analysis of carbohydrate utilization in *Aspergillus nidulans*. *Journal of General Microbiology* 31, 45-8.

ROBERTS, C.F. (1963b). The adaptive metabolism of D-galactose in *Aspergillus nidulans*. *Journal of General Microbiology* 31, 285-95.

ROBERTS, C.F. (1970). Enzyme lesions in galactose non-utilizing mutants of *Aspergillus nidulans*. *Biochimica et Biophysica Acta* 201, 267-83.

ROMANO, A.H. (1973). Properties of sugar transport systems in *Aspergillus nidulans* and their regulation. In *Genetics of industrial microorganisms*, ed. Vanek, Hostalek & Cudlin, Vol. II, 195-212.

ROMANO, A.H., EBERHARD, S.J., DINGLE, S.L. & McDOWELL, T.D. (1970). Distribution of the phosphoenolpyruvate:glucose phosphotransferases in bacteria. *Journal of Bacteriology* 104, 808-13.

ROMANO, A.H. & KORNBERG, H.L. (1968). Regulation of sugar utilization by *Aspergillus nidulans*. *Biochimica et Biophysica Acta* 158, 491-3.

ROMANO, A.H. & KORNBERG, H.L. (1969). Regulation of sugar uptake by *Aspergillus nidulans*. *Proceedings of the Royal Society, London, B* 173, 475-90.

ROSENBERGER, R.F. & KESSEL, M. (1967). Synchrony of nuclear replication in individual hyphae of *Aspergillus nidulans*. *Journal of Bacteriology* 94, 1464-9.

ROWLANDS, R.T. & TURNER, G. (1974). Physiological and biochemical studies of nuclear and extra-nuclear oligomycin-resistant mutants of *Aspergillus nidulans*. *Molecular and General Genetics* 132, 73-88.

RUSSEL, P.J. & SRB, A.M. (1974). L-glutamine D fructose-6-phosphate amidotransferase in *Neurospora crassa*. *Molecular and General Genetics* 129, 77-86.

SAINSARD, A., CLAISSE, M. & BALMEFREZOL, M. (1974). A nuclear mutation affecting structure and function of mitochondria in *Paramecium*. *Molecular and General Genetics* 130, 113-25.

SANWAL, B.D. & STACHOW, C.S. (1965). Allosteric activation of nicotinamide adenine dinucleotide specific isocitrate dehydrogenase of *Neurospora*. *Biochimica et Biophysica Acta* 96, 28-44.

SAUER, B.L., COLVIN, H.J. & MUNKRES, K. (1973a). The effect of nitrogen source upon respiration of *Neurospora crassa*. *Biochemical and Biophysical Research Communications* 55, 389-95.

SAUER, B.L., COLVIN, H.J. & MUNKRES, K. (1973b). Glucose utilization and ethanolic fermentation by wild-type and extrachromosomal mutants of *Neurospora crassa*. *Journal of Bacteriology* 116, 1322-8.

SCARBOROUGH, G.A. (1970). Sugar transport in *Neurospora crassa*. II. A second glucose transport system. *Journal of Bioological Chemistry* 245, 3985-7.

SHARPLESS, T.K. & BUTON, R.A. (1970). An inducible alternate terminal oxidase in *Euglena gracilis* mitochondria. *Journal of Biological Chemistry* 245, 58-70.

SHERALD, J.L. & SISLER, H.D. (1970). Antimycin A resistant respiratory pathway in *Ustilago maydis*. *Plant Physiology* 46, 180-2.

SINGH, J. & WALKER, T.K. (1956). Changes in the composition of fat of *Aspergillus nidulans* with age of the culture. *Biochemical Journal* 62, 286-9.

SKINNER, V.M. & ARMITT, S. (1972). Mutants of *Aspergillus nidulans* lacking pyruvate carboxylase. *Federation of European Biochemical Societies Letters* 20, 16-18.

SMITH, J.E. & BERRY, D.R. (1975). *The filamentous fungi*, Vol. 2. *Biosynthesis and metabolism*. London: Edward Arnold.

SPETH, J.L. & NIEDERPRUEM, D.J. (1976). Enzyme activities associated with arabitol and mannitol biosynthesis and catabolism in *Schizophyllum commune*. *Archives of Microbiology* 107, 81-6.

STAN, H.J. & SCHORMULLER, J. (1968). Regulation of pyruvate carboxylase of *Penicillium camemberti* by the adenylate system. *Biochemical and Biophysical Research Communications*. 32, 289-94.

STRANDBERG, G.W. (1969). D-mannitol metabolism by *Aspergillus candidus*. *Journal of Bacteriology* 27, 1305-9.

SUBIK, J., KUZELA, S., KOLAROV, J., KOVAC, L. & LACHOWICZ, T.M. (1970). Oxidative phosphorylation in yeast. VI. ATPase activity and protein synthesis in mitochondria isolated from nuclear mutants deficient in cytochromes. *Biochimica et Biophysica Acta* 205, 213-8.

SUBIK, J., KOLAROV, J.& LACHOWICZ, T.M. (1972). A mutant of *Saccharomyces cerevisiae* lacking α-ketoglutarate dehydrogenase activity. *Federation of European Biochemical Societies Letters* 97, 81-4.

TRINCI, A.P.J. (1971). Influence of the peripheral growth zone on the radial growth rate of fungal colonies. *Journal of General Microbiology* 67, 325-44.

VALENTINE, B.P. (1975). The isolation and characterization of temperature sensitive mutants of *Aspergillus nidulans* with special reference to cell wall synthesis and mannose utilization. Ph.D. Thesis, University of London.

VALENTINE, B.P. & BAINBRIDGE, B. W. (1975). Mutations affecting the incorporation of mannose into the cell wall of *Aspergillus nidulans*. *Proceedings of the Society for General Microbiology* 11(4), 90.

WATSON, K. & SMITH, J.E. (1967). Respiratory components of *Aspergillus niger* mitochondria. *Journal of Bacteriology* 96, 1547-50.

ZONNEVELD, B.J.M. (1971). Biochemical analysis of the cell wall of *Aspergillus nidulans*. *Biochimica et Biophysica Acta* 249, 506-14.

ZONNEVELD, B.J.M. (1972). Morphogenesis in *Aspergillus nidulans*. The significance of α-1,3-glucan of the cell wall and α-1,3-glucanase for cleistothecium development. *Biochimica et Biophysica Acta* 273, 174-87.

The Regulation of Carbon Metabolism in *Aspergillus nidulans*

H. N. ARST, JR. and C. R. BAILEY

Department of Genetics, University of Cambridge, Cambridge, U.K.

INTRODUCTION

In *Aspergillus nidulans*, at least two forms of regulation have been demonstrated over enzymes and permeases whose role is partially or wholly to provide a source of carbon and energy. These types of regulation are induction and carbon catabolite repression. In most cases there is no evidence that these phenomena affect *de novo* protein synthesis although there is evidence that one or more forms of regulation involve *de novo* synthesis of the following *A. nidulans* enzymes: arginase (Cybis & Weglenski, 1972; Bartnik, Weglenski & Piotrowska, 1973), xanthine dehydrogenase and urate oxidase (Scazzocchio, unpublished data), nitrate reductase (Cove, unpublished data), acetamidase (Dunsmuir & Hynes, 1973; Hynes, 1974), formamidase (Hynes, 1974), and histidase (Hynes, 1974). We shall therefore assume that regulation affects *de novo* protein synthesis. We shall also assume that control is exerted at the level of transcription by extrapolation of the results of Cybis & Weglenski (1972) and Bartnik *et al.* (1973) for arginase, Dunsmuir & Hynes (1973) for acetamidase, and Scazzocchio (unpublished data) for urate oxidase, which indicate that induction of all three enzymes and nitrogen metabolite (i.e. ammonium) repression of arginase probably occur at the level of transcription.

ENZYME INDUCTION

There have been relatively few genetical and biochemical analyses of induction of enzymes involved in carbon metabolism in *A. nidulans*. This is partly due to the fact that the analyses of lactose and galactose utilisation, which have proved so fruitful for regulatory studies in *Escherichia coli*, have been comparatively unrewarding in *A. nidulans*. The ineffectiveness of analogues and selective techniques employed with the *E. coli* systems, the likelihood of a second pathway of galactose catabolism, and the lack of correlation between the ability to utilise lactose and β-galactosidase levels have frustrated attempts to exploit these systems in *A. nidulans* (Paszewski, Chojnacki, Litwinska & Gajėwski, 1970; Gajewski, Litwinska, Paszewski & Chojnacki, 1972; Roberts, 1970; Fantes & Roberts, 1973).

The extensive study of induction of enzymes involved in acetate utilisation has been discussed by Roberts, McCullough & Payton (this volume). However, some mention ought to be made of two other systems, involving the genes *gal*A and *int*A. *gal*A$^-$ mutations result in loss of inducibility of galactokinase and galactose-1-phosphate uridyl transferase whilst not affecting galactose induction of UDP-galactose-4-epimerase (Roberts, 1970). *gal*A$^-$ mutations are both frequent and recessive (Roberts, 1963, 1970). Therefore it might be postulated that the *gal*A product is either a positive-acting regulatory substance required for galactose induction of the first two enzymes of the Leloir pathway or an enzyme which catalyses conversion of galactose to the actual co-inducer. Unfortunately no constitutive mutants have been selected.

The *int*A gene is a positive regulatory gene which mediates the induction of the products of three unlinked structural genes. The three gene products are acetamidase, γ-amino-*n*-butyrate permease, and a transaminase for ω-amino acids (Arst, 1976; H.A. Penfold, C.R. Bailey & H.N. Arst, unpublished). At least the acetamidase is subject to an alternative, parallel induction mechanism involving another probable positive regulatory gene, *amd*A, which appears to be specific for acetamidase regulation. Preliminary evidence (C.R. Bailey & H.N. Arst, unpublished) indicates that, whereas *amd*A might mediate acetamidase induction by acetamide, *int*A probably mediates its induction by γ-amino-*n*-butyrate, β-alanine, and possibly other ω-amino acids. *int*A is an especially interesting regulatory gene because it resembles closely the hypothetical integrator genes of the model for gene regulation proposed by Britten & Davidson (1969, 1971; Davidson & Britten, 1973). This model proposes that positive regulatory genes acting in parallel can elicit the expression of a particular structural gene in a variety of different contexts, on a one 'context'-one positive regulatory gene basis. (Thus, presence of an amide would be the "context" for *amd*A-mediated induction of acetamidase and presence of an ω-amino acid the "context" for *int*A-mediated induction of acetamidase.) Any such positive regulatory gene which affects the concomitant expression of two or more unlinked structural genes, which are also capable of independent expression, is designated an integrator gene. Thus, *int*A integrates expression of the acetamidase with expression of the γ-amino-*n*-butyrate permease and transaminase.

The Britten-Davidson model was developed to account for cell differentiation in higher eukaryotes, but it is not surprising that the first demonstration of an integrator gene should have come from studies of catabolism in *A. nidulans*, an organism of considerable metabolic versatility. We might expect that the most efficient way of obtaining evidence for the existence of integrator genes would be by analysing regulatory mutations affecting inducible enzymes which participate in more than one metabolic pathway. Unfortunately we are unable to suggest at present any role for acetamidase in the catabolism of ω-amino acids and thus to understand why *int*A should regulate acetamidase synthesis. Conveniently, the presence of the gene specifying thiogalactoside transacetylase in the *lac* operon of *E. coli* provides a respectable precedent for mysteries of this kind!

CARBON CATABOLITE REPRESSION

The most extensively investigated form of regulation of carbon metabolism in *A. nidulans* is carbon catabolite repression. This is the ability of certain carbon and energy sources, generally the better ones, to prevent the synthesis of enzymes and permeases required for the utilisation of other, generally poorer carbon sources. Growth in the presence of a repressing carbon source such as glucose has been shown to reduce levels of acetamidase (Hynes, 1970; Hynes & Pateman, 1970a, b); monoamine oxidase (Page, 1971; Page & Cove, 1972); alcohol dehydrogenase (Page, 1971; Page & Cove, 1972; Bailey & Arst, 1975); proline permease and proline oxidase (Arst & MacDonald, 1975; Bailey & Arst, 1975); γ-amino-*n*-butyrate (GABA) transaminase (H.A. Penfold & H.N. Arst, unpublished); NAD-linked glutamate dehydrogenase (Kinghorn & Pateman, 1974; Bailey & Arst, 1975; Parbtani & Cove, 1975); extracellular protease (Cohen, 1973); β-galactosidase (Bailey & Arst, 1975); D-quinate dehydrogenase (Bailey & Arst, 1975); sorbitol dehydrogenase (Bailey & Arst, 1975); isocitrate lyase (Armitt *et al.* 1970); malate synthase (Armitt, Roberts & Kornberg, 1971); malic enzyme (McCullough & Roberts, 1974), and invertase (Andres & Peberdy, 1974). Data in Table 1 show control characteristics of alcohol dehydrogenase, an enzyme showing a strong response to both induction and carbon catabolite repression. Data in Table 2 show the pattern of repression of proline oxidase, an enzyme which can participate in both carbon and nitrogen nutrition. However, carbon catabolite repression is apparent only as a prerequisite for ammonium repression.

Table 1. Alcohol dehydrogenase activities in a wild type and a carbon catabolite depressed strain

		Relative Activities	
Growth conditions	Induction	Wild type (*cre*A$^+$)	(*cre*A^d-1)
derepressing	uninduced	11	10
repressing	induced	2	71
derepressing	induced	100	107

For experimental details, see Bailey & Arst (1975). Activities are normalised with induced wild type value = 100. Both strains carry the biotin auxotrophy, *bi*A-1, and are isogenic apart from the *cre*A alleles specified above.

Table 2. Proline oxidase activities in a wild type and a carbon catabolite depressed strain

Final growth conditions		Relative activities	
Carbon source	Nitrogen source	Wild type (*cre*A$^+$)	(*cre*A^d-1)
D-glucose	L-proline	97	102
D-glucose	L-proline + NH_4^+	22	72
glycerol	L-proline	100	86
glycerol	L-proline + NH_4^+	92	86

Table 2 (cont'd).....

Unpublished data of C.R. Bailey.
For experimental details see Bailey & Arst (1975) and Arst & MacDonald (1975), except that growth was carried out at 37° in appropriately supplemented minimal medium. After 6h growth on medium containing 1% D-glucose as carbon source and 100 mg/l uric acid as nitrogen source, mycelia was transferred for a further 2½h growth to the media listed above. D-glucose and glycerol were present at 1% (w/v), L-proline at 5 mM and NH_4^+ at 10 mM as the (+)-tartrate. Activities are normalised with induced, derepressed wild type = 100. Both strains carry the biotin auxotrophy, *bi*A-1, and are isogenic apart from the *cre*A alleles specified above.

Genetic investigation of carbon catabolite repression demands ways of assessment *in vivo*. We have developed two systems which allow both the measurement of the degree of carbon catabolite repression *in vivo* and the selection of derepressed mutants. One of these systems involves the use of strains carrying an *are*A^r mutation. *are*A^r strains lack a positive-acting regulatory substance necessary for the synthesis of a wide range of ammonium-repressible enzymes and permeases. Consequently, *are*A^r strains are unable to utilise nitrogen sources other than ammonium on glucose minimal medium (Arst & Cove, 1973). It is useful to note an analogy with *crp*$^-$ mutants of *E. coli*. *crp*$^-$ mutants lack an activator protein necessary for the synthesis of a wide range of carbon catabolite-repressible enzymes and permeases, and are consequently unable to utilise carbon sources other than D-glucose (Zubay, Schwartz & Beckwith, 1970) . However, *are*A^r strains can utilise acetamide, L-alanine, L-asparagine, L-asparate, L-arginine, L-glutamate, L-glutamine, GABA, L-ornithine,L-proline, and L-threonine as sole carbon or sole carbon and nitrogen sources (Arst & Cove, 1973; Bailey & Arst, 1975; C.R. Bailey & H.N. Arst, unpublished results). In addition, the presence of certain carbon sources such as (at 1%) glycerol, ethanol, lactose, melibiose, or L-arabinose still allows considerable utilisation of acetamide, proline, GABA, and glutamine as nitrogen sources by *are*A^r strains. Other compounds from the above list, such as alanine, asparagine, and glutamate and non-carbon sources, such as formamide, uric acid and nitrate, cannot be utilised by *are*A^r strains in the presence of glycerol, *etc*. as carbon source. It is unclear whether this division of potential carbon and nitrogen sources reflects a dichotomy of regulatory mechanisms or whether a single mechanism is operating with the effect of replacement of glucose by glycerol, *etc*. being below the limit of detectability for alanine, asparagine, *etc*. We favour the latter possibility for simplicity and because the four nitrogen sources, proline, acetamide, glutamine, and GABA, form a continuum in the degree to which they can be utilised in the presence of glycerol, *etc*. Proline utilisation is clearly enhanced the most by replacement of glucose by glycerol and GABA utilisation the least.

The other system utilises strains carrying a *pdh*A$^-$ mutation (Romano & Kornberg, 1968, 1969), leading to loss of pyruvate dehydrogenase. *pdh*A$^-$ strains require a source of acetate being unable to convert pyruvate to acetyl-coenzyme A. Ethanol, ethylammonium, and acetamide can supplement *pdh*A$^-$ strains when present as sole carbon sources but not in the presence of 1% D-glucose in normal minimal medium. (In this case the nitrogen source is irrelevant). However, replacement of glucose by 1% glycerol, lactose, melibiose, or arabinose allows ethanol, ethylammonium, and acetamide to supplement *pdh*A$^-$ strains (Bailey & Arst, 1975). We have already noted that alcohol dehydrogenase, monoamine oxidase and acetamidase, which are necessary for the conversion to acetate of ethanol, ethylammonium, and acetamide, respectively, are subject to carbon catabolite repression. In practice, we have mostly worked with ethanol supplementation in order to avoid complications arising from the facts that monoamine oxidase (Page, 1971; Page & Cove, 1972) and acetamidase (Hynes, 1970; Hynes & Pateman, 1970a,b) are also subject to ammonium repression.

It is important to decide whether these two systems measure the same thing, namely, carbon catabolite repression. There is good evidence that they do for the extent to which various carbon sources prevent utilisation of proline, acetamide, glutamine, and GABA as nitrogen sources by *are*A^r strains closely parallels the extent to which they prevent ethanol supplementation of *pdh*A$^-$ strains. Thus D-glucose, D-xylose, and sucrose can be classified as strongly carbon catabolite repressing, whilst glycerol, *meso*-inositol, melibiose, lactose, and L-arabinose are derepressing carbon sources. Intermediate between these two extremes are D-mannose, D-galactose, D-sorbitol, D-fructose, and D-mannitol. Acetate and ethanol, which can be tested only in the *are*A^r system, are classified as strongly repressing and strongly derepressing, respectively. When a large number of carbon sources are ranked according to the extent to which they are repressing in the two systems, the rankings are virtually identical. Lowering the concentration of repressing carbon sources from 1% to 0.1 or 0.05% considerably reduces the apparent degree of repression by sugars such as D-glucose, D-fructose, D-xylose, and sucrose in both systems. *In vitro* data, such as that in Table 2, confirm that growth on 1% glycerol, 0.1% D-fructose, or 0.1% D-glucose leads to higher levels of carbon catabolite-repressible enzymes and permeases, as compared to growth on 1% D-glucose (Bailey & Arst, 1975; Arst & MacDonald, 1975).

Both systems have proved suitable for the selection of

mutations leading to carbon catabolite derepression. Such mutations have been obtained by reverting *are*A^r strains on glucose-acetamide and glucose-proline media, and by reverting *pdh*A$^-$ strains on sucrose-glucose-ethanol medium (Arst & Cove, 1973; Bailey & Arst, 1975; C.R. Bailey & H.N. Arst, unpublished results). In *are*A^r strains these mutations result in the ability to utilise one or more of the nitrogen sources acetamide, proline, GABA, and glutamine in the presence of repressing carbon sources, such as D-glucose. In *pdh*A$^-$ strains these mutations result in the ability to be supplemented by ethanol (and usually also ethylammonium and acetamide) in the presence of repressing carbon sources. Carbon catabolite derepressed mutations, selected using the two systems, have been shown to be allelic, mapping in the *cre*A gene, which is tightly linked to *gal*D in linkage group I. Mutations selected using one system generally result in some degree of derepression in the other system when the appropriate double mutants (*cre*A^d; *pdh*A$^-$ or *cre*A^d; *are*A^r) are constructed by crossing. To a first approximation derepressed alleles of the *cre*A gene mimic the replacement of a repressing by a derepressing carbon source. For example, *cre*A^d-1; *are*A^r double mutants utilise acetamide, proline, GABA, and glutamine in the presence of repressing carbon sources, and *cre*A^d-1, *pdh*A-1 double mutants are supplemented by ethanol, ethylammonium, and acetamide in the presence of repressing carbon sources. This conformity to the predicted phenotype for carbon catabolite derepressed mutations was crucial to their recognition in early work (Arst & Cove, 1973; Bailey & Arst, 1975). Subsequent work (Bailey & Arst, 1975 and unpublished results) has revealed a heterogeneity of phenotypes amongst *cre*A^d alleles, a point to which we shall return later.

Tables 1 and 2 show that *cre*A^d-1 partially relieves carbon catabolite repression of alcohol dehydrogenase and proline oxidase. Bailey & Arst (1975) have shown that *cre*A^d-1 has a similar effect on sorbitol dehydrogenase, mannitol dehydrogenase, and NAD-linked glutamate dehydrogenase, but no effect on carbon catabolite repressibility of β-galactosidase or quinate dehydrogenase. We have not yet established whether this is an allele-specific characteristic, in which case other *cre*A^d alleles might derepress these enzymes, or whether another regulatory mechanism is involved. A comparable finding has been made in *Saccharomyces carlsbergensis* (Schamhart, ten Berge & van de Poll, 1975). Arst & MacDonald (1975) have shown that *cre*A^d-1 leads to derepression of the proline permease. Arst & Cove (1973) and Bailey & Arst (1975) have described a number of ways in which carbon catabolite derepression can be observed *in vivo*. These both extend the range of activities over which *cre*A control can be observed and enable scoring of *cre*A^d mutations in strains carrying wild type *pdh*A and *are*A alleles. As has been observed with carbon catabolite derepressed mutations of yeast (Montenecourt, Kuo & Lampen, 1973; Ghosh, Montenecourt & Lampen, 1973; Elliot & Ball, 1973; Schamhart *et al.*, 1975) most *cre*A^d mutations affect colony morphology, resulting in compact colonies. This is not surprising since carbon catabolite repression might be expected to affect, directly or indirectly, synthesis of carbohydrate, glycoprotein and glycolipid components of the cell wall and membrane.

*cre*A^d mutations are recessive and are not difficult to obtain. All of our (unpublished) attempts to obtain 'repressed' *cre*A alleles have failed, suggesting that this type of allele is rare. This indicates that the *cre*A^d phenotypes result from loss of function mutations. We might consider three general hypotheses to explain the role of the *cre*A gene: (i) *cre*A^d mutatants are blocked in the uptake or catabolism of repressing carbon sources, (ii) *cre*A^d mutants are able to take up and catabolise repressing carbon sources normally but lack an enzyme which catalyses the synthesis of an effector needed for catabolite repression (or, if catabolite repression be mediated by a positive control mechanism as in *E. coli*, *cre*A^d mutants lack an enzyme which catalyses the degradation of an effector needed to activate the regulatory substance), and (iii) *cre*A^d mutants lack a repressor macromolecule, with carbon catabolite repression mediated by a negative control system. The available evidence, outlined below, supports the third possibility.

Bailey & Arst (1975) showed that *cre*A^d-1 has no effect on glucose uptake. *cre*A^d mutants show no growth defects on repressing or partially repressing carbon sources (C.R. Bailey and H.N. Arst, unpublished results). There is thus no indication of any catabolic defect being responsible for the *cre*A^d phenotype. Since *cre*A^d mutations relieve carbon catabolite repression on all strongly and partially repressing carbon sources (with the possible exception of acetate, on which their effects are more marginal), any catabolic block or block in the synthesis of an actual co-repressor would have to involve a step common to a wide range of carbon sources, such as sucrose, maltose, D-xylose, D-glucose, D-fructose, and D-mannitol. The significance of acetate as a possible exception is unclear. *A. nidulans* has a unique morphology on acetate media and its physiology may be altered.

If carbon catabolite repression were mediated by a positive control system involving 3'-5' cyclic adenosine monophosphate (cyclic AMP) as in *E. coli*, (see Rickenberg, 1974 for a review) the most likely explanation for recessive mutations leading to carbon catabolite derepression on a wide range of carbon sources would be that they lead to loss of cyclic AMP phosphodiesterase. Alper & Ames (1975) summarise the inconclusive evidence for this being the case in *E. coli* and related bacteria. Therefore we (unpublished results) compared cyclic AMP phosphodiesterase levels in *cre*A^{d}-1 and wild type strains. *cre*A^{d}-1 does not lead to loss of the phosphodiesterase and appears to result in enhanced levels. Polya, Brownlee & Hynes (1975) have shown that cyclic AMP phosphodiesterase of *A. nidulans* also has phosphomonoesterase activity and is subject to nitrogen metabolite repression. They found no evidence for carbon regulation of the enzyme or for an alteration in its level in a *cre* mutant selected in their laboratory.

The evidence most difficult to reconcile with an indirect role of the *cre*A product in carbon catabolite repression, (such as a block in uptake or catabolism, or in the synthesis or degradation of an effector molecule), is the degree of phenotypic heterogeneity found amongst different *cre*A^{d} alleles. This type of evidence also indicated direct regulatory involvement of the products of the *are*A (Arst & Cove, 1973; Arst & Scazzocchio, 1975) and the *int*A genes (Arst, 1976). To illustrate such an argument consider the allelic heterogeneity in two genes involved in carbon catabolite repression in *E. coli*. *cya* is the structural gene for adenylate cyclase, and *crp* is the gene specifying the catabolite activator protein (CAP), which is active when it binds cyclic AMP (see Rickenberg (1974) for a review). *cya*$^{-}$ and *crp*$^{-}$ mutants are both pleiotropically negative for the utilisation of a wide range of carbon sources. However, the two classes of mutant are easily distinguishable because *cya*$^{-}$ mutants can utilise these carbon sources in the presence of exogenous cyclic AMP, whereas *crp*$^{-}$ mutants cannot. There exists a 'hierarchy' of carbon catabolite-prepressible structural genes, some requiring a higher concentration of cyclic AMP-CAP complex than others (summarised and discussed by Rickenberg (1974) and Alper & Ames (1975)). For example, Lis & Schlief (1973) have shown that synthesis of the *lac* operon enzyme β -galactosidase does not require as much cyclic AMP as synthesis of the *ara* operon enzyme ribulokinase. Therefore amongst a sample of *cya*$^{-}$ mutants, some leaky *cya*$^{-}$ alleles might allow lactose but not L-arabinose utilisation, but no *cya*$^{-}$ allele should every show the opposite phenotype, i.e. *lac*$^{-}$ *ara*$^{+}$. On the other hand there is no reason why mutant forms of CAP should not lead to such opposite phenotypes when a number of alleles are compared. Thus, even if *E. coli* were impermeable to cyclic AMP and its derivatives and if it were not possible to detect adenylate cyclase and CAP *in vitro* evidence of this type could distinguish a gene having an indirect regulatory role, *cya*, from a gene having a direct regulatory role, *crp*.

The more aspects of the phenotype that can be screened, the more sensitive will be this method for distinguishing between regulatory genes having an indirect and a direct role. If sensitivity to carbon catabolite repression increases (i.e. if the transcriptional requirement for cyclic AMP-CAP complex increases) for a series of enzymes A....Z, and we find a mutation in a regulatory gene having a (partial) phenotype A^- B^- C^+ .... J^- K^+.... Z^+, then we can conclude that the regulatory gene has a direct role. Of course, even for regulatory genes such as *crp*, having a direct regulatory role, not all phenotypic combinations may be possible. This will depend upon restrictions imposed by the structure of the regulatory gene product and by the structures of the various receptor sites for it. The main point is that whereas a sample of *cya*$^-$ alleles can show only hierarchical heterogeneity we can expect to observe non-hierarchical heterogeneity amongst *crp*$^-$ alleles. Unfortunately, no systematic study of *cya*$^-$ and *crp*$^-$ alleles has been reported so far as we are aware.

Data in Table 3 show that at least three distinct phenotypes exist amongst *cre*A^d alleles when compared for suppression of an *are*A^r mutation. *cre*A^d-2 and *cre*A^d-25 show an opposite phenotype when comparing proline and acetamide utilisation. This can only be classed as non-hierarchical heterogeneity. In double mutants with *pdh*A-1, *cre*A^d-25 allows little ethanol utilisation in the presence of glucose whereas *cre*A^d-1 and *cre*A^d-2 allow considerable utilisation. This correlates with the degree of derepression of alcohol dehydrogenase as shown in Table 4. There are also morphological differences between *cre*A^d alleles. *cre*A^d-2 leads to an extremely compact colony morphology, *cre*A^d-1 to a less compact morphology, and *cre*A^d-25 had virtually no morphological effect. We are collecting further *in vitro* data in order to quantify the non-hierarchical heterogeneity amongst *cre*A^d alleles. At present all of the indications are that the *cre*A gene specifies the repressor molecule (defective or missing in *cre*A^d mutants) of a negative control system mediating carbon catabolite repression (Arst & MacDonald, 1975).

Table 3. Suppression of *are*A^r-1 by various *cre*A^d mutations

Relevant Genotype	Growth on nitrogen source				
	NH_4^+	proline	acetamide	GABA	glutamine
Wild type (*cre*A$^+$; *are*A$^+$	+++	+++	+++	+++	+++
*are*A^r-1	+++	–	–	–	±
*cre*A^d-1; *are*A^r-1	+++	++	++	++	++
*cre*A^d-2; *are*A^r-1	+++	–	++	–	±
*cre*A^d-25; *are*A^r-1	+++	++	±	±	+

Table 3 (cont'd).....

Unpublished data of C.R. Bailey.
Growth tests were carried out at 37° on appropriately supplemented minimal medium containing 1% D-glucose as carbon source and the nitrogen sources listed above at 10 mM (NH_4^+ as the (+)-tartrate, acetamide) or 5 mM (L-proline, GABA, L-glutamine). For further details of growth-testing of *A. nidulans* see Arst & Cove (1969). Growth scores (after 2 days): - = no growth, ± = slight growth, + up to +++ = increasing levels of growth. Growth scores on different media are not necessarily equivalent, but +++ represents the maximum growth observed on any medium.

Table 4. Derepression of alcohol dehydrogenase in strains carrying different *cre*A alleles

Relevant genotype	% Derepression
Wild type (*cre*A$^+$)	0
*cre*A^d-1	59
*cre*A^d-2	80
*cre*A^d-25	10

For experimental details see Bailey & Arst (1975).
% Derepression = Repressed, induced specific activity ÷ Derepressed, induced specific activity x 100.
Repressing medium contained 1% D-glucose and derepressing medium contained 0.1% D-fructose, and both media contained 1% ethanol as inducer. All strains carried the biotin auxotrophy , *bi*A-1, and are isogenic apart from the *cre*A alleles specified above.

The next question we might ask is whether cyclic AMP is involved in carbon catabolite repression in *A. nidulans*. If a negative control system does mediate carbon catabolite repression in *A. nidulans* then its evolutionary origin is probably different from that of the positive control system found in *E.coli* and related bacteria. Thus there is no particular reason to suspect a role forcyclic AMP. In yeasts there is conflicting evidence over an involvement of cyclic AMP in carbon catabolite repression. One of the most recent reports (Schlanderer & Dellweg, 1974) favours such an involvement. We have found no evidence for any involvement, albeit using only indirect experiments. Neither cyclic AMP nor its N^6-2'-O-dibutyryl derivative, when added exogeneously, affect carbon catabolite repression in either the *are*A^{r} or *pdh*A^{-} systems (C.R. Bailey & H.N. Arst, unpublished results). However, there is no indication that these compounds are taken up, for they neither supplement adenine auxotrophies nor serve as nitrogen sources. (Attempts to obtain a mutation allowing cyclic AMP to supplement an adenine auxotrophy by reverting a strain doubly auxotrophic for adenine on cyclic AMP-containing medium failed (D.J. Cove & H.N. Arst, unpublished results). Activators of adenylate cyclase, such as glucagon and isoproterenol and inhibitors of cyclic AMP phosphodiesterase, such as theophyllin aminophylline, and 3-isobutyl-1-methyl xanthine, do not relieve carbon catabolite repression in the *are*A^{r} and *pdh*A^{-} systems when added exogeneously (C.R. Bailey & H.N. Arst, unpublished results). An adenylate cyclase inhibitor, ethacrynic acid, does not show increased toxicity on carbon sources such as ethanol, whose utilisation is subject to strong carbon catabolite repression as compared to glucose. Ethacrynic acid toxicity is also unaffected by *cre*A^{d} mutations (C.R. Bailey & H.N. Arst, unpublished results). Finally, Zonneveld (1976) found no evidence that glucose plays a role in regulating cyclic AMP levels in *A. nidulans*. The conclusions which can be drawn from his work are restricted by the fact that the lowest concentration of glucose he used was 0.5%, a strongly repressing concentration. Also, no comparisons were made to mycelia grown on non-repressing carbon sources. Growing for long periods of time until glucose exhaustion is not necessarily equivalent to log phase growth because other physiological factors might be operating as cultures approach senescence.

If cyclic AMP is not involved, what is the effector? At present, we can only phrase this question in a more modest way: what steps are necessary for repressing carbon sources to exert repression? We can do this by constructing doubly mutant strains carrying an *are*A^{r} or a *pdh*A^{-} mutation and a mutation blocking utilisation of the repressing carbon source. It is relevant to note here that glucose and other repressing sugars metabolised through pyruvate are still repressing in *pdh*A^{-} strains. Bailey & Arst (unpublished results) have used both the *are*A^{r} and *pdh*A^{-} systems to show that maltose is not repressing if the stains carry a mutation blocking maltose uptake and that D-galactose is not repressing in strains lacking glactokinase. Of course, we are limited in that we cannot use mutational blocks resulting in toxicity in the presence of the repressing carbon source. The effects of several blocks in acetate utilisation upon acetate repression have been studied. Acetate is not repressing in strains lacking acetyl-CoA synthetase and only slightly repressing in strains lacking isocitrate lyase and malate synthase. Loss of fructose-1,6-diphosphate phosphatase does not, however, affect acetate repression (C.R. Bailey & H.N. Arst, unpublished results).

Finally, we return to the pattern of regulation of proline

oxidase seen in Table 2, in which we noted that carbon catabolite repression is seen only as a prerequisite for nitrogen metabolite repression. How is such a finding compatible with a negative control model in which the *cre*A product is a repressor? A simple explanation may be that where expression of a structural gene is subject to both carbon catabolite and nitrogen metabolite repression the *cre*A product can act as a repressor only when the *are*A product is inactive (in the absence of the *cre*A product there is no requirement for the *are*A product). This can be envisaged to happen in one of two ways. Firstly, there may be partial or complete overlap of *cre*A and *are*A product receptor sites so that the *cre*A and *are*A products compete with each other. Alternatively, binding of the *are*A product to its receptor site may allosterically prevent the *cre*A product from binding to its receptor site. Such molecular mechanisms are entirely speculative at present, but they provide a plausible explanation of the enzyme data. We do not know whether other enzymes, which are subject to both carbon catabolite and nitrogen metabolite repression, such as acetamidase and GABA transaminase, show a similar pattern of regulation to proline oxidase or whether the two forms of regulation are independent in these cases.

ACKNOWLEDGEMENTS

H.N. Arst is grateful to the Royal Society to be the Smithson Research Fellow. C.R. Bailey thanks the Science Research Council for support through a grant to Professor J.M. Thoday. We thank Claudio Scazzocchio, Keith Rand, Hugh Penfold and Don MacDonald for useful discussions and Mick Pierson for technical assistance.

REFERENCES

ALPER, M.D. & AMES, B.N. (1975). Cyclic 3',5'-adenosine monophosphate phosphodiesterase mutants of *Salmonella typhimurium*. *Journal of Bacteriology* 122, 1081-90.

ANDRES, I. & PEBERDY, J.F. (1974). The production of invertase in *Aspergillus nidulans* with reference to the effects of glucose and sucrose. *Microbios* 10, 15-23.

ARMITT, S., ROBERTS, C.F.& KORNBERG, H.L. (1970). The role of isocitrate lyase in *Aspergillus nidulans*. *Federation of European Biochemical Societies Letters* 7, 231-4.

ARMITT, S., ROBERTS, C.F. & KORNBERG, H.L. (1971). Mutants of *Aspergillus nidulans* lacking malate synthase. *Federation of European Biochemical Societies Letters* 12, 276-8

ARST, H.N., JR. (1976). Integrator gene in *Aspergillus nidulans*. *Nature, London* 262, 231-4.

ARST, H.N., JR. & COVE, D.J. (1969). Methylammonium resistance in *Aspergillus nidulans*. *Journal of Bacteriology* 98, 1284-93.

ARST, H.N., JR. & COVE, D.J. (1973). Nitrogen metabolite repression in *Aspergillus nidulans*. *Molecular and General Genetics* 126, 111-41.

ARST, H.N., JR. & MACDONALD, D.W.(1975). A gene cluster in *Aspergillus nidulans* with an internally located *cis*-acting regulatory region. *Nature, London* 254-31.

ARST, H.N., JR. & SCAZZOCCHIO, C. (1975). Initiator constitutive mutation with an 'up-promoter' effect in *Aspergillus nidulans*. *Nature, London* 254-, 31-4.

ARST, H.N., JR., PARBTANI, A.A.M. & COVE, D.J. (1975). A mutant of *Aspergillus nidulans* defective in NAD-linked glutamate dehydrogenase. *Molecular and General Genetics*138, 165-71.

BAILEY, C. & ARST, H.N., JR. (1975). Carbon catabolite repression in *Aspergillus nidulans*. *European Journal of Biochemistry* 51, 573-7.

BARTNIK, E., WEGLENSKI, P. & PIOTROWSKA, M. (1973). Ammonium and glucose repression of the arginine catabolic enzymes in *Aspergillus nidulans*. *Molecular and General Genetics* 126, 75-84.

BRITTEN, R.J. & DAVIDSON, E.H. (1969). Gene regulation for higher cells: a theory. *Science* 165, 349-57.

BRITTEN, R.J. & DAVIDSON, E.H. (1971). Repetitive and non-repetitive DNA sequences and a speculation on the origins of evolutionary novelty. *The Quarterly Review of Biology* 46. 111-38.

COHEN, B.L. (1973). Regulation of intracellular and extracellular neutral and alkaline proteases in *Aspergillus nidulans*. *Journal of General Microbiology* 79, 311-20.

CYBIS, J. & WEGLENSKI, P. (1972). Arginase induction in *Aspergillus nidulans*. The appearance and decay of the coding capacity of messenger. *European Journal of Biochemistry* 30, 262-8.

DAVIDSON, E.H. & BRITTEN, R.J. (1973). Organization, transcription and regulation in the animal genome. *The Quarterly Review of Biology* 48, 565-613.

DUNSMUIR, P. & HYNES, M.J. (1973). Temperature sensitive mutants affecting the activity and regulation of the acetamidase of *Aspergillus nidulans*. *Molecular and General Genetics* 123, 333-46.

ELLIOT, J.J. & BALL, A.J.S. (1973). Glucosamine resistant mutants of *Saccharomyces cerevisiae*. *Genetics* 74, s71.

FANTES, P.A. & ROBERTS, C.F. (1973). β-galactosidase activity and lactose utilization in *Aspergillus nidulans*. *Journal of General Microbiology* 77, 471,86.

GAJEWSKI, W., LITWINSKA, J., PASZEWSKI, A. & CHOJNACKI, T. (1972). Isolation and characterization of lactose non-utilizing mutants of *Aspergillus nidulans*. *Molecular and General Genetics* 116, 99-106.

GOSH, B.K., MONTENCOURT, B. & LAMPEN, J.O. (1973). Abnormal cell envelope ultrastructure of a *Saccharomyces* mutant with invertase formation resistant to hexoses. *Journal of Bacteriology* 116, 1412-20.

HYNES, M.J. (1970). Induction and repression of amidase enzymes in *Aspergillus nidulans*. *Journal of Bacteriology* 103, 482-7.

HYNES, M.J. (1974). Effects of ammonium, L-glutamate, and L-glutamine on nitrogen catabolism in *Aspergillus nidulans*. *Journal of Bacteriology* 120, 1116-23.

HYNES, M.J. & PATEMAN, J.A.J. (1970a). The genetic analysis of regulation of amidase synthesis in *Aspergillus nidulans*. I. Mutants able to utilize acrylamide. *Molecular and General Genetics* 108, 97-106.

HYNES, M.J. & PATEMAN, J.A.J. (1970b). The genetic analysis of regulation of amidase synthesis in *Aspergillus nidulans*. II. Mutants resistant to fluoroacetamide. *Molecular and General Genetics* 108, 107-16.

KATZ, L. & ENGLESBERG, E. (1971). Hyperinducibility as a result of mutation in structural genes and self-catabolite repression in the *ara* operon. *Journal of Bacteriology* 107, 34-52.

KINGHORN, J.R. & PATEMAN, J.A. (1974). The regulation of NAD L-glutamate dehydrogenase in *Aspergillus nidulans*. *Genetical Research, Cambridge* 23, 119-24.

LIS, J.T. & SCHLEIF, R. (1973). Different cyclic AMP requirements for induction of the arabinose and lactose operons of *Escherichia coli*. *Journal of Molecular Biology* 79, 149-62.

MCCULLOUGH, W. & ROBERTS, C.F. (1974). The role of malic enzyme in *Aspergillus nidulans*. *Federation of European Biochemical Societies Letters* 41, 238-42.

MONTENECOURT, B.S., KUO, S.-C. & LAMPEN, J.O. (1973). *Saccharomyces* mutants with invertase formation resistant to repression by hexoses. *Journal of Bacteriology*114, 233-8.

PAGE, M.M. (1971). Genetic and biochemical studies on the catabolism of amines and alcohols in *Aspergillus nidulans*. Ph.D. Thesis, University of Cambridge.

PAGE, M.M. & COVE, D.J. (1972). Alcohol and amine catabolism in the fungus *Aspergillus nidulans*. *Biochemical Journal* 127, 17P.

PASZEWSKI, A., CHOJNACKI, T., LITWINSKA, J. & GAJEWSKI, W. (1970). Regulation of lactose utilization in *Aspergillus nidulans*. *Acta Biochemica Polonica* 17, 385-91.

POLYA, G.M., BROWNLEE, A.G. & HYNES, M.J. (1975). Enzymology and genetic regulation of a cycylic nucleotide-binding phosphodiesterase-phosphomonoesterase from *Aspergillus nidulans*. *Journal of Bacteriology* 124, 693-703.

RICKENBERG, H.V. (1974). Cyclic AMP in prokaryotes. *Annual Reviews of Microbiology* 28, 353-69.

ROBERTS, C.F. (1963). The genetic analysis of carbohydrate utilization in *Aspergillus nidulans*. *Journal of General Microbiology* 31, 45-58.

ROBERTS, C.F. (1970). Enzyme lesions in galactose non-utilizing mutants of *Aspergillus nidulans*. *Biochemica et Biophysica Acta* 201, 267-83.

ROMANO, A.H. & KORNBERG, H.L. (1968). Regulation of sugar utilization by *Aspergillus nidulans*. *Biochemica et Biophysica Acta* 158, 491-3.

ROMANO, A.H. & KORNBERG, H.L. (1969). Regulation of sugar uptake by *Aspergillus nidulans*. *Proceedings of the Royal Society, Series B*, 173, 475-90.

SCHAMHART, D.H.J., ten BERGE, A.M.A. & van de POLL, K.W. (1975). Isolation of a catabolite repression mutant of yeast as a revertant of a strain that is maltose negative in the respiratory-deficient state. *Journal of Bacteriology* 121, 747-52.

SCHLANDERER, G. & DELLWEG, H. (1974). Cyclic AMP and catabolite repression in yeasts. In *Schizosaccharomyces pombe* glucose lowers both intracellular adenosine 3':5'-monophosphate levels and the activity of catabolite-sensitive enzymes. *European Journal of Biochemistry* 49, 305-16.

ZONNEVELD, B.J.M. (1976). The effect of glucose and manganese on adenosine-3',5'-monophosphate levels during growth and differentiation of *Aspergillus nidulans*. *Archives of Microbiology* 108, 41-4.

ZUBAY, G., SCHWARTZ, D., & BECKWITH, J. (1970). Mechanism of activation of catabolite-sensitive genes: a positive control system. *Proceedings of the National Academy of Science of the United States of America* 66, 104-10.

Nitrogen Metabolism

J. R. KINGHORN and J. A. PATEMAN

Department of Genetics, University of Glasgow, Glasgow, U.K.

INTRODUCTION

In this article, we review studies of the major pathways and transport systems concerned with general nitrogen metabolism in *Aspergillus nidulans* with particular emphasis on the genetical contribution. These somewhat limited studies show that *Aspergillus* is rather similar in this area of metabolism to the closely related but more extensively studied *Neurospora crassa* (for reviews see Esser & Kuenan, 1967; Fincham & Day, 1971; Pateman & Kinghorn, 1976). Unfortunately, although a relatively large number of mutants have been isolated in *Aspergillus nidulans*, perhaps even the majority of these have not been exploited extensively biochemically. Instead, such mutants have been used primarily as a tool for classical genetic work; for example as markers or in the study of forward or/and back mutation. Only recently, has more attention been paid to elucidating their functions and to our understanding of metabolic pathways. Since various symbols are used in the text with respect to gene-enzyme relationships, it is necessary to define these. Where the catalytic function of a gene is known and conditional mutants have been made or even where catalytic activity is lost, a gene is described without any attached symbol. Mutants, which are thought to be impaired in a particular catalytic step, only on indirect but fairly suggestive evidence, such as one or more of the following: growth tests, resistance to certain toxic analogues, accumulation of intermediates, cross feeding, and epistatic relationships are set in brackets. Furthermore, where the enzymic step of a particular mutant is doubtful or where there are a number of possible steps in which a mutant could be blocked, a question mark follows. Another cautionary note which should be stressed is that certain similar mutants have been isolated independently by two or more workers and may or may not have been checked for allelism. Where the mutants are known to be allelic, this is shown by an equivalent sign. Finally, loci located in the same linkage group or chromosome should be taken as unlinked unless it is indicated to the contrary. The work reviewed here provides a general account of the known routes of nitrogen metabolism. We also review the recent studies on sulphur amino acid metabolism since this is of interest to the main theme of the contribution. Subjects dealt with extensively in other chapters e.g. arginine, purine and proline metabolism are only lightly touched upon to give the reader an overall picture on how *Aspergillus* manages its nitrogenous metabolic affairs. Moreover, all regulatory aspects of general nitrogen metabolism have been omitted since it is the subject of the next article.

NITRATE AND NITRITE ASSIMILATION

In *Aspergillus nidulans*, the assimilation of nitrate, via nitrite to ammonium is catalysed by nitrate reductase, (E.C. 1.6.6.3. and nitrite reductase, (E.C. 1.6.6.4.). The evidence for these two systems being sufficient to effect the entire reduction comes from studies of mutant strains. Mutation in at least eighteen genes can prevent or reduce the utilisation of nitrate or/and nitrite as nitrogen source (for reviews see Gross, 1969; Pateman & Kinghorn, 1976; Cove, 1970, 1977).

Nitrate

Strains mutant in the *niaD* gene were initially isolated on the basis of their inability to utilise nitrate as nitrogen source, whilst remaining able to utilise nitrite (Cove & Pateman, 1963; Pateman, Cove, Rever & Roberts, 1964; Cove, 1966; Pateman & Cove, 1967; Pateman, Rever & Cove, 1967; Cove & Pateman, 1969; Cove, 1970, 1976a,b). A number of different *niaD* mutants, isolated in this manner, have been examined for nitrate reductase activity, and all were found to have enzyme levels of lower than those of wild-type strains. Mutants in the *niaD* gene were also made using two other techniques. First, the putrescine starvation method which allows the selection of strains unable to utilise nitrate (or indeed other nitrogen sources) (Herman & Clutterbuck, 1966; Cove, 1976a). Second, by selection for resistance to chlorate (see below).

Several temperature-sensitive *niaD* mutant strains are known (McDonald & Cove, 1974). These mutations restrict growth on nitrate at 37°, whilst permitting growth at 25°. Nitrate reductase has been isolated from mycelium of strains carrying these temperature sensitive alleles and found to be thermolabile. This thermosensitivity of nitrate reductase from temperature-sensitive *niaD* mutant strains shows that the *niaD* gene product is a structural component of the nitrate reductase enzyme molecule.

Among the strains isolated on the basis of their inability to utilise nitrate were a class which retained the ability to utilise nitrite, but which could not utilise hypoxanthine as nitrogen source. These strains were able to grow on urate. This pattern of growth is consistent with impaired activity of both nitrate reductase and xanthine dehydrogenase. These mutants were therefore designated *cnx* and fell into seven complementation groups, namely *cnxA, B,C,E,F,G,* and *H*. Pateman *et al.*(1964) proposed that the *cnx* genes determine the production of a co-factor necessary for both nitrate reductase and xanthine dehydrogenase activity. Nitrate reductase and xanthine dehydrogenase have been shown, in a variety of organisms, to be flavoproteins, which also contain molybdenum and iron as prosthetic groups (Pateman *et al.*,1964). More recent evidence supported the proposition that molybdenum is involved (Arst, MacDonald & Cove,1970). They demonstrated that high concentrations of molybdate added to growing cultures of *cnxE* mutants, restored, at least partially, both nitrate reductase and xanthine dehydrogenase activities and consequently *cnxE* gene is thought to be concerned with the synthesis of a molybdenum containing co-factor, e.g. in the insertion of molybdenum into the co-factor, rather than coding for a cofactor structural component. Furthermore, temperature-sensitive alleles are known for the *cnxE* as well as *cnxF* and *cnxH* genes (MacDonald & Cove, 1974). MacDonald & Cove (1974) have examined the *in vitro* temperature-stability of the nitrate reductase produced by strains carrying certain of these alleles. While nitrate reductase from temperature-sensitive *cnxE* and *cnxF* strains were similar to that from the wild type, temperature-sensitive *cnxH* strains made an enzyme far more thermolabile than wild type enzyme. They concluded that the *cnxH* gene must code for a polypeptide component of the nitrate reductase molecule, whilst the *cnxE* and *cnxF* genes must be involved in co-factor biosynthesis. Attempts to isolate temperature-sensitive *cnxA,B,C or G* mutants have not been successful, despite exhaustive searches, showing that the frequency of temperature sensitive mutants of the *cnxE,B,C* and *G* are lower than that of *cnxE,F* and *H* (MacDonald & Cove, personal communication). The significance of this is not clear and their role is unknown (MacDonald & Cove, personal communication).

A.nidulans grown on nitrate possesses nitrate reductase and nitrite reductase. Mutations in the *niaD* and *cnx* genes lead usually to the loss of nitrate reductase activity and to constitutive synthesis of nitrite reductase. Cove & Coddington (1965) showed that wild type nitrate reductase possesses associated cytochrome-c reductase activity (E.C. 1.6.2.3.), which remains intact in purified preparations of nitrate reductase. *niaD* and *cnx* mutants, which are constitutive with respect to nitrite reductase, are also constitutive for cytochrome-c reductase. More recently, MacDonald, Cove & Coddington (1974a) found evidence in wild type for two species of cytochrome-c reductase with sedimentation coefficients of 13.7s and 7.6s. Only the 7.6s species is thought to be associated with nitrate reductase activity since it, like nitrate reductase, requires NADPH and FAD as cofactors. A third species of cytochrome-c reductase with a sedimentation coefficient of 4.5s was detected in mutant strains *cnxB*, *cnxF*, *cnxG* and *cnxH*. Only the 7.6s species was found in *cnxE* strains. MacDonald *et al.* (1974) suggest that the 4.5s species is a subunit or breakdown product of the 7.6s species since the former is only found in mutant strains and has the same cofactor requirements as the latter. They proposed a model which accounts for some of these and certain other facts (Fig. 1).

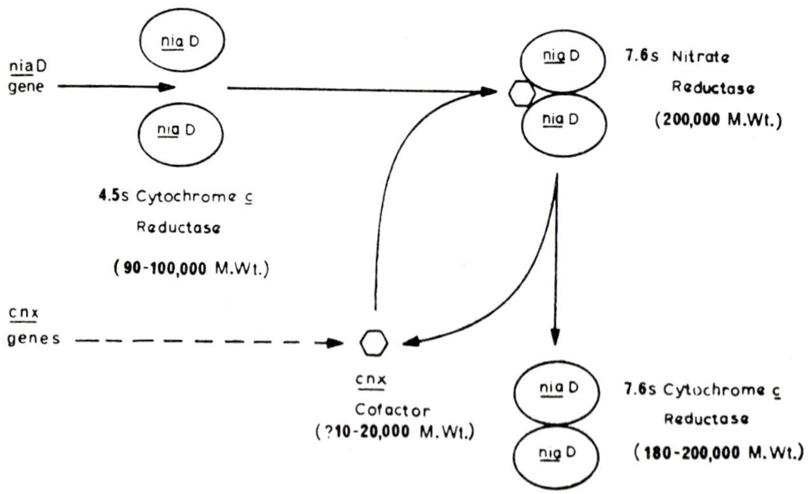

Fig. 1. Suggested model for the role of the *niaD* and *cnx* gene products in specifying the nitrate reductase molecule. The 7.6s nitrate reductase molecule is seen as being made up of two molecules of the 4.5s *niaD* product, together with a small molecular weight cofactor specified by the cnx genes. The 7.6s cytochrome-c reductase would then consist of two 4.5s *niaD* products in the form of a dimer which, however, lacked the cnx cofactor necessary for nitrate reductase activity. Reprinted from MacDonald *et al.* (1974a).

Briefly, 4.5s *niaD* products aggregate to form a 7.6s nitrate reductase together with a low molecular weight co-factor specified by the cnx genes. Only this aggregate has nitrate reductase activity. The co-factor is also required for the dimerisation of two 4.5s. The *cnxB*, *cnxF*, *cnxB* and *cnxH* mutants, which produce the 4.5s species, are unable to produce the co-factor and therefore the 4.5s subunits are unable to form 7.6s molecules. For some reason, still unexplained, the co-factor allows the aggregation of 4.5s subunits in *cnxE* strains to form 7.6s cytochrome-c reductase.

In addition to cytochrome-c reductase, nitrate reductase can show a range of NADPH diaphorase activities. Indeed, the reduction of nitrate may be considered as an electron transport chain confined to a single complex enzyme molecule. The probable situation is represented diagrammatically in Fig. 2. A pair of electrons are donated by NADPH to FAD, then possibly to iron in a cytochrome, then to two molybdenum atoms which are reduced from Mo^{6+} to Mo^{5+} and finally to nitrate which is reduced to nitrite. It is apparent that nitrate reductase can donate electrons from some intermediate point on the chain, probably the flavin, to suitable electron acceptors, and this is the observed NADPH-diaphorase activity. The molecule can also accept electrons from reduced viologen dyes at an intermediate point, probably molybdenum, and thus show benzyl viologen-nitrate oxidoreductase activity (Pateman *et al.*,1967).

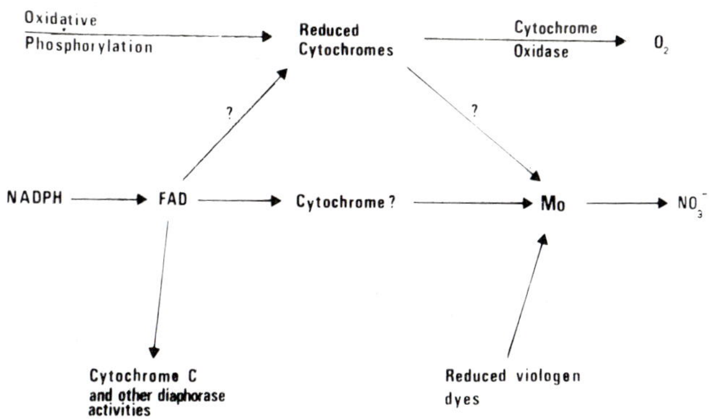

Fig. 2. Electron transfer steps associated with nitrate reduction in fungi. Assimilatory nitrate reduction involves electron transfer on the molecule direct from NADPH to nitrate. The possible involvement of iron in a cytochrome intermediate between FAD and molybdenum is indicated by a query.

In summary, the complete nitrate reductase molecule may be a dimer of two *niaD* gene products and one or more molelcules of the co-factor specified by the *cnx* genes. Only this aggregate of about 200,000 MW possesses nitrate reductase activity in addition to cytochrome-c reductase activity (Fig. 1). In terms of the diagram in Fig. 2 mutation in *niaD* can result in failure of one or more of the electron transfer steps through abnormalities of the *niaD* protein. Mutation in the *cnx* genes only affects the electron transfer mediated by the molybdenum co-factor with the loss of nitrate reduction but production of normal *niaD* protein and its associated NADPH-diaphorase activities.

Chlorate toxicity and nitrate assimilation

As discussed above, mutations in a number of genes lead to inability to utilise nitrate as sole nitrogen source. In most cases, *niaD*, *cnx* and *nirA* (see below) mutations result in resistance to chlorate which is toxic at relatively high concentrations (200 mM) (Cove, 1972, 1976a,b, 1977). The basis of chlorate toxicity and resistance is still an open question. Aberg (1947) proposed that, in a number of organisms, chlorate toxicity is due to the conversion of chlorate to the extremely toxic chlorite, catalysed by nitrate reductase. This may in part account for chlorate toxicity in *A. nidulans*, but Cove (1972, 1976b) has presented evidence that toxicity may also be due to chlorate interference with the breakdown of organic nitrogenous compounds. In this respect chlorate may mimic nitrate which is thought to control certain catabolic pathways. This proposition comes from observations that not all *niaD* and *cnx* mutations lead to chlorate resistance although they abolish nitrate reductase activity. Furthermore, *niaD* mutants are known which result in chlorate resistance but show appreciable nitrate reductase activity and appear normal with respect to growth on nitrate. Although it is less likely these results could be explained by altered nitrate reductase forms, produced by mutant strains, which have higher affinity for nitrate than chlorate and *vice versa*.

Nitrite

Among strains unable to grow on nitrate as sole nitrogen source, is a class which is also unable to grow on nitrite. These are called *niiA* since they produce nitrate reductase, but not nitrite reductase (Pateman *et al*. 1967). The properties of *niiA* mutants demonstrate that nitrite reductase, like nitrate reductase, possesses associated cytochrome-c reductase and other NADPH-diaphorase activities. Pateman *et al*. (1967) proposed that the *niiA* gene codes for the enzyme nitrite reductase, although the evidence for this is circumstantial. *niiA* is the only known locus in which mutations lead to the loss of nitrite reductase and no other activity.

Furthermore, *niiA* is the only gene known in which mutation specifically affects the conversion of nitrite to ammonium. It appears that nitrite reductase must catalyse the complete reduction of nitrite to ammonium although the nature of the reduction products from nitrite is unknown.

Hyponitrite has been suggested as a possible intermediate in the reduction of nitrite to hydroxylamine. Hyponitrite reductase, an enzyme which uses NADH as an electron donor to reduce hyponitrite to ammonium has been demonstrated in *Neurospora crassa* (Medina & Nicholas, 1957a). This enzyme is apparently distinct from the nitrite reductase in *N.crassa* (Medina & Nicholas, 1957b). This may be the case also in *Aspergillus* although there is no evidence. Hydroxylamine is a likely intermediate between nitrite and ammonium since Pateman *et al.* (1967) found relatively high levels of hydroxylamine reductase (1.7.99.1) in mycelium of wild type strains of *A. nidulans* and low levels in strains with *niiA* mutations. This enzyme activity is induced in wild-type strains by the presence of nitrate in the growth medium (see below). In *N.crassa* the enzyme is a flavoprotein inhibited by metal chelating agents and uses NADH as the electron donor to reduce hydroxylamine to ammonium. Manganese or magnesium deficient cells have very low hydroxylamine reductase activity, but neither metal accumulates in purified enzyme, nor does addition of these metals to enzyme preparations increase activity.

It is suggested that the two metals may be necessary for the formation of the enzyme molecule *in vivo*. In *A.nidulans* the electron donor is NADPH not NADH as in *N.crassa*. The enzyme activity *in vitro* is strongly dependent on added FAD, like nitrite reductase activity and to a lesser extent nitrate reductase activity. As described previously, hydroxylamine reductase activity is associated with nitrite reductase activity. The *niiA* gene may be regarded as a structural gene for a protein which catalyses the reduction of both nitrite and hydroxylamine using NADPH as the electron donor. The fact that *riboB* mutants (discussed below) lack normal hydroxylamine reductase activity unless grown with a riboflavin supplement implies that flavin is also a necessary component for hydroxylamine reductase whether or not it is the same protein as nitrite reductase. It is not possible to be sure if *A.nidulans* can use hydroxylamine directly as a nitrogen source since it is extremely toxic and rather unstable. The concentration (10^{-4} M) at which it is not toxic is not sufficient to act as a significant nitrogen source either alone or together with a limiting concentration of another nitrogen source (Pateman, unpublished work).

In an original series of mutation experiments designed to produce mutants unable to utilise nitrate, out of a total of 98 found, one mutant *riboB1* (formerly *rinA* and *nii*, *riboA*) was isolated where growth was poor on nitrate or nitrite in medium without a riboflavin supplement (Pateman *et al.* 1967). This was correlated with the fact that this mutant showed low levels of nitrite reductase (and hydroxylase reductase) activities in the absence of riboflavin supplementation. More recently it was found that mutation in at least three other genes *riboA* (formerly *rinB*) *riboG* (formerly *rinC*) and *riboD* lead to similar but not identical growth patterns (for more detailed discussion - see Cove, 1976c). Mutation in *riboA* and *riboB* can result in an absolute requirement for riboflavin on most nitrogen source whereas all mutations in *riboD* and *riboG* but only some in *riboA* and *riboB* can grow appreciably on most nitrogen except nitrate and nitrite in the absence of riboflavin. These results indicate that nitrite utilisation has a considerable requirement for riboflavin - much higher than for utilisation of any other nitrogen sources. Furthermore, Pateman *et al.* (1967) have shown a high requirement for FAD for nitrite reductase activity *in vitro*. The status of the *ribo* genes at the moment is unclear. There is the possibility that one or more of these genes is concerned with the flavin cofactor of nitrite reductase rather similar to the *cnx* genes for the molybdenum cofactor for nitrate reductase. Alternatively these mutations may result in low intracellular levels of riboflavin and this particularly is severe in the case of nitrite reductase since high levels of FAD is required for activity.

Nitrate and nitrite

Mutants unable to utilise both nitrate and nitrite and lacking nitrate and nitrite reductase are known. One class map in a gene which is concerned with nitrate induction, called *nirA* (formerly *niiB*) (Cove & Pateman, 1963; Pateman *et al.*1964; Pateman *et al.* 1967; Cove, 1970). The *nirA* gene produces a product which is necessary for full expression of the nitrate utilization enzymes. This regulatory gene and its role in the induction of nitrate and nitrite reductase is discussed in the next article. Nitrate and nitrite reductase are also regulated by ammonium repression (see Pateman & Kinghorn, this volume). Two loci *tamA* and *areA* have been identified as playing a role in this control. The product of these two genes are necessary for full repression of nitrate and nitrite reductase (and many other systems) in the presence of ammonium. Mutations in these genes can lead to impaired nitrate and nitrite reductase activity and consequent poor growth on nitrate and nitrite.

Mutants with a phenotype similar to *nirA1* strains (discussed below) but having wild type nitrate reductase activity were isolated (Cove & Pateman - unpublished work). These mutations led to a defect not only in nitrate assimilation but also in the metabolism of pentose sugars (Hankinson, 1974; Hankinson & Cove, 1974). Subsequently, mutants in two genes *pppA* and *pppB* unable to utilise L-arabinose, D-glucuronate or D-xylose as carbon source, and which grew poorly on nitrate and nitrite were isolated. Certain pentose phosphate pathway enzymes are defective in these mutants which are impaired in their ability to reduce $NADP^+$ and consequently grow poorly on nitrate.

AMMONIUM ASSIMILATION

Ammonium has a central position in the metabolism and growth of *Aspergillus* and many other simple eucaryotes. It has been shown in *Aspergillus clavatus* that ammonium is utilised in preference to other nitrogen sources (Robinson, Anthony & Drabble, 1974). To achieve this, systems which are required for the utilisation of other nitrogen sources are repressed by ammonium. This control exerted on a large number of systems in *Aspergillus* by ammonium is discussed in the article on regulation. The various routes by which *A.nidulans* could assimilate ammonium, include NADP and NAD glutamate dehydrogenase, glutamate synthase, aspartase and alanine dehydrogenase. Other related systems for ammonium assimilation include asparagine synthetase, glutamine synthetase, glutamate-oxaloacetate aminotransferase, and alanine-pyruvate aminotransferase. The primary aminating steps together with these secondary systems result in the synthesis of glutamate, glutamine, alanine, aspartate and asparagine, the main starting substances for the biosynthesis of most amino acids. Genetical and biochemical studies have revealed the presence of at least some of these systems in *A.nidulans* (Fig. 3). However, a more detailed scheme awaits the further isolation and biochemical characterisation of mutants defective in some of these systems. Mutants impaired in ammonium assimilation and related or neighbouring systems are discussed below.

Glutamate

The primary amination reaction in *A.nidulans* is catalyzed by NADP L-glutamate dehydrogenase (NADP-GDH) E.C. 1.4.1.4.). Mutation at the *gdhA* locus leads to low NADP-GDH activity and simultaneous derepression of ammonium repressible systems. *gdhA* is the structural gene for the enzyme as shown by the existence of a temperature sensitive allele whose protein is thermolabile; a mutant with abnormally high Michaelis Constants and three *gdhA* mutants which have significantly high levels of immunologically active NADP-GDH material (Kinghorn & Pateman, 1975a).

gdhA mutants grow with a lag response with inorganic nitrogen sources such as ammonium but possess wild type growth with amino acids as sole nitrogen sources. The leaky growth of *gdhA* with ammonium was thought to be due to residual NAD-GDH activity (see glutamate catabolism). However, this supposition is not consistant with the properties of double mutants between *gdhA* and mutants lacking NAD-GDH activity (*gdhB*). The double mutant *gdhA1, gdhB1* grows with a growth lag similar to *gdhA1* alone on ammonium.

Another enzyme which might be responsible for the adaptive growth on ammonium is glutamate synthase (E.C. 2.6.1.53.) (Brown,

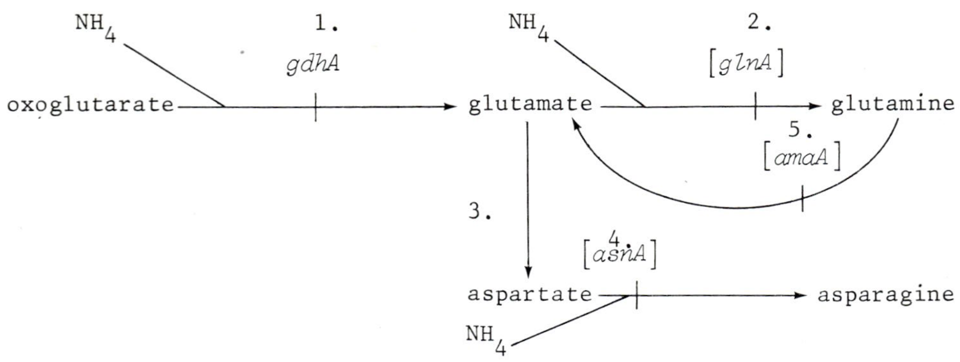

Fig. 3. Ammonium assimilation

Step.	Enzyme system	Mutant locus	Linkage group	References
1.	NADP-glutamate dehydrogenase	*gdhA*	III	Arst & McDonald, 1973: Kinghorn & Pateman, 1973, 1974a, 1975a; Hynes, 1974.
2.	Glutamine synthetase	[*glnA*]		Pateman, 1969; Kinghorn - unpublished work
3.	Glutamate-oxalacetate amino transferase	no mutants		Kinghorn - unpublished work
4.	Asparagine synthetase?	[*asnA*]	III	Arst, Kinghorn & Drainas - unpublished work
5.	Glutamate synthetase?	[*amaA*]		Kinghorn - unpublished work.

gdhA and *asnA* are unlinked.

Burns & Johnston, 1973; Meers, Tempest & Brown, 1970). Two allelic mutants have been selected in *gdhA* strains which stop the leakiness shown by *gdhA* mutants. These map at the *amaA* (ammonium assimilation) locus which is unlinked to *gdhA*. *amaA* mutants are recessive, have wild type NADP-GDH activity and are extremely leaky on ammonium similar to *gdhA* and grow like wild type on amino acids. It is possible that *amaA* mutants are defective in an amminating step such as glutamate synthase. However, such a system has not been detected in the wild type or *gdhA* mutants despite several attempts (Kinghorn - unpublished work; Cove & Hynes - personal communications)

Glutamine

Another important ammonium fixation step is glutamine synthetase (E.C. 6.3.1.2.). Mutations at the *glnA* locus lead to loss of glutamine synthetase activity, *glnA* mutants show wild type growth when supplemented with glutamine and leaky growth with ammonium or glutamate or any other nitrogen source. This phenotype is consistant with the probability that glutamine may result from amino acid catabolism or transamination from asparagine and thus supplement glutamine synthetase defective mutants.

Aspartate

In *Aspergillus* there are two glutamate - oxalacetate aminotransferases (E.C. 2.6.1.1.) and these activities can be separated and detected by electrophoresis. One is probably responsible for aspartate synthesis and the other for aspartate deamination.

Asparagine

Mutants at the *asnA* locus have been isolated which require supplementation with asparagine. These mutants grow extremely poorly with ammonium, glutamate, glutamine or aspartate but as wild type with asparagine. It is likely that *asnA* mutants are defective in some aspect of asparagine synthesis, most likely asparagine synthetase (E.C. 6.5.1.1.), but attempts to find activity in the wild type have failed. The phenotype of *asnA* mutants indicates that only one route is present for asparagine synthesis.

ANABOLIC PATHWAYS

Aspergillus nidulans, in common with most other fungi, can grow on media containing only inorganic nitrogen such as nitrate or ammonium. Under these conditions, all essential organic nitrogen substances are synthesised and the relevant biosynthetic pathways in *Aspergillus* probably closely parallel those found in *Neurospora crassa* (for a review see Esser & Kuenen, 1967; Pateman & Kinghorn

1976). Anablic routes are indispensable where the nitrogen source offered is wholly or mainly inorganic. Under such nutritent conditions, mutationally blocked anabolic pathways lead to auxotrophy for metabolites which are normally synthesised by the fully operational pathways in prototrophs. On the other hand, where Aspergillus is offered organic nitrogenous molecules, biosynthetic pathways may become dispensable and growth of anabolic pathway mutants is not impaired. The biochemistry and genetics of biosynthetic routes for aromatic amino acids, arginine, ornithine, proline, histidine, pyrimidine, purine and polyamine synthesis have been at least partly elucidated in *Aspergillus nidulans* and are discussed below. Relatively little work has been done on the remaining biosynthetic pathways. The auxotrophic mutants known are listed (Table 1).

Table 1. Remaining auxotrophs

Anablic Pathway	Mutant locus	Linkage group	Phenotype	References
Isoleucine	*ileA* ≡*abA*	II	Isoleucine auxotrophy. Defective in threonine dehydratase activity	Pees, 1965, 1966; MacDonald *et al.*, 1974b.
Leucine	*leuA*	I	Leucine auxotrophy	Forbes, 1959a; Dhillion & Garber, 1970; Sinha, 1970.
Lysine	*lysA* *lysB* *lysC* *lysD* *lysE* *lysF*	VI V VII VII V I	Lysine auxotrophy	Pontecorvo *et al.*, 1953; Kafer, 1958; Aspen & Meister, 1962; Pees, 1966; Pollard *et al.*, 1968.

lysB and *lysE* are unlinked.

Aromatic amino acids

Although there are a number of aromatic auxotrophic mutants known for *A. nidulans*, these have been characterised mainly by growth tests and not by detailed biochemical analysis and therefore, with the exception of histidine synthesis, gene-enzyme relationships are in general rather speculative (Fig. 4). Three classes of mutants, defective in the early part of the pathway prior to prephenate have been isolated by Roberts (1969) and Zaudy (1969). *aroA* mutants, which are probably analogous to the *arom* cluster of five genes which

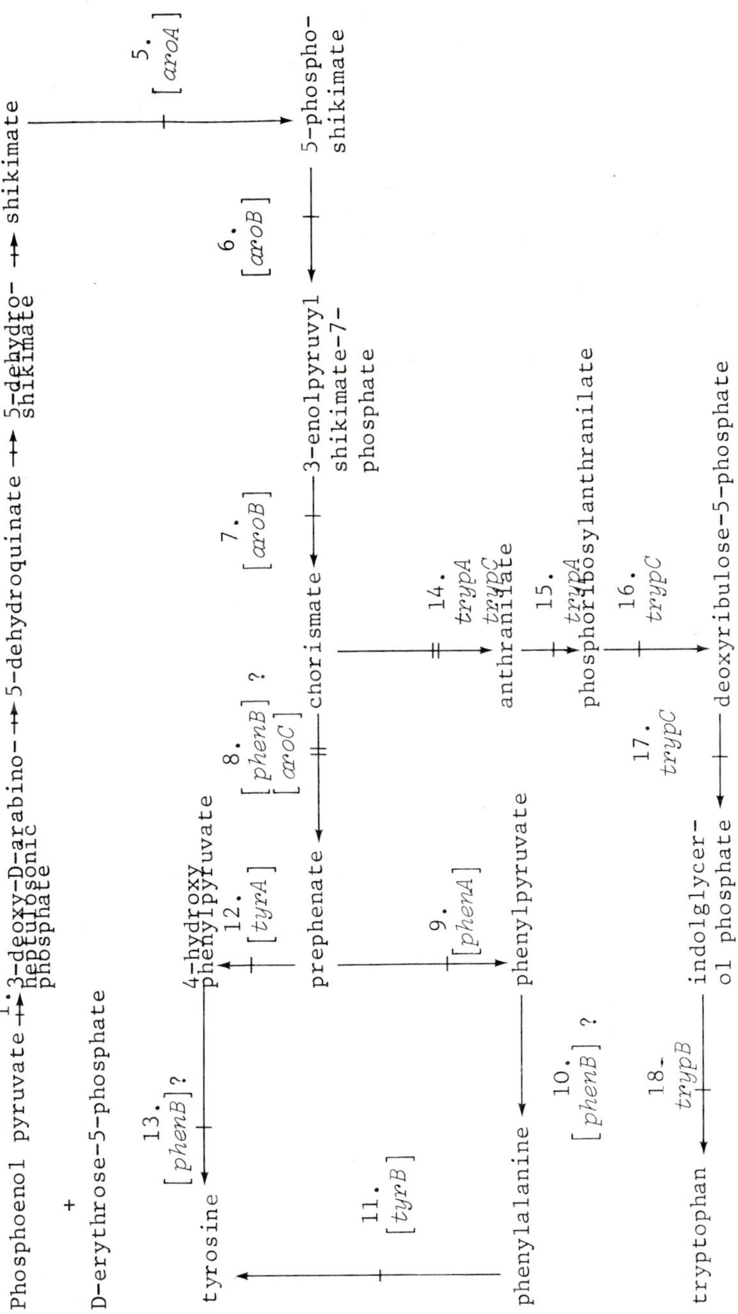

Fig. 4. Aromatic amino acids biosynthesis

Fig. 4. cont'd.

Step	Enzyme system	Mutant locus	Linkage group	References
1.	Deoxy-D-arabinoheptulosonic-7-phosphate synthase	None		-
2.	Dehydroquinate synthetase	[*aroA*]	VIII	Roberts, 1969; Zaudy, 1969.
3.	Dehydroquinate dehydratase	[*aroA*]	VIII	Roberts, 1969; Zaudy, 1969.
4.	Shikimate dehydrogenase	[*aroA*]	VIII	Roberts, 1969; Zaudy, 1969.
5.	Shikimate phosphatase	[*aroA*]	VIII	Roberts, 1969; Zaudy, 1969.
6.	Phenolpyruvyl shikimate 7-phosphate	[*aroA*]	VIII	Roberts, 1969; Zaudy, 1969.
7.	Chorismate synthetase	[*aroB*]	VI	Roberts, 1969; Zaudy, 1969; Noronha, 1970.
8.	Chorismate mutase	[*aroC*]? [*phenB*]?	I VII	Roberts, 1969; Zaudy, 1969; Sinha, 1967.
9.	Prephenate dehydratase	[*phenA*]	III	Sinha, 1967.
10.	Phenylpyruvate aminotransferase	[*phenB*]	VII	Sinha, 1967.
11.	Phenylalanine hydroxylase	[*tyrB*]		Sinha, 1967.
12.	Prephenate dehydrogenase	[*tyrA*] ≡ [*fpaA*]		Warr & Roper, 1965; Sinha, 1967.
13.	Hydroxyphenyl pyruvate aminotransferase	[*phenB*]?	VII	Sinha, 1967.
14.	Anthranilate synthetase	*trypA* ≡ [$anth_i$] *trypC* ≡ [ind_2]	II VIII	Kafer & De Moss, 1973 Roberts, 1967, 1969; Hütter & De Moss, 1967a,b; Sinha, 1967, 1972; Noronha, 1970.

Fig. 4. cont'd.

Step	Enzyme system	Mutant locus	Linkage group	References
15.	Anthranilate PP ribose-phosphoribosyl transferase	*trypD*	II	Kafer & De Moss, 1973; Roberts, 1967, 1969; Hütter & De Moss, 1967a,b; Sinha, 1967, 1972; Noronha, 1970
16.	Phosphoribosylanthranilate isomerase	*trypC* ≡ [ind_2]	VIII	Kafer & De Moss, 1973; Roberts, 1967, 1969; Hütter & De Moss, 1967a,b; Sinha, 1967, 1972; Noronha, 1970.
17.	Indole glycerophosphate synthetase	*trypC* ≡ [ind_2]	VIII	Kafer & De Moss, 1973; Roberts, 1967, 1969; Hütter & De Moss, 1967a,b; Sinha, 1967, 1972; Noronha, 1970.
18.	Tryptophan synthetase	*trypB*	I	Kafer & De Moss, 1973; Roberts, 1967, 1969; Hütter & De Moss, 1967a,b; Sinha, 1967, 1972; Noronha, 1970.

are required for five of the seven steps (Fig. 4 steps 2,3,4,5 & 6) of chorismate synthesis in *N.crassa*(Giles *et al.*,1967; Case & Giles, 1968, 1971). Investigations of the state of aggregation of the five enzymes in *A.nidulans* have shown that the five enzymes sediment together (Ahmed & Giles, 1969). *aroB* mutants may correspond to the *aromB* locus in *N.crassa* controlling chorismate synthetase activity. The enzymic defect of *aroC* is not known although a possible candidate is chorismate mutase activity.

A general pathway of tyrosine and phenylalanine biosynthesis from chorismate in *A.nidulans*, is given by Sinha (1967) and is similar to that of other micro-organisms. A noteworthy feature is that tyrosine may be synthesised by two routes in *A.nidulans*. The more common one in micro-organisms is by the transamination of 4-hydroxyphenyl-pyruvate and second by direct hydroxylation of phenylalanine, the latter route being found mainly in animals.

The pathway, leading to tryptophan biosynthesis, branches from that of tyrosine and phenylalanine at chorismate. The enzyme deficiencies of auxotrophs have been established in this case (Hütter & De Moss, 1967a). These studies show that for this pathway gene-enzyme relationships are fairly complex, but again rather similar to the situation in *Neurospora crassa* (Hütter & De Moss, 1967a,b) (Fig. 4). *trypA* mutants lack anthranilate synthetase, *trypB* mutants lack tryptophan synthetase and *trypD* mutants lack anthranilate PP ribose-phosphoribosyl transferase activity. In *A.nidulans*, as in *N.crassa*, a multienzyme complex catalizes the first, third and fourth steps (Fig. 4,Steps 1, 16 and 17): anthranilate synthetase phosphoribosyl-anthranilate isomerase and indole glycerophosphate synthetase respectively. *trypC* mutants lack all three activities, including anthranilate synthetase, which also depends on the product of the *trypA* gene. Relatively recently, Kafer & De Moss (1973) have found that by mixing extracts of *trypA* and suitable *trypC* mutants, *in vitro* formation of anthranilate activity can be obtained. Furthermore, activity can also be obtained when extracts of *A.nidulans* mutants are mixed with extracts of *N.crassa* mutants of the complementary type indicating that the systems specified by *trypA* and *trypC* in *A. nidulans* and *N.crassa* are similar and perhaps identical. *trypE* mutants were isolated as tryptophan auxotrophs in a *nicB8* strain (Roberts, 1967) and only show a requirement for tryptophan in *nicB* alleles. However, biochemical analyses of these strains showed no defect in the enzymes leading to tryptophan synthesis from chorismate (Hütter & De Moss, 1967a).

Noronha (1970) has described mutants at four loci *anthA* (*anth*$_2$), *anthB*(*anth*$_4$), *anthC*(*anth*$_6$) and *anthD*(*anth*$_5$,*anth*$_7$ and *anth*$_8$) which respond to anthranilate, indole and tryptophan, but not to shikimate suggesting a metabolic block between shikimate and anthranilate. However, *anthA*, *anthC* and *anthD* mutants also respond to nicotinamide. At least one of these may be allelic with a class of known nicotinamide auxotrophic mutants (*nicB*) which also respond to anthranilate and tryptophan. *anthA, C* and *D* mutants are not allelic with *trypE* or *trypA* and their growth responses suggest that they do not belong to *trypC* nor *trypD*. It should also be noted that *nicA* strains respond to nicotinamide and anthranilate but not tryptophan. Therefore *anthA, anthC, anthD, nicA* and *nicB* may be deficient in nicotinamide rather than anthranilate and tryptophan metabolism. This supposition is consistant, at least in *nicA* and *nicB*, with the fact that these mutants respond to extremely low supplementation levels. However, it is unlikely that any of these mutants are simple nicotinamide auxotrophs and deficient in nicotinamide biosynthesis. In connection with this there is a simple nicotinamide auxotroph (*nicC*) which only respons to nicotinamide. There are however possible connections between nicotinamide and tryptophan metabolism which may explain the pleotrophy of the mutants. First, nicotinamide may be a cofactor for some of the tryptophan biosynthetic enzymes. Mutations in genes which are responsible for the synthesis of a nicotinamide-containing apoenzyme may show the above characteristics and be analogous to the *cnx* genes for nitrate reductase activity. Second, there is a possible metabolic route for nicotinamide biosynthesis, direct from tryptophan via kynurenine as has been found in other organisms. It is at least possible that some of the mutants affect the activity and/or control of this tryptophan-nicotinamide pathway.

anthB mutants do not respond to nicotinamide (as does *anthA*, *anthC* and *anthD* mutants) they are not allelic with *trypA, trypB, trypC* or *trypD* mutants. Finally, with respect to tryptophan synthesis, a mutant *indA1* (formerly *ind1*) which is rather similar in growth properties to *trypC* and *trypD* mutants has been isolated; it responds to indole and tryptophan but not to anthranilate, shikimate or nicotinamide (Noronha, 1970). However, *indA1* is not allelic with *trypC* or *trypD* and is located in another linkage group (I).

Histidine

The pathway of histidine biosyntheses is biochemically one of the better characterised of the amino acid *biosynthetic* pathways in *A.nidulans*. Berlyn (1967) isolated two hundred and fifty histidine auxotrophs. These fell into ten groups by complementation analysis: *hisA, B, C, D, E, F, G, H, I* and a complex group, *hisE1D*. Mutations in the *hisE, hisI* and *hisD* loci leads to poor activity of PR-ATP pyrophosphydrolase (Fig. 5, Step 2), PR-AMP cyclohydrolase (Fig. 5 Step 3) and histidinol dehydrogenase (Step 10) (E.C. 1.1.1.23) respectively. Crosses between *hisD, hisE*, and *hisI* showed that these genes are tightly clustered. Furthermore, multiply deficient mutants *hisE1D* were isolated and lacked all three activities. These failed to complement with *hisD*, *hisI* or *hisE* mutants and in crosses between *hisD*, *hisE*, *hisI* and *hisElD*, a low frequency ($> 5 \times 10^{-4}$) of wild type recombinants were recovered showing that all four classes occur in a single region of the chromosome. Furthermore, reversion studies showed that these mutations were not deletions. Thus, the situation is similar, if not identical, to that in *N. crassa* where the *his-3* region specifies a multifunctional protein aggregate which catalyses all three activities and these activities can be individually affected by mutations within the region.

The other mutants map in six unlinked loci in contrast to the single cluster in *Salmonella*. The first catalytic step for histidine biosynthesis, PR-ATP pyrophosphorylase activity, is lost in mutants in the *hisG* gene. The other mutants are biochemically less well characterised, but it is likely from precursor accumulation data that *hisB* and *hisC* mutants lack IGP hydratase (Fig. 5-Step 7) and IAP aminotransferase (Fig.5-Step 8) activities respectively. The enzymic defect in *hisA*, *hisF* and *hisG* mutants is not known for certain. However, since these mutants accumulate imidazole substances (Fig.5-Steps 7,8,9,10) and show normal activity for the first three enzymic reactions (Fig.5-Steps 1,2 and 3), Berlyn (1967) suggests that these may be defective in the intermediate reactions (Fig.5-Steps 4,5 and 6).

Arginine ornithine and proline

The biosynthesis of arginine and ornithine proceed by the acetylated pathway in *Aspergillus nidulans* (Fig. 6). Mutants at the *ornA*, *ornB* and *ornC* loci are repaired by arginine or ornithine, showing that they are blocked prior to ornithine rather than in arginine synthesis. *ornC* mutants lack acetylornithine deacetylase (E.C. 3.5.1.16) while it is suspected that *ornA* and *ornB* lack steps 2 and 3 in Fig. 6.

argA, *argB*, *argC* and *argE* auxotrophies are repaired by arginine but not ornithine indicating that they are defective in systems necessary for the conversion of ornithine to arginine. Mutations in the *argB* gene lead to impaired ornithine carbamyl transferase activity (E.C. 2.1.3.3). Mutations in *argC* and *argE* affect arginosuccinate synthetase activity (E.C.6.3.4.5.) and Cybis and colleague (Cybis *et al.*,1972a) suggest that the situation may be similar to that in yeast where the enzyme is coded for by two structural genes. *argA* has been relatively well characterised and is probably the structural gene for arginosuccinase (E.C. 4.3.2.1.). The breakdown of arginine and ornithine follows another route via a non-acetylated intermediate: glutamate semialdehyde (see catabolic section).

Proline can be synthesised by two routes in *A.nidulans*. The first is the route from glutamate (Fig. 6) and this is present in the wild type. Mutation in two tightly linked loci *proA* and *proB*

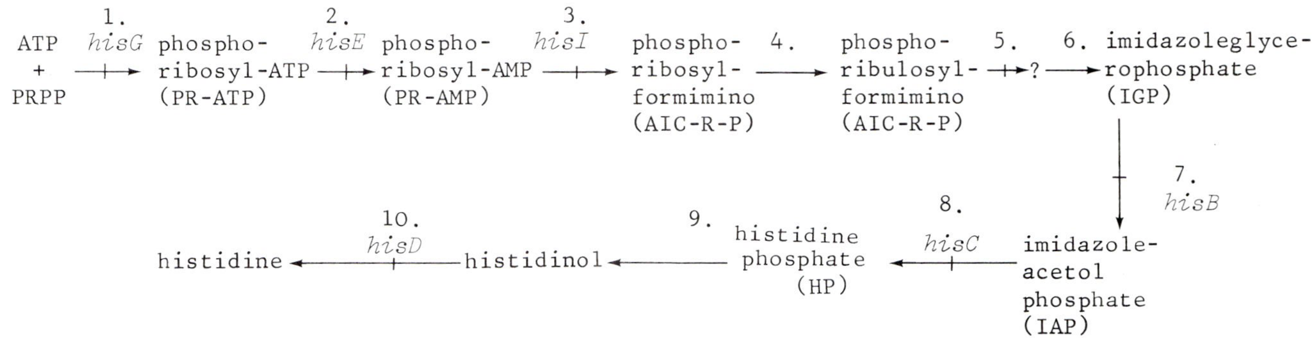

Fig.5. Histidine biosynthesis.

Step	Enzyme system	Mutant locus	Linkage group	References
1.	PR-ATP pyrophosphorylase	*hisG*	II	Berlyn, 1967: Pees, 1966
2.	PR-ATP pyrophosphorylase	*hisE*	VIII	Berlyn, 1967: Pees, 1966
3.	PR-AMP cyclohydrolase	*hisI*	VIII	Berlyn, 1967; Pees, 1966
4.	AIC-R-P isomerase	*hisA* .?	VI	Berlyn, 1967; Pees, 1966
5.	AIC-RIP amidotransferase	*hisH* ?	VIII	Berlyn, 1967; Pees, 1966
6.	Cyclase	*hisF* ?	VII	
7.	IGP dehydratase	*hisB*	I	Berlyn, 1967; Pees, 1966
8.	IAP aminotransferase	*hisC*	VIII	Berlyn, 1967; Pees, 1966
9.	HP phosphatase	No mutants		Berlyn, 1967; Pees, 1966
10.	Histidinol dehydrogenase	*hisD*	VIII	Berlyn, 1967; Pees, 1966

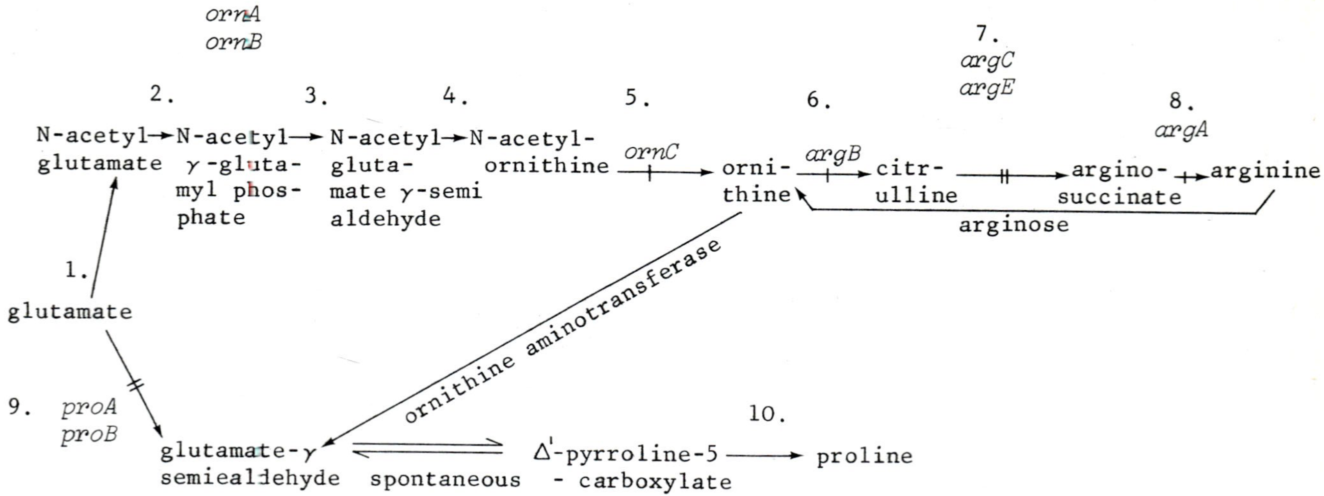

Fig. 6. Arginine, ornithine and proline biosynthesis.

Fig. 6 cont'd

Step	Enzyme system	Mutant locus	Linkage group	References
1.	Glutamate-N-acetyl transferase	None		
2.	N-acetyl-γ-glutamokinase]	*ornA*	IV	Cybis *et al.*,1972a
3.	N-acetylglutamate semialdehyde dehydrogenase]	*ornB*	VIII	Cybis *et al.*,1972a
4.	N-acetyl ornithine aminotransferase	None		Cybis *et al.*,1972a
5.	N-acetyl ornithine deacetylase	*ornC*	?	Cybis *et al.*,1972a
6.	Ornithine carbamyl transferase	*argB*	III	Cybis *et al.*,1970 Cybis *et al.*1972a,b
7.	Arginosuccinate synthetase	*argC* *argE*	VIII	Cybis *et al.*,1972a
8.	Arginosuccinase	*argA*	VI	Bainbridge *et al.*, 1966; Cybis *et al.*, 1972a
9.	Glutamate -γ-semialdehyde system	[*proA*] [*proB*]	I I	Forbes, 1959b; Węgleński, 1966 1967
10.	Pyrroline-5-carboxylate reductase	No mutants		

For further details of arginase and ornithine aminotransferase, see arginine catabolism.

can result in proline auxotrophy. The grown responses of these mutants suggest that they are blocked between glutamate and glutamate-γ-semialdehyde. The mechanism of the conversion of glutamate to glutamyl-γ-semialdehyde is relatively uncertain although it has been called the glutamate-γ-semialdehyde system and may consist of a kinase and a dehydrogenase (for references in *Neurospora* see Pateman & Kinghorn (1976). No mutants have been found for the other proline biosynthetic step, pyrroline-5-carboxylate reductase (E.C. 1.5.1.2.).

The other route uses exogenously supplied or mutationally channelled arginine or ornithine (Fig. 6). *proA* and *proB* mutants respond to exogenous arginine and ornithine and also are subject to suppression by the acquisition of additional mutations which permit endogenous arginine or ornithine to be converted to glutamate semialdehyde. This conversion is catalysed by ornithine aminotransferase and arginase and is discussed with arginine and ornithine catabolism. These locus specific suppressor mutations of *proA* and *proB* mutations have been identified and characterised and are discussed in the section on arginine regulation see also Piotrowska *et al.*,1969; Węgleński, 1966; Cybis *et al.*,1972b.

Pyrimidine

The pathway of pyrimidine biosynthesis has been extensively studied in a number of micro-organisms including *Neurospora crassa*, *Saccharomyces cerevisiae* and *Coprinus lagopus* (for a review see O'Donovan & Neuhard, 1970). Recently, Palmer & Cove (1975) described preliminary studies of pyrimidine biosynthesis and control in wild type and pyrimidine auxotrophic mutants of *Aspergillus nidulans*. Their genetical approach revealed eight complementation groups are involved in the syntheses of uridine-5-phosphate (Fig. 7.). Complementation groups *pyrA*, *pyrB*, *pyrC* map in the cluster along with *pyrN* mutants, the latter do not complement with *pyrA*, *pyrB* or *pyrC*. *pyrA* lack uridine carbomoyl-phosphate synthase (E.C. 2.7. 2.5.) while *pyrB* and *pyrC* mutants lack aspartate carbomoyltransferase activity (E.C.2.1.3.2.), the first two enzymes of the pyrimidine biosynthetic pathway. *pyrN* mutants lack both these activities. The evidence suggests that this complex region codes for products, probably three discrete peptides which associate to form an enzyme aggrega as in *N.crassa*, *S.cerevisiae* abd *C. lagopus*. The enzyme activities of the other classes of pyrimide auxotrophs have not been determined, but on the basis of intermediate accumulation and growth response patterns, Palmer & Cove (1975) concluded that *pyrD* mutants lack dihydro-orotase activity (E.C.3,5,2,3), *pyrE* mutants lack orotate reductase (E.C.1.3.3.1.), *pyrF* mutants lack orotate phosphoribosyl transferase (E.C.2.4.2.10.) and *pyrG* mutants lack orotidine-5'-phosphate decarboxylase (E.C.4.1.1.23).

Purines

Although a large number of adenine auxotrophs have been known for some time in *A.nidulans* (Pontecorvo *et al.*, 1953; Calef, 1957; Kafer, 1958; Dorn, 1967) they have received much less attention than comparable mutants in other micro-organisms, e.g. yeast. Adenine requiring mutants map at nine non-linked or loosely linked loci, *adA*, *B*, *C*, *D*, *E*, *F*, *G*, *H* and *I* (Fig. 8). Only four of these loci have been related to specific functions. Mutation in the *adA* gene leads to defective adenylosuccinase (E.C. 4.3.2.2.) activity (Foley *et al.*,1965; Bainbridge *et al.*,1966). The *adA* is likely to be the structural gene since conditional mutants, which are osmotic remedial or temperature sensitive, map in the *adA* locus,and each have been shown to produce a labile enzyme with varying responses to osmotic and temperature differences.

Mutation in *adB* leads to low levels of AMP synthetase (E.C.6.3.4.9.) activity. *adB* is probably the structural gene although there is no direct evidence of this (Foley *et al.*, 1965). *adA* and *adB* mutants only respond to adenine and differ from adenine auxotrophic mutants at other loci *adC*, *D*, *E*, *F*, *G*, *H* and *I*, in that the latter also respond to hypoxanthine plus histidine and therefore blocked in the synthesis of inosinate.

One mutant at the *adD* locus was able to grow if supplemented with very low concentrations of adenine (Katz & Rosenberger, 1970). Paradoxically, it was found that the cells contained ten times more purine than that supplied in the medium, indicating that adenine was being synthesised *de novo*. Moreover, it was found that growth could be completely restricted by the absorption of CO_2 in the culture flask. This and other evidence e.g. accumulation of AIR suggested that the mutant was altered in the AIR carboxylase and requires CO_2 for activity. A mutant at the *adC* locus also accumulates AIR but does not respond to CO_2. Katz & Rosenberger (1970) suggest that *adC* mutants may also have an altered AIR carboxylase which does not respond to CO_2, *adC* being another cistron for the AIR carboxylase enzyme.

Much less is known about the other adenine loci. *adE*, *F*, *G* and *H* mutants do not accumulate imidazoles and are likely to be blocked before AIR formation (Foley *et al.*,1965). An adenine requiring mutant of *A.nidulans* has been isolated which has a red pigment and therefore may be similar to the red adenine auxotrophs of many other micro-organisms. However, nothing is known about its allelism with other adenine auxotrophs and moreover, many adenine mutants produce a pinkish red pigment (Sadasivam *et al.*,1969).

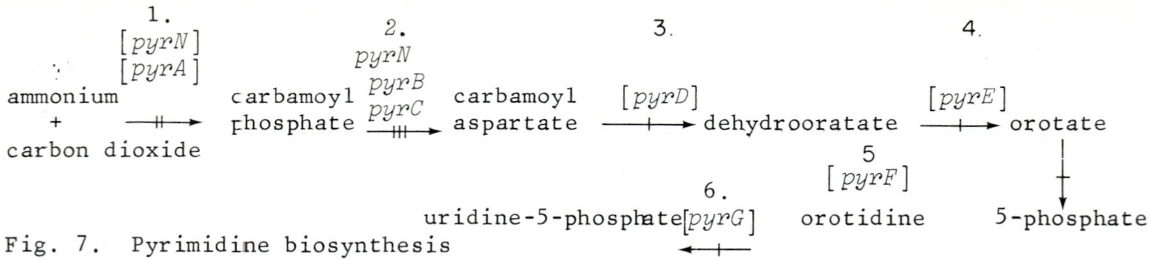

Fig. 7. Pyrimidine biosynthesis

Step	Enzyme system	Mutant locus	Linkage group	References
1.	Carbamoyl phosphate synthase	[*pyrA*]	VIII	Palmer & Cove, 1975
2.	Aspartate carbamoyl transferase	*pyrN*	VIII	
		pyrB	VIII	Palmer & Cove, 1975
		pyrC	VIII	
		pyrN	VIII	
3.	Dihydroorotase	[*pyrD*]	VIII	Palmer & Cove, 1975
4.	Dihydroororatate reductase	[*pyrE*]	I	Palmer & Cove, 1975
5.	Orotate phosphoribosyl transferase	[*pyrF*]	I	Palmer & Cove, 1975
6.	Orotidine-5-phosphate	[*pyrG*]	I	Palmer & Cove, 1975

Polyamines

The metabolic functions of polyamines, such as putrescine, spermine or spermidine, in fungi are unknown although there is some evidence to connect polyamine syntheses with RNA and possibly DNA synthesis and the onset of growth in a number of organisms (for a review see Stevens, 1970). Sneath (1955) isolated a putrescine auxotroph, (*puA*) in *Aspergillus nidulans*. This mutant can also respond to spermidine, but only at much higher concentrations than putrescine. Other polyamines such as butanediol, pyrrolidine etc. or monoamines such as methylammonium do not allow normal growth of the *puA* strain. Stevens and co-workers have investigated polyamine metabolism in relation to germinating conidia. Stevens (1975) found that putrescine is required for germination of conidia of the *puA1* mutant but it may not be necessary for subsequent hyphal growth. In the wild type *A.nidulans* (BWB 272) the putrescine is provided by increased levels of ornithine decarboxylase (E.C. 4.1.1.17.) activity which occurs concurrently with increased RNA synthesis and the putrescine auxotrophy of the *puA*, strains apparently results from defective ornithine decarboxylase activity. The repair of the *puA* mutant by putrescine as well as spermine and spermidine was investigated by Hope & Stevens (1976). They found spermidine and spermine initiate growth at a concentration of ten and one hundred times greater than that of putrescine respectively. They also compared the amount of polyamine taken up by *puA1* cultures growing on the different polyamines. Forty times more putrescine was taken up over a nine hour period by cultures growing on putrescine compared with the spermidine or spermine taken up by comparable cultures. Spathas (personal communication) has found that in twenty-hour cultures of both the wild type and the *puA2* mutant the rate of uptake of putrescine (measured over a five minute period) is approximately three times faster than the rate of uptake of spermidine at the same molarity. Spermine is taken up at a slightly lower rate than spermidine. Uptake of all three polyamines is enhanced by nitrogen starvation and repressed by ammonium or nitrate. In addition, Spathas (personal communication) has selected a mutant (designated *spsA1*) derived from the *puA2* strain which is able to utilize spermidine at comparable concentrations to putrescine. This mutant, however, does not appear to differ significantly from the original *puA2* strain in its uptake of polyamines.

The major polyamines present in *A.nidulans* are spermidine and spermine(Bushell & Bull, 1974; Winther & Stevens, 1976) which is the pattern common to most eukaryotes and in contrast to prokaryotes where putrescine and spermidine usually predominate (Cohen, 1971). Also, the enzyme S-adenosyl methionine (SAM) decarboxylase (E.C. 4.1.1.50.) which catalyzes a step in spermidine and spermine synthesis is activated by putrescine (Winther & Stevens, 1976). In general the eukaryote SAM decarboxylases are putrescine activated

Step	Mutant locus	Linkage group
Ribulose - 5 - phosphate (P-5-P)		
↓		
Phosphoribosyl-1-pyrophosphate (PRPP)	*adE*	I
↓		
Phosphoribosylamine (PRA)	*adF*	I
↓		
Glycinamide ribonucleotide (GAR)	*adG*	I
↓		
Formyl glycinamide ribonucleotide (FGAR)	*adH*	II
↓		
Formyl glycinamidine ribonucleotide (FGAM)		
↓		
Aminoimidazole ribonucleotide (AIR)	[*adC*]	II
↓	[*adD*]	II
Aminoimidazol-N-succinocarboxamide ribonucleotide (SAICAR)		
↓		
Aminoimidoxolecarboxamide ribonucleotide (AICAR)		
↓		
Formamidoimidoxolecarboxamide ribonucleotide (FAICAR)		
↓		
Hypoxanthine ← Inosine ← Inosinate (IMP)		
↓	*adB*	VIII
Adenylosuccinate (AMPS)		
↓	*adA*	I
Adenylate (AMP)		

Fig. 8. Purine biosynthesis

whereas the prokaryote SAM decarboxylases are Mg^{++} dependent but putrescine insensitive (Coppoc, Kallio & Williams-Ashman, 1971). Like ornithine decarboxylase there is also an increase in SAM decarboxylase activity in germinating conidia and this supplies increased levels of spermidine, which appears to be necessary for germination (Stevens *et al.*, 1976). Also the existence of at least seven putrescine auxotrophs (at one locus) makes it unlikely that *A. nidulans* possesses a second pathway to putrescine via arginine decarboxylase such as exists in prokaryotes (Tabor & Tabor, 1972).

TRANSPORT SYSTEMS

Nitrogenous compounds constitute one of the largest and most diverse groups of substances which are transported through the *Aspergillus* cell membrane. Such transport may be partly facilitated diffusion but in most investigated cases is an active transport process. Kinetic and genetic studies have demonstrated the existence of a number of active systems in fungi varying in their specificities and affinities for a range of nitrogenous substances. Transport mutants have been isolated by a variety of methods. Perhaps the most successful is the selection of mutants resistant to toxic analogues such as p-fluorophenylalanine. Many toxic analogues are transported into the cell by one of the normal transport systems. Consequently, a mutation which reduces transport efficiency may confer resistance by keeping the toxic analogue out of the cell. Transport mutants may also be obtained by selecting for poor growth on certain nitrogen or/and carbon sources. The glutamate transport mutant *aauA1*, for example, grows poorly with glutamate as sole nitrogen source and even less well on glutamate as the sole carbon and nitrogen source. This method does not always succeed for a number of reasons: at the concentration of sole nitrogen source required to sustain growth, the compound may simply 'flood' into the cell by passive diffusion; or there may be overlapping transport systems which would all have to be inactivated for poor growth to result. A number of specific transport mutants have been isolated (Table 2) and these, as well as non-specific mutants, are discussed below.

Inorganic nitrogen

Despite their importance, comparatively little work has been done on the transport of nitrate and nitrite in *Aspergillus* or any other fungus. The main reason for this is the lack of suitable radioactive isotopes of nitrogen, which severely limits the resolution of uptake studies. In contrast, there have been a number of reports of ammonium transport in fungi (MacMillan, 1956; Hackette *et al.*, 1970). In *Aspergillus nidulans* an active transport system for ammonium and methylammonium has been characterized using ^{15}N labelled ammonium and ^{14}C labelled methylammonium. Several classes of mutants are known which exhibit abnormal levels of ammonium transport. Mutation in the *meaA* and *mod-meaA* genes result in low transport rates and an altered Km for ammonium and methylammonium, and consequently these genes are thought to determine membrane proteins.

A transport system, which is specific for urea and thiourea, has been reported for *A.nidulans*. Mutants, *ureA*$^-$ (formerly *uruA)*, resistant to the toxic urea analogue, thiourea, have been isolated and these are defective in urea and thiourea transport, but normal for ammonium and amino acid transport. The urea transport system is regulated by ammonium and this is discussed in the article on regulation.

Amino acids

Some genetical studies of amino acid transport have been done in *A.nidulans* and the results are summarized in Table 2. In *Aspergillus nidulans* there are examples of fairly specific transport systems; for example an acidic amino acid system investigated in germinated conidia and young mycelial cells. The specificity of this system is confirmed by a mutation at the *aauA* locus which results in low transport rates for L-glutamate and L-aspartate alone. The *aauA* gene may determine a specific component such as a carrier protein for the acidic amino acid transport site. Other specific amino transport mutants are *prnA* and *gabA* for proline and γ-amino-n butyrate transport respectively. A transport system for basic amino acids has been described but not extensively characterized, although recently a mutant lacking this uptake system has been isolated (Węgleński, personal communication). Mutations have been described in *A.nidjlans* which result in the simultaneous loss of several different amino acid transport activities. These mutations map at the following loci: *aauB*, *aauC*, *aauD* (Kinghorn & Pateman, 1975b) and *fpaD* (Sinha, 1969). Several hypotheses can be advanced to explain these. The *aauB*, *aauC* and *aauD* and *fpaD* loci may determine membrane proteins which are important structural components shared by several or all amino acid transport sites. An alternative explantion is that a general amino acid transport system exists in *Aspergillus* as reported for other fungi including *Neurospora crassa* (Rao & DeBusk, 1975) and one or more of these genes code for this system.

Purines

Mutants, impaired in the transport of purines, have been selected on the basis of resistance to certain toxic purine analogues (Darlington & Scazzocchio, 1967). At least three systems are involved in purine transport as shown by genetical analysis (Table 2). First, mutations in the *azg* gene lead to impaired transport of adenine, hypoxanthine, and guanine. These strains show normal transport of xanthine and urate. Mutants were isolated which were found to be impaired in the transport of xanthine and urate, whilst retaining normal transport of adenine, hypoxanthine and guanine. These map in the *uapA*, *uapB*, *uapC* genes. Finally, a mutation in the *alp* gene leads to defective transport of allantoin, whilst retaining wild type transport of adenine, xanthine, hypoxanthine, quanine and urate.

General aspects of transport

There are a number of possible functions for genes known to be involved in some way in the determination of transport. Many may determine proteins (permeases) which have a specific function in the transport of a single substance, while others will determine proteins with a transport function specific for a group such as the acidic amino acids. But in no case so far has a specific transport function for a membrane protein been directly demonstrated although some work has been directed towards this goal (Pateman, unpublished work). There are other probable functions for genes with apparent involvements in transport, especially those genes in which mutation can result in multiple transport defects. They might determine enzymes concerned with some aspect of metabolism, affecting energy systems coupled with transport, or they might be regulatory genes. Most mutations affecting transport in *Aspergillus* are recessive. However, a few dominant or partially dominant mutants are known such as *aauC*, *aauD* (Kinghorn & Pateman, 1975b) and *fpaD* (Sinha, 1969) and these may be regulatory genes. In an alternative, and perhaps more attractive hypothesis taking into account the number of genes involved, the dominance relationship might be explained by supposing that the genes *aauC*, *aauD* and *fpaD* code not for specific transport proteins, but for structural membrane proteins and these hold the specific transport proteins in the membrane. A heterozygous diploid containing the normal and mutant alleles specifies two forms of structural membrane protein and various combinations of these are built into the membranes. If such a compound membrane was unable to support more than a few of the specific transport proteins this would be apparent as complete or partial dominance of the mutant.

Recent studies in *Aspergillus* have shown that transport systems may be regulated by induction and/or nitrogen and carbon repression. These aspects are discussed in other contributions.

CATABOLIC PATHWAYS

It is only fairly recently that attention has been given to

Table 2. Specific transport mutants in *Aspergillus*

Group	Transport system	Specificity	Mutant locus	Linkage group	References
Inorganic nitrogen	ammonium	Takes up ammonium and methyl-ammonium	*meA* *mod*, *meaA*	IV	Pateman *et al*.1973; Arst & Page, 1973; Pateman *et al*. 1974a
	urea	Takes up urea and thiourea	*ureA*	VIII	Dunn & Pateman, 1972.
Amino acids	L-acidic	Takes up L-glutamate, L-aspartate and L-cysteate	*aauA*	VII	Sinha, 1969; Kinghorn & Pateman 1972; Hynes, 1973; Robinson *et al*.1973,a,b; Pateman *et al*.1974b; Kinghorn & Pateman, 1975b; Kinghorn & Pateman, 1975b
	L-basic	Takes up L-arginine and possibly L-lysine and L-ornithine			Węgleński-personal communication Cybis & Weglenski, 1969
	proline	Takes up proline only	*prnA*	VII	Arst & MacDonald, 1975; *ibid*
	γ-amino-n-butyrate	Takes up γ-amino-n- butyrate only	*gabA*	VI	Arst, 1976; *ibid*
Purines	adenine	Takes up adenine, hypoxanthine, guanine	*azgA*	V	Darlington & Scazzocchio, 1967
	xanthine	Takes up xanthine and urate	*uapA* *uapB* *uapC*	I VII I	Darlington & Scazzocchio, 1967; Scazzocchio & Gordon. *ibid*. Scazzocchio & Gordon *ibid*.
	allantoin	Takes up allantoin only	*alp*	VIII	Darlington & Scazzocchio, 1967

uapA is unlinked with *uapC*

genetical and biochemical aspects of catabolic systems. A number of organic substances can be utilised as nitrogen and/or carbon sources in *A.nidulans* by catabolic routes which are for the most part, regulated by ammonium or carbon control. Catabolic routes may become indispensable when only one or two organic substances are the main or only nitrogen sources. Under these conditions mutations in genes specifying catabolic systems lead to poor growth. Thus, altering the cultural conditions, the organism can be forced to open certain pathways and this has been used as a tool in isolation of catabolic mutants in *A.nidulans*. In *Aspergillus nidulans* a number of breakdown systems have been worked out recently, viz, glutamate, asparagine (discussed below) and purine, proline and arginine which will also be described elsewhere. In addition a start has been made on the elucidation of systems whose functions are necessary for amide aspartate, amine, glutamate, glutamine, histidine and urea catabolism. These are discussed below.

Glutamate

Mutants were isolated which lack NAD-linked glutamate dehydrogenase (NAD-GDH) (E.C.1.4.1.2.) activity but have wild-type levels of NADP-GDH activity (Fig.9-Step 4). These map at the *gdhB* locus, which is not linked to *gdhA*, the structural gene for NADP-GDH, and indeed is located on a different linkage group. Several lines of evidence indicate that *gdhB* specifies a structural component for NAD-GDH. The growth characteristics of mutants lacking NAD-GDH activity are interesting since (i) they indicate the routes by which amino acids are utilised as nitrogen and carbon/energy sources, and (ii) they show that NAD-GDH has a deaminating or catabolic role in fungi, as has been previously suggested on biochemical evidence alone. The utilization of glutamate as the sole nitrogen source is extremely poor in *gdhB* mutants - this is likely to be due to a low ammonium yield from glutamate. In contrast, the wild type grows with glutamate as the sole nitrogen source since ammonium is generated via NAD-GDH activity. Presumably, ammonium is required for such steps as the formation of glutamine by glutamine synthetase NADP-GDH activity does not substitute for NAD-GDH activity in *gdhB* mutants, probably due to either or both of the following reasons. First, glutamate as a nitrogen source results in low repressed levels of NADP-GDH activity. Second, the efficiency of NADP-GDH with respect toglutamate deamination is low as judged by the equilibrium of the reaction, which favours L-glutamate synthesis and the relatively high K_m of NADP-GDH for glutamate. Poor growth of mutant *gdhB1* with aspartate, proline, ornithine, alanine, phenylalanine, or isoleucine as the sole nitrogen source indicates that these amino acids are mainly utilized via glutamate and that ammonium liberation is mainly a result of NAD-GDH activity. Wild-type growth by mutant

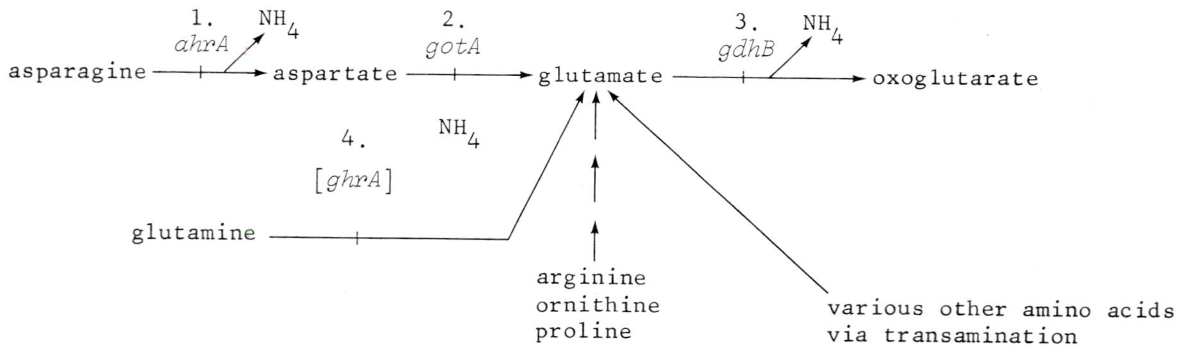

Step	Enzyme system	Mutant locus	Linkage group	References
1.	Asparaginase	*ahrA*	VIII	Drainas, Kinghorn & Pateman, 1976, 1977 Drainas & Pateman, 1977.
2.	Glutamate-oxolacetate amino-transferase	*gotA*	I	Kinghorn - unpublished work.
3.	NAD glutamate dehydro-genase	*gdhB*	IV	Kinghorn & Pateman, 1973, 1974b, 1976. Arst & Cove, 1973.
4.	Glutaminase?	[*ghrA*]	I	Kinghorn - unpublished work

gotA and *ghrA* are unlinked.

Fig. 9. Amino acid catabolism

gdhB1 with the other amino acids and bases tested indicates that ammonium is obtained from these by a preceding step (e.g. glutamine is broken down to glutamate and ammonium by glutaminase) or an alternative route (e.g. glycine is deaminated to glyoxylate and ammonium). The formation of ammonium by these systems allows mutant *gdhB1* to grow like the wild type. The reason mutant *gdhB1* grows normally on certain amino acids but fails to conidiate is unclear. Poor growth of mutant *gdhB1* with glutamate as the carbon and nitrogen source is likely to result from carbon starvation. Similarly, the poor growth of mutant *gdhB1* with aspartate, alanine, proline, ornithine, phenylalanine, isoleucine, glutamine, asparagine, valine, arginine, tyrosine, leucine, or casamino acids indicates that NAD-GDH activity is required for these amino acids to be utilized as the carbon source. This means that the amino group of these amino acids is first transferred to oxoglutarate to form glutamate, and subsequently NAD-GDH activity regenerates oxoglutarate from glutamate. The keto acid derived from the starting amino acid is then utilized as a carbon source in which, eventually, the carbon is utilized via the central metabolic routes.

Aspartate

Our unpublished work on *Aspergillus nidulans* suggests that there are two forms of glutamateoxaloacetateaminotransferase (E.C. 2.6.1.1.) activity (GOT) (Fig. 9-Step 2). One form of GOT is important in synthesis and the other in the breakdown of aspartate. Recessive mutations in the *gotA* gene leads to low activity of the deaminating form as shown by starch gel electrophoresis. *gotA* mutants are unable to utilise aspartate but show wild type growth with glutamate as carbon and nitrogen source.

Asparagine

The major route of asparagine utilisation in micro-organisms is catalysed by asparaginase (E.C.3.5.1.1.) with aspartate and ammonium formation (Fig.9-Step 1). Mutations at the *ahrA* locus isolated on the bases of resistance to a toxic analogue of asparagine, aspartic hydroxamate, lead to low levels of asparaginase activity. The toxic substance is not aspartic hydroxamate itself but hydroxylamine produced by asparaginase activity. *ahrA* mutants are resistant to aspartic hydroxamate but not to hydroxylamine. There is a spectrum of aspartic hydroxamate resistance depending on the nitrogen source and the concentration of aspartic hydroxamate. Furthermore, there is a direct correlation between aspartic hydroxamate resistance and asparaginase activity and this is supported by weakly-resistant mutants which have reduced asparaginase levels. *ahrA* is probably the structural gene for an asparaginase although there is no evidence for this except that all aspartic hydroxamate resistant mutants locate in the *ahrA* gene. Paradoxically, *ahrA* mutants grow as wild type with asparagine as sole nitrogen source indicating that asparagine can be metabolised by another system. Furthermore, this is probably not asparagine synthetase since the double mutant between *ahrA1* and the asparagine auxotroph *asnA2*, which probably lacks asparagine synthetase activity, grows like the single mutant *ahrA1*.

Glutamine

Mutants (designated *ghrA*) resistant to glutamate hydroxamate, a toxic analogue of glutamine, have been isolated (Fig.9-Step 4). *ghrA* mutants are recessive and are able to utilise glutamine as sole nitrogen source. By analogy with *ahrA* mutants *ghrA* mutations possibly lead to low levels of glutaminase activity. However, this proposition has not been tested, since glutaminase activity has not been detected in wild type strains (Hynes, Kinghorn, Drainas, unpublished work). *ghrA* mutants grow like wild type on glutamine as sole nitrogen source.

A second class of glutamic hydroxamate resistant mutants (*ghrB*) were isolated and are cross resistant to aspartic hydroxamate but not to other amino acid toxic analogues, D-serine or p-fluorophenyalanine. It is possible that *ghrB* mutants have poor asparagine and glutamine transport although they grow as wild type with glutamine and asparagine as sole nitrogen or as sole nitrogen and carbon sources.

Histidine

The utilization of histidine as a nitrogen source in *Aspergillus* is mainly through urocanate and ammonium, this step being carried out by histidase (E.C.4.3.1.3.) (Polkinghorne & Hynes, 1975). There is no evidence that histidine is metabolised further than urocanate (Hynes, personal communication). Histidine is extremely poor as a sole nitrogen source for wild type cells and this appears to be due to limited synthesis of histidase. The *areA102* mutation leads to elevated levels of histidase activity and consequent better growth with histidine as sole nitrogen source. Although no structural gene mutants for histidase have been isolated, mutations in the *sarA* and *sarB* loci lead to poor growth of strain *areA102* and this is correlated again with a reduction of the maximal levels of histidase activity under optimum conditions. Furthermore, these mutations also result in poor growth with a number of other amino acids as sole nitrogen sources. Polkinghorne & Hynes (1975), from the pleiotrophy shown by *sar* mutants, favour the hypothesis that *sarA* and *sarB* loci have a regulatory function.

γ-Amino-n-butyrate

γ-amino-n-butyrate (GABA), β-alanine, δ-amino-n-valerate and certain other ω-amino acids are first catabolised by transamination to glutamate plus succinate semi-aldehyde, catalysed by γ-amino-n-butyrate amino-transferase (E.C.2.6.1.19.) (Arst 1976) (Fig. 10). Mutations in the *gatA* gene lead to loss of GABA aminotransferase activity; Arst (1976) suggests that *gatA* is the structural gene for GABA aminotransferase. The lack of GABA aminotransferase activity results in poor growth of mutant *gatA* strains with GABA as sole carbon or nitrogen source and β-alanine δ-amino-n-valerate as sole nitrogen sources. Glutamate formed from GABA aminotransferase activity is catabolised by NAD-GDH while succinic semialdehyde is reduced to succinate catalysed by succinic semialdehyde dehydrogenase (E.C. 1.2.1.16.) activity.

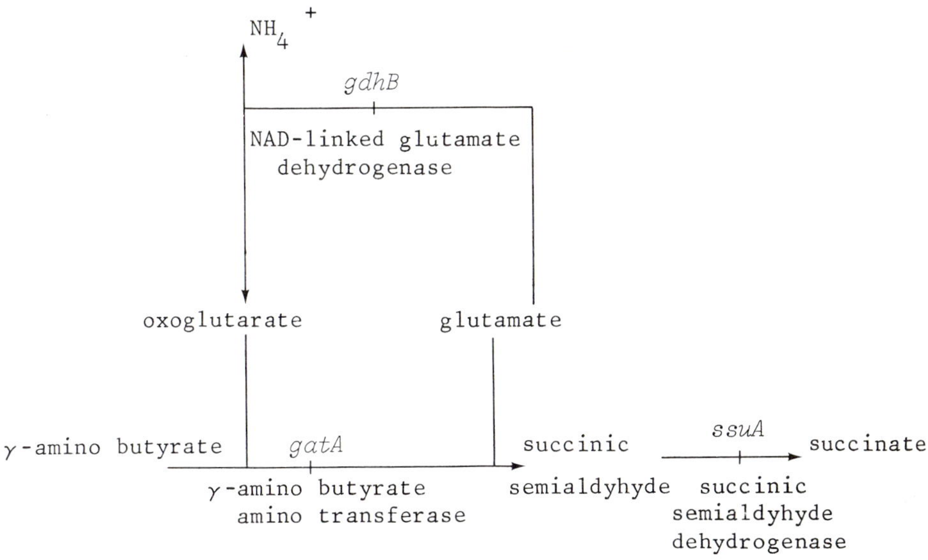

Fig. 10 γ-Amino butyrate catabolism
Modified from Arst (1976).

Arginine, ornithine and proline

The catabolism of arginine and ornithine proceeds as shown in Fig.11. Two enzymes, arginase (E.C..3.5.3.1.) and ornithine-γ-aminotransferase (E.C.2.6.13.) are required for the catabolism of arginine to ornithine and urea, and subsequently of ornithine to glutamate-γ-semialdehyde respectively in *A.nidulans*. Mutations in the *agaA* gene lead to impaired arginase activity and consequent ability to utilise arginine since neither ornithine nor urea is produced (Węgleński, personal communication). Mutations in the gene *otaA*, result in loss of ornithine-γ-aminotransferase activity. These mutants fail to grow on ornithine as a sole nitrogen source but are able to utilise arginine, presumably from the urea formed. In *A.nidulans* arginase and ornithine-γ-aminotransferase activities are inducible by arginine (Cybis & Węgleński, 1972; Stevens & Heaton, 1973; Stevens *et al.*,1976) and repressible by ammonium and/or glucose (Bartnik & Węgleński, 1973, 1974; Bartnick *et al.*, 1973a,b). Proline can be produced from exogenous arginine by the action of arginase and ornithine aminotransferase. The first step in proline catabolism is catalysed by proline oxidase and results in an equilibrium mixture of Δ' pyrroline-5-carboxylate and glutamate-γ-semialdehyde. Mutations in *prnA*, and *prnD* genes lead to impaired proline oxidase activity and consequent poor growth with proline as sole nitrogen source. Δ'pyrroline-5-carboxylate dehydrogenase activity, which catalises the oxidation of Δ'pyrroline-5-carboxylate with glutamate formation, is also abolished in *prnA* and in *prnC* strains (Arst & MacDonald, 1975). Glutamate formed from arginine, ornithine or proline catabolism is deaminated by NAD-linked glutamate dehydrogenase (see glutamate catabolism) while urea is catabolised by urease with ammonium production in both cases.

Purines

The purine degradation pathway is shown in Fig. 12 and for a general review see Vogel & Van Der Drift (1976). Studies of *Aspergillus nidulans* mutants have shown that hypoxanthine is converted to urate by the enzyme xanthine or hypoxanthine dehydrogenase (E.C.1.2.3.2.) which can use both xanthine and hypoxanthine as substrates (Fig. 12-Step 1). Urate is deaminated to ammonium via allantoin, allantoate and urea (Fig. 12). Darlington, Scazzocchio & Pateman (1965) isolated a number of mutant strains able to use ammonium, but unable to use hypoxanthine as nitrogen source, these are discussed below.

Mutants unable to utilise hypoxanthine, but able to utilise urate were isolated. As well as mutants in the *cnx* genes (see Section 2), mutants in two further genes, *hxA* and *hxB*, which differed from *cnx* mutants in being unaffected in their utilisation of nitrate, were made. All these classes of mutants were shown to have abnormally low levels of xanthine dehydrogenase (Scazzocchio, personal communication). It was proposed by Scazzocchio & Darlington (1967) that the *hxA* gene was involved in the induction of xanthine dehydrogenase, while the *hxB* gene was involved structurally in xanthine dehydrogenase. Recent work (Scazzocchio, Holl &

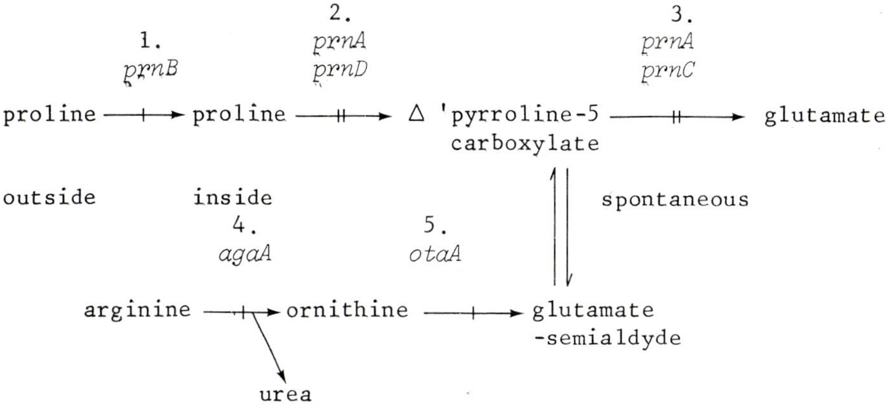

Fig. 11. Arginine, ornithine and proline catabolism

Step	Mutant locus	Linkage group	References
1. Proline transport	*prnB*	VII	Arst & MacDonald, 1975
2. Proline oxidase	*prnA*,	VII	Arst & MacDonald, 1975
	prnD	VII	Arst & MacDonald, 1975
3. Pyrroline-5-carboxylate dehydrogenase	*prnA*,	VII	Arst & MacDonald, 1975
	prnC	VII	Arst & MacDonald, 1975
4. Arginase	*agaA*	VI	(Guzewska - unpublished) Weglenski, 1966, 1967; Piotrowska *et al.*,1969; Cybis & Weglenski, 1972; Cybis *et al.*,1972b; Bartnik *et al.*,1973a,b; Stevens & Heaton, 1973; Bartnik & Weglenski,1973; 1974; Stevens *et al.*1976.
5. Ornithine aminotransferase	*otaA*	VII	(Guzenska - unpublished) Weglenski,1966,1967; Piotrowska *et al.*,1969; Cybis & Weglenski, 1972; Cybis *et al.*,1972b; Bartnik *et al.*,1973a,b; Stevens & Heaton, 1973; Bartnik & Weglenski, 1973; 1974; Stevens *et al.*1976.-

prnA, *prnB*, *prnC* and *prnD* genes are tightly linked.

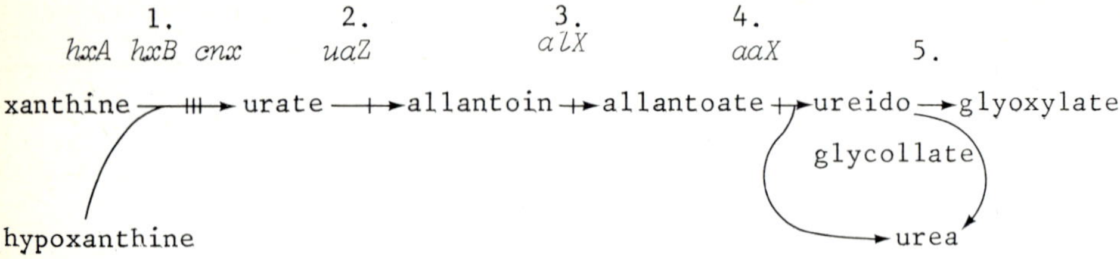

Fig. 12. Purine catabolism

Step	Enzyme system	Mutant locus	Linkage group	References
1.	Xanthine dehydrogenase I	*hxA*	V	Darlington *et al*.1965;
		hxB	VII	Darlington,1966;
	Xanthine dehydrogenase II	*hxB*	VII	Scazzocchio, 1966; Darlington & Scazzocchio, 1967; Scazzocchio & Darlington, 1967, 1968
2.	Urate oxidase	*uaZ*	VIII	Darlington *et al*.1965; Darlington, 1966; Scazzocchio, 1966; Darlington & Scazzocchio, 1967; Scazzocchio & Darlington, 1967, 1968
3.	Allantoinase	*alX*	III	Darlington *et al*.1965; Darlington, 1966; Scazzocchio, 1966; Darlington & Scazzocchio, 1967; Scazzocchio & Darlington, 1967, 1968.
4.	Allantoicase	*aaX*	VIII	Darlington *et al*.1965; Darlington, 1966; Scazzocchio, 1966; Darlington & Scazzocchio, 1967; Scazzocchio & Darlington, 1967, 1968
5.	Ureidoglycollase	no mutants		Darlington *et al*.1965; Darlington, 1966. Scazzocchio, 1966; Darlington & Scazzocchio, 1967; Scazzocchio & Darlington, 1967, 1968

uaZ and *aaX* are unlinked.

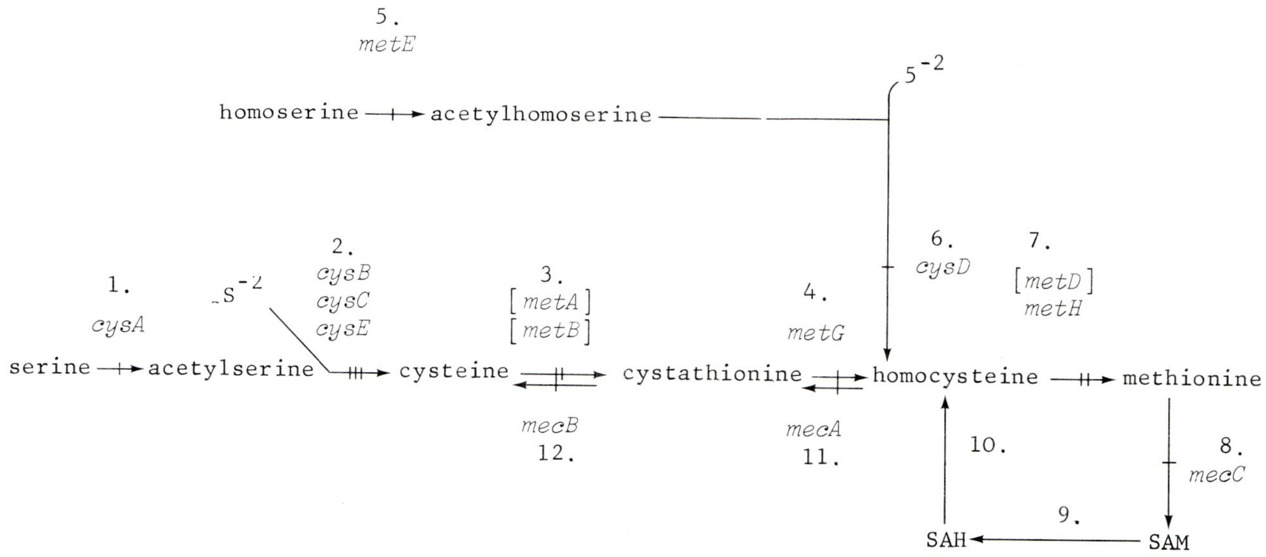

Fig. 13. Sulphur amino acid metabolism

Modified after Stępień *et al.*, (1975).

Fig.13 cont'd

Step	Enzyme system	Mutant locus	Linkage group	References
1.	Serine acetyltransferase	*cysA*	I	Pieniążek *et al.*,1974; Paszewski & Grabski, 1974,1975a; Stępień *et al.*, 1975
2.	Cysteine synthase (Acetylserine sulphydrylase)	*cysB* ≡ *sul02* *cysC* ≡ *sul03* *cysE*		Ayling, 1969; Paszewski & Grabski, 1973, 1975a; Ayling, 1969; Paszewski & Grabski,1973; Pieniążek *et al.*,1973a; Paszewski & Grabski,1975a;Bal *et al.*, 1975; Stępień *et al.*,1975
3.	Cystathionine γ-synthase	[*metA*] [*metB*]	II VI	Gajewski & Litwińska, 1968; Paszewski & Grabski, 1975a; Gajewski & Litwińska, 1968; Paszewski & Grabski, 1975a
4.	Cystathionine β-lyase (β-cystathionase)	*metG* ≡ *metF*	IV	Gajewski & Litwińska,1968; Paszewski & Grabski, 1973, 1975a
5.	Homoserine transacetylase	*metE*	VII	Gajewski & Litwińska,1968; Paszewski & Grabski, 1975a
6.	Acetylhomoserine sulphydrylase (homocysteine synthase)	*cysD*		Paszewski & Grabski,1973,1975a
7.	Tetrahydropteroyltriglutamate methyl transferase	[*metD*] *metH*	III III	Gajewski & Litwińska,1968; Paszewski & Grabski, 1975a,b
8.	Methionine S-adenosyltransferase (SAM synthetase)	*mecC*		Pieniążek *et al.*,1973b
9.	S-adenosylmethionine demethylase	no mutants		Pieniążek *et al.*, 1973b
10.	S-adenosyltransferase	no mutants		Pieniążek *et al.*,1973b
11.	Cystathionine β-synthase	*mecA* = *meth-*i_I	*I*	Pieniążek *et al.*,1973a,b
12.	Cystathionine α-lyase (α-cystathionase)	*mecB*	I	Pieniążek *et al.*,1973a,b, 1974

Mutations in one other locus *metC*I) leads to methionine auxotrophy. *cysA* is unlinked with *mecA* and *mecB*. *MecA* and *mecB* are themselves loosley linked (36 map units).

Foguelman, 1973; Scazzocchio, 1973, 1974) has shown that *A.nidulans* has two xanthine dehydrogenases. It appears that the *hxA* product is required structurally for only one of these activities (xanthine dehydrogenase I), but that the *hxB* product and the co-factor specified by the *cnx* genes is required for both xanthine dehydrogenase I and II.

Two classes of possible structural mutants are impaired in their utilisation of urate whilst remaining normal in their allantoin utilisation. One mutant, designated *uaZ11* lacks urate oxidase (E.C.17.3.3.) but produces xanthine dehydrogenase. Scazzocchio & Darlington (1967) suggested that *uaZ* was the structural gene for urate oxidase. Scazzocchio (personal communication) has isolated five further *uaZ* alleles but none of these is temperature-sensitive.

uaX mutants produce levels of xanthine dehydrogenase and urate oxidase similar to those in wild-type strains grown under similar condition. Scazzocchio (1966)has shown that they are not defective in purine transport. Urate oxidase in several species is part of a particulate system associated with catalase (deDuve & Baudhuin, 1966; Müller & Moller, 1969). If this is so for *A.nidulans* too, then the *uaX* gene product might be involved in the integration of urate oxidase into the multi-enzyme system (Scazzocchio, personal communication).

Mutation in the *alX* gene leads to impaired allantoinase (E.C. 3.5.2.5.) activity. The mutants form a homogenous group which fails to grow on allantoin but grows on allantoate as sole nitrogen source.

The hydrolytic breakdown of allantoate to urea has been reported to proceed through two steps in a number of organisms (Trijbels & Vogels, 1966). By the first step, allantoinase (E.C. 3.5.2.5.) one mole of allantoate yields one mole of urea and one mole of ureidoglycollate, and by the second ureidoglycollase, the ureidoglycollate yields one mole of glyoxylate and another of urea. If a similar situation holds for *Aspergillus* then mutants blocked in the first step would be unable to grow on allantoin. Mutants blocked in the second step would be unimpaired in their utilisation of allantoate, because they would be able to utilise the urea produced by the first step (Scazzocchio, personal communication). Biochemical analysis has shown that both steps are present in *Aspergillus* and that the only mutant, *aaX*, blocked between allantoate and urea lacks allantoinase activity only.

Urea

Urea may be formed from purine and/or arginine catabolism, as described above. Alternatively, urea can be taken up from the surrounding environment by a high affinity specific urea transport system which is lost in *ureA* mutants. Urea is degraded to ammonium possibly by a one step reaction to ammonium by urease (E.C.3.5.1.5.) in *Aspergillus nidulans* (Pateman *et al.*,1964; Scazzocchio & Darlington, 1968 and see Reithel, 1965 for a review). Mutants at three loci *ureB*, *ureC* and *ureD* (formerly *uX*, *uY* and *uZ*) were isolated on the basis of inability to utilise purines as sole nitrogen source. (Pateman *et al.*, 1964; Scazzocchio & Darlington, 1968). Furthermore, growth on all intermediates of the purine degradation pathway including urate, allantoate and urea is poor. In contrast the mutants grow normally with ammonium. All three types of mutants lack urease activity accounting for their poor growth on urea and purines as sole nitrogen sources. A further class of urease deficient mutants,*ureE*, has been isolated (Kinghorn, unpublished work). *ureE3* grows poorer than wild type with urea but significantly better than *ureB*, *ureC* and *ureD* although it lacks *in vivo* and *in vitro* urease activity. The *ureE3* mutation is recessive in heterozygous diploids as are *ureB*, *ureC* and *ureD* mutations. However, *ureE3* differs from *ureB*, *ureC* and *ureD* mutants in urea growth responses and also in selenourea resistance patterns. *ureB*, *ureC* and *ureD* mutants are resistant but *ureE3* is completely sensitive. The roles of these four genes are under investigation. Four loci affecting urease activity are known in *Neurospora crassa* (Kolmark, 1969a,b; Haysman & Branch Howe, 1971).

Amides

The ability of*A.nidulans* to metabolise a large variety of amides as sole nitrogen (and carbon) sources has been extensively studied (Hynes & Pateman, 1970c). There are at least three amidases (E.C.3.5.1.4.) which have been fairly well characterised genetically. First, formamidase whose principal substrate is formamide is non-inducible but subject to ammonium repression and weak carbon repression. Formamide in only used as a nitrogen source. A mutant leading to loss of formamidase (*fmdS1)*has been obtained and it has been suggested that the lesion is in the structural gene. Second, acetamidase which hydrolyses a number of aliphatic amides including acetamide as carbon or as nitrogen sources and is subject to carbon and nitrogen repression and to more than one inducer. Mutants lacking acetamidase have been isolated by selecting for resistance to the toxic analogue fluorocetamide. These mutants map at the *amdS* locus which is likely to be the structural gene for acetamidase since thermolability of the enzyme protein has been demonstrated in an *amdS* mutant carrying a temperature sensitive allele. Moreover, it has been shown that not only does the temperature sensitive *amdS* allele lead to thermolabile enzyme activity but it also renderssynthesis of the enzyme temperature sensitive. They suggest that this mutation affects the ability of subunits to assemble at 40°C but not at 25°C. Third, a general amidase, which has rather similar substrates as acetamidase, is responsible for the utilisation of amides as sole nitrogen sources and consequently is only subject to ammonium repression. Mutation in the gene *gmdS* (unlinked with *amdS* and *fmdS*) leads to loss of the general amidase system and therefore it appears that although the general amidase and acetamidase have common substrates, they are distinct systems specified by different loci and regulated by different means. Finally, a fourth system, capable of breaking down longer chain aliphatic amides such as valeramide and hexamede, may exist although there is no genetical evidence.

Amines

Arst & Cove (1969), Page & Cove (1972) have studied genetical aspects and control of monoamine oxidase (E.C.1.4.3.4.) which uses small *n* - alkylamines, such as methylammonium (methylamine) as substrates. Mutants in 3 loci (*mauA*, *mauB*, *mauC*) were found to have reduced amine catabolism with consequent poor growth on monoamines as sole nitrogen sources. Recessive mutations in the unlinked *mauA* and *mauB* genes lead to impaired monoamine oxidase activity. The *mauA* locus is likely to be a structural gene since mutants carrying temperature sensitive alleles at this locus have increased thermolability of the enzyme. There are no known conditonal defective *mauB* alleles and therefore no definite conclusions can be made about the function of this gene. *mauC* mutants differ from *mauA* and *mauB* in that they show almost wild type monoamine oxidase activity, are semi-dominant and are only expressed in the presence of a putrescine auxotrophic *puA* mutation (Page, 1973).

Polyamines

While *mau* mutants are defective in the catabolism of simple amines such as methylammonium and ethylammonium, these strains can still utilise diamines such as putrescine as sole nitrogen source (Page, 1973; Spathas, personal communication). Putrescine utilization is probably ammonium-repressible since the ammonium repressed *tamA*^{r}*50* mutant does not grow on this nitrogen source. Assays of a diamine oxidase show that this enzyme is lacking in *tamA*^{r}*50* but this may not be the normal route of catabolism of putrescine since other mutants unable to utilize putrescine still possess the diamine oxidase activity (Spathas, personal communication). Spermidine and spermine are not effective nitrogen sources for *Aspergillus nidulans*.

SULPHUR AMINO ACID METABOLISM

Sulphur metabolism in *A.nidulans* has been studied by several groups of workers. These studies reveal the metabolic routes by which inorganic sulphur is initially reduced and amino acids such as methionine, homocysteine and cysteine synthesised and interconverted under various conditions. It is outside the scope of this article to include inorganic sulphur metabolism (for details of the genetics of sulphur reduction see Arst, 1968). Amino acid metabolism has been intensively studied by the Polish groups (see Fig. 13 for references).

Homocysteine

There appear to be two pathways of homocysteine and methionine biosynthesis in fungi. The first is catalysed by cystathionine γ-synthase (E.C. 4.2.99.9.) (Fig.13-Step 3) and cystathionine β-lyase (E.C.4.4.1.8.) (Fig.13-Step 4) and the other by acetylhomoserine sulphydrylase (E.C.4.2.99.7.) (Fig. 13-Step 6). The principle pathway is thought, on genetical evidence, to be the former in *N.crassa* (Kerr & Flavin, 1970) while it is the latter (Cherest *et al.*,1969), in yeast. Paszewski & Grabski (1973) detected cystathionine β-lyase and acetylhomoserine sulphydrylase activities, although not cystathione γ-synthase activity, suggesting the occurrence of both pathways in *Aspergillus nidulans*. The contribution made by both pathways of homocysteine biosynthesis in *Aspergillus* was tested by making mutants lacking these systems (Gajewski & Litwińska, 1968; Putrament, Guzewska & Pieniazek, 1970; Paszewski & Grabski, 1974, 1975a). Mutations in the *cysD* and *metG* genes lead to impaired acetylhomoserine sulphydrylase and cystathionine β-lyase activities respectively. Furthermore, it is likely from accumulation and nutritional data, that *metA* and *metB* mutants lack cystathionine γ-synthase activity. While *metA*, *metB* and *metG* (previously *methA*, *B* and *G*) mutants are auxotrophic (these are repaired by homocysteine or methionine), *cysD* mutants are prototrophic, indicating that although the two pathways exist in *Aspergillus* the main pathway of homocysteine biosynthesis is via cystathionine β-lyase and cystathionine γ-synthase, as in *Neurospora crassa*. Furthermore, the pathway involving acetylhomoserine sulphydrylase is primarily concerned s with cysteine biosynthesis and only found operating in the mutants blocked in the main pathway for cysteine synthesis that is *cysA*, *cysB* and *cysC* (see below). Rather interestingly, homolanthione (a higher homologue of cystathionine) and cystathionine accumulate in methionine auxotrophs (mapping in the *metH* and *metD* genes) which are blocked between homocysteine and methionine (see methionine synthesis) and thereby resembling the situation in similar mutants of *E.coli* (Huang, 1963). The evidence of Paszewski & Grabski (1975b) suggests that homolanthionine is synthesised by the catalytic action of cystathionine γ-synthase activity, as follows:

$$\text{acetylhomoserine + homocysteine} \longrightarrow \text{homolanthionine + acetate}$$

in which homocysteine substitutes for the normal substrate, cysteine.

Methionine

Methionine auxotrophs have been isolated which respond to methionine but not to homocysteine (Gajewski & Litwińska, 1968). These mutations map in the *metD* and *metH* genes. The mutant *metH2* shows lower tetrahydropteroyltriglutamate methyltransferase (E.C. 2.1.1.14.) activity (Fig. 13-Step 7) than the wild type strains. Furthermore, the enzyme from *metH2* is more thermolabile than that from wild type indicating that *metH* is likely to be the structural gene (Paszewski, personal communication). *metD10* lacks the methyl donor (Paszewski & Grabski, 1975a,b). As discussed above, homolanthionine and cystathionine accumulate in these strains.

Cysteine

No less than five routes have been reported for cysteine biosynthesis in micro-organisms. Four of these involve an anabolic route from serine while the fifth channels cysteine from methionine and homocysteine catabolism.

I. serine + acetyl-CoA ⟶ acetylserine + sulphide ⟶ cysteine.

II. serine + sulphide ⟶ cysteine.

III. serine + thiosulphate ⟶ sulphocysteine ⟶ cysteine.

IV. serine + acetyl-CoA ⟶ acetylserine + thiosulphate ⟶ sulphocysteine ⟶ cysteine.

V. homocysteine ⟶ cystathionine ⟶ cysteine

Recently Pieniazek and co-workers (Pieniazek *et al.*, 1973a,b, 1974; Stepien *et al.*, 1975) have been investigating by genetical means the route of cysteine biosynthesis in *Aspergillus*. These detailed studies reveal that probably only two routes operate for cysteine synthesis, one which is anabolic, from serine via acetyserine (route I) and the other catabolic from homocysteine (route V). Other routes have been described in *Aspergillus* namely route II (Pieniązek *et al.*, 1973a) but this reaction was found to be catalized by cystathionine β-synthase like in rat liver. The route III was described by Nakamura & Sato, 1960, 1963, 1965) but the evidence for these is considerably weaker.

Pieniązek and colleagues characterised mutants, mapping at a number of loci, which are defective in the anabolic pathway.

Mutation in the *cysA* gene leads to defective serine acetyltransferase (E.C.2.3.1.30.) activity (Fig. 13-Step 1). *cysB*, *cysC* (isolated by Ayling, 1969) and *cysE* mutants are impaired in cysteine synthase (E.C.4.2.99.8.) activity (Fig.13-step 2); an enzyme which appears to be bound to particles or membranes, (Bal *et al.*, 1975; Cybis, personal communication). *cysB* and *cysC* represent two different loci, while the allelism of *cysE* with respect to *cysB* and *cysC* and the question of structural genes is at the present time unclear (Paszewski & Cybis, personal communication). *cysE1*, the most extensively studied so far, is an extremely interesting mutant and shows a loss of association of the enzyme with particles, an increased Km for acetylserine and other features. This mutant may not affect cysteine synthase itself but rather a subcellular structure to which it is attached (Cybis, unpublished work). This pathway, involving serine acetyltransferase and cysteine synthase, is the main one for cystein synthesis in prototrophic cells. Paradoxically, mutations in genes (*cysA*, *B*, *C* and *E*) leading to the loss of these activities are not auxotrophic for cysteine since they result in consequent low levels of intracellular cysteine which causes derepression of the other route for cysteine synthesis (Paszewski & Grabski, 1975a). This conditional alternative route includes acetylhomoserine sulphydrylase (discussed above) where homocysteine is formed and the so-called trans-sulphurylation pathway in which cysteine is formed from homocysteine via cystathionine. The derepression of acetylhomoserine sulphydrylase in *cysA*,*B*,*C* and *E* mutants not only allows cysteine but also methionine to be made from homocysteine and this accounts for the following observations. First, *cysA*, *B*, *C* and *E* are not auxotrophic neither for cysteine nor for methionine. Second, that they suppress the methionine/homocysteine auxotrophs *metA*, *metB*, and *metG*. Third, *cysA*, *B*, *C* and *E* mutants are only auxotrophic for cysteine in the presence of the mutants, namely *mecA*, *mecB* or *cysD*, which block the cysteine alternative pathway. Mutants at the *mecA* and *mecB* loci result in poor cystathionine β-synthase cystathionine γ-lyase activity respectively (Pieniazek *et al.*, 1973b, 1974; Paszewski & Grabski, 1974. *mec* mutants were isolated on the basis that they do not grow in the presence of methionine and the toxic sulphate analogue, selenate in 48 h test. Since selenate toxicity is reversed by methionine, due to the catabolism of methionine to cysteine, this gave an opportunit to select for mutants with impaired efficiency of methionine breakdown. *mecA* and *mecB* strains are unable to grow on methionine in the presence of selenate and in combination with *cysA*, *cysB*, *cysC* *cysE* alleles, cysteine auxotrophy results, as discussed above. Also methionine can be converted to homocysteine by a three-step reaction (Fig.13-Steps, 8, 9 and 10). A mutant (*mecC13*) defective in S-adeno sylmethionine (SAM) synthetase (E.C. 2.5.1.6.) activity has been isolated.

SAM is an important metabolite, since it can methylate RNA and DNA and it is a precursor for the polyamines spermine and spermidine (Stevens *et al.*,1976). However, it is probably not involved in the regulation of methionine biosynthesis in *A.nidulans* (Paszewski & Grabski, 1974) as has been found in *E.coli* (Holloway *et al.*, 1970) and yeast (Cherest *et al.*,1973).

REFERENCES

ABERG,B. (1947). On the mechanism of the toxic action of chlorate and some related substances upon young wheat plants. *Kungl Lautbrukshögskolans Ann* 15, 37-107.

AHMED,S.I. & GILES,N.H. (1969). Organization of enzymes in the common aromatic synthetic pathway: evidence for aggregation in fungi. *Journal of Bacteriology* 99, 231-7.

ARST Jr.,H.N. (1968). Genetic analysis of the first step of sulphate metabolism in *Aspergillus nidulans*. *Nature* 219, 268-70.

ARST Jr., H.N. (1976). Integrator gene in *Aspergillus nidulans*. *Nature* 262, 231-4.

ARST Jr., H.N. & COVE,D.J. (1969). Methylammonium resistance in *Aspergillus nidulans*. *Journal of Bacteriology* 98, 1284-93.

ARST Jr.,H.N. & COVE,D.J. (1973). Nitrogen metabolite repression in *Aspergillus nidulans*. *Molecular & General Genetics* 126,111-41.

ARST,H.N. & MACDONALD,K. (1973). A mutant of *Aspergillus nidulans* lacking NADP-linked glutamate dehydrogenase. *Molecular & General Genetics* 122, 261-5.

ARST,H.N. & MACDONALD,D.W. (1975). A gene cluster in *Aspergillus nidulans* with an internally located cis-acting regulatory region. *Nature* 254, 26-34.

ARST Jr.,H.N., MACDONALD,D.W. & COVE,D.J. (1970). Molybdate metabolism in *Aspergillus nidulans* *Molecular & General Genetics* 108, 129-45.

ARST Jr.,H.N. & PAGE,M.M. (1973). Mutants of *Aspergillus nidulans* altered in the transport of methylammonium and ammonium. *Molecular & General Genetics* 121, 239-45.

ARST Jr.,H.N., PARBTANI,A.A.M. & COVE, D.J. (1975). A mutant of *Aspergillus nidulans* defective in NAD-linked glutamate dehydrogenase. *Molecular & General Genetics* 138, 165-71.

ASPEN,A.J. & MEISTER,A. (1962). Conversion of α-aminoadipic acid to L-pipecolic acid by *Aspergillus nidulans*. *Biochemistry* 1, 606-12.

BAINBRIDGE,B.W., DALTON,H., & WALPOLE,J.H.(1966). Identification of the arginosuccinase gene. *Aspergillus News Letter* 7, 18.

BAL,J., MALESZKA,R., STEPIEŃ,P. & CYBIS,J.(1975). Subcellular mislocation of cysteine synthase in a cysteine auxotroph of *Aspergillus nidulans*. *Febs Letters* 58, 164-6.

BARTNIK,E., GUZEWSKA,J. & WĘGLEŃSKI,P. (1973a). Mutations simultaneously affecting ammonium and glucose repression of the arginine catabolic enzymes in *Aspergillus nidulans*. *Molecular & General Genetics* 126, 85-92.
BARTNIK,E. & WĘGLEŃSKI,P. (1973). Ammonium and glucose repression of the arginine catabolic enzymes in *Aspergillus nidulans*. *Molecular & General Genetics* 126, 75-84.
BARTNIK,E. & WĘGLEŃSKI,P. (1974). Regulation of arginine catabolism in *Aspergillus nidulans*. *Nature* 250, 591-2.
BARTNIK,E., WĘGLEŃSKI, P. & PIOTROWSKA,M. (1973b). Ammonium and glucose repression of the arginine catabolic enzymes in *Aspergillus nidulans*. *Molecular & General Genetics* 126, 75-84.
BERLYN, U.C. (1967). Gene-enzyme relationships in histidine biosynthesis in *Aspergillus nidulans*. *Genetics* 57, 561-70.
BROWN,C.M.,BURN, V.J. & JOHNSON,B. (1973). Presence of glutamate synthase in fission yeasts and its possible role in ammonia assimilation. *Nature* 246, 115-6.
BUSHELL,M.E., & BULL,A.T. (1974). Polyamine, magnesium and ribonnucleic acid levels in steady-state cultures of the mould *Aspergillus nidulans*. *Journal of General Microbiology* 81, 271-3.
CALEF,E. (1957). Effect on linkage maps of selection of crossovers between closely linked markers. *Heredity* 11, 265-79.
CASE,M.E. & GILES,N.H. (1968). Evidence for nonsense mutations in the arom cluster of *Neurospora crassa*. *Genetics* 60, 49-58.
CASE,M.E. & GILES,N.H. (1971). Partial enzyme aggregates formed by pleiotropic mutants in the *arom* gene cluster of *Neurospora crassa*. *Proceedings of the U.S. National Academy of Science*, 68, 58-62.
CHEREST,H.,EICHLER,F. & DE ROBICHON-SZALMAJSTER,H. (1969). Genetic and regulatory aspects of methionine biosynthesis in *Sacchoromyces cererisiae*. *Journal of Bacteriology* 97, 328-36.
CHEREST,H., SURDIN-KERJAN,Y., ANTONIEWSKI,J. & DE ROBICHON-SZULMAJSTER H. (1973). S-adenosyl methionine-mediated repression of methionine biosynthetic enzymes in *Saccharomyces cerevisiae*. *Journal of Bacteriology* 114, 928-33.
COHEN,S.S. (1971). *Introduction to the polyamides*. New Jersey: Prentice-Hall Incorporating Englewood Cliffe.
COPPOC,G.L.,KALLIO,P., & WILLIAMS-ASHMAN,H.G. (1971). Characteristics of s-adenosyl-l-methionine decarboxylase from various organisms. *International Journal of Biochemistry* 2, 673-81.
COVE,D.J. (1966). The induction and repression of nitrate reductase in the fungus *Aspergillus nidulans*. *Biochimica and Biophysica Acta* 113, 51-6.
COVE,D.J. (1970). Control of gene action in *Aspergillus nidulans*. *Proceedings of the Royal Society, London B* 176, 267-75.
COVE,D.J. (1972). Chlorate toxicity in the fungus *Aspergillus nidulans*. *Proceedings of the Biochemical Society* 127, 19b.

COVE,D.J. (1976a). Chlorate toxicity in *Aspergillus nidulans*. The selection and characterisation of chlorate resistant mutants. *Heredity* 36, 191-203.
COVE,D.J. (1976b). Chlorate toxicity in *Aspergillus nidulans* - studies of mutants altered in nitrate assimilation. *Molecular & General Genetics* 146, 147-61.
COVE,D.J. (1977). Nitrate assimilation in *Aspergillus nidulans*. *Biological Reviews*. Manuscript in preparation.
COVE,D.J. & CODDINGTON,A. (1965). Purification of nitrate reductase and cytochrome c reductase from *Aspergillus nidulans*. *Biochimica et Biophysica Acta* 110, 312-8.
COVE,D.J. & PATEMAN,J.A. (1963). Independently segregating genetic loci concerned with nitrate reductase activity in *Aspergillus nidulans*. *Nature* 198, 262-8.
COVE,D.J. & PATEMAN,J.A. (1969). Autoregulation of the synthesis of nitrate reductase in *Aspergillus nidulans*. *Journal of Bacteriology* 97, 1374-8.
CYBIS,J.,PIOTROWSKA,M. & WĘGLEŃSKI,P. (1970). Control of ornithinetranscarbamylase formation in *Aspergillus nidulans*. *Bulletin de l'Academie Polonaise de Sciences, Serie des sciences biologique*, Cl.II 18, 669-72.
CYBIS,J.,PIOTROWSKA,M. & WĘGLEŃSKI,P. (1972a). The genetic control of the arginine pathways in *Aspergillus nidulans* mutants blocked in arginine biosynthesis. *Acta Microbiologica Polonica Series A* IV(XXI) No.4. 163-9.
CYBIS,J., PIOTROWSKA,M. & WĘGLEŃSKI,P. (1972b). Genetic control of the arginine pathways in *Aspergillus nidulans*. *Molecular & General Genetics* 118, 273-7.
CYBIS,J. & WĘGLEŃSKI,P. (1969). Effects of lysine on arginine uptake and metabolism in *Aspergillus nidulans*. *Molecular & General Genetics* 104, 282-7.
CYBIS,J. & WĘGLEŃSKI,P. (1972). Arginase induction in *Aspergillus nidulans*. The appearance and decay of the coding capacity of messenger. *European Journal of Biochemistry* 30, 262-8.
DARLINGTON,A.J. (1966). Genetic and biochemical studies of purine breakdown in *Aspergillus*. Ph.D. Thesis, University of Cambridge.
DARLINGTON,A.J. & SCAZZOCCHIO,C. (1967). Use of analogues and the substrate-sensitivity of mutants in analysis of purine uptake and breakdown in *Aspergillus nidulans*. *Journal of Bacteriology* 93, 937-40.
DARLINGTON,A.J., SCAZZOCCHIO,C . & PATEMAN,J.A. (1965). Biochemical and genetical studies of purine breakdown in *Aspergillus*. *Nature* 206, 599-600.
DEDUVE,C. & BAUDHUIN,P. (1966). Peroxisomes (microbodies and related particles). *Physiological Reviews* 46, 323.
DHILLON,T.S. & GARBER,E.D. (1970). Methionine-sensitive leucine-requiring mutants of *Aspergillus nidulans*. *Zeitschrift für*

Biologie 116, 349-53.
DORN,G.L. (1967). A revised linkage map of the eight chromosomes of *Aspergillus nidulans*. *Genetics* 56, 619-31.
DRAINAS,C., KINGHORN,J.R. & PATEMAN,J.A. (1976). Aspects of L-asparaginase regulation in *A.nidulans*. *180th Genetical Society Meeting, Heredity* 37, 153.
DRAINAS,C., KINGHORN, J.R. & PATEMAN, J.A. (1977). Aspartic hydroxamate resistance and asparaginase regulation in the fungus *Aspergillus nidulans*. *Journal of General Microbiology* (in press).
DRAINAS, C. & PATEMAN,J.A. (1977). L-asparaginase activity in the fungus *A.nidulans*. *Biochemical Society Transactions* (in press).
DUNN,E. & PATEMAN,J.A. (1972). Urea and thiourea uptake in *Aspergillus nidulans*. *Heredity* 29, 129.
DUNSMUIR,P. & HYNES,M.J. (1973). Temperature sensitive mutants affecting the activity and regulation of the acetamidase of *Aspergillus nidulans*. *Molecular & General Genetics* 123, 333-46.
ESSER,K. & KUENEN,R. (1967). Biosynthetic pathways. *Genetics of Fungi* 356-410.
FINCHAM,J.R.S. & DAY,P.R. (1971). *Fungal Genetics* 3rd Edition, Oxford: *Blackwell Scientific Publications*.
FOLEY,J.M., GILES,N.H. & ROBERTS,C.F. (1965). Complementation at the adenylosuccinase locus in *Aspergillus nidulans*. *Genetics* 52, 1247-63.
FORBES,E. (1959a). Use of mitotic segregation for assigning genes to linkage groups in *Aspergillus nidulans*. *Heredity* 13, 67-80.
FORBES, E. (1959b). Recombination in the *pro* region in *Aspergillus nidulans*. *Microbial Genetics Bulletin* 13, 9-11.
GAJEWSKI,W. & LITWINSKA,J. (1968). Methionine loci and their suppressors in *Aspergillus nidulans*. *Molecular & General Genetics* 102, 210-20.
GILES,N.H., CASE,M.E., PATRIDGE,C.W.H. & AHMED,S.I. (1967). A gene cluster in *Neurospora crassa* coding for an aggregate of five aromatic synthetic enzymes. *Proceedings of the National Academy of Sciences, Washington* 58, 1453-60.
GROSS,S.R. (1969). Genetic regulatory mechanisms in the fungi. *Annual Review of Genetics* 3, 395-424.
HACKETTE,S.L. SKYE,G.E., BURTON,C. & SEGEL, I.H. (1970). Characterization of an ammonium transport system in filamentous fungi with methylammonium-^{14}C as the substrate. *The Journal of Biological Chemistry* 245, 4241-50.
HANKINSON,O. (1974). Mutants of the pentose phosphate pathway in *Aspergillus nidulans*. *Journal of Bacteriology* 117, 1121-30.
HANKINSON,O. & COVE,J.D. (1974). Regulation of the pentose phosphate pathway in the fungus *Aspergillus nidulans*. *Journal of Biological Chemistry* 249, 2344-53.
HAYSMAN,P., & BRANCH-HOWE Jr.,H. (1971). Some genetic and physiological characteristics of urease-defective strains of *Neurospora*

crassa. Canadian Journal of Genetical Cytology 13, 256-69.

HERMAN,C. & CLUTTERBUCK,A.J. (1966). A method for selection of auxotrophs by means of 'spidery' growth. *Aspergillus Newsletter* 7, 13-14.

HOLLOWAY,C.T.,GREENE,R.C. & SU, C.H. (1970). Regulation of S-adenosylmethionine synthase in *Escherichia coli. Journal of Bacteriology* 104, 734-47.

HOPE,J. & STEVENS,L. (1976). The effects of putrescine, spermidine, and spermine on the growth of a polyamine requiring mutant of *Aspergillus nidulans. Biochemical Society Transactions* 4, 1128-30.

HUANG,H.T. (1963). Accumulation of L-homolanthionine by an *Escherichia coli* mutant. *Biochemistry* 2, 296-7.

HÜTTER,R. & DEMOSS,J.A. (1967a). Organization of the tryptophan pathway: a phylogenic study of the fungi. *Journal of Bacteriology* 94, 1896-907.

HÜTTER,R. & DEMOSS,J.A. (1967b). Enzyme analysis of the tryptophan pathway in *Aspergillus nidulans. Genetics* 55, 241-7.

HYNES, M.J. (1970). Induction and repression of amidase enzymes in *Aspergillus nidulans. Journal of Bacteriology* 103,482-7.

HYNES,M.J. (1973). Alterations in the control of glutamate uptake in mutants of *Aspergillus nidulans. Biochemical and Biophysical Research Communications* 54, 685-9.

HYNES,M. J. (1974). The effects of the carbon source on glutamate dehydrogenase activities in *Aspergillus nidulans. Journal of General Microbiology* 81, 165-70.

HYNES M.J. (1975). Amide utilization in *Aspergillus nidulans:* evidence for a third amidase enzyme. *Journal of General Microbiology* 91, 99-109.

HYNES,M.J. & PATEMAN, J.A. (1970a). The genetical analysis of regulation of amidase synthesis in *Aspergillus nidulans*. I. Mutants unable to utilise acrylamide. *Molecular and General Genetics* 108, 97-106.

HYNES,M.J. & PATEMAN,J.A. (1970b). The genetic analysis of regulation of amidase synthesis in *Aspergillus nidulans. Molecular & General Genetics* 108, 107-16.

HYNES,M.J. & PATEMAN,J.A. (1970c). The use of amides as nitrogen sources by *Aspergillus nidulans. Journal of General Microbiology* 63, 317-24.

KAFER,E. (1958). An eight-chromosome map of *Aspergillus nidulans Advances in Genetics* 9, 105-45.

KAFER, E. & DEMOSS, J.A. (1973). Formation of hybrid anthranilate synthetase *in vitro* from components of *Aspergillus* and *Neurospora. Biochemical Genetics* 9, 203-11.

KATZ,D. & ROSENBERGER,R.F. (1970). The effect of CO_2 on the purine requirement of *Aspergillus nidulans* and 3 mutants. *Biochimica et Biophysica Acta* 224, 279-81.

KERR,D.S. & FLAVIN,M. (1970). The regulation of methionine synthesis and the nature of cystathionine α-synthase in *Neurospora*. *The Journal of Biological Chemistry* 245, 1812-55.

KINGHORN, J.R. & PATEMAN,J.A. (1972). Regulation of glutamate transport in *Aspergillus nidulans*. *Heredity* 29, 128.

KINGHORN, J.R. & PATEMAN,J.A. (1973). NAD and NADP L-glutamate dehydrogenase activity and ammonium regulation in *Aspergillus nidulans*. *Journal of General Microbiology* 78, 39-46.

KINGHORN,J.R. & PATEMAN,J.A. (1974a). The effect of the carbon source on ammonium regulation in *Aspergillus nidulans*. *Molecular & General Genetics* 128, 95-8.

KINGHORN,J.R. & PATEMAN,J.A. (1974b). The regulation of NAD L-glutamate dehydrogenase in *Aspergillus nidulans*. *Genetical Research, Cambridge* 23, 119-24.

KINGHORN, J.R. & PATEMAN,J.A. (1975a). The structural gene for NADP L-glutamate dehydrogenase in *Aspergillus nidulans*. *Journal of General Microbiology* 86, 294-300.

KINGHORN,J.R. & PATEMAN,J.A. (1975b). Mutations which affect amino acid transport in *Aspergillus nidulans*. *Journal of General Microbiology* 86, 174-84.

KINGHORN,J.R. & PATEMAN,J.A. (1976). Mutants of *Aspergillus nidulans* lacking nicotinamide adenine dinucleotide-specific glutamate dehydrogenase. *Journal of Bacteriology* 125, 42-7.

KOLMARK,H.G. (1969a). Genetic studies of urease mutants in *Neurospora crassa*. *Mutation Research* 8, 51-63.

KOLMARK,H.G. (1969b). Urease defective mutants in *Neurospora crassa*. *Molecular & General Genetics* 104, 219-34.

MACDONALD,D.W., ARST Jr.,H.N. & COVE,D.J. (1974b). The threonine dehydratase structural gene in *Aspergillus nidulans*. *Biochimica et Biophysica Acta* 362, 60-5.

MACDONALD, D.W. & COVE, D.J. (1974). Studies on temperature-sensitive mutants affecting the assimilatory nitrate reductase of *Aspergillus nidulans*. *European Journal of Biochemistry* 47, 107-10.

MACDONALD, D.W., COVE,D.J. & CODDINGTON,A. (1974a). Cytochrome-c reductase from wild-type and mutant strains of *Aspergillus nidulans*. *Molecular & General Genetics* 128, 187-99.

MACMILLAN, A. (1956). The entry of ammonia into fungal cells. *Journal of Experimental Botany* 7, 113-26.

MEDINA, A. & NICHOLAS,D.J.D. (1957a). Hyponitrite reductase in *Neurospora*. *Nature* 179, 533-4.

MEDINA,A. & NICHOLAS, D.J.D. (1957b). Metallo enzymes in the reduction of nitrite to ammonia in *Neurospora*. *Biochimica et Biophysica Acta* 25, 138-41.

MEERS, J.L., TEMPEST,D.W. & BROWN, C.M. (1970). Glutamine (amide): 2-oxoglutarate amino transferase oxidoreductase (NADP), an enzyme involved in the synthesis of glutamate by some bacteria. *Journal of General Microbiology* 64, 187-94.

MÜLLER, M. & MØLLER, K.M. (1969). Urate oxidase and its association with peroxisomes in *Acanthamoeba sp. European Journal of Biochemistry* 9, 424.

NAKAMURA, T. & SATO, R. (1960). Cysteine-S-sulphonate as an intermediate in microbial synthesis of cysteine. *Nature* 185, 163-4.

NAKAMURA, T. & SATO, R. (1963). Synthesis from sulphate and accumulation of S-sulphocysteine by a mutant strain of *Aspergillus nidulans*. *Biochemical Journal* 86, 328-35.

NAKAMURA, T. & SATO, R. (1965). Assimilation of inorganic sulphur compounds by microorganisms. *Journal of Japanese Biochemical Society* 37, 1.

NORONHA, L. (1970). Genetic investigation of tryptophan-requiring mutants of *Aspergillus nidulans*. *Indian Journal of Experimental Biology* 8, 298-301.

O'DONOVAN, G.A. & NEUHARD, J. (1970). Pyrimidine metabolism in microorganisms. *Bacteriological Reviews* 34, 278-343..

PAGE, M.M. (1973). Genetics and biochemical studies on the catabolism of amines and alcohols in *Aspergillus nidulans*. *Ph.D. Dissertation, University of Cambridge*.

PAGE, M.M. & COVE, D.J. (1972). Alcohol and amine catabolism in the fungus *Aspergillus nidulans*. *Biochemical Journal* 127, 17.

PALMER, L.M. & COVE, D.J. (1975). Pyrimidine biosynthesis in *Aspergillus nidulans*. *Molecular & General Genetics* 138, 243-55.

PASZEWSKI, A. & GRABSKI, J. (1973). Studies on β-cystathionase and O-acetylhomoserine sulphydrylase as the enzymes of alternative methionine biosynthetic pathways in *Aspergillus nidulans*. *Acta Biochimica Polonica* 20, 159-68.

PASZEWSKI, A. & GRABSKI, J. (1974). Regulation of S-amino acids biosynthesis in *Aspergillus nidulans*. *Molecular & General Genetics* 132, 307-20.

PASZEWSKI, A. & GRABSKI, J. (1975a). Enzymatic lesions in methionine mutants of *Aspergillus nidulans:* Role and regulation of an alternative pathway for cysteine and methionine synthesis. *Journal of Bacteriology* 124, 893-904.

PASZEWSKI, A. & GRABSKI, J. (1975b). Homolanthionine in fungi: Accumulation in the methionine-requiring mutants of *Aspergillus nidulans*. *Acta Biochimica Polonica* 22, 263-8.

PATEMAN, J.A. (1969). Regulation of synthesis of glutamate dehydrogenase and glutamine synthetase in microorganisms. *Biochemical Journal* 115, 769-75.

PATEMAN, J.A. & COVE, D.J. (1967). Regulation of nitrate reduction in *Aspergillus nidulans*. *Nature* 215, 1234-7.

PATEMAN, J.A., COVE, D.J., REVER, B.M. & ROBERTS, D.B. (1964). A common co-factor for nitrate reductase and xanthine dehydrogenase which also regulates the synthesis of nitrate reductase. *Nature*, 201, 58-60.

PATEMAN, J.A. & KINGHORN, J.R. (1976). Nitrogen metabolism. In *The Filamentous Fungi, Vol.2.* pp.159-237. Edited by J.E.Smith and D.R.Berry, London: Edward Arnold.

PATEMAN, J.A., DUNN, E., KINGHORN, J.R. & FORBES, E.C. (1974a). The transport of ammonium and methylammonium in wild type and mutant cells of *Aspergillus nidulans. Molecular & General Genetics* 133, 225-36.

PATEMAN, J.A., KINGHORN, J.R., & DUNN, E. (1974b). Regulatory aspects of L-glutamate transport in *Aspergillus nidulans. Journal of Bacteriology* 119, 534-42.

PATEMAN, J.A., KINGHORN, J.R., DUNN, E. & FORBES, E.(1973). Ammonium regulation in *Aspergillus nidulans. Journal of Bacteriology* 114, 943-50.

PATEMAN, J.A., REVER, B.M. & COVE, D.J. (1967). Genetic and biochemical studies of nitrate reduction in *Aspergillus nidulans. Biochemical Journal* 104, 103-11.

PEES, E. (1965). Polarized negative interference in the lys-51 region of *Aspergillus nidulans. Experientia* 21, 514-15.

PEES, E. (1966). Lysine histidine and isoleucine mutants. *Aspergillus Newsletter* 7, 11-12.

PIENIĄŻEK, J., BAL, J., BALBIN, E. & STĘPIEŃ, P.P. (1974). An *Aspergillus nidulans* mutant lacking serine transacetylase: evidence for two pathways of cysteine biosynthesis. *Molecular & General Genetics* 132, 363-6.

PIENIĄŻEK, J., KOWALSKA, I.M. & STĘPIEŃ, P.P. (1973a). Deficiency in methionine adenosyltransferase resulting in limited repressibility of methionine biosynthetic enzymes in *Aspergillus nidulans. Molecular & General Genetics* 126, 367-74.

PIENIĄŻEK, N.J., STĘPIEŃ, P.P. & PASZEWSKI, A. (1973b). An *Aspergillus nidulans* mutant lacking cystathionine β-synthase. *Biochimica et Biophysica Acta* 297 37-47.

PIOTROWSKA, M., SAWICKI, M. & WĘGLEŃSKI, P. (1969). Mutants of the arginine-proline pathway in *Aspergillus nidulans. Journal of General Microbiology* 55, 301-5.

POLKINGHORNE, M. & HYNES, M.J. (1975). Effect of L-histidine on the catabolism of nitrogenous compounds in *Aspergillus nidulans. Journal of General Microbiology* 87, 185-7.

POLLARD, R., KAFER, E. & JOHNSTON, M.T. (1968). Influence of translocations on meiotic and mitotic nondisjunction in *Aspergillus nidulans. Genetics* 60, 743-57.

PONTECORVO, G., ROPER, J.A., HEMMONS, L.M., MACDONALD, K.D. & BUFTON, A.W.J. (1953). The genetics of *Aspergillus nidulans. Advances in Genetics* 5, 141-238.

PUTRAMENT, A., GUZEWSKA, J. & PIENIĄŻEK, D. (1970). Further genetic characteristics of methionine mutants and their suppressors in *Aspergillus nidulans. Molecular & General Genetics* 109, 209-18.

RAO,E. & DE BUSK,A.G. (1975). A mutant deficient in the general (PmG) amino acid transport system. *Neurospora Newsletter* 22, 12-13.

REITHEL, F.J. (1971). Ureases. In *The Enzymes* Vol.4, pp.1-21. London: Academic Press.

ROBINSON,J.H., ANTHONY,C. & DRABBLE, W.T. (1973a). The acidic amino-acid permease of *Aspergillus nidulans*. *Journal of General Microbiology* 79, 53-63.

ROBINSON, J.H., ANTHONY, C. & DRABBLE, W.T. (1973b). Regulation of the acidic amino-acid permease of *Aspergillus nidulans*. *Journal of General Microbiology* 79, 65-80.

ROBERTS, C.F. (1967). Complementation analysis of the tryptophan pathway in *Aspergillus nidulans*. *Genetics* 55, 233-9.

ROBERTS, C.F. (1969). Isolation of multiple aromatic amino-acid mutants in *Aspergillus nidulans*. *Aspergillus News Letter* 10, 19.

ROBINSON, J.H., ANTHONY, C. & DRABBLE, W.T. (1974). The utilization of nitrogen sources by *Aspergillus clavatus*. *Journal of General Microbiology* 85, 23-8.

SADASIVAM, S., SHANMUGASUNDARAM, R. & SHANMUGASUNDARAM, F.R.B. (1969). The pinkish red pigment produced by an adenineless mutant of *Aspergillus nidulans*. *Indian Journal of Biochemistry & Biophysics* 6, 237.

SCAZZOCCHIO, C. (1966). Studies on the genetic control of purine oxidation in *Aspergillus nidulans*. *Ph.D.Thesis, University of Cambridge*.

SCAZZOCCHIO, C. (1973). The genetic control of molybdoflavoproteins in *Aspergillus nidulans*. *Molecular & General Genetics* 125, 147-55.

SCAZZOCCHIO, C. (1974). The genetic determination of molybdoflavo-enzymes in *Aspergillus nidulans*. *Proceedings of the Climax First International Conference on the Chemistry and Uses of Molybdenum, London.*

SCAZZOCCHIO, C. & DARLINGTON,A.J. (1967). The genetic control of xanthine dehydrogenase and urate oxidase synthesis in *Aspergillus nidulans*. *Memoires* 1503-8.

SCAZZOCCHIO, C. & DARLINGTON, A.J. (1968). The induction and repression of the enzymes of purine breakdown in *Aspergillus nidulans*. *Biochimica et Biophysica Acta* 166, 557-68.

SCAZZOCCHIO, C.,HOLL,F.B. & FOGUELMAN,A.I. (1973). The genetic control of molybdoflavoproteins in *Aspergillus nidulans*. Allopurinol-resistant mutants constitutive for xanthine-dehydrogenase. *European Journal of Biochemistry* 36, 428-45.

SINHA, U. (1967). Aromatic amino acid biosynthesis and para-fluorophenylalanine resistance in *Aspergillus nidulans*. *Genetical Research, Cambridge* 10, 261-72.

SINHA, U. (1969). Genetic control of the uptake of amino acids in *Aspergillus nidulans*. *Genetics* 62, 495.

SINHA, U. (1970). Competition between leucine and phenylalanine and its relation to p-fluorophenylalanine resistant mutations in *Aspergillus nidulans*. *Archiv für Mikrobiologie* 72, 308-17.

SINHA, U. (1972). Studies with p-fluorophenylalanine-resistant mutants of *Aspergillus nidulans*. *Beiträge zur Biologie Pflanzenbau* 48, 171-80.

SNEATH, P.H.A. (1955). Putrescine as an essential growth factor for a mutant of *Aspergillus nidulans*. *Nature* 175, 818.

STĘPIEŃ, P.P., PIENIĄŻEK, N.J., BAL, J. & MORZYCHA, F. (1975). Cysteine biosynthesis in *Aspergillus nidulans*. *Acta Microbiologi Polonica Series A*. Vol. VII (24) 201-10.

STEVENS, L. (1970). The biochemical role of naturally occurring polyamines in nucleic acid synthesis. *Biological Reviews* 45,1-27.

STEVENS, L. (1975). Ornithine decarboxylase activity in germinating conidia of *Aspergillus nidulans*. *FEBS LETTERS* 59, No.1 80-2.

STEVENS, L. & HEATON, A. (1973). Induction, partial purification and properties of ornithine transaminase from *Aspergillus nidulans*. *Biochemical Society Transactions, 537th Meeting, Canterbury* 1, 749-51.

STEVENS, L., MCKINNON, I.M. & WINTHER, M. (1976). Polyamine and ornithine metabolism during the germination of conidia of *Aspergillus nidulans*. *Biochemical Journal* 158, 235-41.

TABOR,H. & TABOR, C.W. (1972). Biosynthesis and metabolism of putresine, spermidine, spermine, and related amines. *Advances in Enzymology*. 36, 203-68.

TRIJBELS, F. & VOGELS, G.D. (1966). Allantoicase and ureidoglycolase in *Pseudomonas* and *Penicillium* species. *Biochimica et Biophysica Acta* 118, 387-95.

VOGELS, G.D. & VAN DER DRIFT, C. (1976). Degradation of purine and pyrimidine by microorganisms. *Bacteriological Reviews* 40, 403-68.

WARR, J.R. & ROPER, J.A. (1965). Resistance to various inhibitors in *aspergillus nidulans*. *Journal of General Microbiology* 40, 273-81.

WĘGLEŃSKI,P. (1966). Genetical analysis of proline mutants and their suppressors in *Aspergillus nidulans*. *Genetical Research Cambridge* 8, 311-21.

WĘGLEŃSKI, P. (1967). The mechanisms of action of proline suppressors in *Aspergillus nidulans*. *Journal of General Microbiology* 47, 77-85.

WINTHER, M. & STEVENS, L. (1976). Synthesis of polyamines in the germinating conidia of *Aspergillus nidulans*. *Biochemical Society Transactions* 4, 1126-8.

ZAUDY,G.(1969). The location of some multiply aromatic mutants in *Aspergillus nidulans*. *Aspergillus Newsletter* 10, 22.

Genetic Regulation of Nitrogen Metabolism

J. A. PATEMAN and J. R. KINGHORN

Department of Genetics, University of Glasgow, Glasgow, U.K.

INTRODUCTION

The study of the regulation of gene action in such lower eukaryotes as *Neurospora* and *Aspergillus* has lagged far behind that of such prokaryotes as *Escherichia coli*. An important reason for such a difference is the lack of viral and plasmid agents in fungi similar to those which provide such powerful techniques for genetic and biochemical analysis in the prokaryotes. A second important reason is the unfortunate accident that *Neurospora crassa* and various yeast species were used for most of the earlier biochemical and genetical work. There is little doubt that if *A. nidulans*, with its well defined parasexual cycle and stable diploids, had been exploited in the same intensive fashion as *N. crassa* our knowledge of genetic regulation in simple eukaryotes would be much greater.

One of the few important generalisations that can be made about gene action and organisation in lower eukaryotes is that genes coding for related or sequential functions are seldom clustered on the chromosomes. This is a direct contrast to *E. coli* where something like 50% of the known structural genes are linked in groups of two or more related to their metabolic functions. A few examples of the clustering of functionally related genes are known in *A. nidulans*; the *niaD* and *niiA* genes for nitrate and nitrite reductases, the *prnA*, *prnB*, *prnC* and *prnD* genes in proline catabolism; the *ureB* and *ureA* genes for urease and urea uptake. It is not clear why there should be such a difference in the frequency of gene clusters in prokaryotes and lower eukaryotes, apart from the pleasible assumption that it is connected with the more elaborate organisation of the eukaryote chromosome. A related puzzle in the lower eukaryotes to date has been the scarcity of control regions - particularly cis-acting ones - closely linked to structural genes. As some gene clusters have at last been recognised, so have control regions associated with them been defined by mutations with regulatory effects; notably in the proline catabolism and urea utilization systems in *A. nidulans*. The first regulator genes to be discovered in *A. nidulans* showed some degree of dominance, implying a positive action of the regulator gene product. We are far from fully understanding the reasons for the apparent low frequency of gene clusters and cis-acting control regions, and the preponderance of positive acting regulator genes in *A. nidulans*. However, in the light of recent work, much of it reported in various contributions to this symposium, the situation now appears less paradoxical and is discussed in the final section of this article.

The range of nitrogen metabolism regulatory systems about which something is known in *A. nidulans* is too great for each to be treated exhaustively here. Some topics are dealt with in some detail because of their importance and the author's special interest. A summary will be given of most of the other areas.

NITRATE UTILIZATION

In *A. nidulans* the enzyme activities necessary for nitrate reduction are nitrate reductase and nitrite reductase. These two enzyme activities and the associated hydroxylamine reductase activity are induced by nitrate and nitrite and repressed by ammonium (Cove & Pateman, 1963; Pateman *et al.*, 1964; Pateman & Cove, 1967). It is probable that both nitrate induction and ammonium repression act by determining the rate of enzyme synthesis, but it is not known if this is at the transcription or translation level or both. There is also some evidence that nitrate reductase *in vivo* may be inactivated in the absence of nitrate and/or the presence of ammonium, with consequent loss of activity (Hynes, 1973a; Dunn-Coleman & Pateman, 1975). The genes known to play some role in nitrate reduction are; *nia-D*, the structural gene for nitrate reductase; *niiA* , the probable structural gene for nitrite and hydroxylamine reductase; the five *cnx* (co-factor, nitrate reductase, xanthine dehydrogenase) genes *cnxABC*, *cnxE*, *cnxF*, *cnxG*, *cnxH* concerned with a co-factor common to nitrate reductase and xanthine dehydrogenase; *nirA* (previously *niiB*), probably a regulator gene concerned with the induction of nitrate and nitrite reductases. Recessive mutations in the *nia-D* and *cnx* genes result in the failure to grow on nitrate; recessive mutations in the *niiA*, and *nirA* genes result in failure to grow on nitrate or nitrite. A number of genes which affect the regulation of the pathway by ammonium are described later.

Regulation of nitrate reductase activity by NADPH

The level of nitrate reductase activity in *A. nidulans* is dependent upon an activation/inactivation mechanism, involving the enzyme co-factor NADPH. To reduce nitrate to ammonium a great deal of NADPH is used. The cell compensates for this by elevating the levels of the enzymes of the Pentose Phosphate Pathway (P.P.P.), which generate NADPH from glucose breakdown. It has been shown by Hankinson & Cove (1974) that both nitrate and the *nirA* gene are involved in this elevation of the PPP enzymes. When wild type cells are carbon starved rapid inactivation of nitrate reductase results, concurrent with a rapid decrease in the NADPH concentration. *In vitro* studies with cell extracts have shown that nitrate reductase activity is dependent on the concentration of NADPH (Dunn-Coleman & Pateman, 1975). Extracts of a mutant with altered levels of the PPP enzymes and reduced levels of NADPH lose their nitrate reductase activity more rapidly than wild type extracts. This *in vitro* loss of nitrate reductase activity can be reversed only by the addition of NADPH. These results suggest that the concentration of NADPH *in vivo* plays an important role in the regulation of nitrate reductase activity; inactivation of the enzyme, by oxidation, resulting when the level of NADPH is low. This inactivation mechanism would provide a rapid means of regulating nitrate reductase activity and enable the cell to conserve NADPH. Thus it seems likely that nitrate reductase in *A. nidulans* is subject to two levels of regulation. The synthesis of the enzyme is determined by nitrate induction and ammonium repression. The activity of the enzyme molecules is determined by the NADPH concentration, a mechanism which provides a fast-acting and reversible control analogous to the 'feed-back' inhibition found in many prokaryote systems.

The regulatory role of nitrate reductase

The majority of mutations in the *nia-D* and *cnx* genes not only result in the loss of nitrate reductase activity, they also result in the constitutive synthesis of nitrite reductase in the absence of nitrate. Furthermore, some *nia-D* gene mutants and all *cnx* mutants retain the cytochrome c reductase activity possessed by the nitrate reductase molecule. This cytochrome c reductase activity, which serves as a measure of the synthesis of nitrate reductase protein, is also synthesised constitutively in the absence of nitrate. Thus, when nitrite reductase activity is lost as a result of mutation in any one of six different genes, the synthesis of nitrate reductase itself - as represented by its cytochrome c reductase moiety - and of nitrate reductase becomes constitutive. There are two main hypotheses to account for this remarkable situation:

1) The apparent constitutive synthesis of the two nitrate induced enzyme proteins might be due to the accumulation of a low concentration of nitrate in cells that lack nitrate reductase activity. This nitrate might come from the oxidation of ammonium or the concentration of trace amounts from the medium. There are two main arguments against this: (a) at low concentrations nitrite is a comparable inducer to nitrate (Pateman & Cove, 1967) and *niiA* mutants which lack nitrite reductase activity would convert the nitrate to nitrite. Thus, as a result of nitrite induction, *niiA* mutants should appear constitutive for nitrate reductase activity. However, all *niiA* mutants are fully nitrate inducible for nitrate reductase; (b) in the normal prototrophic strains ammonium repression is more effective than nitrate induction in that the presence of equimolar amounts of ammonium and nitrate results in low levels of nitrate and nitrite reductase. Also the constitutive enzyme synthesis of the *nia-D* and *cnx* mutants is repressed by ammonium. Thus it is possible to observe the effect of various combined nitrate and ammonium concentrations in the medium on nitrite reductase levels in the prototroph and the mutants. If the *nia-D* and *cnx* mutants were constitutive because of intracellular nitrate accumulation then the mutants should respond to nitrate in the presence of ammonium by as much or more enzyme synthesis as the prototroph. However, the mutants have significantly lower levels of enzyme in these conditions (Cove & Pateman, 1969). This is strong evidence that the *nia-D* and *cnx* mutants are not constitutive because of the intracellular concentration of nitrate.

2) The alternative and favoured hypothesis is that nitrate reductase has two functions, one catalytic and the other regulatory It is based on the observations that mutations in the *nia-D* and *cnx* genes (but not in the *nii* gene) can simultaneously result in: i) loss of nitrate reductase activity; ii) constitutive synthesis of all the enzyme proteins of the nitrate reduction pathway in the absence of ammonium; iii) altered response to nitrate induction in the presence of ammonium.

The regulatory gene nirA

A common class of recessive mutations in the *nirA* gene result in the loss of all the enzyme activities necessary for the reduction of nitrate. A second type of mutation can occur in the *nirA* gene and result in constitutive synthesis of the nitrate and nitrite reductases in the absence of nitrate. The constitutive mutant, $nirA^c$, possesses the following characteristics: 1) The $nirA^c$ mutation is the same gene as the $nirA^-$ mutations; 2) the $nirA^c$ mutation affects the level of all the enzymes necessary for nitrate reduction; 3) the $nirA^c$ is semi-dominant (i.e. intermediate in phenotype) with respect to the induction of nitrate reductase in the heterozygous diploid genotype, $nirA^c/nirA^+$; 4) the constitutive enzyme synthesis in $nirA^c$ is repressed by ammonium in a similar fashion to the ammonium repression of the nitrate induced enzyme synthesis in $nirA^+$. In order to explain these results it was proposed that the *nirA* gene produced a regulator product which was necessary for the synthesis of the nitrate reducing enzymes.

A possible model for the regulation of nitrate reduction

A model basically similar to that proposed by Cove & Pateman (1969) which illustrates possible mechanisms for the control of nitrate and nitrite reductase is given in Fig. 1. Although it is rather complex it is the simplest type of model which takes into account the involvement of nitrate, nitrate reductase and the *nirA* gene product in the regulation of the pathway and the further complication of ammonium repression (see later section). There is no direct biochemical evidence as yet concerning the mechanisms of the postulated regulatory functions of nitrate reductase and the *nirA* gene product. Therefore, the relationships suggested in the model between the various control components are speculative. They represent just one of a number of possible ways of explaining the experimental observations.

There are several points which are worth further discussion. The postulated requirement of the *nirA* gene product (i.e. positive action) for synthesis of the enzyme proteins is based on the lack of all enzyme activity in the recessive $nirA^-$ mutants and the semi-dominance of the $nirA^c$ mutant. The $nirA^-$ mutants would produce either no regulatory protein or an abnormal form unable to mediate induction of enzyme synthesis. The $nirA^c$ mutant would produce a regulatory protein whose positive function in enzyme induction was not prevented by the nitrate reductase molecule and the haploid $nirA^c$ would thus be constitutive. The semi-dominance of the $nirA^c$ mutant in the heterozygous diploid $nirA^c/nirA^+$ might be explained by a repressor function (i.e. negative action) of the *nirA* regulatory protein after interaction with nitrate reductase or ammonium. Alternatively, it may be that the *nirA* regulatory protein has only a positive inducing function and the semi-dominance of *nirA*^{c} is a gene dosage effect, as suggested by Cove & Pateman (1969) and Cove (1970). The postulated positive action of the *nirA* gene product is analogous to that of the C gene regulatory protein in the L-arabinose system in *Escherichia coli* (Englesberg *et al*., 1965) and in direct contrast to the repressor function (negative action) of the *i* gene regulatory protein in the lactose system of *E. coli*.

The nature of the control system, if one exists, for the synthesis of the molybdenum co-factor determined by the *cnx* genes is an intriguing but unresolved question. It is certainly possible that some form of genetic regulation exists since the requirement for the co-factor would increase at least a hundred-fold after the induction of nitrate reductase and xanthine dehydrogenase by nitrate and purines respectively. Garrett & Cove (1977) and unpublished have evidence which suggests that the synthesis of the molybdenum co-factor is subject to ammonium repression, but not to nitrate induction.

There is no direct evidence that the site of action of the product of the *nirA* gene is a region close to the *niaD* and *niiA* structural genes as indicated in Fig. 1. However, it is known that at least one such closely linked control region exists (see following section) and the interaction of the *nirA* product with this region is the most plausible hypothesis. Fig. 1 also indicates that ammonium regulates the expression of the *niaD* and *niiA* genes through the interaction of the products of the *tamA* and *areA* genes with a control region closely linked to the structural genes. The reasons for this form of model are given in the following section and the general account of ammonium regulation.

The nia region

The *tamA*^{r}*50* allele results in general ammonium repression (see next section) and consequent inability to grow on any nitrogen source except ammonium. A number of secondary mutants were obtained after nitrosoguanidine mutagenesis in a *tamA*^{r}*50* strain by selecting for growth on nitrate (Pateman & Kinghorn, unpublished results). One of these mutations provisionally named nia 22 allows growth of *tamA*^{r}*50* on nitrate alone, that is it specifically suppresses *tamA*^{r}*50* on nitrate. The *tamA*^{r}*50* nia 22 double mutant remains repressed for all the ammonium regulated functions except nitrate reductase and nitrite reductase and can only grow on either ammonium or nitrate. The nia 22 mutant is unlinked to the *tamA* gene and closely linked about 2 cM to the *niiA* gene. Thus it is a specific suppressor mutation of *tamA*^{r}*50* which maps in a region close to the *niaD* and *niiA* structural

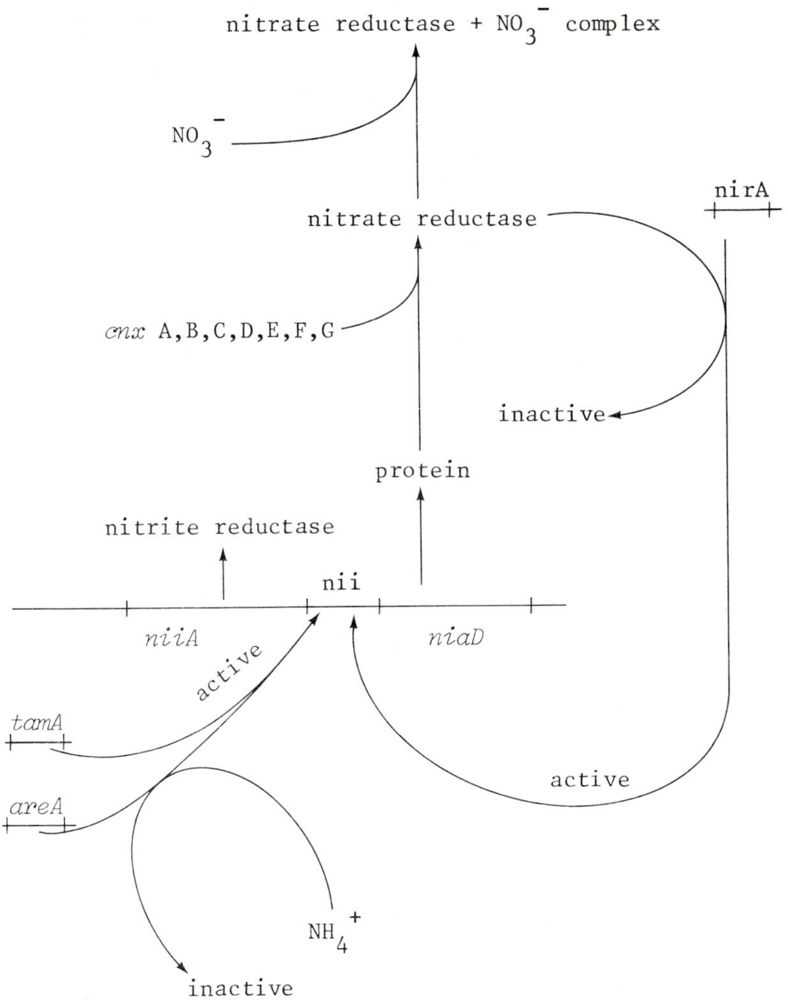

Fig. 1. Hypothetical model for the action of the *nirA*, *areA* and *tamA* regulator genes

The *nirA* gene product performs a necessary function, at the nii control region adjacent to the *niaD* and *niiA* genes, to allow expression of those structural genes. When not complexed with NO_3^-, nitrate reductase inactivates the *nirA* product. The *areA* and *tamA* gene products also perform a function, in some way mediated through the

Fig. 1 cont'd.....

nii control region, necessary for the expression of the *niaD* and *niiA* structural genes. The presence of ammonium outside the cell results indirectly in the inactivation of the *areA* and *tamA* gene products and consequent ammonium repression.

---

genes. The nia 22 mutation has a remarkable property when separated from *tamA*^{r}*50*. The *tamA*$^+$ nia 22 genotype grows on nitrite but not on nitrate. Thus there appears to be an interaction between the *tamA*^{r}*50* gene product and the nia 22 allele which allows expression of the *niaD* structural gene for nitrate reductase, and the *niiA* structural gene for nitrite reductase in the presence of nitrate. Conversely the normal *tamA*$^+$ gene product does not interact with the nia 22 allele to allow expression of *niaD* in the presence of nitrate. The simplest explanation is that the *tamA* and *areA* gene products regulate the transcription of the *niaD* and *niiA* genes by interacting with a control region analogous to a promotor and/or operator next to the structural genes. The interaction between the *tamA*^{r}*50* product and the nia 22 allele allows some transcription of the *niaD* and *niiA* genes in the presence of nitrate whereas the normal *tamA*$^+$ product cannot interact with the altered nia 22 region in the presence of nitrate to allow expression of the *niaA*. Thus the nia 22 mutation is behaving as a promotor or operator mutation analogous to the hyper-repressible operator-type mutant in the sulphate permease gene of *A. nidulans* (Lukaszkiewicz & Paszewski, 1976). Since the nia 22 mutation has a differential effect on the two genes it is necessary to suppose that a part of the control region is specific to the regulation of each structural gene and probably a part is concerned with both. To accommodate this dual requirement the control region is thought to lie between *niaD* and *niiA* and is shown in that position in Fig. 1, although of course this position needs to be established by further genetic analysis.

The nii region

In a similar fashion to that described above mutants were obtained in a *tamA*^{r}*50* strain by selecting for growth on nitrite. One of these mutations named nii 14 allows growth of *tamA*^{r}*50* on nitrite. The *tamA*^{r}*50* nii 14 double mutant remains repressed for all the ammonium regulated functions except nitrite reductase and can only grow on ammonium and nitrite. The nii 14 mutation maps close to, ca 2 cM, to the *niaD* gene. Thus like nia 22, the nii 14 mutation is a specific suppressor of *tamA*^{r}*50* in a region close to the *niaD* and *nii4* structural genes. It is of course probable that nia 22 and nii 14 are different mutations in a single control region. The *tamA*+ nii 14 genotype grows on nitrate. Therefore their appears to be an interaction between the *tamA*^{r}*50* product and the nii 14 allele which allows expression of the *niiA* structural gene in the presence of nitrite, but does not allow expression of *niaD* in the presence of nitrate. The nia 22 and nii 14 have differential effects on the expression of the *niaD* and *niiA* genes and the product of the *nirA* regulatory plays some role in nitrate and nitrite induction. Therefore the simplest hypothesis is that the products of the *tamA*, *areA* and *nirA* regulatory genes all act at a common region and consequently are so depicted in Fig. 1. The further genetic and biochemical analysis of this region, particularly the dominance relationships of nia 22 and nii 14, should give a new insight into a mechanism of eukaryote control.

AMMONIUM REGULATION

A number of enzyme and uptake systems in *A. nidulans* are regulated by ammonium. The level of all these systems is minimal if prototrophic (normal) cells are grown, or held, in the presence of ammonium. It is not known whether this effect of ammonium is on protein synthesis, is an inhibition of activity, or is a combination of both, however, it is usually referred to as Ammonium repression. Repression by ammonium is widespread and very important in the whole area of inorganic and simple organic nitrogen metabolism in an analogous fashion to catabolite repression in the carbon metabolism of many prokaryotes. Systems repressed by ammonium include nitrate reduction (Pateman & Cove, 1967); purine degradation (Scazzocchio & Darlington, 1968); amide degradation (Hynes & Pateman, 1970a,b); proline utilization (Arst & MacDonald, 1975); amino acid transport systems (Pateman *et al.*, 1974b); urea transport (Dunn & Pateman, 1972 Pateman *et al.*, 1973); purine transport (Arst & Cove, 1969); ammonium transport (Pateman *et al.*, 1973; Pateman *et al.*, 1974a); L-arginine (Bartnik *et al.*, 1973; Bartnik & Węgleński, 1973, 1974) and L-histidine catabolism (Polkinghorne & Hynes, 1974); extracellular protease (Cohen, 1972). Mutations in a number of genes can effect the ammonium regulation of these systems. The relevant properties of these mutants will be described in separate sections.

meaA and mod meaA

It was shown by Arst & Cove (1969) that mutations in two genes *meaA* and *meaB* can result in simultaneous resistance to methylammonium and derepression for several ammonium repressed activities. The *meaA* gene probably determines some membrane component which affects ammonium and methylammonium transport (Pateman *et al.*, 1974a).

Mutations in a modifier gene *mod meaA*, when combined in a double mutant *meaA mod meaA* (previously called DER-3), result in altered K_ms for both ammonium and methylammonium transport (Pateman *et al.*, 1974a). This strongly indicates that the two genes *meaA* and mod *meaA* in some way determine a structural component(s) of the ammonium transport sites. It may be of significance that both *meaA* and the double mutant *meaA* mod *meaA* also efflux ammonium considerably faster than the normal strain under conditions which result in a high intracellular concentration of ammonium (Pateman unpublished work). It is easy to understand why a reduction in transport efficiency should confer resistance to the toxic analogue methylammonium. It is less clear why mutation affecting a membrane component should also result in ammonium derepression.

meaB

Mutations in *meaB* result in methylammonium resistance and some ammonium derepression. The function of *meaB* is unknown; it may determine a membrane component or have some function in sugar metabolism.

amrA gene

A mutant *amrA1* was obtained by selection for poor growth on ammonium as a sole nitrogen source (Pateman *et al.*, 1973). This mutation in the *amrA* gene simultaneously results in ammonium derepression for some but not all ammonium-regulated systems, but it does not confer methylammonium resistance. The *amrA1* possesses a low level of ammonium transport and of methylammonium transport under certain conditions (Pateman *et al.*, 1974a). It also effluxes ammonium faster than the normal strain under certain conditions (Pateman, unpublished work). The function of the *amrA* gene is unknown: one possibility is that it determines a membrane component, since the *amrA1* mutant shows altered transport properties; another possibility is that *amrA* has a regulatory function in glucose metabolism affecting the supply of NADPH (Dunn-Coleman, unpublished data).

gdhA

Mutations in the *gdhA* gene result in the loss of NADP glutamate dehydrogenase activity (Kinghorn & Pateman, 1973), and it has been shown that *gdhA* is the structural gene for the enzyme (Kinghorn & Pateman, 1975b). The *ghdA* mutants are derepressed for a number of ammonium regulated systems. They are not resistant to methylammonium but rather more sensitive than the wild type. They are sensitive to high external ammonium concentrations (200 mM), a fact which has been utilized in selection experiments. The basis of this sensitivity to ammonium is not known - it does not appear to be simply due to the loss of NADP glutamate dehydrogenase activity.

The *gdhA* mutants are unique in that they are the only class of mutants exhibiting abnormalities of ammonium regulation where the primary gene function - determination of the NADP glutamate dehydrogenase protein - is known. There are two main hypotheses to explain the involvement of NADP glutamate dehydrogenase in ammonium regulation. It might be that ammonium itself is not the true effector whose concentration determines the level of the various enzymes and uptake systems. Instead, the true effector might be some metabolite, e.g. L-glutamate or L-glutamine produced from ammonium via glutamate dehydrogenase activity. Thus, lack of glutamate dehydrogenase activity would result in low levels of the true effector and consequent derepression. There are two powerful arguments against this hypothesis. The only likely candidates for the role of effector are L-glutamate and L-glutamine and both are readily taken up by cells, yet neither results in effective repression of all the systems under discussion. This is difficult to understand if they themselves, or anything derived from them, were the true effector. Furthermore, it appears that it is the extracellular concentration of ammonium, not the intracellular concentration, which determines the level of many of the ammonium regulated systems (Pateman *et al.*, 1973 and see later section). This too is difficult to understand if the true effector were some product derived from intracellular ammonium.

The alternative hypothesis is that the NADP glutamate dehydrogenase protein has some kind of function in ammonium regulation in addition to its catalytic activity. We consider this the more probable state of affairs, and a possible model for ammonium regulation which incorporates a regulatory role for NADP glutamate dehydrogenase is given later. In *A. nidulans* as in many other fungi, there is a NAD glutamate dehydrogenase in addition to the NADP specific enzyme. The NAD glutamate dehydrogenase does not appear to be involved in ammonium regulation and *gdhB* mutants which lack NAD glutamase dehydrogenase activity are in no way abnormal with regard to ammonium regulation (Kinghorn & Pateman, 1973, 1976).

tamA

The toxic effects of the ammonium analogue methylammonium, the urea analogue thiourea (Dunn & Pateman, 1972) and the asparagine analogue aspartic hydroxamate (Drainas, Kinghorn & Pateman, 1976) are like those of chlorate relieved by ammonium. Consequently, there is a possibility that simultaneous resistance to two or more of such toxic analogues might result from mutation altering the properties of a regulatory molecule concerned with ammonium repression. A number of resistant mutants were obtained by selecting for simultaneous resistance to the three analogues chlorate, thiourea and aspartic hydroxamate (Kinghorn & Pateman, 1973; Kinghorn & Pateman, 1975a). These triply resistant mutants were genetically complex, each strain carrying at least two and often three mutations. Genetic analyses of these mutants showed the existence of basically two classes of triply resistant mutants, and these map in the *tamA* (thiourea, asparate hydroxamate and methylammonium resistance) and *areA* (see below) loci. In addition, enhancers of specific resistances, e.g. chlorate, were found but the studies of these have not been pursued.

The *tamA* resistant mutants *tamA*^{r}*101*, *tamA*^{r}*103*, *tamA*^{r}*105* and *tamA*^{r}*119* are resistant to thiourea, aspartic hydroxamate, methylammonium, chlorate with alanine as a sole nitrogen source, and are recessive in the heterozygous diploid. These mutants are partially repressed, i.e. about 25% of wild-type activity with respect to nitrate reductase, ammonium transport, urea transport and asparaginase. The *tamA*^{r}*119* mutant shows partially repressed activity of other ammonium control systems such as L-glutamate transport and arginase. Partially repressed (i.e. resistant) mutants, in contrast to fully repressed alleles, e.g. *tamA*^{r}*50* (see below) are able to grow on all nitrogen sources, and this is probably attributable to the low level of various metabolic activities being sufficient for growth. On the other hand, the loss of 75% of activity protects partially repressed strains against analogue toxicity. The partially resistant strains have two further unique characteristics - they efflux ammonium particularly when using amino acids as nitrogen sources, and they have low levels (about 10%) of NADP glutamate dehydrogenase activity.

tamA^{r}*50* is a fully ammonium repressed mutant mapping in the *tamA* locus. This mutant is only able to use ammonium as a sole nitrogen source (Pateman & Kinghorn, 1976). *tamA*^{r}*50* was found to have extremely low levels of ammonium regulated systems, e.g. less than 1% wild type nitrate reductase levels. These low transport and enzyme levels presumably account for its poor growth characteristics. *tamA*^{r}*50* is recessive to *tamA*$^{+}$.

A mutation mapping in *tamA*, designated *tamA*^{d}*1* (Pateman & Kinghorn, 1976) leading to ammonium derepression of a number of ammonium repressible activities, was obtained by direct selection for derepression in a *tamA*^{r}*119* strain. *tamA*^{d}*1* is semi-dominant to *tamA*$^{+}$ and leads to relief of repression of a number of systems but does not restore wild-type NADP-GDH activity. The existence of alleles at the *tamA* locus which can determine either repression or derepression indicates that *tamA* is a regulatory gene involved in ammonium regulation.

areA

Arst & Cove (1973) obtained two chlorate resistant strains which were the result of mutations in a gene which they called *areA*. These two mutants, *areA*^{r}*-1* and *areA*^{r}*-2*, grow poorly on most nitrogen sources including nitrate, nitrite, amino acids, purines and amides. The growth characteristics of the *areA*^{r} strains are recessive in heterokaryons and heterozygous diploids with the *areA*$^+$ allele. The *areA*^{r}*-1* strain possesses low levels of nitrate and nitrite reductase and of urate oxidase, the enzyme necessary for nitrate and uric acid utilization. These properties of the *areA*^{r} strains suggest that the *areA* gene is a gene involved in ammonium regulation. The mutant alleles would result in the gene product repressing the ammonium regulated enzymes and uptake systems even in the absence of ammonium, and thus growth would only occur when ammonium itself was the nitrogen source. Three other fully ammonium repressed alleles of *areA* are known: *areA*^{r}*550* and *areA*^{r}*551*, Kinghorn & Pateman, unpublished results; and *areA*^{r}*217* (Hynes, 1975). All of these are repressed for many ammonium controlled functions and are unable to grow normally on any nitrogen source except ammonium.

Partially repressed *areA* alleles (*areA*^{5}*500-510*) similar to *tamA* partially repressed alleles, were isolated on the basis of triple resistance to chlorate, thiourea and aspartic hydroxamate with nitrate as sole nitrogen source (Kinghorn & Pateman, 1975). Unlike partially repressed *tamA*^{r} strains, partially repressed *areA* (e.g. *areA*^{r}*500*) alleles are not resistant with alanine as sole nitrogen source. Again, similar to *tamA*^{r}*119*, partially repressed *areA* alleles grow on all nitrogen sources, but in contrast have normal NADP-GDH and ammonium efflux activity.

The *xprD1* (*areA*^{d}*D1*) mutant was obtained by Cohen (1972); it is ammonium derepressed for a number of metabolic activities including extracellular protease production. This mutant is rather difficult to characterise since it grows poorly even on ammonium, it is not certain if its extreme characteristics are due to the nature of the allele or to modifying mutatations. It is almost unique amongst *areA* and *tamA* alleles in that it is fully derepressed for extracellular protease; although some *tamA* alleles which suppress *sarA* and *sarB* (Polkinghorne & Hynes, unpublished results) also affect extracellular protease regulation. Another general ammonium derepressed allele *areA575* is known (Kinghorn & Pateman, unpublished results). It shows ammonium derepression for all ammonium regulated activities except for extracellular protease production. The existance of *areA* alleles which result in ammonium repression and others which result in ammonium derepression indicates that *areA* is a regulatory gene involved in ammonium regulation.

A model for ammonium regulation

It is evident from the brief survey of the more important characteristics of the numerous genes involved in ammonium regulation that the system is extraordinarily complex. In such a situation it is both instructive and necessary to consider the simplest set of assumptions concerning the functions of the genes which could plausibly explain the experimental observations. There are four main features which need to be incorporated into a general scheme of ammonium regulation: (1) the differential effect of extracellular and intracellular ammonium on ammonium repression. The thiourea and glutamate uptake systems are fully derepressed in cells which have been nitrogen starved (Pateman *et al.*, 1973). In contrast, cells held in 10 mM ammonium are fully repressed, with the specific activities of the two uptake systems some ten to twenty-fold lower. The concentration of the intracellular ammonium pool is similar, about 100 to 200 μM, after the two treatments. Therefore normal cells can be either fully repressed or fully derepressed for the two uptake systems and yet have the same low level of intracellular ammonium. This indicates that, in some way, normal cells monitor and respond to the extracellular ammonium concentration with respect to the regulation by ammonium of glutamate and thiourea uptake. When the external concentration of ammonium is high (50 mM) the intracellular pool is also high and the level of methylammonium uptake is low. A high intracellular ammonium pool derived from urea by urease activity also results in a low level of methylammonium uptake (Pateman *et al.*, 1973). This indicates that it is the concentration of the intracellular ammonium pool which determines the level of the methylammonium uptake system, which is in fact the uptake system for ammonium itself. However, the *gdhA* mutants are considerably derepressed with respect to regulation by intracellular ammonium of methylammonium uptake. The *gdhA* mutants - which lack NADP glutamate dehydrogenase - are unique in their response to intracellular ammonium, unlike the normal strain or any of the other mutants; (2) the role of NADP glutamate dehydrogenase in ammonium regulation; (3) the roles of the regulatory genes *areA* and *tamA*; (4) the roles of the various other classes of ammonium derepressed mutants.

The main features of our hypothesis (illustrated in Fig. 2) are: (a) Normal cells of *A. nidulans* monitor separately the extracellular and intracellular ammonium concentrations with respect to ammonium regulation; (b) the extracellular ammonium concentration determines the level of a number of enzyme and uptake systems, notably nitrate reductase, urea uptake and glutamate uptake; (c) the intracellular ammonium concentration determines the level of the ammonium uptake system; (d) NADP glutamate dehydrogenase located in an ammonium monitoring site in the cell membrane can complex with extracellular but not intracellular ammonium. The concentration of this complex indirectly determines the level of nitrate reductase, urea uptake and glutamate uptake, etc.; (e) NADP glutamate dehydrogenase can complex with intracellular ammonium. The concentration of this second type of complex determines the level of the ammonium uptake system; (f) a number of cell membrane components are essential both for the function of a number of metabolically unrelated transport systems such as urea and glutamate, and also for the function of the external ammonium monitoring site. Mutation in the genes determining such structural membrane components could simultaneously result in the malfunction of a number of transport systems and the ammonium monitoring sites. Thus genes such as *meaA* and *amrA*, which show both transport and ammonium regulation abnormalities, would be interpreted as structural genes for cell membrane proteins; (g) a control pathway converts information concerning the external and internal concentration of ammonium - which is in the form of the NADP glutamate dehydrogenase complexes - into action with respect to synthesis of the regulated enzyme and uptake systems. The regulatory genes *areA* and *tamA* determine proteins which mediate later steps in this control pathway.

The role of NADP-GDH in monitoring the extracellular ammonium concentration

It seems clear from the intracellular ammonium pool data and the derepression characteristics of the *gdhA* mutants that cells of *A. nidulans* distinguish between external and internal ammonium with respect to regulation. A monitoring mechanism located in the cell membrane itself is the simplest type that can be envisaged for distinguishing between the external and internal concentrations of a metabolite. An alternative possibility is that NADP-GDH plays some role in the internal compartmentation of ammonium between a membrane vesicle and the cytosol and that only the cytosol ammonium pool is significant with respect to regulation. There is a situation of this kind in arginine metabolism in *Neurospora crassa* (Weiss, 1976). However, this latter alternative makes it less easy to understand the transport characteristics of *meaA* and other genes. The role suggested for NADP glutamate dehydrogenase explains why recessive mutations in the *gdhA* structural gene can result in ammonium derepression

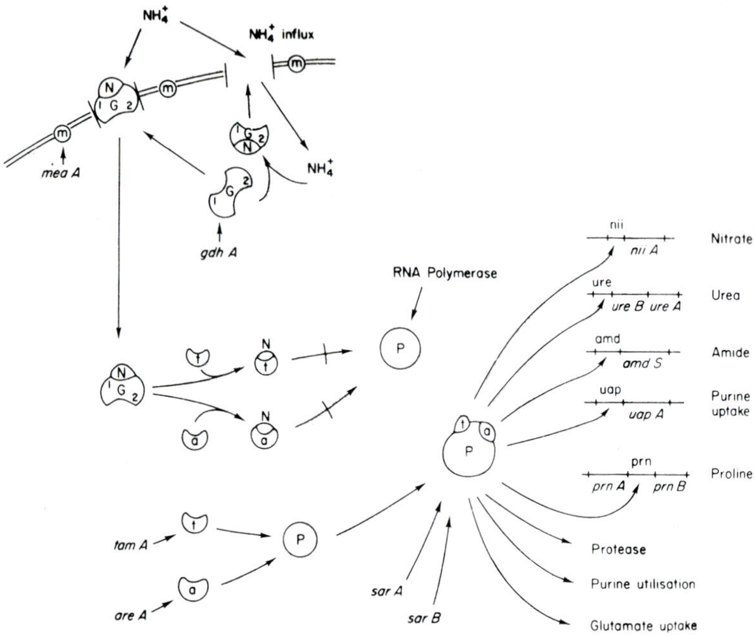

Fig. 2. Hypothetical model for the action of the *ghdA*, *areA* and *tamA* gene products in ammonium repression

The NADP-GDH molecule in a regulatory site in the cell membrane complexes with external ammonium. The GDH-ammonium complex inactivates the *areA* and *tamA* gene products and this results in ammonium repression. In the absence of external ammonium the *areA* and *tamA* products with a DNA-dependant RNA polymerase form a complex which is capable of transcribing the various structural genes which are subject to ammonium repression. Thus the *areA* and *tamA* products possess a positive regulatory action mediated through cis-acting control regions by conferring specificity on a core RNA polymerase.

in addition to loss of catalytic activity. The altered or absent enzyme protein no longer interacts with ammonium and the regulatory system behaves as if ammonium were absent.

The roles of the tamA and areA genes

Mutations which can cause either general ammonium repression or general ammonium derepression can occur in both the *tamA* and *areA* genes. The repressed alleles are recessive and the derepressed alleles all show some degree of dominance. The only simple explanation of these characteristics is that *tamA* and *areA* are regulatory genes which code for products - presumably protein - which have a positive action that is a necessary function in the expression of many ammonium regulated genes. When the ammonium status of the cell is low the *tamA* and *areA* products are in a form which performs some essential role in the expression of many genes. When the ammonium status of the cells is high the *tamA* and *areA* products are in a form which does not promote the expression of the ammonium regulated genes.

Since there are a number of genes affecting ammonium regulation the epistatic relationships between alleles in different genes might provide information about the temporal sequence of action of the gene products in the control pathway. Accordingly various double mutants have been made and their ammonium regulation characteristics determined. The properties of the majority of the relevant double mutants are given in Table 1. There are two points to be noted concerning these epistatic relationships: (a) Only the phenotype with respect to ammonium control is relevant. It is possible that some of the gene products have other functions in addition to their role in ammonium control. The effect of NADP-GDH levels of all known mutations in *tamA* may be an example. The expression of such characters in double mutants is irrelevant with respect to ammonium regulation. (b) It is only possible to score for epistacy in double mutants carrying alleles determining opposite characteristics, that is one gene must carry a repressed and the other a derepressed allele.

A number of general conclusions can be drawn from the data in Table 1 and from the general characteristics of the mutants. First, the repressed alleles of both *tamA* and *areA* are epistatic to the derepressed alleles in *meaA*, *meaB*, *amrA*, and *gdhA*. This indicates that the action of *meaA*, *meaB*, *amrA* and *gdhA* with respect to ammonium regulation is earlier than that of *tamA* and *areA*. This is in line with our hypothesis that these four genes are in some way concerned with monitoring the internal and external ammonium status of the cell. Second, all the alleles of *meaA*, *meaB*, *amrA* and *gdhA* just cause general ammonium derepression without appreciable differential effects on the levels of the

Table 1. Genes involved in ammonium regulation

Gene	Phenotype	Dominance	Epistacy	Reference
meA8 IV	General ammonium derepressed Low ammonium transport	Semi-dominant to *meaA*+	*areA*^{r}*1*, *tamA*^{r}*50*; *tamA*^{r}*119*; all epistatic to *meA8*	Arst & Cove, 1969 Arst & Page, 1973 Pateman *et al.*, 1974
gdhA1 III	General ammonium derepressed. Low NADP-GDH activity. Structural gene for NADP-GDH	Recessive to *gdhA*+	*areA*^{r}*1*; *tamA*^{r}*50*; *tamA*^{r}*119*; all epistatic to *gdhA1*	Arst & Macdonald, 1973 Kinghorn & Pateman, 1973; 1975
tamA^r*119* VI *tamA*^{r}*105*	Partial ammonium repression Low NADP-GDH activity	Recessive to *tamA*+	*areA*^{d}*575* epistatic to *tam*^{r}*119* and *tam*^{r}*105*	Kinghorn & Pateman, 1975 Polkinghorne & Hynes (unpublished)
tamA^r*50*	Full ammonium repression. Low NADP-GDH activity	Recessive to *tamA*+	*areA*^{d}*575*; ure-1; nia 22; nii 14, ASP20; all epistatic to *tamA*^{r}*50*	Pateman & Kinghorn, 1976
tamA^d*1*	General ammonium derepression Low NADP-GDH activity	Semi-dominant to *tamA*+	*areA*^{r}*500*; *areA*^{r}*550*; *nirA1*; all epistatic to *tamA*^{d}*1*	Pateman & Kinghorn, 1976
areA^r*500* III	Partial ammonium repression	Recessive to *areA*+		Kinghorn & Pateman, 1975
areA^r*1* *areA*^{r}*2*	Full ammonium repression	Recessive to *areA*+	*prn*^{d}*20*; *amdA*; *intA*^{c}*2*; *creA*^{d}*1* all epistatic to *areA*^{r}*1*	Arst & Cove, 1973 Arst, 1976
areA^r*19* (amdT19)				Hynes, 1972; 1975 Arst & Cove, 1973

Table 1 cont'd...

Gene	Phenotype	Dominance	Epistacy	Reference
areA^{r}*550*	Full ammonium repression.	Recessive to *areA*$^{+}$	ure-1 epistatic to *areA*^{r}*550*	Pateman & Kinghorn, 1976
areA^{d}*101*	General ammonium derepression	Semi-dominant to *areA*$^{+}$	*nirA1* epistatic to *areA*^{d}*101*	Arst & Cove, 1973
areA^{d}*575*	General ammonium derepression	Semi-dominant to *areA*$^{+}$		Pateman & Kinghorn, 1976
areAD1 (xpD1)	General ammonium derepression	Semi-dominant to *areA*$^{+}$		Cohen, 1973 Arst & Cove, 1973
areA102 (amdT102)	Acetamidase derepression; ammonium repression of formamidase, uric acid transport and xanthine transport	Semi-dominant to *areA*$^{+}$	*sarA3; sarB33* epistatic to *areA102*	Hynes & Pateman, 1970; Hynes, 1972, 1975; Arst & Cove, 1973; Arst & Scazzocchio, 1975
ure1 (UR1) VIII	Suppresses *tamA*^{r}*50* on urea only and ammonium derepressed for urea utilization only	Cis-dominant to *ure*$^{+}$		Pateman & Kinghorn, 1976
nii14	Suppresses *tamA*^{r}*50* on nitrite only			Pateman & Kinghorn, (unpublished)
nia22	Suppresses *tamA 50* on nitrate only			Pateman & Kinghorn, (unpublished)

Table 1 cont'd...

Gene	Phenotype	Dominance	Epistacy	Reference
amd9 (amd19)	Suppresses *areA*^{r}*217* on acetamide only	Cis-dominant to *amd*$^+$ only		Hynes, 1975
sarA3 VII	Suppresses *areA102*	Semi-dominant to *sarA*$^+$	*tamA*^{r}*105* epistatic to *sarA3*	Polkinghorne & Hynes, 1975 and unpublished
sarB33 VII	Suppresses *areA102* Reduced utilization many nitrogen sources	Recessive to *sarB*$^+$	*tamA*^{r}*105* epistatic to *sarB33*	Polkinghorne & Hynes, 1975 and (unpublished)
ASP20	Suppresses *tamA*^{r}*50* on asparagine only			Pateman & Drainas (unpublished)
prn 20 VII	Suppresses *areA*^{r}*1* on proline only	Cis-dominant to *prn*$^+$		Arst & MacDonald, 1975
uap100 I	Suppresses *areA*^{r}*102* on uric acid	Cis-dominant to *uap*$^+$		Arst & Scazzocchio, 1975
amdA^d*301* VII	Suppresses *areA*^{r}*1* on acetamide only	Semi-dominant		Arst, 1976
intA (amdR) II	Controls expression of genes for acetamide GABA transaminase and GABA permease			Arst, 1976

Table 1 cont'd...

Gene	Phenotype	Dominance	Epistacy	Reference
amdR$^-$	Low acetamidase and/ or low formamidase	Recessive to *amdR*$^+$ and *amdR*^{c}		Hynes & Pateman, 1970a,b
amdR^c	Constitutive acet- amidase	Semi-dominant to *amdR*$^+$		Hynes & Pateman, 1970a,b

Gene designation in brackets denotes previously used gene symbol

Roman numerals indicate linkage group of each gene

various ammonium regulated activities. In contrast some alleles of both *tamA* and *areA* have much more marked effects on the levels of some metabolic activities than on others. This also implies that the products of *tamA* and *areA* act more directly on the expression of the ammonium controlled genes. Third, there is no clear cut unilateral relationship between all the alleles of *tamA* and *areA*, some alleles of *tamA* are epistatic to some alleles of *areA* and *vice versa* although in the majority of cases *areA* alleles are epistatic to *tamA* alleles. The products of the *sarA* and *sarB* genes interact in some way with the products of the *tamA* and *areA* genes since *sarA* and *sarB* alleles can be epistatic to *areA* alleles and hypostatic to *tamA* alleles. Fourth, there are several strong reasons for supposing that the products of the *tamA* and *areA* genes act at the level of transcription: (a) The great range of control effects determined by alleles of both *tamA* and *areA* often with differential effects on various metabolic activities; (b) five examples are now known in which a cis-dominant mutation in a region next to a structural gene can suppress the repressive action of a *tamA*^{r} or *areA*^{r} allele on the expression of the structural gene; (c) the nii 14 mutation, which maps close to *niiA* the structural gene for nitrite reductase, allows expression of *niiA*$^+$ but not of *niaD* in the presence of the *tamA*^{r}*50* allele. Thus it is probable that the product of the *tamA* and *areA* genes interact with control regions adjacent to structural genes to regulate the transcription of those structural genes.

Britten & Davidson (1969) have put forward an appealing general model for gene regulation in eukaryotes. Its essential features are that products of 'integrator' genes interact with 'receptor' genes to control the transcription of batteries of 'producer' genes. It is argued above that the products of the *areA* and *tamA* genes act at the level of transcription. If this is so then the *areA* and *tamA* are clearly examples of 'integrator' genes, the closely linked cis-dominant regions such as ure^{d}1 and prn^d are 'receptors' and the various structural genes such as *ureA*, *ureB* and *prnA*, *prnB* are 'producer' genes. Britten & Davidson make no assumptions about the transcription control mechanism other than that the usual state of the eukaryote genome is some form of histone-mediated repression which requires specific activation for gene expression. There are two main possibilities concerning the mode of action of the *areA* and *tamA* products. They might be involved in specifically 'opening up' or in some way making available for transcription those regions of the genome carrying genes specifying ammonium regulated functions. These regions would presumably be masked by histones or some general repressive mechanism unless the ammonium status of the cell were low and the *areA* and *tamA* products were in the active form with respect to their unmasking action. We now know that the various classes of histones are present in *A. nidulans* (Morris, these proceedings) and presume that *Aspergillus* chromosomes have a complex structure similar to that of other eukaryotes, but there is no direct evidence for any such general repressive mechanism at present. A second possibility is that the *tamA* and *areA* products play a necessary role, together with RNA polymerase, in transcription by interaction with cis-dominant promoter regions next to the structural genes. There are well known examples of this type of action in prokaryotes. The expression of the L-arabinose operon in *E. coli* B/T requires cAMP, the cAMP receptor protein (CRP) and the action of the product of the *araC* regulator gene interacting with the *araI*, *araP*, *araCRP* and *araO* sites which are next to the *araB* structural gene (Heffernan, Bass & Wilcox, 1976). The precise mode of action of the *araC* gene product is unknown but there are obvious similarities between the characteristics of the *araC* and the *tamA* and *areA* genes. In several bacteriophage systems, T4 in *E. coli* and SP01 in *Bacillus subtilis* (Fox, 1976; Tjian & Pero, 1976) specific transcription of phage genes is controlled by phage coded regulatory proteins that interact with the core bacterial RNA polymerase. The late genes of phage SP01 can only be transcribed by bacterial core RNA polymerase that is bound to host delta protein and the products of the phage regulatory genes 33 and 34. That is the products of regulatory genes 33 and 34 act as sub-units of a specific form of RNA polymerase which is essential for the transcription of the late phage genes. This type of positive control mechanism would accommodate very well all the known characteristics of the *areA* and *tamA* genes. Of course there is no experimental evidence for such a mechanism in *A. nidulans*, but for ease of presentation and understanding this mode of action has been attributed to the *tamA* and *areA* genes in the purely speculative model outlined in Fig. 2.

The sequence of events which occur during ammonium regulation, outlined in Fig. 2, may be described as follows: (1) Extracellular ammonium can enter the cell through active transport ammonium influx sites into the intracellular pool. Intracellular ammonium can interact in a specific way with NADP-GDH molecules, this NADP-GDH ammonium complex in some way controls the rate of ammonium inflex through the ammonium transport sites. Extracellular ammonium can interact in a second specific fashion with NADP-GDH molecules held in special ammonium monitoring sites in the cell membrane. It is this special regulatory form of the NADP-GDH ammonium complex which interacts with the *tamA* and *areA* gene products to prevent the expression of all ammonium regulated metabolic functions except ammonium influx. Thus common recessive mutations in the *gdhA* gene which determine abnormal NADP-GDH molecules would result in ammonium derepression because of the absence of the normal NADP-GDH regulatory function. The *meaA* gene would determine a membrane structural protein which is necessary for the correct functioning of both the ammonium monitoring sites and various transport sites such as that for ammonium. Thus common recessive mutations in the *meaA* gene could result in both ammonium derepression and abnormal ammonium transport. (2) The complex of extracellular ammonium and NADP-GDH interacts with the products of *tamA* and *areA* genes in such a way that these gene products are no longer active in promoting the expression of ammonium regulated systems. In Fig. 2 this interaction is represented by the transfer of ammonium to the *tamA* and *areA* products and their subsequent failure to interact with the RNA polymerase. Alternatively if NADP-GDH is not complexed with ammonium it does not interact with the *tamA* and *areA* products and they are free to associate with a core RNA polymerase. Thus the *tamA* and *areA* products act as sub-units in a specialised form of RNA polymerase which alone is capable of effectively transcribing the structural genes of the ammonium regulated metabolic functions. It is easy to see how mutation in *tamA* or *areA* could alter their products in such a way as to result in either ammonium repression or derepression. Also, such mutationally altered regulator gene products might have differential effects on the transcription of the various structural genes depending on the precise relationship between the RNA polymerase and the cis-dominant control regions. Furthermore, such a function for the *tamA* and *areA* products could explain why some *areA* alleles are epistatic to some *tamA* alleles and *vice versa*. Epistacy in this situation would reflect the final effect of mutationally altered sub-units on the functional properties of the multimeric polymerase protein. A similar possibility is that the *tamA* and *areA* genes alone determine a specialised RNA polymerase which has the unique function of transcribing the ammonium regulated genes. Finally, there is the question of how the interaction of carbon and nitrogen regulation might be fitted into such a system. In general, in *A. nidulans* it seems that when a metabolite may be utilised as both a carbon and nitrogen source, carbon control is epistatic to nitrogen control. This obviously reflects the fact that in the short term an energy supply is more important than nitrogen for cell survival. It may be that regulatory genes such as *creA* and *sarA* and *sarB* also determine sub-units of the RNA polymerase. If such a carbon regulating sub-unit could substitute in whole or in part for the function of the *tamA* and *areA* products it would allow carbon control to take precedence over nitrogen control. This kind of possibility is indicated in Fig. 2 by arrows suggesting an interaction of *sarA* and *sarB* products with the RNA polymerase. In summary it must be emphasised that the hypothesis outlined is based on a great range of experimental observations concerning the characteristics and interactions of various regulatory genes. However, the endeavour to add some flesh to the bones by way of suggesting the more plausible molecular mechanisms is purely speculative. The future will show how wild such guessing can be.

UREA UTILIZATION

There is an active transport uptake system for urea, which also transports thiourea, in *A. nidulans* (Dunn & Pateman, 1972 and unpublished results). Mutations in the *ureA* gene (formerly *uruA*) on chromosome VIII result in defective urea and thiourea transport and consequent protection from thiourea toxicity. Inside the cell urea is converted to carbon dioxide and ammonium by urease. Urea transport and urease activity are regulated by ammonium repression. Four genes are known in which mutation can result in lowered activity: *ureB* (formerly *uX*) on chromosome VIII; *ureC* (formerly *uY*) on chromosome VIII; *ureD* (formerly *uZ*) on chromosome VII; *ureE* on chromosome VIII. The status of these genes with respect to urease is still unclear, but *ureB* is probably the structural gene for urease (Kinghorn & Pateman unpublished results). The *ureA* and *ureB* genes are closely linked; crosses between *ureA1* and *ureB2* gave eleven recombinants in a total of about 2640 progeny, that is about 0.4% recombination.

The *tamA*^{r}*50* mutant cannot grow on urea. A mutant ure 1 (formerly UR-1) was obtained in a *tamA*^{r}*50* strain by selection for growth on urea. The ure 1 allele suppresses *tamA*^{r}*50* specifically with respect to urea utilisation and the *tamA*^{r}*50* ure 1 double mutant is able to grow only on ammonium or urea. The *tamA*^{r}*50* ure 1 and the *tamA*$^{+}$ ure 1 genotypes are both ammonium derepressed specifically for urea transport and urease activity. The ure 1 mutation is closely linked to *ureA1* with a recombination frequency of about 1.3% from a total progeny of 610. Thus the probable structural genes *ureA* and *ureB* and the control mutation ure 1 form a gene cluster. The position of ure 1 relative to the structural genes is important and some information is available. In a cross of ure 1, *ureB*$^{+}$, *ureA*$^{+}$ X ure^{+} *ureB*$^{-}$*2* *ureA*$^{-}$*1* a total of 5 *ureB*$^{+}$ *ureA*$^{-}$*1* recombinants were obtained. All five recombinants were ure 1 *ureB*$^{+}$ *ureA*$^{-}$*1*. This indicates that ure 1 is next to *ureB* and thus either between *ureB* and *ureA* or at the end of the cluster next to *ureB*. The above data and the recombination frequencies between ure 1 and *ureA1* of 1.3%, between ure 1 and *ureB2* of 0.2%, and between *ureB2* and *ureA1* of 0.4% suggest ure 1 *ureB2* *ureA1* as the most probable order of the genes in the cluster.

Table 2. Dominance relationships of *ure*^{d}*1*

Diploid genotype	Dominance of ure^{d}1 allele	Phenotype with respect to growth on urea and to ammonium control of urease and urea transport
tamA^{r}*50 ure*^{d}*1 ureB*$^{+}$ *ureA*$^{+}$ / *tamA*^{r}*50 ure*$^{+}$ *ureB*$^{+}$ *ureA*$^{+}$	Dominant	Grows on urea. Derepressed for urease and urea transport.
tamA^{+} *ure*^{d}*1 ureB*$^{+}$ *ureA*$^{+}$ / *tamA*$^{+}$ *ure*$^{+}$ *ureB*$^{+}$ *urA*$^{+}$	Dominant	Derepressed for urease and urea transport.
tamA^{+} *ure*^{d}*1 ureB*$^{+}$ *ureA*$^{-}$ / *tamA*$^{+}$ *ure*$^{+}$ *ureB*$^{+}$ *ureA*$^{+}$	Not trans-dominant	Repressed for urea transport
tamA^{+} *ure*^{d}*1 ureB*$^{+}$ *ureA*$^{+}$ / *tamA*$^{+}$ *ure*$^{+}$ *ureB*$^{+}$ *ureA*$^{-}$	Cis-dominant	Derepressed for urea transport.
tamA^{r}*50 ure*$^{+}$ *ureB*$^{+}$ *ureA*$^{+}$ / *tamA*^{r}*50 ure*$^{+}$ *ureB*$^{+}$ *ureA*$^{+}$	Not trans-dominant	Unable to grow on urea.
tamA^{r}*50 ure*^{d}*1 ureB*$^{+}$ *ureA*$^{+}$ / *tamA*^{r}*50 ure*$^{+}$ *ureB*$^{-}$ *ureA*$^{+}$	Cis-dominant	Grows on urea.

tamA, regulator gene for ammonium repression; ure^{d}1, ammonium derepressed allele in ure control region; *ureB*, probable structural gene for urease; *ureA*, probable structural gene for urea uptake protein.

The dominance relationships of the ure 1 mutation and the *ureA* and *ureB* genes were determined in appropriate heterozygous diploids (Pateman & Kinghorn, 1976 and unpublished results).

The properties of some of these heterozygous diploids are given in Table 2. The urea 1 mutation is cis-dominant with respect to the expression of both the urease and the urea transport genes. The cis-dominance is clear cut with respect to the suppression of the repressive action of the *tamA*^{r}*50* alleles and also with respect to ammonium derepression in the presence of the *tamA*$^{+}$ alleles. Thus a gene for urease, a gene determining urea transport and cis-acting control region are all tightly linked. This is likely to be a particularly interesting case for further investigation, since it is the only instance so far in *A. nidulans* of a gene cluster where a single type of control, namely ammonium repression, operates. In other systems, such as nitrate utilization or proline utilization, induction and/ or carbon regulation are confounded with repression. It is of particular importance to determine the position of the urea control region, if this is at one end of the cluster it will imply the existence of a polycistronic message from the *ureB* and *ureA* genes, and at present there is no definite evidence for this in any other eukaryotic system.

NADP AND NAD LINKED L-GLUTAMATE DEHYDROGENASES

A. nidulans can synthesize two glutamate dehydrognases, one NADP linked (NADP-GDH) and the other NAD linked (NAD-GDH). A good deal is known about the biochemistry and genetics of these enzymes but little about their regulation. NADP-GDH is of particular interest since there is strong evidence that it may act as a regulatory protein in ammonium control in addition to its catalytic function.

NADP-GDH

The role of this enzyme is primarily an animating one. There is a single structural gene, *gdhA*, in *A. nidulans* (Kinghorn & Pateman, 1973). NAD-GDH activity does not substitute for NADP-GDH activity in *gdhA*$^{-}$ mutants. The level of NADP-GDH activity is markedly affected by both the nitrogen and carbon sources utilized by the cells. If *A. nidulans* is grown on glucose as a carbon source, the synthesis of NADP-GDH is high with inorganic nitrogen sources such as ammonium and is low when L-glutamate is the source of nitrogen (Pateman, 1969). If glucose is replaced as a carbon source by L-glutamate or other amino acids, or the cells are carbon starved, there is a rapid loss of NADP-GDH activity which can be prevented by the addition of the protein inhibitor cycloheximide (Hynes, 1974a; Kinghorn & Pateman, 1974a). This suggests that in some way glucose is a prerequisite both for NADP-GDH activity and for glutamate repression of the enzyme. It may be that this loss of NADP-GDH activity in the absence of carbon is caused by an inactivating protein, possibly a protease, and glucose prevents the synthesis of this protein.

Pateman & Kinghorn (unpublished work) have made 'control' mutants for NADP-GDH. The experimental strategy was based on the following observations: wild-type *A. nidulans* growing on L-glutamate as a carbon source has high NAD-GDH and neglibible NADP-GDH activity. Mutants at the *gdhB* locus lack NAD-GDH activity and are unable to grow on L-glutamate as a carbon source (see later section. Consequently 'suppressor' mutations which allowed a *gdhB* mutant to grow on L-glutamate as a carbon source might include a class in which the carbon regulation of NADP-GDH had been relaxed. A large number of suppressors which allowed growth of *gdhB1* on L-glutamate as a carbon source were isolated. Four of these were found to have NADP-GDH activity with L-glutamate as carbon source (unlike the wild type, which has no activity), and this activity allows *gdhB1* to grow on L-glutamate, i.e. NADP-GDH substitutes for NAD-GDH activity. The genetics of these mutants is not yet well characterised, but preliminary data show that at least two new genes are involved. The nature of these suppressor mutations is obviously of interest and importance.

NAD-GDH

The primary role of NAD-GDH is a daminating one. The structural gene for NAD-GDH in *A. nidulans* is probably *gdhB* which is unlinked to *gdhA*. Recent observations (Kinghorn & Pateman, 1974) suggest that NAD-GDH activity in *A. nidulans* is at least partly regulated by glucose repression. The main facts which support this argument are as follows: first, carbon starvation after growth on glucose results in appreciable NAD-GDH activity. This activity is maximal after 3 h; second, maximal NAD-GDH activity is found in cells in which the carbon status is low, but probably sufficient for protein synthesis. This is the case when either acetate, L-glutamate or L-asparate or casamino acids is the main or only carbon source. Since NAD-GDH activity is found in cells after a short period of carbon starvation and in cells grown on acetate, the system is probably regulated by glucose repression alone or by repression and induction, induction being subordinate to repression. This seems similar to the situation in *N. crassa* (Strickland, 1971) but different to that in *Saccharomyces cerevisiae*, where ammonium represses NAD-GDH activity (Hierholzer & Holzer, 1963).

A mutant *gdhC1* has been isolated which can grow on L-glutamate as sole nitrogen source better than the wild type (Kinghorn

& Pateman, 1974b). The increased level of growth in this mutant (*gdhC1*) is due to appreciable derepressed levels of NAD-GDH activity when grown in the presence of glucose. The *ghdC* locus is not linked to *gdhA* or to *gdhB*. Since *gdhC1* is semi-dominant in the heterozygous diploid, a model is favoured where *gdhC* codes for a regulatory product which, on interaction with a carbon metabolite effector, represses synthesis, i.e. a positive control system. However, if this were so, one should find amongst strains lacking NAD-GDH activity common recessive mutations, in the *gdhC* gene, which result in repressed NAD-GDH activity and indeed any other system *gdhC* controls. Paradoxically, such a class of mutants has not so far been detected (Kinghorn & Pateman, unpublished results).

PURINE UTILIZATION

The uptake systems and enzymes responsible for purine utilization are subject to induction controls and to ammonium repression in *A. nidulans*. No attempt to summarise the main findings will be made here since the article by Scazzocchio, C. & Gordon, D. in this volume covers the area.

ARGININE UTILIZATION

In *A. nidulans* the enzymes responsible for arginine breakdown are subject to a complex series of controls which include ammonium repression. The article by Węgleński in this volume provides a comprehensive and succinct account.

PROLINE UTILIZATION

The uptake system and enzymes necessary for proline utilization in *A. nidulans* are subject to controls which include ammonium repression. A paper was presented at the meeting by Arst, H. & MacDonald, D. but they were unable to prepare a paper for this volume. The following paragraph is a statement provided by Arst, H. & MacDonald, D. for inclusion here.

An analysis of L-proline catabolism in *A. nidulans* is being carried out by Arst & MacDonald (1975 and unpublished results). They have shown that the activities necessary for degradation of proline are specified by a gene cluster in linkage group VII with the structure *prnA*-*prn*D-control region-*prn*B-*prn*C. Mutations in *prn*B eliminate a proline permease. Mutations in *prn*A and *prn*D eliminate proline oxidase. Mutations in *prn*A and *prn*C eliminate Δ^1-pyrroline-5-carboxylate dehydrogenase. Unfortunately, data bearing on the control of this cluster are not amenable to presentation at the moment, and Drs. Arst and MacDonald were therefore unable to prepare a paper for this volume.

HISTIDINE UTILIZATION

L-histidine is a poor nitrogen source for wild type *A. nidulans* and is not an appreciable carbon source for any known genotypes. The level of histidase appears to be the limiting factor in the utilization of histidase. No histidase structural genes are known. Histidase levels are controlled by ammonium repression and there may also be a regulatory effect of glutamate and glutamine (Polkinghorne & Hynes, 1975). Mutations in several genes can result in alterations in histidase levels:

areA

The *areA19* allele (formerly *amdT19*) results in poorer growth on histidine and a four-fold decrease in histidase levels compared with *areA*$^+$ in the absence of ammonium. The *areA102* (formerly *amdT102*) allele results in better growth on histidine as a nitrogen source and histidase levels are increased ten-fold compared with *areA*$^+$ in the absence of ammonium, but histidase production is still repressed by ammonium. The action of the *areA102* allele is semi-dominant with respect to histidase as it is for acetamidase. *areA102* is a unique allele of the *areA* gene in that ammonium derepressed alleles *areA*^{d} do not result in higher levels of histidase, nor do ammonium derepressed alleles of *tamA* have this effect (see section on ammonium regulation).

tamA

Ammonium repressed alleles, *tamA*^{r}, prevent growth on histidine presumably as the result of low histidase levels (Pateman & Kinghorn, unpublished results). The partially ammonium derepressed mutant *tamA*^{r}*105* (Kinghorn & Pateman, 1975) grows as *tamA*$^+$ on histidine, but the double mutant *areA102*, *tamA*^{r}*105* has much lower acetamidase activities than *areA102* alone (Hynes & Polkinghorne unpublished results). Thus both *areA* and *tamA* play some role in the ammonium control of histidase and the action of the *tamA*^{r}*105* allele is epistatic to that of the *areA102* allele.

sarA and *sarB*

In an attempt to obtain histidase structural gene mutants, Polkinghorne & Hynes (1975) selected for mutations which suppressed the enhanced growth of *areA102* on histidine. They did not obtain any histidase structural gene mutants, but did isolate suppressor mutations of *areA102* which mapped in two loci designated *sarA* and *sarB*. In all the *areA102 sarA* and *areA102 sarB* double mutants the histidase levels are correlated with their degree of growth on histidine and are below those of *areA102* alone. In strains carrying either *sarA* or *sarB* alleles histidase levels are still subject to ammonium repression. The *sarA* alleles are semi-dominant to *sarA*$^+$ with respect to histidase levels and histidine utilization, while the *sarB* alleles are recessive to *sarB*$^+$ for both properties in heterozygous diploids. The *sarA* and *sarB* alleles affect the utilization of many nitrogen sources resulting in poorer growth on amino acids such as phenylalanine, citrulline, glutamate, valine and serine, and on urea. Some *sarA* and *sarB* also affect milk clearing ability, i.e. extracellular protease activity.

Polkinghorne (unpublished results) has selected for mutations resulting in enhanced growth on histidine of a double mutant *areA102 sarA*. Third stage mutations suppressing the suppressive action of *sarA* on *areA102* were obtained. The great majority of this new class of suppressor mutations are *tamA* alleles. This common class of *tamA* alleles are epistatic in function to the *sarA* allele, which itself is epistatic in function to the *areA102* allele.

It is clear that the regulatory genes *areA* and *tamA* are concerned in ammonium control histidase. In addition, it seems likely that the *sarA* and *sarB* genes have a regulatory function both with respect to histidase and some other aspect of amino acid catabolism. A strange feature in this system is that with the number of regulatory components already uncomfortably large, the histidase structural gene is not known nor has the subsequent enzyme activity, urocanase, been detected in *A. nidulans*

AMIDE UTILIZATION

A. nidulans can grow on acetamide as both a carbon and nitrogen source and can grow on formamide as a nitrogen source. There are at least three amidases responsible for amide utilization (Hynes, 1975b). The general amidase appears only to be involved in the utilization of amides as nitrogen sources; whereas the acetamidase is responsible for the utilization of aliphatic amides as either carbon and/or nitrogen sources. The synthesis of the three enzymes is subject to a variety of regulatory controls. The formamidase and acetamidase are induced by formamide and acetamide and repressed by ammonium. The acetamidase is also repressed by carbon metabolites derived from glucose and acetate. Carbon derepression resulting from growth on acetate or acetamide as a carbon source overrides ammonium repression (Hynes, 1970).

Mutations in a number of genes can result in regulatory effects:

amdR

The *amdR* gene is on linkage group II. *amdR*$^-$ mutants grow poorly on acetamide. There are two types of *amdR*$^-$ allele; one produces low acetamidase under all conditions, the other type gives low acetamidase and formamidase on glucose but normal enzyme levels on carbon derepression. *amdR*$^-$ alleles are recessive to *amdR*$^+$ and *amdR*^{c} alleles. *amdR*^{c} alleles produce high constitutive acetamidase levels on acrylamide, which is a substrate for the enzyme, but does not induce acetamidase in the wild type *amdR*$^+$. *amdR*^{c} alleles are semi-dominant to *amdR*$^+$ alleles (Hynes & Pateman, 1970a,b). The *amdR* gene is also involved in the control of γ-amino-n-butyric acid metabolism (GABA) and has been renamed *intA* (Arst, 1976).

amdA

The *amdA* gene is in linkage group VII; it apparently controls the expression of *amdS*, the structural gene for acetamidase. *amdA*^{d} alleles are ammonium derepressed for acetamidase and suppress the ammonium repressed phenotype of *areA*^{r} mutations (see section on ammonia regulation) on acetamide (Arst, 1976).

amdI

The *amd19* mutation results in increased induction by acetamide of acetamidase and better growth on acetamide as a carbon or nitrogen source. The *amd19* mutation is closely linked to the acetamidase structural gene *amdS* and is cis-dominant to *amdI*$^+$ for enhanced growth on acetamide and elevated acetamidase activities (Hynes, 1975). It seems that the *amdI* is a control region analogous to a promotor or operator involved in the induction by acetamide of the expression of the *amdS* gene. The *amdI* region has been renamed and in Table 1 in line with the nomenclature for such control regions.

amdT (now known to be *areA* and mentioned as such in other sections)

The *amdT* gene is in linkage group III. The allele *amdT*^{c}*102* allows elevated levels of acetamidase on glucose in the absence of acetamide and permits growth on acrylamide. The *amdT*^{c}*102* allele is semi-dominant to the *amdT*$^+$ allele in heterozygous diploids. A possible explanation is that *amdT* (i.e. *areA*) codes for a positive acting regulatory product which is involved in carbon metabolite repression of acetamidase in addition to its role in ammonium regulation, so that *amdT*^{c}*102* is in one sense a carbon derepressed mutant (Hynes & Pateman, 1970b). The *amdT*^{c}*102* allele is still subject to ammonium repression with respect to acetamidase. The *amdT*^{c}*102* allele also results in poor growth on formamide and low levels of formamidase synthesis, yet formamide is not a carbon source and there is no indication of any form of carbon control of formamidase. So with respect to the control of formamidase, *amdT*^{c}*102* is behaving as an ammonium repressed allele (*areA*^{r}) of the *areA* gene.

areA and *tamA*

Mutations which result in ammonium repression, *areA*^{r} and *tamA*^{r}, and mutations which result in ammonium derepression, *areA*^{d} and *tamA*^{d} for all three amidases, are known. The characteristics of the *areA* and *tamA* alleles confirm that amidase synthesis in *A. nidulans* is subject to the same general ammonium regulation as many other metabolic functions and an *areA*^{r} ammonium repressed mutant has been used to obtain two other classes of regulatory mutants, *amd19* and *amdA*^{d}. The further properties of the *areA* and *tamA* genes are discussed in the section on ammonium regulation.

In this brief account no mention has been made of possible control effects of such metabolites as glutamate and glutamine (Hynes, 1974). The situation is very unclear, due partly to the very real problem of deciding which metabolites are the true effectors *in vivo* in both nitrogen and carbon control. At present we can do no more than bear in mind the possible existence of interdependent controls involving glutamate and glutamine in addition to those involving ammonium and carbon metabolites.

GENERAL CONCLUSIONS

It can be seen from the various sections of this and accompanying articles that our knowledge at the descriptive level of genetic regulation in *A. nidulans* has increased greatly in the last few years. There is now a plethora of regulatory mutants including cis-dominant types and there are several instances of gene clusters. There is also a rather overwhelming impression of complexity in the best known systems both with respect to the number of genes determining regulatory components and their interactions. It is apparent that control of metabolic functions can operate in some or all of the following ways: reversible enzyme inactivation, substrate induction of enzyme synthesis and the repression of enzyme synthesis by nitrogen and carbon metabolites. The nitrate utilization, arginine utilization and amide utilization systems strikingly illustrate these various complexities. Although our knowledge of regulation in all systems is still fragmentary a few broad principles seem to be emerging. We put forward our ideas as providing some sort of framework for thinking about regulation in *Aspergillus*, tenuous as these ideas may be at the present time.

(1) It is becoming increasingly probable that induction and repression act at the level of transcription in many control systems. This has not been proved in any system, but a number of cis-dominant control mutations closely linked to structural genes are now known and this makes transcriptional control the most likely hypothesis.

(2) There are two instances where an enzyme is thought to play some role in regulation; nitrate reductase in nitrate utilization and NADP glutamate dehydrogenase in ammonium repression. The likely hypothesis in these cases is that the enzyme is acting as a sensor for the co-effector, which is one of its substrates. The regulation system responds to the concentration of the enzyme-co-effector complex as the initial control signal rather than the concentration of the unbound co-effector. This results in autoregulation when an enzyme has this regulatory role for a metabolic pathway of which it is itself a component enzyme. This is the case for nitrate reductase but probably not for glutamate dehydrogenase. Proteins with this dual role of catalysis and regulation may prove to be a not uncommon feature of regulatory systems. It is easy to see how they might be of advantage from an evolutionary point of view.

(3) A two tier system of control is common; with specific induction for single pathways interacting with a more general repression which regulates a number of metabolically related pathways. This permits economy of control components in achieving the metabolically efficient objective of repression by a centrally important metabolite (such as ammonium) over-riding specific induction of peripheral pathways.

(4) The majority of regulatory genes known appear to determine positive acting regulatory products. The reason for this is unclear; it may be that it allows greater possibilities for a hierarchy of interactions between regulatory systems. Such interactions are important and clear cut; ammonium repression overrides specific inductions such as nitrate, but carbon derepression overrides ammonium repression. This allows the utilization of such metabolites as glutamate as carbon sources in the presence of ammonium and absence of glucose.

(5) The requirement for interactions between control systems may also explain the existence of several regulatory genes in some systems. The *tamA* and *areA* genes in the ammonium repression system appear to have very similar properties, but it may be that they have different specificities with respect to some of the induction systems or to carbon control. Two regulatory proteins

with some overlapping properties may be necessary to allow a wide range of interactions between specific inductions and general nitrogen and carbon control.

(6) There are good examples now in *A. nidulans* of gene clusters, cis-acting control regions analogous to operator/promotor regions, and regulatory genes. However, there is still no definate example of an operon structure with a cis-acting control region determining the production of a polycistronic message. Four genes concerned with proline utilization are closely linked in a cluster with some kind of control region in the centre. However, at the present time the regulation of these proline genes is unclear. There is evidence for cis-acting control mutations affecting the expressions of only one gene, *prnC*, of the four genes *prnA*, *prnB*, *prnC* and *prnD* in the cluster (H. Arst, this symposium). The structural genes for the nitrate and nitrite reductases are closely linked to each other and to the control mutations nia 22 and nii 14 (Pateman & Kinghorn, this article and unpublished results). But the position of the control regions relative to the structural genes is not yet known. In the case of the urea utilization gene cluster a cis-dominant mutation, ure 1, determines the expression of two closely linked genes *ureA* and *ureB*. The genetic evidence at present strongly favours the gene order ure 1 *ureB ureA* in the cluster. This order would indicate a polycistronic message. However, the appropriate delection mutants are required in this and the other *Aspergillus* gene clusters to provide definate evidence for operon structures if these exist.

The molecular biology of *Aspergillus* lags far behind the genetic and physiological studies. Such studies were an essential preliminary for our understanding of regulatory processes, but now the scene is set for investigations at the molecular level. A combination of molecular biological techniques and the fascinating variety of *Aspergillus* regulatory mutants would surely provide unique insights into regulation in eukaryotes.

REFERENCES

ARST, Jr., H.N. (1976). Integrator gene in *Aspergillus nidulans*. *Nature* 262, 231-4.

ARST, Jr., H.N. & COVE, D.J. (1969). Methylammonium resistance in *Aspergillus nidulans*. *Journal of Bacteriology* 98, 1284-93.

ARST, Jr., H.N. & COVE, D.J. (1973). Nitrogen metabolite repression in *Aspergillus nidulans*. *Molecular and General Genetics* 126, 111-41.

ARST, H.N. & MACDONALD, K. (1973). A mutant of *Aspergillus nidulans* lacking NADP-linked glutamate dehydrogenase. *Molecular and General Genetics* 122, 261-5.

ARST, Jr., H.N. & MACDONALD, B.W. (1975). A gene cluster in *Aspergillus nidulans* with an internally located cis-acting regulatory region. *Nature* 254, 26-34.

ARST, Jr., H.N. & PAGE, M.M. (1973). Mutants of *Aspergillus nidulans* altered in the transport of methylammonium and ammonium. *Molecular and General Genetics* 121, 239-45.

ARST, Jr., H.N. & SCAZZOCCHIO, C. (1975). Initiator constitutive mutation with an 'up-promotor' effect in *Aspergillus nidulans*. *Nature* 254, 31-34.

BARTNIK, E., GUZEWSKA, J. & WĘGLEŃSKI, P. (1973). Mutations simultaneously affecting ammonium and glucose repression of the arginine catabolic enzymes in *Aspergillus nidulans*. *Molecular and General Genetics* 126, 85-92.

BARTNIK, E. & WĘGLEŃSKI, P. (1973). Ammonium and glucose repression of the arginine catabolic enzymes in *Aspergillus nidulans*. *Molecular and General Genetics* 126, 75-84.

BARTNIK, E. & WĘGLEŃSKI, P. (1974). Regulation of arginine catabolism in *Aspergillus nidulans*. *Nature* 250, 591-2.

BRITTEN, R.J. & DAVIDSON, E.H. (1969). Gene regulation for higher cells: A theory. *Science* 165, 349-57.

COHEN, B.L. (1972). Ammonium repression of extracellular protease in *Aspergillus nidulans*. *Journal of General Microbiology* 71, 293-9.

COHEN, B.L. (1973). Regulation of intracellular and extracellular neutral and alkaline proteases in *Aspergillus nidulans*. *Journal of General Microbiology* 79, 311-20.

COVE, D.J. (1970). Control of gene action in *Aspergillus nidulans*. *Proceedings of the Royal Society, London, B* 176, 267-75.

COVE, D.J. & PATEMAN, J.A. (1963). Independently segregating genetic loci concerned with nitrate reductase activity in *Aspergillus nidulans*. *Nature* 198, 262-3.

COVE, D.J. & PATEMAN, J.A. (1969). Autoregulation of the synthesis of nitrate reductase in *Aspergillus nidulans*. *Journal of Bacteriology* 97, 1374-8.

DRAINAS, C., KINGHORN, J.R. & PATEMAN, J.A. (1976). Aspects of L-asparaginase regulation in *A. nidulans*. *180th Genetical Society Meeting, Heredity* 37, 153.

DUNN, E. & PATEMAN, J.A. (1972). Urea and thiourea uptake in *Aspergillus nidulans*. *Heredity* 29, 129.

DUNN-COLEMAN, N.S. & PATEMAN, J.A. (1975). The regulation of nitrate reductase in the fungus *A. nidulans*. *Biochemical Society Transactions* 3, 531.

ENGLESBERG, E., IRR, J., POWER, J. & LEE, N. (1965). Positive control of enzyme synthesis by gene C in the L-arabinose system. *Journal of Bacteriology* 90, 946-57.

FOX, T.D. (1976). Identification of phage SP01 proteins coded by regulatory genes 33 and 34. *Nature* 262, 748-52.
GARRETT, R.H. & COVE, D.J. (1977). Formation of NADP-nitrate reductase activity *in vitro* from *Aspergillus nidulans niaD* and *cnx* mutants. *Molecular and General Genetics* (In press).
HANKINSON, O. & COVE, D.J. (1974). Regulation of the pentose phosphate pathway in the fungus *Aspergillus nidulans*. *Journal of Biological Chemistry* 249, 2344-53.
HEFFERNAN, L., BASS, R. & ENGLESBERG, E. (1976). Mutations affecting catabolite repression of the L-arabinose regulon in *Escherichia coli* B/r. *Journal of Bacteriology* 126, 1119-31.
HIERHOLZER, G. & HOLZER, H.(1963). Repression der synthese von DPM-abhangiger glutaminsaure dehydrogenase in *Saccharomyces cerevisiae* durch ammoniumionen. *Biochemische Zeitschrift* 339, 175-82.
HYNES, M.J. (1970). Induction and repression of amidase enzymes in *Aspergillus nidulans*. *Journal of Bacteriology* 103, 482-7.
HYNES, M.J. (1972). Mutants with altered glucose repression of amidase enzymes in *Aspergillus nidulans*. *Journal of Bacteriology* 111, 717-22.
HYNES, M.J. (1973a). Pleiotropic mutants affecting the control of nitrogen metabolism in *Aspergillus nidulans*. *Molecular and General Genetics* 125, 99-107.
HYNES, M.J. (1973b). Alterations in the control of glutamate uptake in mutants of *Aspergillus nidulans*. *Biochemical and Biophysical Research Communications* 54, 685-9.
HYNES, M.J. (1974a). The effects of the carbon source on glutamate dehydrogenase activities in *Aspergillus nidulans*. *Journal of General Microbiology* 81, 165-70.
HYNES, M.J. (1974b). Repression of enzymes of nitrogen catabolism by methylammonium and caesium chloride in strains of *Aspergillus nidulans* insensitive to ammonium repression. *Molecular and General Genetics* 132, 147-52.
HYNES, M.J. (1974c). Effects of ammonium, L-glutamate and L-glutamine on nitrogen catabolism in *Aspergillus nidulans*. *Journal of Bacteriology* 120, 1116-23.
HYNES, M.J. (1975a). Studies on the role of the *areA* gene in the regulation of nitrogen catabolism in *Aspergillus nidulans*. *Australian Journal of Biological Science*, 28, 301-13.
HYNES, M.J. (1975b). Amide utilization in *Aspergillus nidulans*: evidence for a third amidase enzyme. *Journal of General Microbiology* 91, 99-109.
HYNES, M.J. & PATEMAN, J.A. (1970a). The genetic analysis of regulation of amidase synthesis in *Aspergillus nidulans*. *Molecular and General Genetics* 108, 97-106.
HYNES, M.J. & PATEMAN, J.A. (1970b). The genetic analysis of regulation of amidase synthesis in *Aspergillus nidulans*. *Molecular and General Genetics* 108, 107-16.

KINGHORN, J.R. & PATEMAN, J.A. (1973). NAD and NADP L-glutamate dehydrogenase activity and ammonium regulation in *Aspergillus nidulans*. *Journal of General Microbiology* 78, 39-46.

KINGHORN, J.R. & PATEMAN, J.A. (1974a). The effect of the carbon source on ammonium regulation in *Aspergillus nidulans*. *Molecular and General Genetics* 128, 95-8.

KINGHORN, J.R. & PATEMAN, J.A. (1974b). The regulation of NAD L-glutamate dehydrogenase in *Aspergillus nidulans*. *Genetical Research Cambridge* 23, 119-24.

KINGHORN, J.R. & PATEMAN, J.A. (1975a). Studies of partially repressed mutants at the *tamA* and *areA* loci in *Aspergillus nidulans*. *Molecular and General Genetics* 140, 137-47.

KINGHORN, J.R. & PATEMAN, J.A. (1975b). The structural gene for NADP L-glutamate dehydrogenase on *Aspergillus nidulans*. *Journal of General Microbiology* 86, 294-300.

KINGHORN, J.R. & PATEMAN, J.A. (1976). Mutants of *Aspergillus nidulans* lacking nicotimamide adenine dinucleotide-specific glutamate dehydrogenase. *Journal of Bacteriology* 125, 42-7.

LUKASZKIEWICZ, Z. & PASZEWSKI, P. (1976). Hyper-repressible operator-type mutant in sulphate permease gene of *Aspergillus nidulans*. *Nature* 259, 337-8.

PATEMAN, J.A. (1969). Regulation of synthesis of glutamate dehydrogenase and glutamine synthetase in microorganisms. *Biochemical Journal* 115, 769-75.

PATEMAN, J.A. & COVE, D.J. (1967). Regulation of nitrate reduction in *Aspergillus nidulans*. *Nature* 215, 1234-7.

PATEMAN, J.A., COVE, D.J., REVER, B.M. & ROBERTS, D.B. (1964). A common co-factor for nitrate reductase and xanthine dehydrogenase which also regulates the synthesis of nitrate reductase. *Nature* 201, 58-60.

PATEMAN, J.A., DUNN, E., KINGHORN, J.R. & FORBES, E.C. (1974a). The transport of ammonium and methylammonium in wild type and mutant cells of *Aspergillus nidulans*. *Molecular and General Genetics* 133, 225-36.

PATEMAN, J.A.& KINGHORN, J.R. (1976a). Nitrogen metabolism. In *The Filamentous Fungi*, Vol. 2 159-237. Edited by J.E. Smith and D.R. Berry, London; Edward Arnold.

PATEMAN, J.A. & KINGHORN, J.R. (1976b). Ammonia regulation in *Aspergillus nidulans*. *Proceedings of the Society for General Microbiology* Vol. III Part 4 p. 187.

PATEMAN, J.A.,KINGHORN, J.R. & DUNN, E. (1974b). Regulatory aspects of L-glutamate transport in *Aspergillus nidulans*. *Journal of Bacteriology* 119, 534-42.

PATEMAN, J.A., KINGHORN, J.R., DUNN, E. & FORBES, E. (1973). Ammonium regulation in *Aspergillus nidulans*. *Journal of Bacteriology* 114, 943-50.

PATEMAN, J.A., REVER, B.M. & COVE, D.J. (1967). Genetic and biochemical studies of nitrate reduction in *Aspergillus nidulans*. *Biochemical Journal* 104, 103-11.

POLKINGHORNE, M. & HYNES, M.J. (1975). Mutants affecting histidine utilization in *Aspergillus nidulans*. *Genetical Research Cambridge* 25, 119-35.

SCAZZOCCHIO, C. & DARLINGTON, A.J. (1968). The induction and repression of the enzymes of purine breakdown in *Aspergillus nidulans*. *Biochimica et Biophysica Acta* 166, 557-68.

STRICKLAND, W.N. (1971). Regulation of glutamate dehydrogenase in *Neurospora crassa* as a response to carbohydrate and amino acids in the media. *Australian Journal of Biological Sciences* 24, 905-15.

TJIAN, R. & PERO, J. (1976). Bacteriophage SP01 regulatory proteins directing late gene transcription *in vitro*. *Nature* 262, 753-7.

WEISS, R.L. (1976). Compartmentation and control of arginine metabolism in *Neurospora*. *Journal of Bacteriology* 126, 1173-9.

Regulation of Arginine Catabolism in *Aspergillus nidulans*

E. BARTNIK, J. GUZEWSKA, J. KLIMCZUK,
M. PIOTROWSKA and P. WEGLENSKI

Department of Genetics, University of Warsaw, Warsaw, Poland

INTRODUCTION

Arginine catabolism in *Aspergillus nidulans* provides a good system for studying genetic regulation. The pathway has been worked out biochemically, the enzymes involved are easy to assay and depending on growth conditions show differences in their rates of synthesis sufficient to follow kinetics of induction and the effects of carbon and nitrogen catabolite repression. Methods were developed to select mutants either derepressed or non-inducible for the arginine catabolic enzymes and a number of such mutants were obtained.

In this paper the properties of several mutants are described and the evidence is presented for the existence of both positive and negative control of the arginine catabolism. The inter-relationships between specific and general (i.e. responsible for carbon and nitrogen catabolite repression) regulatory mechanisms are also discussed.

CATABOLISM OF ARGININE

A scheme for arginine metabolism in *A. nidulans* and its relation to metabolism of proline is given in Fig. 1. The levels of the first two enzymes of the arginine catabolic pathway, arginase and ornithine δ-transaminase (OTAse), are low when the wild type mycelium is grown in the standard minimal medium (MM). Addition of arginine, as well as of several related compounds (Cybis & Weglenski, 1969), leads to the 10-15 fold induction of both enzymes. Increase in the arginase and OTAse levels is due to de novo synthesis of the enzymes; there is no evidence for the post-transcriptional regulation in this system.

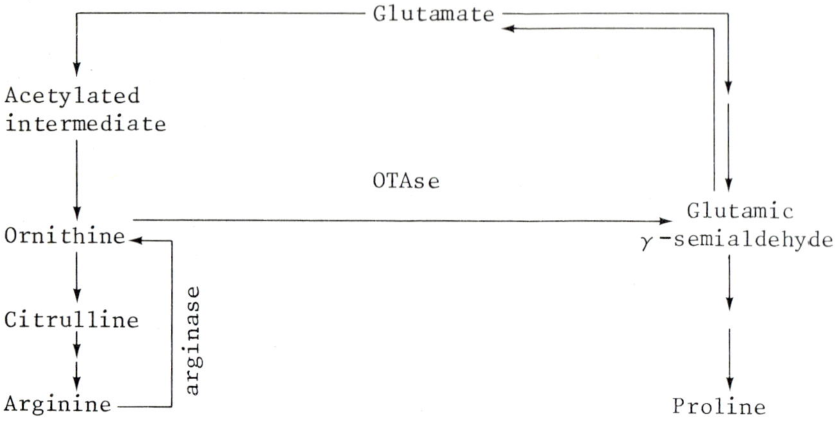

Fig. 1. The scheme of the arginine and proline metabolism in *Aspergillus nidulans*.

The half-life of arginase messenger in mycelium grown at 30°C with doubling time of 3.9 h was calculated for 2.7 min (Cybis & Weglenski, 1972).

Carbon or nitrogen starvation of the wild type mycelium results in derepression of the OTAse and has no effect on the arginase level (Bartnik, Piotrowska & Weglenski, 1973), however, the rate of induced synthesis of both arginase and OTAse is decreased in the presence of glucose and nitrogen sources (other than arginine) in the growth medium (Fig. 2). Induction of synthesis of both enzymes by arginine is prevented by high concentrations of ammonium (Fig. 2).

When arginine regulation mutants are isolated it is necessary to check them for three properties: (1) the level of the enzymes and their induction by arginine; (2) ability of the OTAse synthesis to derepress under carbon and nitrogen limitation, and (3) sensitivity of the process of induction of both enzymes by arginine to the inhibitory effect of ammonium. As will be shown, these properties are controlled by different sets of genetic elements.

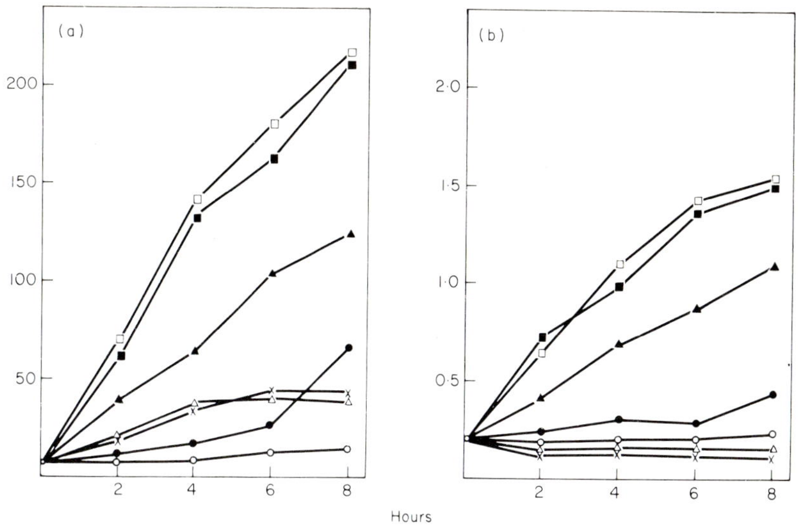

Fig. 2. Induction and derepression of OTAse (A) and arginase (B). Mycelium of the wild type strain was grown in the minimal medium containing 1% glucose as a carbon source and 10 mM sodium nitrate as a nitrogen source. After 16 h (time 0 on the graphs) mycelium was transferred to the media listed below and the activity of enzymes was measured in samples withdrawn every 2 h. Media: MM (○——○), MM + 5 mM arginine (▲——▲), MM + 5 mM arginine + 5 mM ammonium tartrate, (●——●), MM with no glucose with (■——■) or without (x——x) 5 mM arginine, MM with no nitrogen source with (□——□) or without (△——△) arginine. Units of arginase and OTAse activity are that amount of enzyme which forms 1 μmol of urea and 1 μmol of glutamic γ-semialdehyde per min per mg protein respectively.

SELECTION OF MUTANTS

Mutants in which arginase and OTAse are non-inducible by arginine were obtained as those which are unable to utilise arginine as a source of proline or as a sole source of nitrogen. By the same methods, mutants in the structural genes for arginase and OTAse were selected, as well as mutants defficient in transport of basic amino acids (Piotrowska, unpublished results). The OTAse and arginase structural genes are not linked to each other.

Mutants with high levels of arginase and OTAse were selected as suppressors of proline mutants blocked in one of the first two steps of the proline biosynthetic pathway (see Fig. 1). Several hundreds of proline suppressor mutants were obtained and analysed both biochemically and genetically (Weglenski, 1967; Klimczuk & Weglenski, 1974). They were assigned to 22 genetic loci. All these suppressors act by opening the pathway of proline synthesis from the arginine. The majority of suppressor mutants show high or elevated levels of the arginine catabolic enzymes, what indicates the direct or indirect involvement of the suppressor genes in regulation of the arginase and OTAse synthesis. Although at the moment it is not possible to establish functions of all these genes, properties of several mutants can be explained by a model of regulation which is discussed below.

POSITIVE AND NEGATIVE CONTROL OF THE ARGINASE AND OTASE INDUCTION

Mutants in one locus, designated *arcA*, result either in non-inducibility of the arginase and OTAse synthesis (*arcA*^{r} phenotype) or in the 10-fold increase in the level of both enzymes (*arcA*^{d} phenotype). Mutation of the *arcA*^{r} type are recessive, while *arcA*^{d} are semidominant over the wild type and *arcA*^{r} alleles (Bartnick & Weglenski, 1974). The frequency of recombination between the *arcA*^{r} and *arcA*^{d} alleles is less than 0.3 per cent, and some of the reversions of the *areA*^{r}1 mutation were found to map within the *arcA* locus and to have the *arcA*^{d} phenotype.

The simplest interpretation of the properties of the *arcA*^{r} and *arcA*^{d} mutants is that the *arcA* gene specifies a product (inducer) which is necessary for expression of the arginase and OTAse structural genes. In wild type strains inducer requires activation by a coinducer - arginine, whereas in the *arcA*^{d} mutant the inducer is active even in the absence of arginine. In the *arcA*^{r} mutants the inducer is inactive or is not produced.

Mutations in any of six proline suppressor loci designated *suApro*, *suDpro*, *suEpro*, *suHpro*, *suJpro* and *suLpro* result in approximately 10-fold increase in the arginase and OTAse levels. All mutations in these loci are recessive and their effects on the enzyme level are not additive. Experiments with double mutants *arcA*^{d} *suApro* and *arcA*^{r} *suApro* has shown that the effects of the *arcA*^{d} and *suApro* mutations are additive and the *arcA*^{r} mutation is epistatic to *suApro* (Table 1) as well as to all other suppressor mutants belonging to the *suApro* group. It can be assumed that *suApro* and the remaining genes of this type, participate in formation of a repressor which is responsible for maintaining of the low level of the arginine catabolic enzymes in the absence of arginine. Repression exerted by the products of these genes can be overcome by an inducer (the product of the *arcA* gene) activated by arginine. When due to mutations in one of the suppressor genes of the *suApro* type the repression

Table 1. OTAse activity in wild type and mutant strains of *A. nidulans*

No.	Relevant genotype	MM	MM arg	MM arg NH_4	-N	-N arg
1	*w.t.*	7	110	30	38	180
2	*arcA*^{r}*1*	5	7		5	11
3	*arcA*^{d}*47*	51	105	95	110	190
4	*suA25pro*	70	130	120		
5	*suD25pro*	75	140	135	140	210
6	*suF105pro*	36	104	46	38	160
7	*AniA6*	7	7		34	220
8	*areA*^{r}*1*	7	60	10	9	12
9	*areA*^{d}*102*	9	50	22	33	140
10	*xprD1*	7	110	115	40	190
11	*gdhA10*	8	140	130	42	180
12	*arcA*^{r}*1 suA25pro*	10	22	16	12	26
13	*arcA*^{r}*1 suD25pro*	6	9	7	7	8
14	*arcA*^{r}*1 suF105pro*	6	7		6	12
15	*arcA*^{r}*1 xprD1*	7	8		7	10
16	*arcA*^{r}*1 gdhA10*	7	6		7	14
17	*arcA*^{d}*47 suA25pro*	180	190			190
18	*arcA*^{d}*47 suD25pro*	200	200			220
19	*arcA*^{d}*47 areA*^{r}*1*	44	140			

Table 1 cont'd...

No.	Relevant genotype	MM	MM arg	MM arg NH_4	-N	-N arg
20	*suD25pro areA^{r}1*	84	140			
21	*suD25pro suF72pro*	145	160			190
22	*suD19pro AniA6*	85	90			210
23	*AnIA6 gdhA10*	7	8			210
24	*AniA6 xprD1*	9	12			200

Strains were grown in minimal medium (MM) containing sodium nitrate (10 mM) as nitrogen source and glucose (1%) as a carbon source. In the case of strains No. 8, 19 and 20, ammonium tartrate (1.25 mM) was added to MM. After 16 h mycelia were collected, washed and transferred to the following media: MM, MM + 5 mM arginine, MM + 5 mM arginine + 5 mM ammonium tartrate, MM with no nitrogen source (-N) and -N + 5mM arginine. The activities of OTAse were measured 6 h after transfer. The units of the OTAse activity are given in Fig. 2.

is abolished, arginine is not necessary for induction of arginase and OTAse, but the presence of the *arcA* product is still obligatory.

CATABOLITE REPRESSION OF THE ARGINASE AND OTASE SYNTHESIS

As it was mentioned earlier, under nitrogen or carbon starvation approximately 4-fold derepression of the OTAse synthesis is observed in the wild type strain, while the arginine level remains constant (Fig. 2). Among proline suppressors mutants were found and mapped in the *suFpro* locus, which have low arginase and the OTAse derepressed to the level characteristic for the wild type strain starved for carbon or nitrogen (Bartnick, Guzewska & Weglenski 1973). Moreover, starvation of the *suFpro* for carbon or nitrogen does not result in further derepression of the OTAse synthesis, while in mutants of the *suApro* type which have higher OTAse levels than *suFpro*, derepression occurs (Table 1). Experiments with double mutants *arcA^r suFpro* has shown that the *arcA^r* mutations are fully epistatic to *suFpro*. It means that the presence of the *arcA* gene product is necessary not only for induction but also for derepression of OTAse. It means also, that the control systems for induction and derepression of the OTAse synthesis are partially overlapping.

Selecting for the arginase and OTAse non-inducible mutants a class of particularly interesting mutants was obtained. These are dominant mutants mapping in the *AniA* locus (Bartnik, Guzewska & Weglenski, 1973). It appeared that the arginase and OTAse in the *AniA* strains are induced by arginine only when arginine is a sole nitrogen source, or when glucose is withdrawn from the growth medium (or replaced by a non-repressive carbon source, e.g. lactose). Mutations in the *AniA* locus apparently result in the increase in sensitivity of the arginine catabolic enzymes synthesis to carbon and nitrogen catabolite repression.

Since in the *AniA* mutants derepression of the OTAse synthesis under conditions of nitrogen or carbon starvation occurs as in the wild type strain (Bartnik, Guzewska & Weglenski, 1973), it seems that the product of this gene is directly or indirectly involved in the control of the process of induction. On the other hand, mutantions in the *suFpro* gene conceal effects of the *AniA* mutations completely, while the double mutants of the *suApro AniA* type retain properties of both mutants: the level of the arginase and OTAse is high like in the *suApro* strain, but no further induction of the enzymes by arginine in the presence of glucose and any nitrogen source in the medium is possible (Table 1).

Although no fully satisfying interpretation of the *AniA* gene function can be offered, it is possible that the product of this gene acts in tandem with the product of the *suFpro* gene, furnishing "repressor" which is responsible for slowing down the rate of the OTAse synthesis in the presence of a good carbon and nitrogen source. The *suFpro* mutations abolish functions of the repressor, while the *AniA* mutations change the repressor in such a way that it cannot be removed by the inducer (product of the *arcA* gene) in the process of induction by arginine. Carbon or nitrogen starvation inactivates repressor as well in the wild type strain as in the *AniA* mutants.

Whatever the functions of the *suFpro* and *AniA* genes are, it is worth pointing out, that single mutation in either one of these genes affects sensitivity of the enzyme synthesis to carbon and nitrogen repression simultaneously. This may indicate the existence of a common link in the control mechanisms for these two processes. There is no evidence so far for effects of the *suFpro* and *AniA* genes on any other metabolic system except arginine catabolism.

Two genes, *areA* (Arst & Cove, 1973) and *gdhA* (Kinghorn &

Pateman, 1973) have been suggested to play an essential role in the control of the nitrogen metabolism in *A. nidulans*. Mutation of the *areA*^{r} type causes the inability to utilise many nitrogen sources including arginine, and the reason is the non-inducibility of the appropriate catabolic enzymes and uptake systems. Other mutations in the same gene (*areA*^{d}) result in a relief from ammonium repression of several metabolic systems, but generally *areA*^{d} mutations are less pleiotrophic than *areA*^{r} ones (Arst & Cove, 1973). The effects of the *areA* mutations on the arginine catabolism are as follows: there is no induction of arginase and OTAse in the *areA*^{r}*1* mutant under conditions where arginine is a sole nitrogen source or when ammonium is present in the medium at concentrations higher than 2 mM (Bartnik, Klimczuk, Kowalska & Weglenski, 1976). In the *areA*^{r}*1* mycelium grown in the medium containing 5 mM arginine and 1.25 mM ammonium, induction of arginase and OTAse occurs, although at a slower rate than in the wild type strain. Two *areA*^{d} mutations (*areA*^{d}*101* and *areA*^{d}*102*) were found to have no effect on arginine catabolism. Only in the case of the *xprD1* mutation (an allele of the *areA* isolated by Cohen, 1972) ammonium, contrary to the situation in the wild type strain, does not prevent induction of arginase and OTAse by arginine (Table 1). Mutations in the *gdhA* gene have the same effect. Both *xprD1* and *gdhA10* mutations have no effect on the basal level of the OTAse. Several combinations of the *gdhA* and *areA* alleles with mutations in *arcA*, *suA* and other proline suppressor loci were tested for the level and inductivity of the arginine catabolic enzymes. The results (Table 1) show that the *arcA*^{d}, *suApro* and *suDpro* mutants are epistatic to *areA*^{r}. The double mutants *arcA*^{r}*1*, *gdhA10* and *arcA1 xprD1* have the *arcA*^{r} phenotype (Table 1). This indicates that both *areA* and *gdhA* mutations influence synthesis of the arginine catabolic enzymes only when the presumably more specific system of genetic control of this pathway is intact.

GENERAL CONCLUSIONS

The graphical summary of the results of experiments performed with mutants affecting arginine catabolism is given in Fig. 3. Three classes of genes involved in regulation of this pathway can be distinguished. The first class represented by the *arcA* gene is responsible for the positive regulation of the system. Judging by the dominance relationships between the *arcA*$^+$, *arcA*^{r} and *arcA*^{d} alleles and taking into account the full epistasy of the *arcA*^{r} mutations to all other regulatory mutations, the product of the *arcA* gene is obligatory for the synthesis of arginase and OTAse at a rate higher than that which is sufficient to maintain the

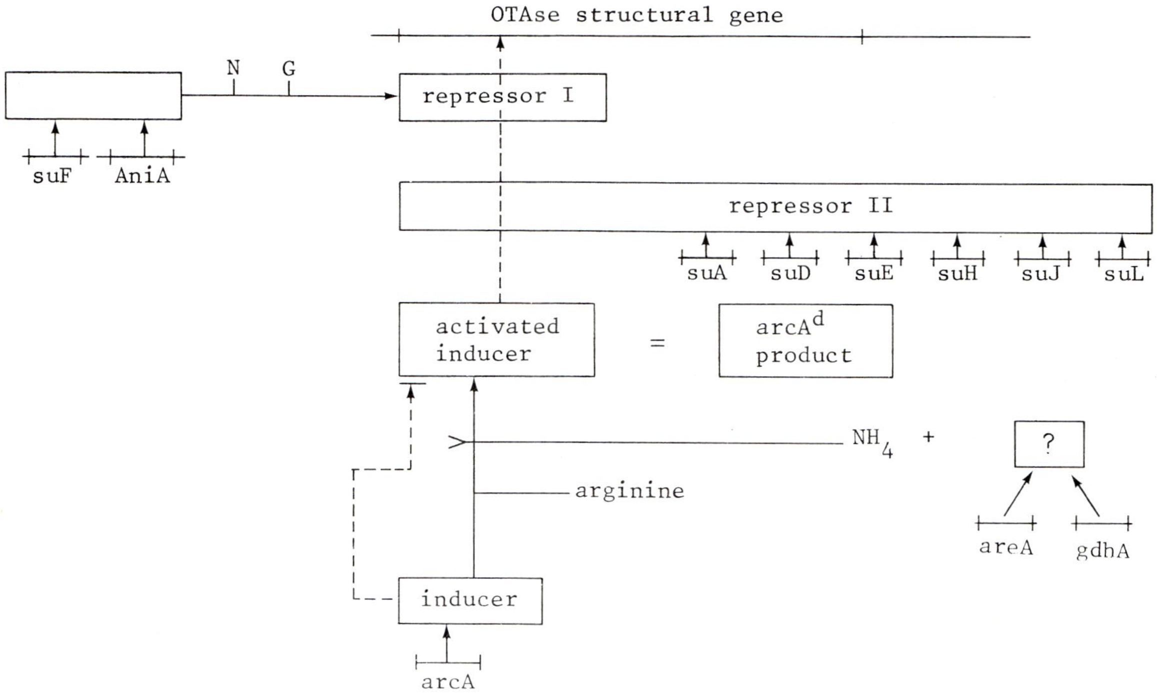

Fig. 3. The scheme of regulation of expression of the OTAse structural gene. The symbols N and G designate unknown effectors derived from nitrogen and glucose metabolism respectively. For the remaining explanations see text.

basal levels of these enzymes.

The second class consists of six genes of the *suApro* type which are responsible for the negative control of the pathway. Nothing is known on the nature of the products of these genes. One of several possibilities is, that products of these genes directly participate in formation of a complex repressor molecule which blocks structural genes for OTAse and arginase. Another possibility is that they participate in formation of a corepressor of an unidentified repressor. It seems noteworthy that mutations in genes belonging to this class affect also synthesis of the ornithine transcarbamylase (Cybis, Piotrowska & Weglenski, 1972), nitrate reductase (Bartnik, unpublished results) and suppress some effects of the *areA*^{r}*1* mutation enabling growth of the *areA*^{r}*1* strain on several (but not all) nitrogen sources (Arst & Cove, 1973). We have no good explanation for the pleiotrophy of the *suApro* and the remaining suppressor genes belonging to this class, although one can speculate that these genes are responsible for synthesis or functioning of some elements of the chromatine which unspecifically block expression of many unrelated genes.

The third class of genes involved in regulation of the arginine catabolism consists of the *suFpro* and *AniA* genes. The products of these genes contribute to the system which ensures normal (i.e. that which is observed in wild type strain) response of the arginase and OTAse synthesis to carbon and nitrogen catabolite repression. From the formal point of view, *suFpro* and *AniA* genes specify a repressor which is active only in the presence of the unknown effectors derived from the carbon and nitrogen metabolism. The levels of the OTAse in the *suFpro*, *suDpro* and *suFpro suDpro* strains (Table 1) indicate that the mechanisms of repression exerted by the products of the *suFpro* and *suDpro* genes are independent.

The physiological meaning of the regulatory mechanism specified by the *suFpro* and *AniA* genes is to moderate expression of the arginase and OTAse genes in response to the availability of carbon and nitrogen in the growth medium. Thus the function of the *suFpro-AniA* system is the same as that which is commonly ascribed to general regulatory mechanisms involved in carbon and nitrogen catabolite repression. However, the functions of the *suFpro* - *AniA* system seem to be limited only to arginine catabolism, since the synthesis of several enzymes which are under carbon or nitrogen repression, including acetamidase, nitrate reductase and β-galactosidase, is not affected by either *suFpro* or *AniA* mutations.

The effects of the *suFpro* and *AniA* mutations on the arginine catabolism are apparently different from that of the *areA* and *gdhA*. The last two genes seem to influence the response of the arginase and OTAse synthesis to ammonium. Ammonium affects synthesis of the arginine catabolic enzymes in two ways: (1) prevents derepression of the OTAse as any other good nitrogen source; and (2) prevents induction of arginase and OTAse by arginine. This conclusion is based on the fact, that in the *suFpro* mutants in which OTAse is already derepressed ammonium prevents induction of this enzyme by arginine. Lack of effects of the *areA*^{d}/or *xprD*/ and *gdhA* mutations on the basal level of the OTAse indicates, that both genes are involved only in the mechanism(s) by which ammonium interferes with arginase and OTAse induction. One of the possible targets of ammonium (or its derivative), whenever it acts directly or its effect is mediated through the products of the *areA* and *gdhA* genes, is the process of activation of the inducer by arginine. This supposition can be drawn from the observations on the effect of the *arcA*^{d} mutation. In the *arcA*^{d} mutant, ammonium does not prevent induction of the arginase and OTAse by arginine, and *arcA*^{d} mutants are epistatic to *areA*^{r} ones.

Despite considerable effort, we were unable to select mutants for arginase and OTAse of the operator or "promotor-up" type. This is surprising, especially in the view of the fact that operator constitutive mutants for these enzymes were obtained in yeasts (Wiame 1971). Getting mutants of this type would be very helpful to test several predictions concerned with arginase and OTAse regulation. It seems however that for verification of the proposed model of regulation, the products of at least some of the postulated regulatory genes should be isolated and their functions tested at the molecular level.

REFERENCES

ARST, H.N. & COVE, D.J. (1973). Nitrogen metabolite repression in *Aspergillus nidulans*. *Molecular and General Genetics* 126, 111-41.

ARST, H.N. & MACDONALD, D.W. (1973). A mutant of *Aspergillus nidulans* lacking NADP-linked glutamate dehydrogenase. *Molecular and General Genetics* 122, 261-5.

COHEN, B.L. (1972). Ammonium repression of extracellular protease in *Aspergillus nidulans*. *Journal of General Microbiology* 71, 293-9.

CYBIS, J., PIOTROWSKA, M. & WEGLENSKI, P. (1972). Genetic control of the arginine pathways in *Aspergillus nidulans*. Common regulation of anabolism and catabolism. *Molecular and General Genetics* 118, 273-7.

CYBIS, J. & WEGLENSKI, P. (1969). Effects of lysine on arginine uptake and metabolism in *Aspergillus nidulans*. *Molecular and General Genetics* 104, 282-7.

CYBIS, J. & WEGLENSKI, P. (1972). Arginase induction in *Aspergillus nidulans*. The appearance and decay of coding capacity of messenger. *European Journal of Biochemistry* 30, 262-8.

BARTNIK, E., GUZEWSKA, J. & WEGLENSKI, P. (1973). Mutations simultaneously affecting ammonium and glucose repression of the arginine catabolic enzymes in *Aspergillus nidulans*. *Molecular and General Genetics* 126, 85-92.

BARTNIK, E., KLIMCZUK, J., KOWALSKA, I. & WEGLENSKI, P. (1976). Effect of the *areA* gene on regulation of arginine catabolism in *Aspergillus nidulans*. *Acta Microbiologica Polonica* In Press.

BARTNIK, E., PIOTROWSKA, M. & WEGLENSKI, P. (1973). Ammonium and glucose repression of the arginine catabolic enzymes in *Aspergillus nidulans*. *Molecular and General Genetics* 126, 75-84.

BARTNIK, E. & WEGLENSKI, P. (1974). Regulation of arginine catabolism in *Aspergillus nidulans*. *Nature* 250, 590-2.

KINGHORN, J.R.& PATEMAN, J.A. (1973). NAD and NADP L-glutamate dehydrogenase activity and ammonium regulation in *Aspergillus nidulans*. *Journal of General Microbiology* 78, 39-46.

KLIMCZUK, J. & WEGLENSKI, P. (1974). New proline suppressor mutants in *Aspergillus nidulans*. *Acta Microbiologica Polonica* 6, 93-100.

PONTECORVO, G., ROPER, J.A., HEMMONS, L.M., MACDONALD, R.D. & BUFTON, A.W.J. (1953). The genetics of *Aspergillus nidulans*. *Advances in Genetics* 5, 141-238.

WEGLENSKI, P. (1967). The mechanism of action of proline suppressors in *Aspergillus nidulans*. *Journal of General Microbiology* 47, 77-85.

WIAME, J.M. (1971). The regulation of arginine metabolism in *Saccharomyces cerevisiae*. Exclusion mechanisms. *Current topics of cellular regulation* 4, 1-38.

The Regulation of Purine Breakdown

C. SCAZZOCCHIO and D. GORTON

Department of Biology, University of Essex, Colchester, U.K.

INTRODUCTION

The pathway of purine breakdown (Fig. 1) is a dispensable pathway where the structural and regulatory genes are scattered throughout the genome (Table 1). The enzyme activities catalysing the different steps are subject to both specific induction and repression by ammonia (Scazzocchio & Darlington, 1968), and are coded by genes constituting a system of at least two regulons subject to semisequential induction.

ENZYME ACTIVITIES

No biochemical work has been done on any of the enzymes of purine degradation except for the enzyme catalysing the first step of Fig. 1, xanthine dehydrogenase I. Xanthine dehydrogenase I is similar to other xanthine dehydrogenases and oxidases in its substrates, inhibitor specificities and molecular weight; NAD seems to be the physiological electron acceptor (Lewis, Hurt & Scazzocchio, unpublished results). The *cnx* mutations, defective in the synthesis of a molybdenum containing co-factor (Pateman *et al.*, 1964; Arst *et al.*, 1970), result not only in loss of xanthine dehydrogenase I activity Scazzochio *et al.*, 1973; Scazzochi 1973, 1974) but also in the appearance of a protein, immunologically indistinguishable from xanthine dehydrogenase I and with half its molecular weight (Lewis & Scazzocchio, submitted for publication. As in the case of nitrate reductase (Macdonald *et al.*, 1974) the *cnx* cofactor also plays a role in the maintainance of the correct quaternary structure.

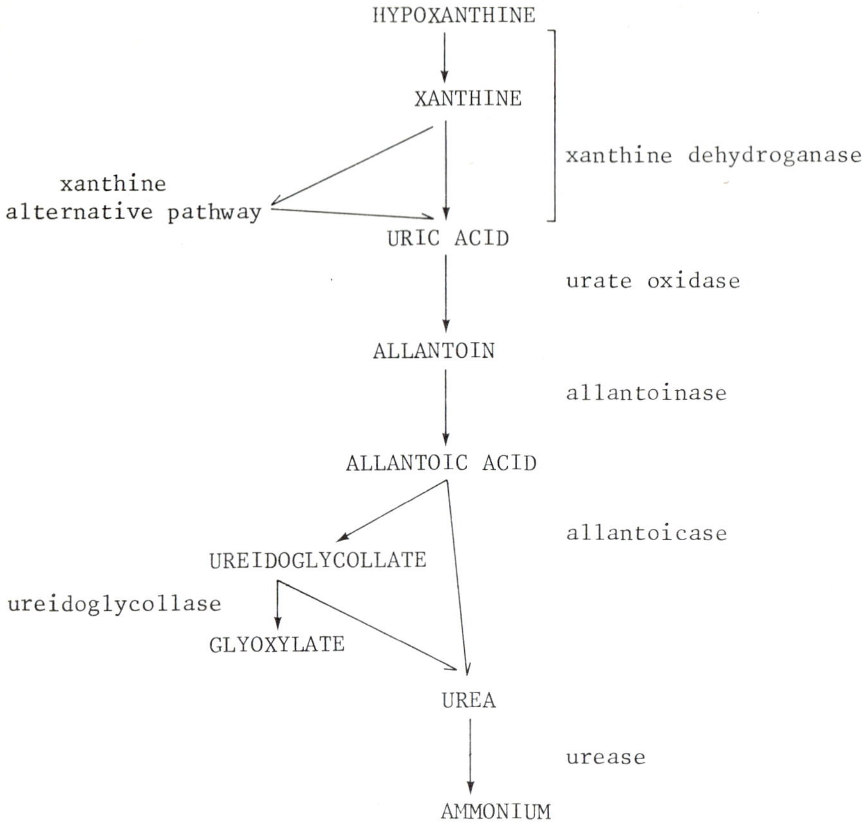

Fig. 1 The purine breakdown pathway

The enzyme that we have called xanthine dehydrogenase II (Scazzocchio, 1974; Scazzocchio, 1973) can accept hypoxanthine as a substrate, but is not induced by purines, is under the control of the *apl*A regulatory gene, and is physiologically the first enzyme of nicotinate breakdown. It has no physiological role in the breakdown of purines. It is only relevant here because mutations at the *hx*B gene eliminate both xanthine dehydrogenase I and II activities. We have presented evidence that the protein coded by *hx*B is inducible (Scazzocchio *et al.*, 1973; Scazzocchio & Darlington, 1967); this contrasts with *cnx*. Our prejudice, based on complementation data, was that the synthesis of the cofactor is constitutive, and this has recently been shown to be the case by Garrett & Cove (1976). We have presented evidence that another pathway, distinct from the xanthine dehydrogenases, exists for the conversion of xanthine to uric acid (Darlington & Scazzacchio, 1968). No evidence as to the nature of this pathway has been forthcoming, nor have defective mutants been isolated.

UPTAKE OF PURINES AND THEIR DERIVATIVES

Darlington & Scazzocchio have presented evidence for two uptake systems, one responsible for the uptake of adenine, guanine and hypoxanthine, and the analogues azaguanine and purine, and a second one responsible for the uptake of xanthine and uric acid and their thio-analogues (Darlington & Scazzocchio, 1967). Unpublished work by Arst & Scazzocchio show that the situation is more complex. While mutations at the *azg*A gene (Darlington & Scazzocchio, 1967), resulting in resistance to azaguanine, grow leakily on hypoxanthine, an *are*A-102 *azg*A^{-} double mutant is non-leaky. Since *azg*A is probably not under *are*A control (Arst & Cove, 1973) this suggests that there is another permease in the wild type that is not present in *are*A-102. Moreover there are at least two genes involved in uric acid and xanthine uptake, *uap*A (Darlington & Scazzocchio, 1967; Arst & Scazzocchio, 1975) and *uap*B (which also affects allantoin uptake. They seem to code for permeases with partially overlapping specificities. And there is also evidence for a third permease under the control of another gene, *uap*C. Mutations resulting in lack of allantoin uptake map in *al*P (Darlington & Scazzocchio, 1967) and the urea uptake system is specified by *ure*A (Kinghorn & Pateman, 1976).

GENES INVOLVED IN PURINE DEGRADATION

The genes involved in purine uptake have been dealt with previously, as most of the evidence for the uptake systems is genetical. The other genes of purine degradation in Table 1 have been identified by absence of the corresponding enzyme activity (Scazzocchio & Darlington, 1968). To the information already published (Scazzocchio & Darlington, 1968) we should add that we have shown formally that *hx*A is the structural gene for xanthine dehydrogenase I, as mutants with modified electrophoretic mobility can be obtained as second site revertants of *hx*A-1 (Scazzocchio & Lewis, unpublished data).

*hx*B was thought to code for a common component of xanthine dehydrogenase I and II (Scazzocchio 1974; Scazzocchio *et al.*, 1973); it now seems more likely to code for a post-translational modification common to both enzymes, but the problem is still open.

Table 1. Genes of purine degradation and others mentioned in the text

Locus	Linkage group	Linkage to other genes involved in purine degradation and uptake, where tested	Presumed function
*azg*A	V		hypoxanthine, adenine, guanine permease (Darlington & Scazzocchio 1967)
*uap*A	I	unlinked to *uap*C	uric acid-xanthine permease (Darlington & Scazzocchio 1967) Arst & Scazzocchio 1975
*uap*B	VII		uric acid-xanthine-allantoin permease (Arst & Scazzocchio unpublished data)
*uap*C	I		control of a uric acid-xanthine permease. (Arst, unpublished data)
*a*lP	VIII		allantoin permease (Darlington & Scazzocchio 1967)
*uru*A	VIII	tightly linked to *u*X	urea permease (Kinghorn & Pateman 1976)
*hx*A	V		known to code for xanthine dehydrogenase I structural component (Scazzocchio & Darlington 1968; Lewis & Scazzocchio unpublished data)

Table 1 cont'd....

Locus	Linkage group	Linkage to other genes involved in purine degradation and uptake, where tested	Presumed function
*hx*B	VII	unlinked to *cnx*F and to *u*Y	affects both xanththine dehydrogenases I and II; either codes for a common subunit or determines a common post-translational modification (Scazzocchio & Darlington 1968, Scazzocchio 1974; Scazzocchio *et al.* 1973; Scazzocchio & Darlington 1967)
cnx A,B,C	VIII	the *cnx*ABC complex locus is 9cM from *ua*Y	all involved in synthesis of molybdenum containing cofactor of nitrate reductase and xanthine dehydrogenases I and II (Pateman, *et al.* 1964; Arst, *et al.* 1970; Scazzocchio, 1974)
*cnx*E	II		
*cnx*F	VII		
*cnx*G	VI	not tightly linked to *apl*A	
*cnx*H	III	unlinked to *are*A	
*ua*Z	VIII	36cM from *ua*Y	urate oxidase structural gene (Scazzocchio & Darlington, 1968)
*ua*X	VI	unlinked to *aa*X and to *apl*A	see Text (Scazzocchio & Darlington, 1968)
*al*X	III	unlinked to *are*A	allantoinase structural gene (Scazzocchio & Darlington, 1968)
*aa*X	VI		allantoinase structural gene (Scazzocchio & Darlington, 1968)

Table 1 cont'd....

Locus	Linkage group	Linkage to other genes involved in purine degradation and uptake, where tested	Presumed function
*u*X	VIII	unlinked to *u*Z, and to *ua*Y	urease structural gene? (Scazzocchio & Darlington, 1968; Kinghorn & Pateman, 1976)
*u*Y	VII		urease structural gene? (Scazzocchio & Darlington, 1968)
*u*Z	VIII		urease structural gene? (Scazzocchio & Darlington, 1968)
*ua*Y	VIII		uric acid induction of number of purine breakdown and uptake enzymes (Scazzocchio & Darlington, 1968; Scazzocchio *et al.*; 1963 Scazzocchio & Darlington, 1967; Scazzocchio 1966 and Philippides & Scazzocchio, in preparation)
*apl*A	VI		regulation of xanthine dehydrogenase II (Scazzocchio *et al.* 1973, Scazzocchio & Darlington, 1967)
*are*A	III		regulatory gene for nitrogen metabolite repression (Arst & Cove, 1973)

From: Scazzocchio *et al.* 1973; Arst & Cove, 1973; Kinghorn & Pateman, 1976; Scazzocchio, 1966; Darlington, 1966; and H.N. Arst, unpublished results.

---

*ua*X is quite a mysterious gene. The three alleles isolated result in slow growth on uric acid as sole nitrogen source, but the *in vitro* urate oxidase levels are normal (Scazzocchio & Darlington, 1968). It has been reported for several organisms including the yeast *Saccharomyces cerevisiae* (Parish, 1975), that urate oxidase is associated with catalase in peroxisomes. The oxidation of uric acid produces hydrogen peroxide, toxic unless broken down by catalase. *ua*X mutations might result in defective integration of urate oxidase in the subcellular particle, or even be defective in the catalase itself.

We have isolated allele-specific suppressors of *al*X-4; their nature is as yet unknown. The specific functions of the three genes involved in urease activity is not clear (but see Kinghorn & Pateman, this volume, for recent work on urease). Until recently it seemed that none of the genes involved in purine degradation fell into clusters. The exception appears to be *u*X and *ure*A, closely linked in chromosome VIII (Kinghorn & Pateman, 1976).

REGULATORY PATTERNS

All enzyme activities are to some extent repressed by ammonia (Scazzocchio & Darlington, 1968). Induction proceeds in a semisequential way. Uric acid induces xanthine dehydrogenase I, urate oxidase (Scazzocchio & Darlington, 1968; Scazzocchio, 1973), allantoinase and allantoicase (Scazzocchio & Darlington, 1968), and at least one uric acid-xanthine permease (Arst & Scazzocchio, 1975; Scazzocchio, 1966). But allantoinase and allantoicase are also and additively induced by allantoin (Scazzacchio & Darlington, 1968). Ureidoglycollase and urease seem to be constitutive (Scazzacchio & Darlington, 1968).

*The regulatory gene ua*Y

Mutations mapping in the *ua*Y gene result in non-inducibility of all activities induced by uric acid. *ua*Y-mutations are co-dominant in diploids (Scazzocchio, 1966 and Philippides & Scazzocchio, in preparation). This gives no evidence as to whether *ua*Y acts through positive or negative control. However the frequency of non-inducible mutations, and the phenotype of a deletion show *ua*Y to be a positive control gene (Philippides & Scazzocchio, in preparation). This is meant in the most general sense - the product of the *ua*Y gene is necessary for induction to take place, and does not prejudge at what level the *ua*Y product is acting. Unpublished experiments are compatible with the induction of both xanthine dehydrogenase I and urate oxidase proceeding through *de novo* synthesis, possibly at the level of transcription.

We have considered previously the possibility that the *ua*Y gene coded for an enzyme responsible for the formation of an uric acid derivative, directly involved in the induction process, or even coded for an uric acid uptake system at the level of the nuclear membrane (Scazzocchio & Darlington, 1967; Scazzocchio, 1970). The whole idea of an indirect role of *ua*Y was due to the fact that *apl*A^{c} mutations suppressed *ua*Y mutations for xanthine dehydrogenase activity (Scazzocchio *et al.*, 1973; Scazzocchio & Darlington, 1967). When we showed that the xanthine dehydrogenase present in strains carrying *apl*A^{c} mutations was a different enzyme from the one induced by uric acid any positive evidence for an 'induction pathway' role of the *ua*Y gene collapsed. Indeed a search for a uric acid derivative has been negative (Philippides & Scazzocchio, in preparation), and data of Scazzocchio & Arst show that a functional *ua*Y gene is necessary for the expression of the *uap*-100 up-promotor initiator constitutive mutation. While this is difficult to understand, it surely points to a quite direct involvement of the *ua*Y product in the induction process.

There are rare mutations with a different phenotype mapping in or very near the *ua*Y gene. They are resistant to oxipurinol toxicity, and constitutive at a low level for xanthine dehydrogenase I and urate oxidase. One allele, *oxp*^{r}-5, has been studied in detail. If formation of uric acid is prevented by introducing an *hx*A or *hx*B mutation, the double mutant is inducible, just as the wild type (Cove, 1970 and Philippides & Scazzocchio, in preparation). This is consistent with a tightly binding mutant *ua*Y product with a lower dissociation constant for uric acid. However, only a fine structure map could determine whether *oxp*^{5}-5 maps *within* the *ua*Y gene, as, if two closely linked genes were involved in uric acid-mediated induction, other interpretations of *oxp*^{r}-5 function should be considered.

The co-dominance of *ua*Y mutations indicate that one *ua*Y gene makes enough product to elicit expression of only one set of structural genes. This dose effect could also explain the lack of complementation in heterocaryons between *ua*Y and the structural genes under its control (Scazzocchio & Darlington, 1967 and Cove, this volume). Philippides & Scazzocchio (in preparation) have recently identified a protein likely to be coded by the *ua*Y gene. It binds strongly to phospho-cellulose and DNA-cellulose columns. It binds both uric acid and the gratuitous inducer 2-thiouric acid, to the physiological inducer with a dissociation constant of the order of 10^{-6}. And finally the binding activity is missing in a strain carrying a small deletion of the *ua*Y gene.

It can be predicted that another regulatory gene, involved in allantoin-mediated induction, should be found. Mutations in this gene would not have been found by Darlington (1966) and Scazzocchio (1966) in their search for purine non-utilising mutants, as they would grow on all purines. A simple negative control mechanism would override a positive control, in the regulation of the same gene - therefore, since allantoinase and allantoicase are induced additively by uric acid and allantoin, the regulatory gene concerned with allantoin induction is most probably a positive control gene. A search is in progress for mutants in this proposed regulatory gene.

Cis-dominant mutations

As is commonly the case in catabolic pathways most, if not all, of the genes involved in purine degradation or uptake need for their expression two regulatory signals acting simultaneously, the nitrogen metabolite repression (ammonia repression) signal, and the specific induction signal. But it should be noticed that *hx*B (under the control of both *ua*Y and *apl*A (Scazzocchio *et al.* 1973), *al*X, and *aa*X should respond to two alternative inducing signals, and one presumes their cis-dominant regulatory regions will reflect this complexity.

Kinghorn & Pateman (1976) have described a mutation, *UR*-1, (*ure* 1) tightly linked to the *u*X-*ure*A cluster, and resulting in ammonium derepression of both urea permease and urease. This result is interesting, taking into account that two other genes (*u*Y, *u*Z) outside the cluster have some unspecified role in the production of urease activity.

Arst & Scazzocchio have described in detail *uap*-100 (Arst & Scazzocchio, 1975) which is closely linked to *uap*A. It results in constitutivity of the permease specified by the *uap*A gene and has been shown to act only in *cis*. Some considerations are merited here on the nature of this mutation. It was selected as a revertant of *are*A-102, an *are*A allele resulting in lack of the uric acid-xanthine permease activities. However it does not bypass the need for an *are*A product, as it does not bypass *are*A-1, (preliminary in vivo evidence), an *are*A allele resulting in loss of all activities subject to ammonium repression. Also *uap*-100 does not result in ammonium derepression. Nor does it bypass the need for the *ua*Y protein. So what does it do? (1) It can work with a modified *are*A product. (2) It does not need the coinducer while still needing the apo-inducer (the *ua*Y product). (3) It has an up-promoter effect. It is not a classical cis-cominant, initiator constitutive mutation. It is hardly surprising that in eukaryote, with DNA organised in chromatin on the bead-interbead scheme (Morris, 1976), we should find unprecedented cis-dominant control regions.

REFERENCES

ARST, H.N.,Jr. & COVE, D.J. (1973). Nitrogen metabolite repression in *Aspergillus nidulans*. *Molecular and General Genetics* 126, 111-41.

ARST, H.N., Jr., MACDONALD, D.W. & COVE, D.J. (1970. Molybdate metabolism in *Aspergillus nidulans* I. Mutations affecting nitrate reductase and/or xanthine dehydrogenase. *Molecular and General Genetics* 108, 128-45.

ARST, H.N., Jr. & SCAZZOCCHIO, C. (1975). Initiator constitutive mutation with an 'up promotor' effect in *Aspergillus nidulans*. *Nature* 254, 31-4.

COVE, D.J. (1970). Control of gene action in *Aspergillus nidulans*. *Proceedings of the Royal Society of London, Series B* 176, 267-75.

DARLINGTON, A.J. (1966). Genetic and biochemical studies of purine breakdown in *Aspergillus*. *Ph.D. thesis, University of Cambridge*.

DARLINGTON, A.J. & SCAZZOCCHIO, C. (1967). Use of analogues and the substrate-sensitivity of mutants in the analysis of purine uptake and breakdown in *Aspergillus nidulans*. *Journal of Bacteriology* 93, 937-40.

DARLINGTON, A.J. & SCAZZOCCHIO, C. (1968). Evidence for an alternative pathway of xanthine oxidation in *Aspergillus nidulans*. *Biochimica et Biophysica Acta* 166, 569-71.

GARRETT, R.H. & COVE, D.J. (1976). Formation of NADPH-nitrate reductase activity *in vitro* from *Aspergillus nidulans nia*D and *cnx* mutants *Molecular and General Genetics* 149, 179-86.

KINGHORN, J.R. & PATEMAN, J.A. (1976). A gene cluster in a simple eukaryote. *Heredity* 37, 154.

MORRIS, N.R. (1976). Nucleosome structure in *Aspergillus nidulans*. *Cell* 8, 357-64.

MACDONALD, D.W., COVE, D.J. & CODDINGTON, A. (1974). Cytochrome-c reductases from wild type and mutant strains of *Aspergillus nidulans*. *Molecular and General Genetics* 128, 187-99.

PARISH, R.W. (1975). The isolation and characterization of peroxisomes (microbodies) from baker's yeast, *Saccharomyces cerevisiae*. *Archives of Microbiology* 105, 187-92.

PATEMAN, J.A., COVE, D.J., REVER, B.M. & ROBERTS, D.B. (1964). A common cofactor for nitrate reductase ans xanthine dehydrogenase which also regulates the synthesis of nitrate reductase. *Nature* 201, 58-60.

SCAZZOCCHIO, C. (1966). Studies on the genetic control of purine oxidation in *Aspergillus nidulans*. *Ph.D. thesis, University of Cambridge*.

SCAZZOCCHIO, C. (1970). Nuclear compartmentalisation in the control of gene action in *Aspergillus nidulans*. *Heredity* 25, 683.

SCAZZOCCHIO, C. (1973). The genetic control of molybdoflavoproteins in *Aspergillus nidulans*. II: Use of the NADH dehydrogenase activity associated with xanthine dehydrogenase to investigate substrate and product induction. *Molecular and General Genetics* 125, 147-55.

SCAZZOCCHIO, C. (1974). The genetic determination of molybdoflavoenzymes in *Aspergillus nidulans*. *Journal of Less Common Metals* 36, 461-4.

SCAZZOCCHIO, C. & DARLINGTON, A.J. (1967). The genetic control of xanthine dehydrogenase and urate oxidase synthesis in *Aspergillus nidulans*. *Bulletin de la Societe de chimie biologique* 49, 1503-8.

SCAZZOCCHIO, C. & DARLINGTON, A.J. (1968). The induction and repression of the enzymes of purine breakdown in *Aspergillus nidulans*. *Biochimica et Biophysica Acta* 166, 557-68.

SCAZZOCCHIO, C., HOLL, F.B. & FOGUELMAN, A.I. (1973). The genetic control of molybdoflavoproteins in *Aspergillus nidulans*. Allopurinol-resistant mutants constitutive for xanthine dehydrogenase. *European Journal of Biochemistry* 36, 428-45.

The *Aspergillus* Nucleus: Histones Chromatin and Tubulin

N. R. MORRIS, R. A. FELDEN, M. A. GEALT, R. V. NARDI, G. SHEIR-NEISS and M. M. SANDERS

Department of Pharmacology, C.M.D.N.J.—Rutgers Medical School, Piscataway, New Jersey, U.S.A.

INTRODUCTION

Aspergillus nidulans promises to become an important model system for studying the molecular biology of the eukaryotic nucleus. It is one of the simplest organisms with a true nucleus and has one of the smallest genomes of any differentiating eukaryote. The small size of its genome, about 2 x 10^7 base pairs (Bainbridge,1971), its well developed genetics, the availability of mutations affecting regulation, development, and mitosis (see Clutterbuck, 1976; Orr & Rosenberger, 1976; Morris, 1976a) and the initial biochemical characterization of its histones, chromatin and tubulin (see below) make *A.nidulans* unique in its suitability and potential usefulness for studies of nuclear structure and function.

The morphology of the *Aspergillus* nucleus and the changes it undergoes during the cell cycle have been described by Robinow & Caten (1969). The nucleus is bounded by a membrane and has a true nucleolus. At metaphase the nucleolus disappears and the nuclear chromatin condenses into densely-staining chromosome-like structures which orient along a microtubular spindle. At anaphase the chromosomes move to the ends of the spindle, the nucleus divides and the chromatin becomes diffuse. These resemblances between the morphological events of mitosis in *Aspergillus* and in higher organisms suggest that the biochemical organization of the nucleus may be similar in *Aspergillus* and higher eukaryotes. Two features of mitosis in *Aspergillus* differ from those in higher eukaryotes. The ends of the spindle terminate in spindle plaques rather than centrioles and the nuclear membrane persists throughout metaphase.

Although the morphology of the nucleus is well known, the biochemistry of the nucleus in *Aspergillus* has not previously been described. For this reason we have initiated a biochemical characterization of some of the major features of the *Aspergillus* nucleus. In order to accomplish this a method has been developed for preparation of nuclei from vegetative mycelia. The histones have been isolated and characterized, the chromatin structure has been studied by digestion with nucleases and *Aspergillus* tubulin has been co-precipitated with brain tubulin and characterized by electrophoresis and isoelectric focussing. As anticipated from the morphological similarities, the histones, chromatin and tubulin of *A.nidulans* are not dissimilar in their general attributes from those of higher eukaryotes. They do however differ sufficiently to provide some useful and interesting comparisons.

EXPERIMENTAL STUDIES

Preparation of Nuclei

The major difficulty in preparing nuclei from fungi is to develop a method for disruption of the tough cell wall which does not damage the nuclei. Although it is possible to make protoplasts from *A.nidulans* (Peberdy & Gibson, 1971), we were unable to make protoplasts on a scale sufficiently large to generate large amounts of nuclei and therefore we developed a method based on large-scale physical disruption of frozen mycelia which does not cause excessive nuclear damage (Gealt, Sheir-Neiss & Morris, 1976). The method used was derived from a procedure for preparation of mitochondria from *Neurospora crassa* (Minssen & Munkres, 1973). Mycelia at late log phase were frozen rapidly in liquid nitrogen and were disrupted by grinding in liquid nitrogen in a large Waring blender. The homogenate was then thawed rapidly to $4^{o}C$. The nuclei were released by homogenization in a large volume of isosmotic buffer (in order to dilute nucleases and proteolytic enzymes) and purified by filtration and differential centrifugation through sucrose solutions (Gealt, *et al.*, 1976; Morris, 1976b). Phenylmethylsulfonyl fluoride (1mM) was included in all solutions to inhibit proteolysis. All steps in the procedure were monitored by fluorescence microscopy. The criteria used for nuclear purification were: DNA recovery, the absence of intact cells in the final nuclear fraction; morphological similarity of *in situ* and isolated nuclei and the presence of histones. The yield of nuclei obtained from wild type mycelium by this method was about 20% based on the recovery of DNA. The DNA from these nuclei was of large molecular weight (>20 x 10^6 daltons) and the ratio of DNA/RNA/protein was 1/2.8/8.7. The yield and DNA/RNA/protein ratios were similar to those described for nuclei from other fungi (Doi & Doi, 1974: Wintersberger, Smith & Letnansky, 1973; Bhargava & Halvorson, 1971; Hsiang & Cole, 1973) and the morphology of the isolated nuclei by fluorescence and electron microscopy was similar to that seen *in situ*. These nuclei were used as the starting material for isolation of *A.nidulans* histones and for studies of chromatin.

Histones

Early work on fungal nuclear proteins either failed to reveal any histone-like proteins (Dwivedi, Dutta & Bloch, 1969; Leighton, Dill, Stock & Philips, 1971) or demonstrated a partial complement of histone-like proteins (Tonino & Rozijn, 1966; Wintersberger, Smith & Letnansky, 1973; Franco, Johns & Navlet, 1974). One possible explanation of the failure to find a full complement of histones in fungi was that in these studies histones had been lost by proteolysis. Since we paid special attention to preventing proteolysis during nuclear isolation, we decided to look for histones in *A.nidulans* using purified nuclei as starting material (Felden, Sanders & Morris, 1976). Nuclei were washed with salt to remove non-histone proteins and then extracted with 0.2 M HCl. The acid-soluble proteins were fractionated first on BioGel P-10 columns and then subfractionated on BioGel P-60 and Sephadex G-100 columns. Five histone-like proteins were identified (Fig. 1). Two were indistinguishable from calf thymus histones H3 and H4 in their chromatographic elution from BioGel P-10 and their electrophoretic mobilities on acid-urea and SDS gels. Amino acid analysis of *Aspergillus* H4 (Table 1) showed it to resemble calf thymus H4 in being arginine rich and in its overall amino acid composition; however it was somewhat less basic. *Aspergillus* H3 like H3 from other sources formed dimers on standing which could be cleaved to monomers by treatment with reducing agents. *Aspergillus* H2A and H2B eluted from BioGel P-10 in the same position as calf thymus H2A and H2B and amino acid analysis of these proteins demonstrated that they were lysine rich and resembled calf thymus H2A and H2B in overall composition but were also less basic (Table 1). A fifth basic protein that we believe to be H1 eluted from BioGel P-10 in the position of calf thymus H1 and has electrophoretic properties similar to calf thymus H1 appears to be but smaller and/or less basic than the calf thymus protein. Because of the possibility that ribosomal basic proteins could be confused with histones, ribosomal basic proteins were prepared from *A.nidulans* and compared with the

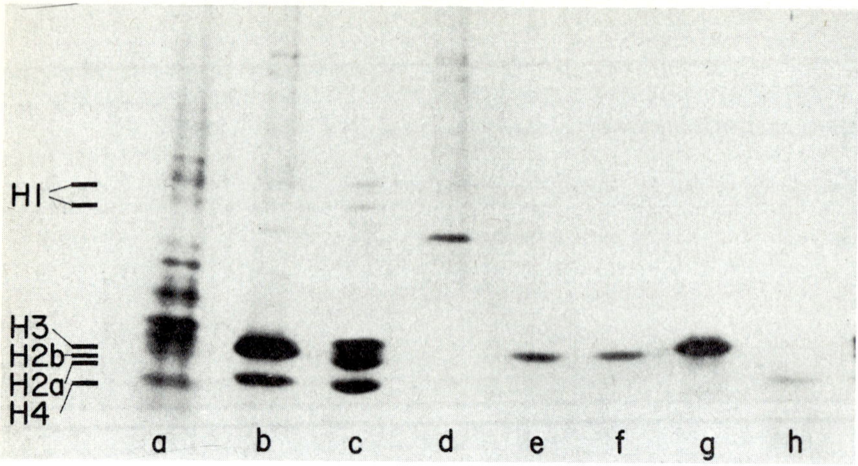

Fig. 1. Discontinuous SDS polyacrylamide gel electropherogram of *A.nidulans* unfractionated histones, *A.nidulans* purified histones, *A.nidulans* ribosomal basic proteins, and calf thymus histones. (a) *A.nidulans* ribosomal basic proteins (19μg); (b) *A.nidulans* unfractionated histones (15μg); (c) calf thymus histones (15μg); (d) *A.nidulans* H1; (e) *A.nidulans* H2a; (f) *A.nidulans* H2b; (g) *A.nidulans* H3; (h) *A.nidulans* H4. The order of migration of the calf thymus histones is shown to the left of the figure. Electrophoresis was from top to bottom.

---

histones. The *Aspergillus* histones and ribosomal basic proteins were clearly two different groups of proteins (Fig.1).

Until recently, the presence of histones in fungal nuclei was a subject of some controversy. However, the almost simultaneous description of histones in the nuclei of *Aspergillus nidulans* in chromatin from *Neurospora crass* (Goff, 1976), and in nuclei of *Saccharomyces cerevisiae* (Thomas & Furber, 1976) strongly suggests that histones are a general feature of the nuclear organization of most fungi.

Chromatin Structure

The association of histones with DNA in higher eukaryotes makes the DNA partially resistant to degradation by nucleases. Moreover, the pattern of nuclease resistance is very characteristic and conveys information about the structural organization of chromatin. The unit of structure in the chromatin of higher eukaryotes is the nucleosome (Olins & Olins, 1974; Oudet, Gross-Bellard & Chambon,

Table 1. Amino Acid Composition*

	A.nidulans (H2a)‡	Calf thymus (H2a)Ø	*A.nidulans* (Hab)$	Calf thymus (H2b)✱	*A.nidulans* (H4)//	Calf thymus (H4)⊕
Aspartic acid	9.0	6.6	5.9	4.8	5.9	4.9
Threonine ψ	4.4	3.9	7.6	6.4	8.3	6.9
Serine ψ	6.6	3.4	9.8	11.2	4.7	2.0
Glutamic acid	10.6	9.8	9.8	8.0	7.4	5.9
Proline	4.9	4.1	5.2	4.8	0	1.0
Glycine	12.5	10.8	8.3	5.6	18.7	16.7
Alanine	12.9	12.9	14.4	10.4	8.2	6.9
Cysteine	0	0	0	0	0	0
Valine	5.5	6.3	5.8	7.2	7.1	8.8
Methionine	0	0	0	1.6	0.5	1.0
Isoleucine	4.0	3.9	4.9	4.8	5.8	5.9
Leucine	11.2	12.4	6.1	4.8	7.0	7.8
Tyrosine**	0	2.2	0	4.0	0	3.9
Phenylalanine	1.5	0.9	1.6	1.6	2.8	2.0
Lysine	8.6	10.2	14.8	16.0	9.2	10.8
Histidine	2.1	3.1	1.2	2.4	2.3	2.0
Arginine	6.5	9.4	5.0	6.4	12.1	13.7
%Basic+	17.2	22.5	21	24.8	23.6	26.5
Lysine/arginine	1.32	1.08	2.96	2.50	0.76	0.79

*Expressed as mole percent
‡ Average from two 24-h hydrolyses
Ø Data taken from amino acid sequences (9)
$ Results of a single 24-h hydrolysis
ψ Not corrected for hydrolytic losses
** No tyrosine was recovered in any of the hydrolysates, probably because the samples were stored in dilute HCl
+ Sum of lysine, histidine and arginine

1975) a bead-like structure containing histones H2A, H2B, H3 and H4. The nucleosomes of rat liver have a regular repeat of about 200 base pairs which can be demonstrated by the fact that gentle digestion with micrococcal nuclease, which first attacks the relatively unprotected DNA linker between nucleosomes, generates a series of DNA fragments which are approximate multiples of 200 base pairs (Noll, 1974a). More extensive digestion with this enzyme degrades the higher multiples down to monomers and 'nibbles in' from the ends to give a transient limit digest fragment 140 base pairs long representing the well-protected nucleosome core (Shaw *et al.*

1976; Noll & Kornberg, in press). Additional information about nucleosome structure can be obtained with another enzyme, DNAse I. DNAse I degrades rat liver chromatin to single-stranded fragments which are multiples of 10 bases in length. This digestion pattern has been thought to reflect some regular organizational feature of the internal structure of the nucleosome (Noll, 1974b).

Gentle digestion of *Aspergillus* nuclear chromatin with micrococcal nuclease produced a regular series of DNA fragments similar to those produced by digestion of rat liver nuclei under the same conditions (fig.2).

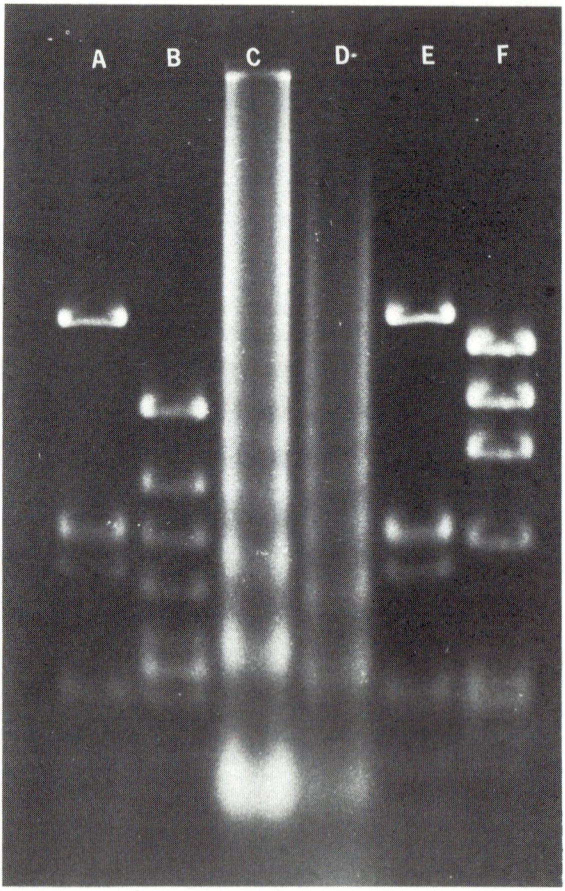

Fig. 2. Micrococcal Nuclease Digestion of Nuclei from Rat Liver and *A.nidulans*. (Methodology on next page).

A.nidulans nuclei were prepared by the method of Gealt *et al.*(1976). Rat liver nuclei were prepared by the method of Hewish & Burgoyne (1973). Nuclei were digested in 0.5 M sucrose, 0.01 M Pipes (pH 6.9), 5 mM CaCl, 5 mM $MgSO_4$ at a concentration of approximately 500 μg/ml chromatin DNA with 250 units per ml of micrococcal nuclease for 1 min at 37°C. The incubations were terminated by addition of EDTA to a final concentration of 20 mM, treated with 0.2 mg/ml of DNAse-free RNAase A for 10 min at 37°C, and deproteinized by shaking with a mixture of chloroform:isoamyl alcohol (24:1) in the presence of 1% sodium dodecylsulfate and 400 mM NaCl. The DNA samples were subjected to electrophoresis on 17 cm slab gels in 2% agarose, 40 mM Tris-acetate buffer (pH 7.8), 1 mM EDTA (TAE buffer) for approximately 14 h at 30 V. The gels were stained with ethidium bromide and photographed under ultraviolet light. Slots A-F contained the following: (A) Hha digest of X174 RF; (B) Hind II digest of X174 RF; (C) micrococcal nuclease digest of rat liver nuclei; (D) micrococcal nuclease digest of *Aspergillus* nuclei; (E) same as (A); (F) Hae III digest of X174 RF.

This result indicated that *A.nidulans* chromatin like rat liver chromatin, had a regular nucleosome-like structure; however, the repeat length was observed to be shorter than that of rat liver (Fig.2). By comparison with restriction fragments of known sizes, the nucleosome repeat length of *A.nidulans* chromatin was calculated to be 154±9 base pairs compared to about 200 base pairs for chromatin from rat liver (Morris, 1976b). The shorter repeat length for *A.nidulans* suggested that either the DNA linker between nucleosome cores was shorter than that of rat liver chromatin or that the cores themselves contained less DNA. Extensive digestion of *Aspergillus* chromatin with micrococcal nuclease generated a 140 base pair fragment which was identical with the core fragment produced from rat liver chromatin (Fig.3) (Morris, 1976b), leading to the conclusion that the nucleosome core was the same as rat liver but the linker was shorter. DNAse I was used to provide additional information about the structure of the core. Digestion of *Aspergillus* nuclear chromatin with DNAse I produced a pattern of fragments at multiples of 10 bases that was similar to that produced from rat liver chromatin. This result indicates that not only the size (140 base pairs) but also the regular structure of the nucleosome core of *A.nidulans* chromatin is very similar to that seen in higher eukaryotes. Similar results have been obtained with other fungi. In both *Neurospora crassa* (Noll, 1976) and *Saccharomyces cerevisiae* (Lohr & Van Holde, 1975; Thomas & Furber, 1976), the size and structure of the nucleosome core are similar to that of higher eukaryotes, but the nucleosome repeat is shorter. Thus the general pattern in fungi seems to be that the core is conserved with respect to higher eukaryotes but the linker between cores is shorter.

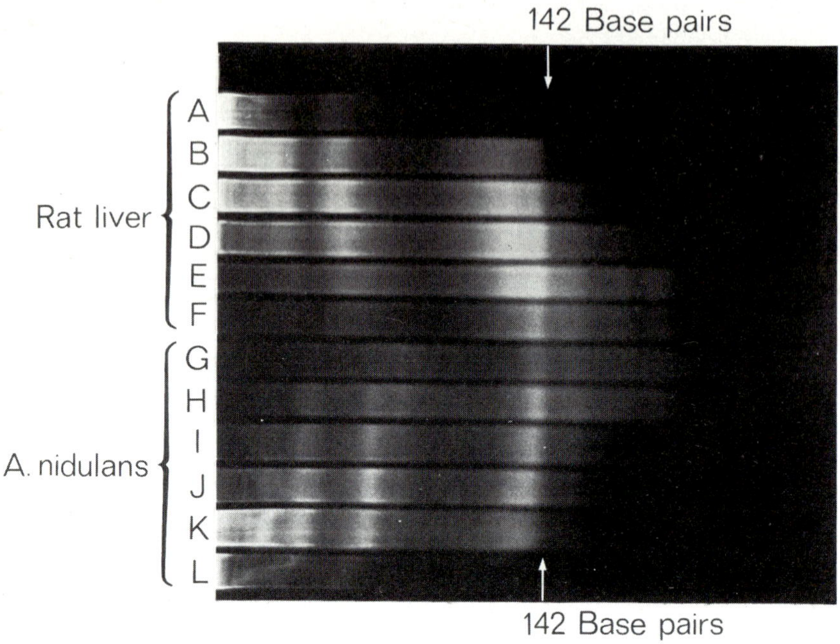

Fig. 3. Comparison of Smaller DNA Fragments from Digests of Rat liver and *A.nidulans* chromatin with micrococcal nuclease. Nuclei were prepared and digested as specified in Fig. 1 with 250 units per ml of enzyme for 0.25, 1,2,4,8 and 16 min. The digests were treated with RNAase, deproteinized as described in Fig. 1, and subjected to electrophoresis in 6% acrylamide with TAE buffer for 12 h at 40 V.

The fungi provided the first demonstration that the nuceosome repeat length was not universal for all eukaryotes. The importance of this finding was that it suggested that the nucleosome repeat length might vary even among higher eukaryotes. Moreover, inspection of the data suggested a reason for variation in the length of the linker. In higher eukaryotes histone H1 has been shown to be associated with the linker DNA between cores (Shaw *et al.*, 1976; Noll & Kornberg, in press). In *N.crassa* as well as *A.nidulans* histone H1 differs substantially from the H1's of higher eukaryotes and appears to be less basic (Goff, 1976; Felden *et al.*,1976). It was therefore suggested that the structure of H1, possibly the net positive charge on the molecule, determines the length of the linker and consequently the length of the nucleosome repeat (Morris, 1976b; Noll, 1976). This hypothesis has been tested for a few higher eukaryotes with very basic H1-like histones. In chicken erythrocyte vs chicken liver (Morris, 1976c) and sea urchin sperm vs sea urchin gastrula (Spadafora, Bellard Compton & Chambon, submitted for publication) the tissue with the more basic H1 has the longer repeat. The 140 base pair core is conserved in these tissues. Thus the variation in nucleosome repeat length must be a consequence of changes in the length of the DNA linker between cores, and the data support the hypothesis that H1 determines the spacing of nucleosome cores.

Tubulin

Electron micrographs demonstrate that *A.nidulans* resembles other eukaryotes in having a microtubular mitotic spindle (Robinow & Caten, 1969). We have identified tubulin, the major protein component of microtubules, in *A.nidulans* by copolymerization with purified porcine brain tubulin (Sheir-Neiss, Nardi, Gealt & Morris, 1976). *A.nidulans* was radioactively labelled by 24 h of growth in medium containing ^{35}S. The (^{35}S)-mycelium was then disrupted, extracted with buffer and centrifuged at 100,000 x g for one hour. Purified porcine tubulin was added to the extract at a final concentration of 4-5 mg of protein per ml and was polymerized at 37° for 20 min in the presence of GTP and glycerol. The microtubules were then collected by centrifugation at 100,000 x g for one hour at 30°, depolymerized at 0° and the polymerization cycle repeated to obtain a purified pellet of copolymerized porcine and (^{35}S)-*Aspergillus* microtubular protein. Approximately 1% of the total acid precipitable radioactivity of the crude extract was found in the final product. Coelectrophoresis of the (^{35}S) *Aspergillus* tubulin with ^{3}H-labelled chick brain tubulin in a discontinuous polyacrylamide gel containing SDS demonstrated two *Aspergillus* components, one comigrating with chick tubulin 1, the other migrating slightly faster than chick tubulin 2 (Fig.4). Coelectrophoresis of ^{35}S *Aspergillus* tubulin with ^{3}H chick brain tubulin in an isoelectric focussing system also gave similar but non-identical profiles (Fig.5). The apparent molecular weights of the *Aspergillus* proteins were 55,000 and 60,000. No other labelled components were found. Davise has also partially purified *Aspergillus* tubulin by copolymerization and by conventional methods and has shown that mutants demonstrating resistance and increased sensitivity to the spindle poison, methylbenzimidazole carbamate have tubulins with altered ability to bind this drug (manuscript submitted).

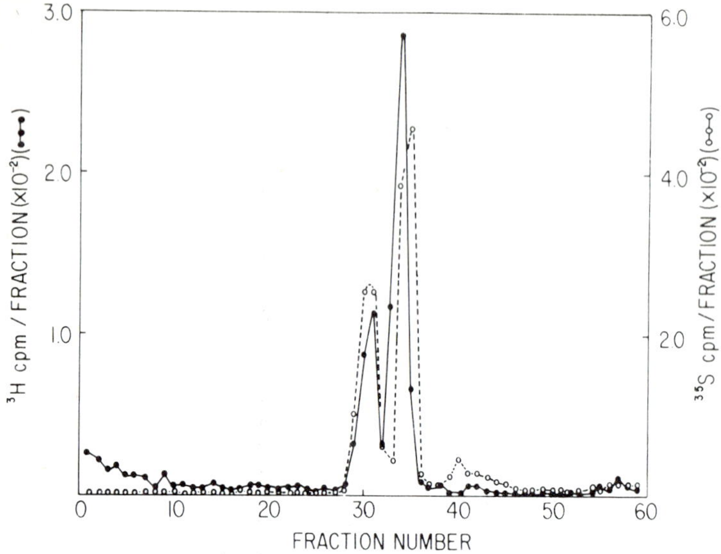

Fig. 4. Coelectrophoresis of [^{3}H] chick tubulin and [^{35}S] *A. nidulans* copolymerized protein in a discontinuous polyacrylamide gel. ^{3}H chick tubulin (20 μg) and [^{35}S] *A.nidulans* copolymerized protein (20 μg) were coelectrophoresed for 6 h through a 10% polyacrylamide gel. The gel was sliced into 1 mm sections, and the slices were counted in a liquid scintillation spectrometer to determine the presence of [^{3}H](•) and [^{35}S] (o). The anode is on the right.

CONCLUSIONS

Our initial studies of the nucleus of *A.nidulans* clearly demonstrate that *Aspergillus* resembles other eukaryotes in the major features of its nuclear biochemistry. *Aspergillus* chromatin has histones, its chromatin is organized into nucleosomes, and *Aspergillus* tubulin is not dissimilar from the tubulin of higher eukaryotes. Differences between the chromatin of *A.nidulans* and the chromatin of higher eukaryotes have focussed attention on previously unnoticed details of chromatin structure in the higher eukaryote notably the variability in the length of the linker between cores, the invariability of the core and the relationship between the structure of H1 and the length of the linker and its effect on the nucleosome repeat length. Although our knowledge of the organization of chromatin in *A.nidulans* is still very limited, the fact

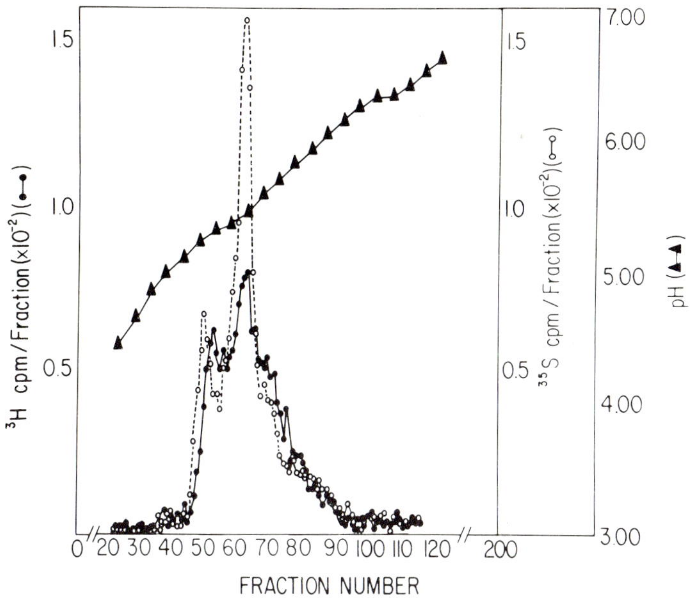

Fig. 5. Coelectrophoresis of [^{3}H] chick tubulin and [^{35}S] *A. nidulans* copolymerized protein in an isoelectric focussing gel. [^{3}H] chick tubulin (40 μg) and [^{35}S] *A.nidulans* copolymerized protein (20 μg) were subjected to isoelectric focussing for 36 h. The gel was sliced into 1 mm sections, and the slices were counted in a liquid scintillation spectrometer to determine the presence of [^{3}H] (o) and [^{35}S] (●). The anode is on the right. A second gel, run without protein samples, was used to determine the pH gradient. This gel was sliced into 5 mm sections. Each section was soaked in 1 ml of distilled water and the pH (▲) of each section was determined.

that *A.nidulans* chromatin resembles that of higher eukaryotes sufficiently to provide a simple model system for experimentation, yet is different enough to provide interesting comparisons, suggests that the important role played by *A.nidulans* in the genetic analysis of eukaryotic regulatory systems may find a parallel in the use of *A.nidulans* for the analysis of chromatin structure and function. Similarly, the characterization of tubulin in an organism in which there are mutants affecting the organization and function of the mitotic spindle promises to be useful in understanding the molecular events of mitosis.

REFERENCES

BAINBRIDGE,B.W.(1971). Macromolecular composition and nuclear division during spore germination in *Aspergillus nidulans*. *Journal of General Microbiology* 66, 319-25.

BHARGAVA,M.M. & HALVORSON,H.O. (1971). Isolation of nuclei from yeast. *Journal of Cell Biology* 49, 423-9.

CLUTTERBUCK,A.J. (1974). *Aspergillus nidulans* in *Handbook of Genetics* Vol.1, ed. R.C. King, Plenum Press.

DOI,K. & DOI,A. (1974). Isolation of nuclei from a tetraploid strain of *Saccharomyces cerevisiae*. *Journal of Biochemistry* 75, 1017-26.

DWIVEDI,R., DUTTA, S. & BLOCK,D. (1969). Isolation and characterization of chromatin from *Neuraspora crassa*. *Journal of Cell Biology* 43, 51-8.

FELDEN,R.A., SANDERS,M.M. & MORRIS,N.R. (1976). Presence of histones in *Aspergillus nidulans*. *Journal of Cell Biology* 68, 430-9.

FRANCO,J.E.W., JOHNS, E.W. & NAVLET,J.M. (1974). Histones from bakers yeast. Isolation and characterization. *European Journal of Biochemistry* 45, 83-9.

GEALT,M.A., SHEIR-NEISS,G. & MORRIS,N.R.(1976). The isolation of nuclei from the filamentous fungus *Aspergillus nidulans*. *Journal of General Microbiology* 94, 204-10.

GOFF,C.G. (1976). Histones of *Neurospora crassa*. *Journal of Biological Chemistry* 251, 4131-8.

HEWISH,D.R. & BURGOYNE,L.A. (1973). A digestion of chromatin DNA at regularly spaced sites by a nuclear deoxyribonuclease. *Biochemical and Biophysical Research Communications* 52,504-10.

HSIANG,M.W. & COLE,R.D. (1973). The isolation of histones from *Neurospora crassa*. *Journal of Biological Chemistry* 248, 2007-13.

LEIGHTON,T.J., DILL,B.C., STOCK,J.J. & PHILIPS,H. (1971). Absence of histones from the chromosomal proteins of fungi. *Proceedings National Academy of Science, U.S.A.* 68, 677-80.

LOHR,D. & VAN HOLDE,K.E. (1975). Yeast chromatin subunit structure *Science* 188, 165-6.

MINSSEN,M. & MUNKRES,K.D. (1973). Preparation of mitochondrial membrane proteins from *Neurospora crassa*. Prevention of lipid autoxidation by an antioxidant. *Biochimica et Biophysica Acta* 291, 398-410.

MORRIS,N.R. (1976a). Mitotic mutants of *Aspergillus nidulans*. *Genetical Research Cambridge* 26, 237-54.

MORRIS,N.R. (1976b). Nucleosome structure in *Aspergillus nidulans*. *Cell* 8, 357-64.

MORRIS,N.R. (1976c). A comparison of chicken liver and chicken erythrocyte chromatin. *Cell*, in press.

NOLL, M. (1974a). Subunit structure of chromatin. *Nature, London* 251, 249-51.
NOLL, M. (1974b). Internal structure of the chromatin subunit. *Nucleic Acids Research* 1, 1573-8.
NOLL, M. (1976). Differences and similarities in chromatin structure of *Neurospora crassa* and higher eukaryotes. *Cell* 8, 349-55.
OLINS,A.L. & OLINS,D.E. (1974). Spheroid chromatin units (ν bodies). *Science* 183, 330-1.
OUDET,P., GROSS-BELLARD,M. & CHAMBON, P. (1975). Electron microscopic and biochemical evidence that chromatin structure is a repeating unit. *Cell* 4, 281-300.
ORR,E. & ROSENBERGER,R.F. (1976). Initial characterization of *Aspergillus nidulans* mutants blocked in the nuclear replication cycle. *Journal of Bacteriology* 126, 895-902.
PEBERDY,J.F. & GIBSON,R.K. (1971). Regeneration of *Aspergillus nidulans* protoplasts. *Journal of General Microbiology* 69,325-30.
ROBINOW,C.F. & CATEN,C.E. (1969). Mitosis in *Aspergillus nidulans*. *Journal of Cell Science*. 5, 403-37.
SHAW,B.R., HERMAN,T.M., KOVACIC,R.T., BEAUDREAU,G.S. & VAN HOLDE,K.E. (1976). Analysis of subunit organization in chicken erythrocyte chromatin. *Proceedings National Academy of Science U.S.A.* 73, 505-9.
SHEIR-NEISS,G., NARDI,R.V., GEALT,M.A. & MORRIS,N.R. (1976). Tubulin-like protein from *Aspergillus nidulans*. *Biochemical and Biophysical Research Communications* 69, 285-90.
THOMAS,J.O. & FURBER,V., (1976). Yeast chromatin structure. *FEBS Letters* 66, 274-80.
TONINO,G. & ROZIJN,T. (1966). On the occurence of histones in yeast. *Biochimica et Biophysica Acta* 124, 427-9.
WINTERSBERGER,V., SMITH,P. & LETNANSKY, K. (1973). Yeast chromatin. Preparation from isolated nuclei, histone composition and transcription capacity. *European Journal of Biochemistry* 33, 123-30.

The Proteases of *Aspergilli*

B. L. COHEN

Department of Genetics, University of Glasgow, Glasgow, U.K.

INTRODUCTION

Protease studies are one of today's growth areas in biology (Reich, Rifkin & Shaw, 1975). This interest stems from technological uses (Blain,1975; Keay, 1971) and from the varied roles of proteases in cell biology. Protease function is characterized by selectivity and irreversibility. Proteases have been implicated in nutrition, in the production by limited proteolysis of active hormones and enzymes, in the control by limited proteolysis of developmental processes and of the assembly of structural proteins, in the responses of tissues to injury or of multicellular organisms to foreign agents and in intracellular communication and interaction. Whilst some of these functions may be irrelevant to fungi, the importance of others has been demonstrated or may be confidently predicted. A corollary of the irreversibility of protease action is that synthesis and control of the proteases themselves must take place in such a way that the cellular environment is protected by superimposed control of protease activity. Thus protease studies demand simultaneous investigation of those intracellular proteins which inhibit protease activity (Fritz, Tschesche, Greene & Truscheit, 1974).

WHAT PROTEASES DO ASPERGILLI MAKE?

Enzyme types

Even to the biochemical taxonomist the classification of proteolytic enzymes presents difficulties. Broadly, however, the Commission on Enzyme Nomenclature (1972) recognizes three groups. Exo-peptidases (EC.3.4.11 - 15) participate in, but cannot themselves complete, the digestion of proteins. Little is known about them in Aspergilli and they will therefore not be further considered here. Dipeptide hydrolases (EC.3.4.13 - 14) have apparently not been studied at all in Aspergilli, although they may be important (Johnson & Brown, 1974). Thus attention will be confined to the third group, the peptide endo-hydrolases (EC.3.4.21 - 24), conveniently called 'proteases'.

The further classification of proteases is principally in terms of catalytic mechanism, usually determined by reactivity towards inhibitors that react with particular residues in the active site region. Using these criteria, endo-proteases fall into four broad classes: alkaline proteases (EC.3.4.21) with a reactive serine residue; thiol proteases (EC.3.4.22); acid proteases (EC.3.4.23) and neutral proteases (EC.3.4.24) generally dependent on a metal ion. From extensive reviews of microbial proteases (Matsubara & Feder, 1971; Nakagawa, 1970; Sodek & Hoffman, 1970), it is evident that the proteases of many Aspergilli fit into this classification. Not all species, however, are known to produce enzymes of each class but this may reflect incompleteness of research. Thus, for example, the *niger* group, including *A.awamori*, *A.niger* and *A.saitoi*, are chiefly known as active producers of extracellular acid protease only. However, *A.niger* alone has also been reported to produce what seems to be an alkaline protease (Bossman, 1973).

Patchy knowledge of the range of proteases produced by Aspergilli is well-demonstrated in the *flavus-oryzae* group which is, paradoxically, the most intensively-studied. Taking all strains and reports together, there is evidence for proteases of all four classes (Davidson,Gertler & Hoffman, 1975; Feinstein & Gertler, 1973; Klapper, Jameson & Meyer, 1973*a*, Kundu, Manna & Pal, 1968; Kundu & Manna, 1975; Nordwig & Jahn, 1968; Subramanian & Kalnitsky, 1964; Turkova,Mikes,Gancev & Boublik, 1969). It seems likely that many apparent inter-strain differences arise from the wide diversity of culture conditions and biochemical methods employed by different investigators and from the fact that, for many of them, the starting-material is a crude or partially-purified industrial preparation from which some protease activities may already have been lost. Furthermore true comparisons are precluded by the use of 'private' or unspecified strains. Systematic comparisons of the purified alkaline proteases of *A.flavus*, *A.oryzae*, *A.sojae* and *A.sulphureus* have been made (Bretschneider, Nordwig, Mikes & Turkova, 1973; Turkova, *et al.* 1972). The results were consistent with known taxonomic relations (Raper & Fennell, 1965), the enzymes of *A.flavus* and *A.oryzae* being indistinguishable.

In view of the incomplete state of knowledge in this field I am making a comparison of many species and strains of Aspergilli using standardized culture and analytical methods and cultures obtained from the Commonwealth Mycological Institute, Kew, or the Northern Regional Research Centre, Peoria. Preliminary results indicate that whilst some of the inter-strain differences suggested by the literature may be real, others are artefactual. Most of the species so far examined produce several distinguishable neutral and alkaline proteinases (and therefore conform to expectations) but strains of *A.fumigatus* seem to produce only one such enzyme. There is considerable divergence in specific activity of culture filtrates (in relation to the mass of producing mycelium) but no correlation can yet be discerned between activity and habitat of isolation.

Intracellular and extracellular proteases

A biologically important distinction may be drawn between proteases that are strictly intracellular and those which are capable of being secreted into periplasmic or extracellular spaces (Pollock, 1962; Holtzer, Betz & Ebner, 1975). For most reports of Aspergillus proteases this distinction cannot easily be made. Thus if, as is often the case, cultures were maintained for many days on complex liquid media or on undefined solid nutrients (e.g. wheat bran) it is difficult to know whether an enzyme found in the medium was secreted or simply escaped as a consequence of autolysis. Moreover, since genuinely extracellular enzymes may be found within cells or in transit through the walls during a phase of active secretion, presence in a mycelial extract does not adequately define an intracellular enzyme. Despite these difficulties I suggest that enzymes obtained in high yield from mouldy bran or stationary phase liquid cultures may provisionally be accepted as extracellular. Cases satisfying more stringent criteria for extracellularity include the acid proteases of *A.niger* (Tomonaga, Ohama & Yanagita, 1964) and the neutral and alkaline proteases of *A.nidulans*. Only protease β of *A.nidulans* (Cohen, 1973*a*) is reasonably certain to be an intracellular enzyme although here it must be said that the distinction between periplasmic, intramural and truly intracellular enzyme can only be made *via* protoplast studies, but these have not been attempted.

Subcellular packaging and secretion mechanism

In *Neurospora*, small vesicles rich in hydrolytic enzymes including protease, and considered to be analogous to lyosomes, have been described (Matile, 1969) and it has been proposed, but not confirmed (Heiniger & Matile, 1974) that secretion of such enzymes is by a process of reverse micro pinocytosis. In *Saccharomyces*, similar vesicles have been found to correspond with the vacuoles (Matile & Wiemken, 1967). No similar studies have been reported in Aspergilli, although preliminary gradient centrifugation results indicate that the proteases of *A.nidulans* may be similarly located (unpublished observations). The whole problem of subcellular packaging and secretion could benefit from a genetic approach, making use of mutants defective in exoenzyme secretion. Some relevant mutants already exist in *A.nidulans* (Cohen, 1973*a,b*) and indeed the very existence of an extracellular proteaseless strain (*xpr*C) that is unable to use protein as sole carbon or nitrogen source, together with the unpublished results mentioned above, is *a priori* evidence for the existence of at least two classes of protease-containing vesicles in this species. Similarly, *A. nidulans* is well-placed amongst fungi to serve as an experimental tool for genetic analysis of situations analogous to the lysosomal storage diseases of man (Hers & van Hoof, 1973).

THE REGULATION OF PROTEASE BIOSYNTHESIS

Extracellular enzymes

The control of synthesis of extracellular hydrolases poses some interesting biological problems in organisms like fungi which have rigid and to some extent semi-permeable walls (Trevithick & Metzenberg, 1966) surrounding the plasmalemma. If synthesis is positively regulated (i.e. substrate-inducible) how is the cell able to detect macromolecular substrate molecules that may or may not be able to diffuse through the wall? Alternatively, if synthesis is negatively regulated (i.e. constitutive, but repressible), what metabolites will act as repression-effectors, and is there any form of internal 'autoinduction' (Shinmyo, Tahara, Matsumoto & Terui, 1971)? Is it possible for particular enzymes to be both positively and negatively controlled? How does the method of control relate to the ecology of the organism in the wild and what are the underlying genetic mechanisms? How is the regulation of extracellular enzyme synthesis related to that of other cell constituents? This last question takes on particular point in the light of the fact that protease synthesis may be highly preferential, S-starved *A.niger* synthesizing extracellular proteases exclusively (Tomonaga, 1966). As will be seen, few of these questions can be answered with any security.

Regulation studies of Aspergillus proteases have been hampered by the fact that initial interest in these enzymes stems from their roles in Japanese traditional food processing (Blain, 1975). In consequence, most studies have used complex or undefined media. However, pioneer investigations (Tomonaga, Ohama & Yanagita, 1964) showed convincingly that the acid protease of *A.niger* is not inducible, but is made *de novo* in response to sulphur-limited culture, i.e. that protease synthesis is constitutive but sulphur- (and also nitrogen-) repressible. Not until much later were similar methods applied to another species, demonstrating that the extracellular neutral and alkaline proteases of *A.nidulans* are also non-inducible but are repressed by carbon-, nitrogen- and sulphur-metabolites in such a way that limitation for any one nutrilite derepressed protease production (Cohen, 1973*b*). *De novo* synthesis in response to carbon catabolite derepression has also been reported in *A.oryzae* (Klapper, Jameson & Meyer, 1973*b*). Reports that protease synthesis in some Aspergilli and in certain other fungi may be inducible are mostly suspect, either because the inducing protein was presented under conditions where nutrilite limitation was likely (O'Sullivan & Mathison, 1971; Shinmyo & Terui, 1970; Shinmyo *et al.*, 1971), or because the term was being used loosely (Gealt & Axelrod, 1974). In *Neurospora crassa*, however, extracellular protease synthesis is definitely both induced by substrate protein and repressed by carbon-, nitrogen- and sulphur-nutrilites (Cohen, Morris & Drucker, 1975). In *A.niger*, one report (Shinmyo *et al.*, 1971) claims that an intracellular inducer was extracted from protease-secreting mycelium which was nutrilite-limited but not from mycelium growing on complex medium and not secreting protease. Confirmation of this finding and its extension to another species would be valuable.

In an attempt to improve knowledge of protease regulation in *Aspergilli* I am in the process of a comparative study, using established methods (Cohen, 1973*a*,*b*) and strains obtained from the Commonwealth Mycological Institute, Kew, Surrey (CMI) or the U.S. Department of Agriculture, Northern Regional Research Center, Peoria, Illinois (NRRL). Results so far obtained from 2 strains of *A.nidulans* (CMI 57395 & 158753), one strain of *A.flavus* (CM 114929 2 strains of *A.fumigatus* (CMI 16152 and 158754), 3 strains of *A.oryzae* (CMI 52144 & 52148; NRRL 2160), one strain of *A.tamarii* (CMI 91888) and one strain of *A.terreus* CMI 45543) are broadly similar. All strains produced substantial levels of extracellular neutral and alkaline protease activity in response to nitrogen starvation but did not produce protease in the presence of ammonium. In no strain was protease production induced by substrate protein in the presence of ammonium. Thus the general rule amongst Aspergilli may be for extracellular protease production to be constitutive but repressible and it therefore seems likely that the differences in protease regulation between *A.nidulans* and *N.crassa* have a phylogenetic basis.

The genetics of enzyme synthesis regulation by carbon-, nitrogen- and sulphur-metabolite repression have been investigated chiefly in *A.nidulans*, *S.cerevisiae* and *N.crassa*. Only in *A.nidulans* (Arst & Cove, 1973; Cohen, 1973*b*; Pateman, Kinghorn, Dunn & Forbes, 1973) and in *N.crassa* (Cohen,Morris & Drucker, 1975; Hanson & Marzluf, 1973, 1975) do results bear on protease regulation, but in neither case is enough known to provide a coherent picture of an undoubtedly complex situation; for example, the model put forward by Hanson & Marzluf (1975) does not make any provision for induction of protease although this is well-established (Drucker, 1972).

Intracellular proteases and specific protease inhibitors

Very little is known about the regulation of synthesis and activity of strictly intracellular proteases in Aspergilli or in any other organism. Indeed it may safely be stated that the whole range of such enzymes in any one species has not yet been discovered. This is,however, an active field of research in *Saccharomyces* studies (and in mammalian biochemistry) where it has become clear that investigation of intracellular proteases must proceed in parallel with the study of protease inhibitors (Holtzer, Betz & Ebner, 1975). These proteins, which specifically inhibit intracellular proteases, may be regarded as one important element in the fine control of the time and place of protease activity. Very little is yet known about the regulation of these inhibitors in fungi except that in yeast, evidence for carbon catabolite repression of inhibitor level has recently been obtained (Betz, 1975). The inhibitors of *A.nidulans* appear to be under both carbon- and nitrogen-repression control (unpublished results).

In yeast it has also been established that the intracellular proteases are localized in membrane-bound vesicles whilst the inhibitors are apparently free in the cytosol (Lenney, Matile, Wiemken, Schellenberg & Meyer, 1974; Matern, Betz & Holzer, 1974). It follows that studies of regulation of inhibitor and cell protease can only be made under conditions where this compartmentation is allowed for in experimental designs. My own results on the apparent regulation of the intracellular protease of *A.nidulans* (Cohen, 1973*b*) must therefore be interpreted with caution until they have been repeated using isotonic cell-fractionation methods. The properties of the unstable β1 and β2 forms I described are quite consistent with them being protease-inhibitor complexes formed perhaps as a result of compartment rupture during mycelial extract preparation.

Similarly, reservations must be applied to any reports in which extracellular proteases are obtained from mycelial extracts, since these enzymes too may interact with the naturally-occurring protease inhibitors (Cohen, 1973*b* and unpublished observations).

STRAIN IMPROVEMENT

Whatever developments may actually have been made within industrial concerns using Aspergilli for protease or other enzyme production, no systematic use of genetic methods has been reported. Indeed, only four reports on strain improvement are known to me (Brotskaya, 1960, 1961; Ikeda, Nakamura, Uchida & Ishitani, 1957; Sekine, Nasumo & Iguchi, 1969) and exploitation of regulatory mutations such as *xpr*D (Cohen, 1973*b*), although promising, has not yet been reported in the scientific literature. Increased exo-enzyme production has been reported to result from a mutation in *Neurospora crassa* affecting wall composition, but there is no report of proteases being affected (Gratzner, 1972). For laboratory studies, an 'improved' strain might be one with reduced or absent proteases. So far as extracellular proteases go, such a strain exists (Cohen, 1973*a*), but it has not yet been exploited by those studying other metabolically regulated enzyme systems. Repeated attempts to isolate mutants defective in intracellular protease have not been successful (unpublished results), presumably because of the existence of multiple activities.

PROTEASES, DEVELOPMENT AND DIFFERENTIATION

Only a few, unrelated studies can be reported under this heading. Yanagita & Nomachi (1967) have discussed which part of the mycelium of *A.niger* is active in protease secretion, concluding that it is the region behind the hyphal tip cell. Ohama, Tomonaga & Yanagita (1966) reported an effect of physiological age on both capacity for and specificity of protease secretion. Likewise, Gealt & Axelrod (1974) have found an effect of physiological age: in *A.nidulans* capacity for extracellular protease and intracellular nitrate reductase synthesis both decline at the time when mycelium becomes competent to initiate conidiophore development. A mutant defective in extracellular protease production grows and differentiates normally, demonstrating that these enzymes are dispensable (Cohen, 1973*a*).

So far, none of the investigations of events in the growing hyphal tip have encompassed the intracellular proteases, nor have any studies on germination in Aspergilli paid attention to the possible roles of these enzymes although they have been implicated in spore germination in *Microsporum* (Page & Stock, 1974). The importance of limited proteolytic cleavage in morphogenetic processes has, however, been indicated through studies of chitin synthetase activation in yeast budding (Cabib & Farkas, 1971).

WHAT DOES EVERY ASPERGILLUS WORKER NEED TO KNOW ABOUT PROTEASES?

(i) Anyone who cultures Aspergilli under conditions where an essential nutrilite (carbon, nitrogen, phosphorus or sulphur) may become limiting should be aware that there will be a spectacular increase in protease activities.

(ii) Anyone who ruptures cells with the intention of isolating or characterizing any protein should be aware of the Pandora's Box of protein degradation that he is opening. This subject has recently been reviewed for yeast (Pringle, 1975) and the advice given there should be widely accepted. It should be particularly noted that protease may co-purify with other proteins even through multi-stage procedures and that appreciable protease activity may survive during sample preparation for SDS-PAGE electrophoresis.

Practical steps that may be taken to minimize proteolytic damage include (a) choice of strain, (b) choice of growth conditions to avoid derepression, (c) use of isotonic media and appropriate centrifugation in the preparation of cell extracts, (d) maintenance of an adequate concentration of protease inhibitors (e.g. phenylmethyl-sulphonyl chloride and 1,10-phenanthroline) throughout all biochemical procedures, (e) fast work at pH7, (f) use of powerful separation methods at early stages of purification, (g) addition of intrinsic protease inhibitors as boiled cell extracts.

REFERENCES

ARST,H.N. & COVE,D.J. (1973). Nitrogen metabolite repression in *Aspergillus nidulans*. *Molecular and General Genetics* 126, 111-41.

BETZ,H. (1975). Levels and turnover of the proteinase B inhibitors in yeast. *Biochimica et Biophysica Acta* 404, 142-51.

BLAIN,J.A. (1975). Industrial enzyme production. In *The Filamentous Fungi*, Vol.1,Eds. Smith,J.E. & Berry,D.R., London: Arnold.

BOSSMAN,H.B.(1973). Protein catabolism II. Identification of neutral and acidic proteolytic enzymes in *Aspergillus niger*. *Biochimica et Biophysica Acta* 293, 476-89.

BRETSCHNEIDER,G., NORDWIG,A., MIKES,O. & TURKOVA,J. (1973). Properties of proteases from *Aspergillus oryzae* and *Aspergillus flavus*. A comparative study. *Hoppe-Seyler's Zeitschrift für Physiologische Chimie* 352, 1372-6.

BROTSKAYA,S.Z. (1960). Effect of UV irradiation in varying dose on production of *Aspergillus nidulans* mutants that produce active proteases. *Mikrobiologiya* (English translation) 29, 264-6.

BROTSKAYA,S.Z. (1961). The effect of various doses of UV radiation on the generation of Aspergillus mutants that produce active proteases. *Akad.Nauk. SSSR*10, 120-8.

CABIB,E. & FARKAS,V. (1971). The control of morphogenesis: an enzymatic mechanism for the initiation of septum formation in yeast. *Proceedings of the National Academy of Sciences, Washington* 68, 2052-6.

CHANG,P.L.Y. & TREVITHICK,J.R. (1974). How important is secretion of exoenzymes through apical cell walls of fungi? *Archives of Microbiology* 101, 281-93.

COHEN,B.L.(1973*a*). The neutral and alkaline proteases of *Aspergillus nidulans*. *Journal of General Microbiology* 77, 521-8.

COHEN,B.L. (1973*b*). Regulation of intracellular and extracellular neutral and alkaline proteases in *Aspergillus nidulans*. *Journal of General Microbiology* 79, 311-20.

COHEN,B.L., MORRIS,J.E. & DRUCKER, H. (1975). Regulation of two extracellular proteases of *Neurospora crassa* by induction and by carbon-, nitrogen- and sulfur-metabolite repression. *Archives of Biochemistry and Biophysics* 169, 324-30.

COMMISSION ON BIOCHEMICAL NOMENCLATURE: RECOMMENDATIONS (1972). *Enzyme Nomenclature*, pp228-250. Amsterdam: Elsevier.

DAVIDSON,R., GERTLER,A. & HOFFMAN,T.(1975). *Aspergillus oryzae* acid proteinase: purification and properties and formation of π-chymotrypsin. *Biochemical Journal* 147, 45-53.

DRUCKER,H. (1972). Regulation of exocellular proteases in *Neurospora crassa:* induction and repression of enzyme synthesis. *Journal of Bacteriology* 110, 1041-9.

FEINSTEIN,G. & GERTLER,A.(1973). Isolation of alkaline proteinases from *Aspergillus oryzae* by one-step affinity chromatography on ovoinhibitor-Sephadex column. *Biochimica et Biophysica Acta* 309, 196-202.

FRITZ,H., TSCHESCHE,H., GREENE,L.J. & TRUSCHEIT,E. Editors (1974). *Proteinase Inhibitors*. Proceedings of the Second International Research Conference. Berlin, Heidelberg and New York: Springer-Verlag.

GEALT,M.A. & AXELROD,D.E. (1974). Coordinate regulation of enzyme inducibility and developmental competence in *Aspergillus nidulans*. *Developmental Biology* 41, 224-32.

GRATZNER,H.B. (1972). Cell wall alterations associated with the hyperproduction of extracellular enzymes in *Neurospora crassa*. *Journal of Bacteriology* 111, 443-6.

HANSON,M.A. & MARZLUF,G.A. (1973). Regulation of a sulfur-controlled protease in *Neurospora crassa*. *Journal of Bacteriology* 116,785-9.

HANSON,M.A. & MARZLUF,G.A. (1975). Control of the synthesis of a single enzyme by multiple regulatory circuits in *Neurospora crassa*. *Proceedings of the National Academy of Sciences, Washington* 72, 1240-4.

HEINIGER,U. & MATILE,P. (1974). Protease secretion in *Neurospora crassa*. *Biochemical and Biophysical Research Proceedings* 60, 1425-32.

HERS,H.G. & Van HOOF,F. (1973). Lysosomes and storage diseases, New York & London: Academic Press.

HOLTZER,H., BETZ,H. & EBNER,E. (1975). Intracellular proteinases in microorganisms. In *Current Topics in Cell Regulation*, edited by B.L. Horecker, New York & London: Academic Press.

IKEDA,Y., NAKAMURA,K., UCHIDA,K. & ISHITANI, C. (1957). Two attempts upon improving an industrial strain of *Aspergillus oryzae* through somatic recombination and polyploidization. *Journal of General and Applied Microbioloau* 3. 93-101.

JOHNSON,G.L. & BROWN,J.L. (1974). Partial purification and characterization of two peptidases from *Neurospora crassa*. *Biochimica et Biophysica Acta* 370, 530-40.

KEAY,L. (1971). Microbial proteases. *Process Biochemistry* 6,17-21.

KLAPPER,B.F., JAMESON,D.M. & MAYER,R.M. (1973*a*). The purification and properties of an extracellular protease from *Aspergillus oryzae* NRRL 2160. *Biochimica et Biophysica Acta* 304, 505-12.

KLAPPER,B.F., JAMESON,D.M. & MAYER,R.M. (1973*b*). Factors affecting the synthesis and release of the extracellular protease of *Aspergillus oryzae* NRRL 2160. *Biochimica et Biophysica Acta*, 304, 513-9.

KUNDU,A.K., MANNA,S. & PAL,N. (1968). Extracellular proteinases of *Aspergillus oryzae*. *Applied Microbiology* 16, 1799-801.

KUNDU,A.K., & MANNA,S. (1975). Purification and characteristics of extracellular proteinases of *Aspergillus oryzae*. *Applied Microbiology* 30, 507-13.

LENNEY,J.F., MATILE,P., WIEMKEN,A., SCHELLENBERG,M. & MEYER,J.(1974). Activities and cellular localization of yeast proteases and their inhibitors. *Biochemical and Biophysical Research Communications* 60, 1378-83.

MATERN,H., BETZ,H. & HOLZER,H. (1974). Compartmentation of inhibitors of proteases A and B and carboxypeptidase Y in yeast. *Biochemical and Biophysical Research Communications* 60, 1051-7.

MATILE,P. (1969). Plant lysosomes. pp.406-430 in *Lysosomes in Biology and Medicine*, edited by J.T. Dingle & H.B. Fell, Amsterdam: North-Holland.

MATILE,P. & WIEMKEN,A. (1967). The vacuole as the lysosome of the yeast cell. *Archiv für Mikrobiologie* 56, 148-55.

MATSUBARA,H. & FEDER,J. (1971). Other bacterial, mold and yeast proteases. In *The Enzymes*, 3rd edition, pp.721-795, edited by P.D.Boyer. New York & London: Academic Press.

NAKAGAWA,Y. (1970). Alkaline proteinases from *Aspergillus*. pp.581-91 in *Methods in Enzymology*, Vol.XIX, edited by G.E.Perlmann & L.Lorand. New York & London: Academic Press.

NORDWIG,A. & JAHN,W.F. (1968). A collagenolytic enzyme from *Aspergillus oryzae*. *European Journal of Biochemistry*, 3,519-29.

OHAMA,H.,TOMONAGA,G. & YANAGITA,T. (1966). Proteinase formation in relation to physiological activities of growing mycelia of *Aspergillus niger*. *Journal of Biochemistry* (Tokyo) 60, 713-7.

O'SULLIVAN,J. & MATHISON,G.E. (1971). The localization and secretion of a proteolytic enzyme complex by the dermatophytic fungus *Microsporum canis*. *Journal of General Microbiology* 68, 319-26.

PAGE,W.J. & STOCK,J.J. (1974). Sequential action of cell wall hydrolases in the germination and outgrowth of *Microsporum gypseun* macroconidia. *Canadian Journal of Microbiology*,20,483-9.

PATEMAN,J.A., KINGHORN,J.R., DUNN,E. & FORBES,E.C. (1973). Ammonium regulation in *Aspergillus nidulans*. *Journal of Bacteriology*, 114, 943-50.

POLLOCK,M.R. (1962). pp.121-178 in *The bacteria* edited by I.C. Gunsalus & R.Y. Stanier. New York & London: Academic Press.

PRINGLE,J.R. (1975). Methods for avoiding proteolytic artefacts in studies of enzymes and other proteins from yeasts. Chapter 9 in *Methods in Cell Biology*, Vol.XIX, edited by D.M. Prescott. New York & London: Academic Press.

RAPER,K.B. & FENNELL,D.I. (1965). *The Genus* Aspergillus. Baltimore U.S.A.: Williams & Wilkins Co.

REICH,E., RIFKIN,D.B. & SHAW,E. (1975). Editors, *Proteases and Biological Control. Cold Spring Harbor Conference on Cell Proliferation*, Vol.2. Cold Spring Harbor Laboratory.

SEKINE,H., NASUMO,S. & IGUCHI,N. (1969). Isolation of highly proteolytic mutants of *Aspergillus sojae*. *Agricultural and Biological Chemistry*, 33, 1477-82.

SHINMYO,A. & TERUI,G. (1970). Adipic acid as inducer for acid protease formation by *Aspergillus niger*. *Proceedings X International Congress of Microbiology, Mexico City*, Abstracts p.30.

SHINMYO, A., TAHARA,T., MATSUMOTO,H. & TERUI,G. (1971). Kinetic studies on enzyme production by microbes, VIII. Autoinduction of acid protease in *Aspergillus niger*. *Journal of Fermentation Technology* 49, 535-43.

SODEK,J. & HOFFMAN,T. (1970). Microbial acid proteinases.pp.372-87 in *Methods in enzymology*, Vol.XIX, edited by G.E.Perlmann & L.Lorand. New York & London: Academic Press.

SUBRAMANIAN,A.R. & KALNITSKY,G. (1964). The major alkaline proteinase of *Aspergillus oryzae*, Aspergillopeptidase B. *Biochemistry*. 3, 1861-7.

TOMANAGA,G. (1966). Preferential synthesis of extracellular protease by *Aspergillus niger* in sulphur deficiency. *Journal of General and Applied Microbiology*. 12, 267-76.

TOMANAGA,G., OHAMA,H. & YANAGITA,T. (1964). Effect of sulphur compounds on the protease formation by *Aspergillus niger*. *Journal of General and Applied Microbiology* 10, 373-85.

TREVITHICK,J.R. & METZENBERG,R.L. (1966). Molecular sieving by *Neurospora* cell walls during secretion of invertase isozymes. *Journal of Bacteriology* 92, 1010-5.

TURKOVA,J., MIKES,O., GANCEV,K. & BOUBLIK,M. (1969). Isolation and characterization of alkaline proteinase of *Aspergillus flavus*. *Biochimica et Biophysica Acta* 178, 100-111.

TURKOVA, J., MIKES,O., HAYASHI,K., DANNO,G. & POLGAR,L. (1972). Alkaline proteinases of the genus *Aspergillus*. *Biochimica et Biophysica Acta* 257, 257-63.

YANAGITA,T. & NOMACHI,Y. (1967). Kinetic analysis of the region of protease formation in the hypha of *Aspergillus niger*. *Journal of General and Applied Microbiology* 13, 227-35.

Chromosome Aberrations in *Aspergillus nidulans*

J. A. BIRKETT and J. A. ROPER

Department of Genetics, University of Sheffield, Sheffield, U.K.

INTRODUCTION

Most microorganisms are not amenable to detailed cytological analysis and so are poor candidates for studies of chromosome aberrations. However, in *A. nidulans*, certain aberrations - in the widest sense of that term - have played a critical role in developing the genetics of the species and the resulting, versatile genetic system has been exploited in the discovery of other aberrations. This survey summarizes the ways in which some aberrations are detected and, more important, outlines their genetic consequences.

DIPLOIDS AND ANEUPLOIDS

Diploidy, the first aberration to be detected (Roper, 1952) has been reviewed previously (Pontecorvo *et al.*, 1953; Pontecorvo, 1956; Roper, 1966; Clutterbuck, 1974) and will be discussed only briefly here.

As a rare event, pairs of unlike haploid nuclei fuse in heterokaryotic hyphae to give diploids which are fairly stable at mitosis. The uninucleate conidia of the heterokaryon are mainly parental haploids but those with a heterozygous diploid nucleus can be selected by using haploid parents with different nutritional markers. During vegetative growth the diploids undergo two independent and infrequent processes, mitotic crossing over and haploidization, which lead to recombination. Isolation of the products of these mitotic events has been made easy by the colonial growth of *A. nidulans*, its uninucleate conidia with their distinctive wild type and mutant colours, and by selective techniques. The series of processes - heterokaryosis, diploidy, mitotic crossing over, haploidization - was termed 'the parasexual cycle' (Pontecorvo, 1954); apart from use in *A. nidulans*, it has opened the way to the genetics of some imperfect species.

Mitotic crossing over has been used to order genes relative to each other and to their centromere on individual chromosome arms. The process is like that first described in *Drosophila* (Stern, 1936); it involves the mitotic segregation of chromatids following exchange at the 4-strand stage (Rober & Pritchard, 1955). Exchanges at mitosis and meiosis differ in both frequency and relative distribution; in some chromosomes the former occur disproportionately towards the centromeres (Pontecorvo & Käfer, 1958).

Haploidization, which is effectively like meiosis without crossing over (Pontecorvo, Tarr-Gloor & Forbes, 1954), provides an invaluable tool in a variety of analyses. It is, in fact, a multistep process; following non-disjuction, which usually gives trisomic (2n + 1) and monosomic (2n - 1) nuclei (Pontecorvo & Käfer, 1958), the hypodiploids lose chromosomes at subsequent divisions until a stable haploid state is reached. Käfer (1961) showed the sequential process, and observed some of the highly unstable intermediates, using well-marked diploids in which loss of a chromosomes exposed recessive alleles on the remaining homologue.

Aneuploids, mainly disomics (n + 1), occur in conidia of haploids (Upshall, 1966) and in ascospores. Their frequency in both is usually 0.1 to 0.3 per cent, but the meiotic frequency is increased substantially in crosses heterozygous for a translocation, mainly by disomics for the linkage groups involved in the aberration (Pollard, Käfer & Johnston, 1968; Upshall & Käfer, 1974). All possible disomics and trisomics have been isolated and most can be identified specifically; they grow less well than wild type and produce vigorous euploid sectors (Käfer & Upshall, 1973). Particular strains or treatments may give disomics of mainly one kind; conidia of one unstable strain gave frequent chromosome III disomics (Azevedo & Roper, 1970) and acriflavine treatment of metabolizing, haploid conidia gave up to 25% chromosome IV disomics (Ball & Roper, 1966).

INVERSIONS AND TRANSLOCATIONS

A. nidulans meiotic chromosomes are too small for easy observation of inversion heterozygotes (Elliott, 1960); at present detection depends on linkage studies and evidence for the existence of inversions is equivocal. Roberts (personal communication) has reported one probable case involving the left arm of chromosome I. Käfer-Boothroyd (1964) suggested a chromosome I inversion in one lineage but later (Käfer, 1966) offered other explanations including a possible diffuse centromere. It would be interesting to study the stability of a diploid heterozygous for an inversion as a breakage-fusion-bridge cycle might arise by mitotic crossing over and so produce vegetative instability.

Translocations are detected easily through meiosis, as outlined above, or through haploidization of a diploid formed between the tested haploid and a well-marked standard. A heterozygote for a reciprocal translocation gives only parental haploids for the chromosomes involved in the aberration. A non-reciprocal translocation heterozygote gives one viable recombinant class but, as these carry a duplicate segment and grow poorly, they are rarely isolated. Käfer (1962, 1965) first used this method and described previously unsuspected translocations, probably of X-ray and ultraviolet origin, in the Glasgow stocks.

Most of the isolated translocations produce no detectable phenotypic effect but two show phenotypic instability with similarities to position effect variegation (Ball, 1967; Clutterbuck, 1970).

DUPLICATIONS AND DELETIONS

There is no firm evidence for tandem duplications, but non-tandem duplications are known and have been the major source of information on deletions. Pritchard (1956) isolated the first duplication strain during selection for spontaneous adenine-independent revertants of a strain with *adE*20, a leaky allele on the right arm of chromosome I. One apparent revertant had an extra chromosome I segment, with *ade*20, terminally attached to chromosome II. Two doses of the leaky allele conferred selective advantage on adenine-free medium.

All other duplicaction strains have been obtained *via* non-reciprocal translocations. Bainbridge & Roper (1966) found that crosses between certain morphologically-normal parents produced progeny of which one third were abnormal. A pedigree search, based on Käfer's analyses, showed that these crosses always involved one parent with a particular translocation, later shown to be a non-reciprocal translocation of a segment of chromosome III to VIII. Detailed analyses explained the unusual 2 : 1 ratio as: the two parental types; one inviable class, presumably lacking the translocated segment; one morphologically-abnormal type with a duplicate chromosome III segment. Because of the unexpected properties of the first two duplication strains, a search has been made for others; this has required irradiation to produce translocations, parasexual analysis to detect these, and crosses to obtain duplication progeny from non-reciprocal translocations (Nga, 1968; Azevedo, personal communication; Ortori, unpublished). Some seven or eight different duplication strains have been found; all have apparently similar behaviour, but only those shown in Fig. 1 are defined sufficiently for detailed study.

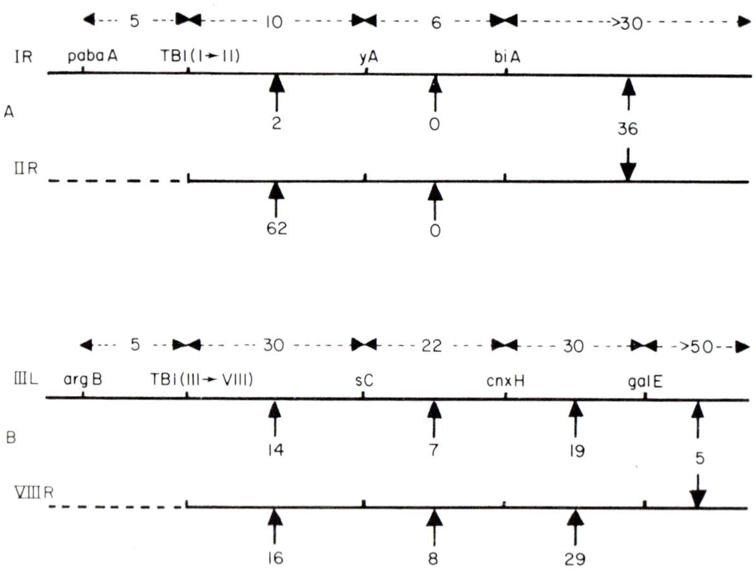

Fig. 1. Duplication strains A and B with relevant markers and approximate map distances in centi-morgans (dotted lines). Mutant alleles (detailed by Clutterbuck, 1974) were used in various arrangements within the duplicate segments. Numbers and unbroken arrows show breaks, as a percentage of the total detected, foreach marked interval of the two segments. In some instances break positions could not be assigned with certainty to a particular interval; in these cases multiple arrows from one number show the possible interval or intervals involved. Most deletions were indistinguishable from single-break events involving the loss of all distal material.

Duplication strains grow slowly and have morphological features characteristic for each duplicate segment. All are unstable and the most frequently observed consequences are sectors with growth rate and morphology nearer, or equal, to wild type. Sectors of strain B were the first analyzed (Bainbridge & Roper, 1966) and, in essence, were similar to those from other duplication strains. The sectors had lost a variable part of either duplicate segment and so, with reduced imbalance, had selective advantage over their duplication parent.

Many sectors have been analysed; in a few cases, with sectors which still showed instability, this has been extended to the products of second and further events of instability. The position and extent of deletion in some sectors can be deduced directly from phenotypic changes involving heterozygous markers in the duplicate segments though, as a necessary check, several representatives of each type have been analyzed genetically. These analyses, of which one aspect is summarized in Fig. 1, reveal a number of features of the deletion process. Almost all sectors kept the parental arrangement of undeleted alleles in the duplicate segments. Thus, although mitotic crossing over is known to occur infrequently in such duplicate segments, it is not implicated in what appears to be an intrachromosomal process of deletion. A definitive check for interstitial deletions would require terminal markers and these are not available. With that important reservation, most losses cannot be distinguished from single-break, terminal deletions. A new duplication strain (Ortori, unpublished), not detailed here, provides a notable exception in that at least 8 per cent of deletions must be interstitial. In strain A most breaks occur in the translocated segment, within a few meiotic units of the attachment point. Strain B behaves quite differently; breaks in this are more nearly equal in the translocated and untranslocated segments and the distribution is not so disproportionate to the meiotic measures of the marked intervals.

Deletions in strain A are between 1 in 10^4 and 1 in 10^6 per mitosis (Birkett, unpublished). There can be no measure of deletions in a standard haploid to compare with this, so the deletion frequency reveals nothing about the possible role in instability of the duplication. Diploids formed between A and a standard haploid are far more unstable than A itself. They show frequent loss of the translocated segment, without genetic change in the rest of the diploid genome. Nga & Roper (1969) incorrectly interpreted this as indicating instability provoked by, and initially confined to, segments present in triplicate in an otherwise diploid genome; they extended this argument also to the duplicate segments of duplication haploids. Using strains with more markers, Case & Roper (1975) showed that most of the extra instability in such unbalanced diploids, compared with unbalanced haploids, was due to mitotic crossing over; loss of the terminally-attached segment resulted from exchange (and appropriate strand segregation) between the duplication-bearing arm and its normal homologue. To exclude this form of instability, irrelevant to the present issue, Case & Roper (unpublished) studied diploids homozygous for a strain A type duplication; mitotic crossing over in these could not eliminate excess genetic material. However, such diploids were extremely unstable. Analyzed sectors, phenotypically representative of the majority, arose by simultaneous deletions from both untranslocated chromosome I segments; that is, they were quantitatively near-diploid and had a translocation homozygote configuration. Other deletions occurred at rates roughly like those found in the duplication haploid. Strain A and its corresponding diploids differ, then, both in instability, and in the relative distribution of breaks. Although this comparison raises new questions, it again implicates excess genetic material as a factor in its own instability.

Clues to the more precise processes of instability come from two sources. The first involves conditions in which DNA repair is probably reduced. Some mutations which confer sensitivity to ultraviolet light increase the instability of strain A (Burr, unpublished; Burr, Palmer & Roper, 1971), and caffeine, at concentrations causing no detectable deletions in diploids, also unselectively increases instability in that strain (Roper, Palmer & Watmough, 1972). The interpretation was that deletions arise *via* DNA lesions, perhaps single-strand gaps, most of which are repairable; reduction of repair, by caffeine or by mutations, would then give a more accurate impression of the incidence of initial errors.

The second line of evidence is obtained from products of instability other than those discussed so far. In addition to deletion products with improved morphology and decreased instability, strains A & B yield "deteriorated" variant sectors, some with sharply increased instability. A large number of deteriorated sectors from strain A have been analyzed and they represent a bizarre picture of instability. Many isolates accumulated successive mutations, affecting morphology and stability, in all linkage groups; one produced frequent disomics; a number, when combined in diploid with a standard haploid, generated recessive lethals; others showed what may have been transpositions of genetic elements from one chromosomal site to another. Only very tentative explanations of these events could be advanced (Nga & Roper, 1968; Azevedo & Roper, 1970); the particular difficulties in this were the apparent diversity of processes and uncertainty about the nature of most of the initial events of instability. However, certain deteriorated sectors of strain B have been more rewarding and it has been possible to define both the initial event and to show the relationship between the highly unstable types of individual lineages (Birkett & Roper, unpublished). One deteriorated sector, typical of many, produced a mixture of conidial types from which isolates were analyzed. The essential results from one analysis are summarized in Fig. 2.

Nga & Roper (1968) proposed crossing over within intrachromosomal loops to explain deletions and, more tentatively, as a source of transposable elements. The idea can be extended for the above situation in which the "sticky" locus may well be a modified telomere, perhaps produced by crossing over between a palindromic telomeric sequence and an internal sequence. Further modification of the telomere, to produce different states of the sticky locus, could occur by a similar process. This model, which will be elaborated elsewhere, can account plausibly for most of the instability events in duplication strains.

CONCLUSIONS

Despite the absence of detailed cytology, chromosome aberrations have been a fruitful field of study in *A. nidulans*. Diploids provided an important tool in the study of gene action; further, through mitotic crossing over and haploidization, they revealed new processes of recombination which have been exploited in *A. nidulans* and have been the only route to formal genetic analysis in some imperfect species. The total genetic system of *A. nidulans*, together with other favourable features of this species, have permitted elucidation of other chromosome aberrations. Of these, duplications - produced mainly *via* non-reciprocal translocations - have been among the most interesting. They add a novel tool in mutation studies (Cooke, Roper & Watmough, 1970), and in exploration of gene dosage effects in unbalanced genomes (Birkett, unpublished); but the prime application so far has been in the study of their genetic instability. Two duplication strains, and preliminary investigations of one not detailed here, reveal the apparent diversity of events involved in instability. This could reflect, in one species, some of the apparent complexity and diversity which stands out in the overall view of genetic instability. A few features of the causes and effects of instability in *Aspergillus* are being clarified and further progress may bring some much-needed unity to the wider picture of genetic instability.

ACKNOWLEDGEMENT

The authors are greateful to colleagues who have made unpublished results freely available.

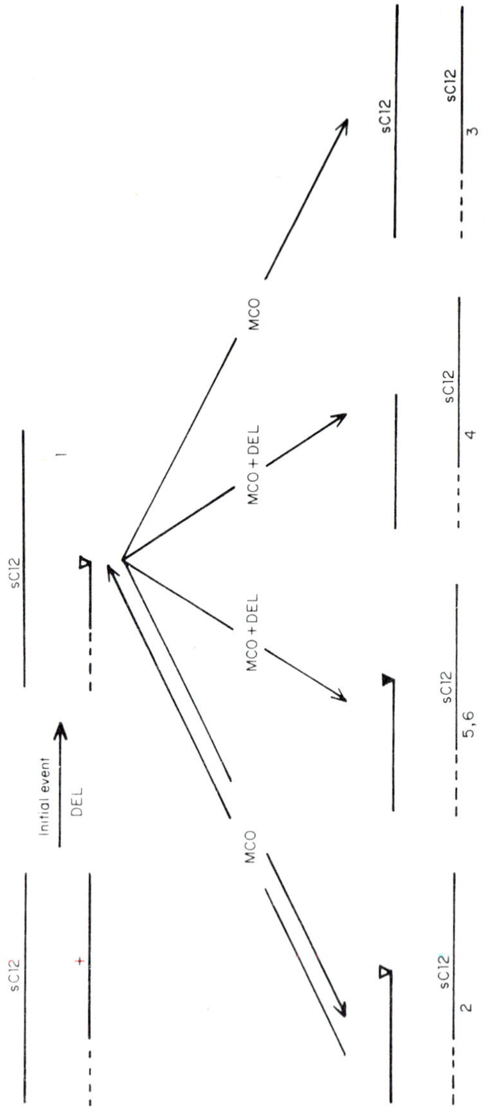
sC12
1
Initial event
DEL
MCO
MCO + DEL
MCO + DEL
MCO
sC12
3
sC12
4
sC12
5,6
sC12
2

Fig. 2. The principal genotypes isolated directly from a single deteriorated sector of strain B. Chromosomes III and VIII are shown by unbroken and dotted lines respectively. *sC* 12, a mutant allele determining thiosulphate requirement; +, its wild type allele. Arrows (except for the initial event) indicate the origin of one type from another established by study of isolates.

The initial event was a deletion (DEL) leaving a sticky locus (∇) which provoked mitotic crossing over (MCO). In type 2 the sticky locus was unchanged, as it could regenerate type 1, but its expression was less pronounced. Type 3 was the relatively stable, full duplication. Type 4, stable and non-deteriorated, had lost the sticky locus, presumably by a further small deletion. Types 5 and 6 arose by crossing over and, perhaps as a result of further deletions, had nodified states or expressions of the sticky locus (▼); in meiotic and mitotic analyses 5 and 6 showed, in less extreme form, various properties of type 2.

REFERENCES

AZEVEDO, J.L. & ROPER, J.A. (1970). Mitotic non-conformity in *Aspergillus nidulans*: successive and transposable genetic changes. *Genetical Research* 16, 79-93.

BAINBRIDGE, B.W. & ROPER, J.A. (1966). Observations on the effects of a chromosome duplication in *Aspergillus nidulans*. *Journal of General Microbiology* 42, 417-24.

BALL, C. (1967). Chromosome instability related to gene suppression in *Aspergillus nidulans*. *Genetical Research* 10, 173-83.

BALL, C. & ROPER, J.A. (1966). Studies on the inhibition and mutation of *Aspergillus nidulans* by acridines. *Genetical Research* 7, 207-21.

BURR, K.W., PALMER, H.M. & ROPER, J.A. (1971). Mitotic non-conformity in *Aspergillus nidulans* : the effects of reduced DNA repair. *Heredity* 27, 487.

CASE, B.L. & ROPER, J.A. (1975). Genetic imbalance and mitotic non-conformity in *Aspergillus nidulans*. *Aspergillus Newsletter* 13, 8-11.

CLUTTERBUCK, A.J. (1970). A variegated position effect in *Aspergillus nidulans*. *Genetical Research* 16, 303-16.

CLUTTERBUCK, A.J. (1974). *Aspergillus nidulans*. In Handbook of Genetics, ed. R.C. King, Vol. 1, pp 447-510.

COOKE, P., ROPER, J.A.& WATMOUGH, W. (1970). Trypan blue-induced deletions in duplication strains of *Aspergillus nidulans*. *Nature, London* 226, 276-77.

ELLIOTT, C.G. (1960). The cytology of *Aspergillus nidulans*. *Genetical Research* 1, 462-76.

KÄFER, E. (1961). The processes of spontaneous recombination in vegetative nuclei of *Aspergillus nidulans*. *Genetics* 46, 1581-609.

KÄFER, E. (1962). Translocations in stock strains of *Aspergillus nidulans*. *Genetics* 33, 59-68.

KÄFER-BOOTHROYD, E. (1964). Evidence for a paracentric inversion from mitotic crossing over in *A. nidulans*. *Aspergillus Newsletter* 5, 11.

KÄFER, E. (1965). The origins of translocations in *Aspergillus nidulans*. *Genetics* 52, 217-32.

KÄFER, E. (1966). Multiple mitotic crossing over in the centromere region or diffuse centromeres? *Aspergillus Newsletter* 7, 14-15.

KÄFER, E. & UPSHALL, A. (1973). The phenotypes of the eight disomics and trisomics of *Aspergillus nidulans*. *Journal of Heredity* 64, 35-8.

NGA, B.H. (1968). Production and analysis of new translocations in *A. nidulans*. *Aspergillus Newsletter* 9, 16-17.

NGA, B.H. & ROPER, J.A. (1968). Quantitative intrachromosomal changes arising at mitosis in *Aspergillus nidulans*. *Genetics* 58, 193-209.

NGA, B.H. & ROPER, J.A. (1969). A system generating spontaneous intrachromosomal changes at mitosis in *Aspergillus nidulans*. *Genetical Research* 14, 63-70.

POLLARD, R., KÄFER, E. & JOHNSTON, M. (1968). Influence of translocations on meiotic and mitotic non-disjunction in *Aspergillus nidulans*. *Genetics* 60, 743-57.

PONTECORVO, G. (1954). Mitotic recombination in the genetic system of filamentous fungi. *Caryologia (Supplement)* 6, 192-200.

PONTECORVO, G. (156). The parasexual cycle. *Annual Review of Microbiology* 10, 393-400.

PONTECORVO, G. & KAFER, E. (1958). Genetic analysis based on mitotic recombination. *Advances in Genetics* 9, 71-104.

PONTECORVO, G., ROPER, J.A., HEMMONS, D.W., MACDONALD, K.D. & BUFTON, A.W.J. (1953). The genetics of *Aspergillus nidulans*. *Advances in Genetics* 5, 141-238.

PONTECORVO, G., TARR-GLOOR, E. & FORBES, E. (1954). Analysis of mitotic recombination in *Aspergillus nidulans*. *Journal of Genetics* 52, 226-37.

PRITCHARD, R.H. (1956). A genetic investigation of some adenine-requiring mutants of *Aspergillus nidulans*. Ph.D. Thesis, The University, Glasgow.

ROPER, J.A. (1952). Production of heterozygous diploids in filamentous fungi. *Experientia* 8, 14-15.

ROPER, J.A. (1966). The parasexual cycle. In *The Fungi* ed. G.C. Ainsworth & A.S. Sussman, Vol. 2, pp 589-617. London : Academic Press.

ROPER, J.A., PALMER, H.M. & WATMOUGH, W.A. (1972). Mitotic non-conformity in *Aspergillus nidulans:* The effects of caffeine. *Molecular and General Genetics* 118, 125-33.

ROPER, J.A. & PRITCHARD, R.H. (1955). The recovery of the complementary products of mitotic crossing over. *Nature London* 175, 639.

STERN, C. (1936). Somatic crossing over and segregation in *Drosophila melanogaster*. *Genetics* 21, 625-730.

UPSHALL, A. (1966). Somatically unstable mutants of *Aspergillus nidulans*. *Nature, London* 209, 1113-5.

UPSHALL, A. & KÄFER, E. (1974). Detection and identification of translocations by increased specific non-disjunction in *Aspergillus nidulans*. *Genetics* 76, 19-31.

The Genetics of Conidiation in *Aspergillus nidulans*

A. J. CLUTTERBUCK

Department of Genetics, University of Glasgow, Glasgow, U.K.

INTRODUCTION

THE DECISION TO CONIDIATE

At the point of initiation of conidiation a wild-type colony of *Aspergillus nidulans* has a number of possible courses of development open to it. It may continue with purely mycelial growth, it may concentrate on sexual reproduction or it may conidiate. I present these as alternatives because it is evident from the behaviour of mutants and also from variants found in the wild that an excess of any one of these features inevitably means a reduction in the others.

The point of decision as far as conidiation is concerned is not reached until a fixed amount of colony growth is completed. Investigations by Axelrod (1972) and Axelrod, Gealt & Pastushok (1973) have shown that this point is reached by colonies grown from single conidia on agar, and also by mycelium grown *en mass* in liquid culture in about 20 h at 36°. Before this time, no exposure to a stimulus will produce conidiation while at any time after this the mycelium is competent to receive a stimulus which will lead to the formation of conidia approximately seven hours later. Gealt & Axelrod (1974) isolated a mutant which achieved competence approximately two hours earlier than the wild type, and similar mutants have been isolated by the author; these latter mutants have the additional peculiarity that the conidia they produced have only transient viability (unpublished material). By far the most efficient conidiation stimulus in *Aspergillus nidulans* is exposure to a gaseous environment, but carbon or nitrogen starvation may also lead to conidiation in submerged culture (Saxena & Sinha, 1973; Martinelli, 1976). A mature colony growing on agar conidiates approximately 2mm behind the hyphal tips and since this is close to the distance the tips extend in six or seven hours there would appear to be no latent period before competence is achieved by hyphae in a mature colony.

The nature of competence is unknown, but Gealt & Axelrod (1974) have shown that its acquisition is accompanied by a significant reduction in the induced levels of nitrate reductase and extracellular protease: two enzymes deliberately chosen because they have no obvious connection with conidiation. It may also be relevant that in surface cultures on agar and in submerged cultures the cellular level of adenosine 3',5'-cyclic monophosphate (cAMP) falls from a high level in early growth and reaches a stable low level at about 20 h (Clutterbuck, 1975). In *Neurospora* premature conidiation of *crisp-i* mutants appears to depend on a defective adenyl cyclase and hence a deficiency of cAMP (Terenzi, Flawiá, Tellez-Iñón & Torres, 1976), however, attempts to induce premature conidiation in *A.nidulans* by administration of adenyl cyclase inhibitors has given only inhibition of conidiation (Clutterbuck unpublished). This may be due to the lack of specificity of the inhibitors some of which in *Neurospora* may also inhibit the phosphodiesterase that destroys cAMP (Scott & Solomon, 1975).

If cAMP does play any part in the control of conidiation it is natural to speculate that it may operate through the regulation of carbohydrate metabolism. Zonneveld (1974) has examined a number of mutants varying in density of cleistothecium production, some of which showed inversely related changes in conidiation. These mutants had levels of α-1,3 glucan and α-1,3 glucanase correlated with their levels of cleistothecium formation, a finding in keeping with Zonneveld's previous conclusions (Zonneveld, 1972 and this volume) that this glucan is built up during early growth and then is metabolized by the glucanase during sexual reproduction. Zonneveld (1974) also investigated three mutants initially selected as defective in later stages of conidiation (Clutterbuck 1969). The *bristle* mutant (Fig. 1b) which produced indefinitely growing conidiophores but no conidia appeared to use up its carbohydrate reserves in this process and stored little glucan, thus illustrating competition between two developmental processes for carbohydrate reserves. The *stunted* mutant, on the other hand, produces no cleistothecia or Hülle cells and stores little glucan, while its conidiophores are stunted and lacking their normal thickening (Fig.1e); this suggests a joint defect in the production of two polysaccharides. Finally, the *medusa* mutant is normal in glucan metabolism but produces conidia only on top of multiseriate sterigmata (Fig. 1d) and has only sterile masses of Hülle cells in place of cleistothecia; it may have a block in cleistothecium formation later than the stage at which glucan utilization is relevant.

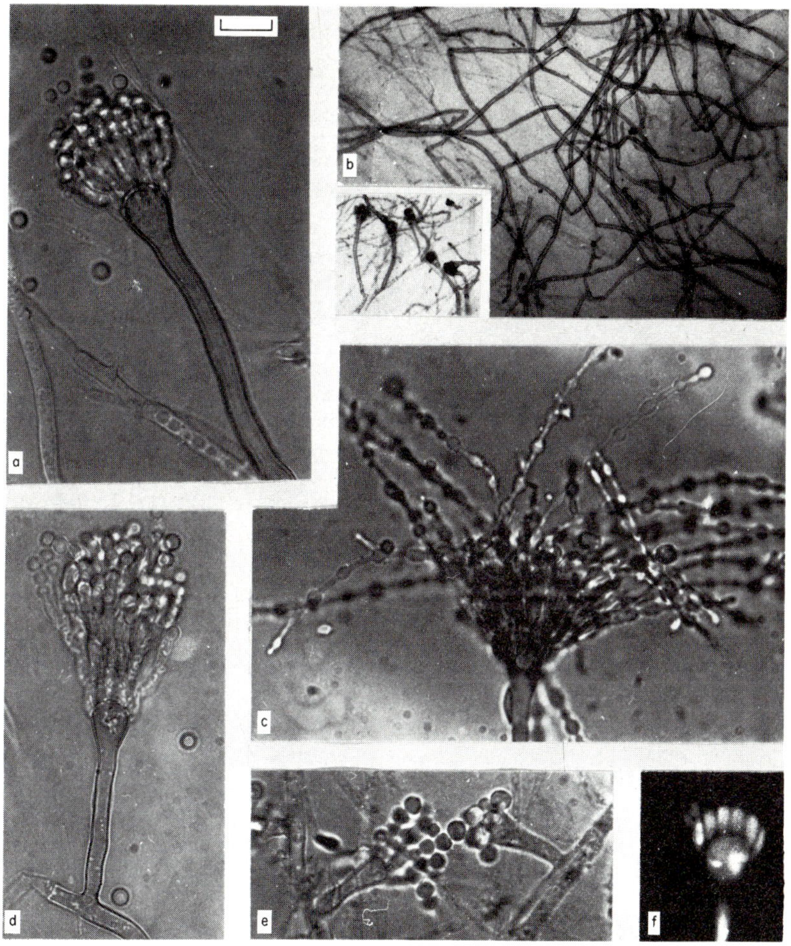

Fig. 1. Wild-type and mutant conidial heads of *Aspergillus nidulans*. a. Wild-type; b. *bristle* mutant (inset: wild-type at same scale); c. *abacus* mutant; d. *medusa* mutant; e. *stunted* mutant; f. *aps* mutant stained by acridine orange fluorescence to show anucleate primary sterigmata failing to develop further while one nucleated sterigma has produced secondary sterigmata.

The scale bar represents 10 μm in a, c, d, e, f, and 70 μm in b.

Mutants diverting the colony's resources into sexual reproduction or excessive mycelial growth have been studied more frequently as examples of cytoplasmic inheritance than as tools in analysis of development. A number of studies of wild variants (Jinks, 1954, 1956; Arlett, 1960; Mahoney & Wilkie, 1962; Croft, 1966) have shown that segregation of conidial and cleistothecial types during vegetative growth is best attributed to cytoplasmically inherited factors although the nature of these and their mode of variation is not known. In the Glasgow strains used for many genetic studies this variability is still sometimes evident but is largely masked by the *velvet* mutation which is present in most of these strains and leads to the predominance of the conidial form in colonies grown on standard media.

"fluffy" mutants producing excessive aerial mycelium occur spontaneously in *Aspergillus* with high frequency especially in cultures grown from old conidia (Grose & Clutterbuck, 1975). These mutants are phenotypically similar to those described by Roper (1958) as 'mycelial'. When first isolated, Roper's mutants could be transferred in heterokaryons to other strains but, unlike most cytoplasmic mutants the transfer was not reciprocal, the original mycelial parent retaining its abnormal morphology. Furthermore, the mycelial character was not transferred freely to the recipient strain, but only intermittently and with a frequency declining with time. Eventually the mycelial strains appeared to have lost their ability to transfer (Roper, personal communication). Roper was able to map two chromosomal genes either of which would allow expression of the putative cytoplasmic factor. This pattern of irregular transfer is clearly very different from mitochondrial cytoplasmic inheritance (see Turner & Rowlands, and this volume) and might now perhaps be more readily explained by a transmissible plasmid (cf. Peterson, 1970), mappable as a nuclear gene in its chromosomal state.

Fluffy mutants isolated by Dorn (1970) were also capable of genetic analysis as chromosomal genes although Dorn, Martin & Purnell (1967) had previously reported that the fluffy condition could be transferred to another strain in a heterokaryon. These mutants also had the interesting property of being phenotypically repaired by various nucleic acid and protein synthesis inhibitors, different inhibitors being specific for different mutants. In addition these strains were invasive of normal colonies which may suggest that the mutants are insensitive to a normal mechanism of self-inhibition which prevents unregulated mycelial growth, but, on the other hand, it may be that invasion occurs merely through overgrowth by the luxuriant aerial hyphae.

THE STEPS TO CONIDIATION

Martinelli & Clutterbuck (1971), in a survey of conidiation

mutants, attempted to classify the mutants according to the stage of development primarily affected and also made estimates of the total number of loci involved in conidiation by comparing mutation rates with those for auxotrophs and other known loci. The great majority of mutants defective in conidiation were also defective in hyphal growth and if these were included in the estimates of conidiation loci a figure of 300 to 500 genes would be reached. However, it is possible to reason that if such mutants are defective in hyphal growth the genes mutated are not specifically concerned with conidiation and could in fact be concerned with many different areas of general metabolism which happened to be of greater importance for conidia than for hyphae. Many auxotrophs fall into this category as is evident from their nearly normal mycelial growth but total failure to conidiate on poorly supplemented media. Exclusion of all slow-growing mutants reduces the estimates of conidiation loci to the more manageable levels of 45 to 150 loci. The great majority of these loci, moreover, appeared to be concerned with the complex of processes involved in the initial choice of developmental pathways which have already been discussed. Most of these mutants were not totally unable to conidiate but could better be described as oligosporogenous mutants, again suggesting that they lead to a shift of a delicate balance of functions. The unusually high proportion of these mutants which were temperature-sensitive may also mean that the functions of many of these loci are only required during growth at high temperatures.

An earlier collection of mutants (Clutterbuck, 1969) had concentrated on those that were able to start normally on the road to conidiation, that is they produced conidiophores, but they were completely blocked before conidia themselves were formed. Despite an intensive search for mutants in this well-defined area of development, only two phenotypes were found and complementation analysis of over 35 mutants in each class showed that each of these two mutant phenotypes resulted from mutation at a single locus. One of these, the *bristle* locus, gives mutants which produce conidiophores elongating to more than ten times the normal length but failing to develop vesicles or any subsequent structures (Fig. 1b). Analysis of the behaviour of leaky mutants and of heterokaryons between *bristle* and normal strains suggest that the same mutational defect results in the inability to form both vesicles and sterigmata, while recent unpublished studies with a temperature-sensitive *bristle* mutant suggest that the function of the *bristle* locus is required at these stages but not once conidial buds have started to appear. The requirement of this gene function for both vesicles and sterigmata - superficially very different structures - could imply that this locus is a master switch for a number of functions, however there is a possibility that vesicles and sterigmata are not as unlike as they look: sterigmata grow and bud like yeast cells so the function of the *bristle* locus could be to operate a hyphal-yeast dimorphism switch (reviewed by Romano, 1966). In this case it is not difficult to imagine that the first step in the change of a broad conidiophore towards a yeast-like structure might be the formation of a swollen tip from which a number of sterigmata could bud.

The second locus identified in this study was named "*abacus*" since its mutants produced rod-like structures with swellings at intervals which were interpreted as consisting of repeated secondary sterigmata (Fig.1c). Similar mutants have been isolated in other *Aspergillus* species (Yuill & Yuill, 1938; Raper & Fennell, 1953; Vujičić & Muntanjola-Cvetković, 1973). The *abacus* mutants were unable to complete the change from the production of sterigmata which grow apically, each from the tip of the last, to conidia, each of which buds from the end of the secondary sterigma (phialide), below the previous conidium. This function also involves the rounding off and isolation of the conidium from its growth point. In this case, studies with a temperature-sensitive mutant show that a switch to the nonpermissive temperature after the start of conidiation produces a shift back to the mutant form, so the *abacus* gene function is required throughout the production of conidia.

These two loci, which could be described as "*morphogenetic*" or perhaps "*tactical*" loci are peculiar in that they appear to determine the next morphological step to be taken in development, but they do not affect the growth of the structures involved: conidiophores producing *bristle* or *abacus* structures continue to grow as healthily as normal conidiophores, but only by maintaining or repeating the previous step in development instead of moving on to the next one. In contrast to these loci there are a great many others which may be described as "*support*" loci whose functions are the supply of materials for development and whose mutants show a failure of growth as well as morphological progression of the developing structures. In its widest sense this class should include the slow growing conidiation mutants already discussed in which the material supplied is necessary for hyphal as well as conidial growth; it would also include loci supplying the specific requirements of conidiation. In addition to the growth criterion other features distinguish mutants at these loci from those at tactical loci: the majority of support locus mutants are leaky and it is often difficult to define precisely the stage at which development stops, so that a gradual exhaustion of supplies is suggested.

Following the military analogy, loci concerned with the choice of hyphal growth, sexual reproduction or conidiation could aptly be termed "*strategic*" loci, while a final class of loci, concerned with modifications of the basic structure, could be called "*auxiliary*" loci. To this class belong the *stunted* and *medusa* loci whose mutants produce conidia rather inefficiently on an abnormal conidial apparatus (Fig. 1, d and e), spore and conidiophore pigment loci, and presumably a number of loci whose mutants were detected by Martinelli & Clutterbuck (1971) as producing nonviable spores.

The order of action of developmental genes

As a rule the stage of blockage in developmental mutants can be deduced from the morphology of the final stage reached. In many mutants at *support* loci, as mentioned above however, this final stage is variable and Barbata, Valdes & Sermonti (1973) have proposed an alternative method of ordering mutant blockage points based on the tendency for conidia of one component to predominate in heterokaryons between two morphological mutants. The basis of this method is not entirely clear: it is true that repair by complementation of late mutants may not be possible since the defects will occur in sterigmata or conidia which are uninucleate and therefore cannot be heterokaryotic, but mutants with defects at earlier stages should all repair each other reciprocally. Apart from this distinction between heterokaryon-reparable and irreparable mutants, it is likely that unequal conidial ratios reflect unequal nuclear ratios as suggested by Hannan (1975) for similar results in *A.niger*. Since Barbata *et al.* (1973) were using slow-growing mutants, unequal nuclear ratios would be expected if the hyphal defects in different mutants were not of equal severity.

brlAi2 AND *aps* MUTANTS

Two of the mutants found in collections of aconidial mutants are proving interesting for reasons not specifically connected with conidiation. One of these is the *brlAi2* mutant which has the *bristle* mutant phenotype and does not complement with other mutants at the *brlA* locus, but is unusual in that it is variegated, that is, colonies bear a few more or less normal conidial heads scattered among the undeveloped bristle-like conidiophores. The suspicion that this might be a variegated position effect mutant was confirmed by finding in this strain a translocation whose breakpoint mapped very close to the *brlA* locus (Clutterbuck, 1970). While the detection of heterochromatin in fungi is difficult because of the small size of the nuclei, the implications of a variegated position effect mutation are that the nucleus contains inactive chromatin capable of spreading its inactivation to adjacent regions, and this, together with Morris's demonstration of histones in *A.nidulans* (this volume) confirms the complex eukaryotic chromosome structure in this fungus. Further studies of this mutant are being made by the isolation of suppressors with a diversity of properties which it is hoped may reveal something of the basis of modifications of chromosome structure.

The second type of mutant referred to above raises another aspect of *Aspergillus* physiology which might not otherwise be suspected. Mutants of this type are designated *aps* since they have *a*nucleate *p*rimary *s*terigmata which fail to grow or develop further; they map at two loci: *apsA* and *apsB* (see Clutterbuck, 1974 for map positions). It appears that in the wild type a specific function is required to organize the movement of nuclei from the vesicle, one into each primary sterigma bud, and this function is missing in the *aps* mutants. However, some conidia are formed by all *aps* mutants and it appears that a few sterigmata may obtain nuclei, perhaps by the random movements of nuclei in the vesicle, after which they can develop normally without further requirement for the missing function. This is illustrated by the conidial head in Fig.1f. While there is no evidence for long distance nuclear migration in *A.nidulans*, the *aps* mutants provide evidence that short distance nuclear movements are essential for normal conidiation as well as for maintainance of a uniform distribution of nuclei in the hyphae. Like the *nu*clear *d*istribution (*nud*) mutants described by Morris (1975) the *aps* mutants have hyphal nuclei which are irregularly distributed and are often concentrated near to septa, however, while the *nud* mutants are temperature-sensitive lethals, the *aps* mutants grow only slightly more slowly than the wild type.

CONTROL OF DEVELOPMENTAL ENZYMES

One of the goals of a developmental study is to learn something about the regulation of developmental enzymes. At present no strictly morphogenetic enzymes are known, but two enzymes with auxiliary functions specific to conidiation are under investigation. The first one to be studied is a laccase (*p*-diphenol oxidase) concerned with the oxidation of the yellow precursor of the spore pigment to the normal green form. This enzyme is missing in yellow-spored mutants and is coded for by the *yA* locus (Clutterbuck, 1973). It is only produced at the time of formation of conidia (see Fig. 2) and while *abacus* mutants do produce the enzyme, *bristle* mutants do not. Inhibitor studies still in progress have so far shown that the translation stage of enzyme synthesis does coincide with the appearance of enzyme activity.

The second enzyme is also a phenol oxidase and from its substrate specificity may conveniently be called a 'cresolase' (Clutterbuck, 1973 It appears to be responsible for the grey-brown pigment of the conidiophores and sterigmata. This pigment is laid down in the outer walls of these structures (Oliver, 1972) where it is preceded by the presence of sulphydryl groups (Oliver, 1974). The chemical constitution of both the conidial and the conidiophore pigments is still uncertain, but the two phenol oxidases have substrate specificities which readily distinguish them from each other and from the phenol oxidase found in the mycelium (Bull & Carter, 1973; Martinelli & Bainbridge, 1974). The cresolase is again only produced during the relevant developmental stages (Fig. 2) and is absent from nonleaky *bristle* mutants

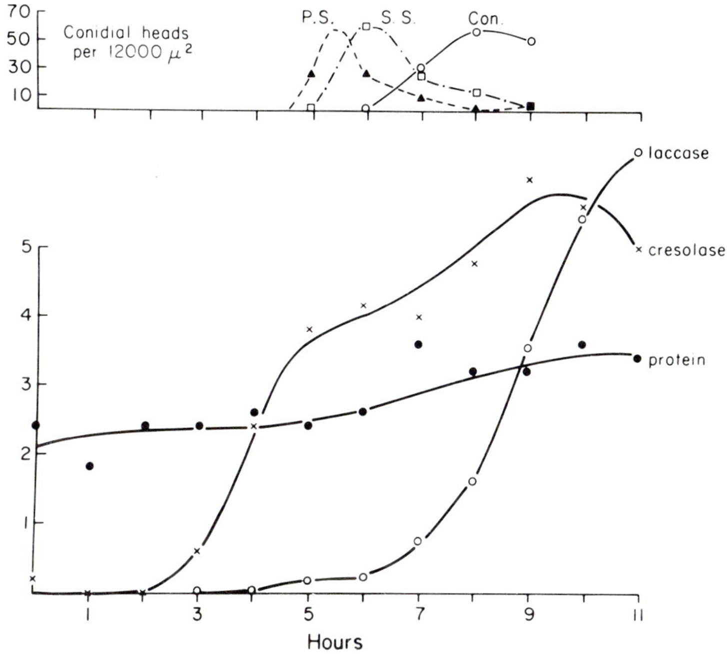

Fig. 2. Enzymes appearing during conidiation in *Aspergillus nidulans*. Cultures were grown as top layers on agar from 10^6 conidia per plate. A cellulose membrane convered the culture for the first 24 h of incubation and was then removed (at 0 h in the Fig.) to allow the initiation of conidiation. Mycelium from single dishes was harvested at intervals and assayed for protein, cresolase and laccase: these are expressed in arbitrary units per culture. The upper graph shows the frequency per 12000 μm^2 culture surface of conidiophores bearing primary sterigmata (p.s.), secondary sterigmata (s.s.) or conidia (con.). Bare conidiophores preceded the primary sterigma stage by one or two hours.

However, an interesting feature of the regulation of this enzyme is its presence in apparently normal amounts in all bristle mutants showing any sign of leakiness as judged by their morphology. Other mutants which lack the pigment but develop normally have been called *ivory* mutants and map at two loci: *ivoA* and *ivoB*. Mutants of the second locus have no cresolase but accumulate a phenol, as yet unidentified, while mutants at the first locus are deficient in this phenol which is presumed to be the pigment precursor.

At present it looks as if the *bristle* locus may play a key role in regulation of these activities. Since laccase is not produced by any *bristle* mutant unless it is so leaky as to produce some conidia, it seems that the *bristle* locus activity has to be completed before laccase is switched on. There might however, be some intervening steps between the actions of the two genes although no mutants for these steps have been found. In the case of the *ivoB* locus, producing the cresolase enzyme, it seems likely that the *bristle* gene product itself or some immediate product of its activity acts as inducer or co-inducer of cresolase synthesis. Furthermore, while all leaky *bristle* mutants produce cresolase, as mentioned above, not all are highly pigmented, so the *bristle* locus may also switch on, in a more gradual manner the synthesis of the pigment precursor (perhaps by the *ivoA* locus).

It has already been mentioned that *bristle* mutants have reduced storage of α-1,3 glucan, although this may be the result of carbohydrate utilization rather than regulatory action. It is nevertheless tempting to speculate that the *bristle* locus may represent a multiple switch point: at the same time that laccase synthesis begins a number of known, and probably many more unknown gene activities will be required; these include other spore pigment genes such as *white*, *fawn* and *wet-white* (see Clutterbuck (1974) for a list of *A.nidulans loci)* as well as the *abacus* locus and presumably many genes concerned with the function rather than the appearance of conidia. If one locus was involved in the regulation of these activities, the *bristle* locus would be a good candidate for the job, in which case its effect on morphology might also be through regulatory action rather than by coding for a morphogenetic protein. The graded morphological series of leaky *bristle* mutants is probably consistent with either explanation. Whatever is the case, there are many good reasons to try to find out more about the function of this gene.

REFERENCES

ARLETT,C.F. (1960). A system of cytoplasmic variation in *Aspergillus nidulans*. *Heredity* 15, 377-88.

AXELROD,D.E. (1972). Kinetics of differentiation of conidiophores and conidia by colonies of *Aspergillus nidulans*. *Journal of General Microbiology* 73, 181-4.

AXELROD,D.E., GEALT,M. & PASTUSHOK,M. (1973). Gene control of developmental competence in *Aspergillus nidulans*. *Developmental*

Biology 34, 9-15.
BARBATA,G., VALDES,L. & SERMONTI,G. (1973). Complementation among developmental mutants in *Aspergillus nidulans*. *Molecular and General Genetics* 126, 227-32.
BULL,A.T. & CARTER,B.L.A. (1973). The isolation of tyrosinase from *Aspergillus nidulans* cell walls, its kinetic and molecular properties and some consideration of its activity *in vivo*. *Journal of General Microbiology* 75, 61-73.
CLUTTERBUCK,A.J. (1969). A mutational analysis of conidial development in *Aspergillus nidulans*. *Genetics* 63, 317-27.
CLUTTERBUCK,A.J. (1970). A variegated position effect in *Aspergillus nidulans*. *Genetical Research, Cambridge* 16, 303-16.
CLUTTERBUCK,A.J. (1972). Absence of laccase from yellow-spored mutants of *Aspergillus nidulans*. *Journal of General Microbiology* 70, 423-35.
CLUTTERBUCK,A.J. (1973). Inter-relations between development and pigmentation during conidiation of *Aspergillus nidulans*. *Genetics* 74, s50.
CLUTTERBUCK,A.J. (1974). *Aspergillus nidulans*. In *Handbook of genetics*, ed. R.C. King, Vol.1, 447-510. Plenum Press: New York.
CLUTTERBUCK,A.J. (1975). Cyclic AMP levels during growth and conidiation. *Aspergillus News Letter* 12, 13-15.
CROFT,J.H. (1966). A reciprocal phenotypic instability affecting development in *Aspergillus nidulans*. *Heredity* 21, 565-79.
DORN,G.L. (1970). Genetic and morphological properties of undifferentiated and invasive variants of *Aspergillus nidulans*. *Genetics* 66, 267-79.
DORN,G.L., MARTIN,G.M. & PURNELL,D.M. (1967). Genetic and cytoplasmic control of undifferentiated growth in *Aspergillus nidulans*. *Life Sciences*, 6, 629-33.
GEALT,M.A. & AXELROD,D.E. (1974). Coordinate regulation of enzyme inducibility and developmental competence in *Aspergillus nidulans*. *Developmental Biology* 41, 224-32.
GROSE,S. & CLUTTERBUCK,A.J. (1975). "Fluffy" colonies obtained from Aspergillus conidia after ageing and other treatments. *Aspergillus News Letter* 13, 17-18.
HANNAN,M.A. (1975). Isolation and analysis of conidiation defective mutants of *Aspergillus niger*. *Molecular and General Genetics* 142, 333-40.
JINKS,J.L. (1954). Somatic selection in fungi. *Nature, London* 174, 409-10.
JINKS,J.L. (1956). Naturally occurring cytoplasmic changes in fungi. *Comptes Redn. Trav. Lab. Carlsberg* 26, 183-203.
MAHONEY,M. & WILKIE,D. (1962). Nucleo-cytoplasmic control of perithecial formation in *Aspergillus nidulans*. *Proceedings Royal Society, London Ser.B* 156, 524-32.

MARTINELLI, S.D. (1976). Conidiation of *Aspergillus nidulans* in submerged culture. *Transactions of the British Mycological Society* 67, 121-8.

MARTINELLI, S.D. & BAINBRIDGE, B.W. (1974). Phenoloxidases of *Aspergillus nidulans*. *Transactions of the British Mycological Society* 63, 361-70.

MARTINELLI, S.D. & CLUTTERBUCK, A.J. (1971). A quantitative survey of conidiation mutants in *Aspergillus nidulans*. *Journal of General Microbiology* 69, 261-8.

MORRIS, N.R. (1975). Mitotic mutants of *Aspergillus nidulans*. *Genetical Research, Cambridge* 26, 297-306.

OLIVER, P.T.P. (1972). Conidiophore and spore development in *Aspergillus nidulans*. *Journal of General Microbiology* 73, 45-54.

OLIVER, P.T.P. (1974). Localization of free sulphydryl groups by ferricyanide reduction in developing conidiophores of *Aspergillus nidulans*. *Histochemical Journal* 6. 319-25.

PETERSON, P.A. (1970). Controlling elements and mutable loci in maize: their relationship to bacterial episomes. *Genetica* 41, 33-56.

RAPER, K.B. & FENNELL, D.I. (1953). Heterokaryosis in *Aspergillus*. *Journal of Elisha Mitchell Scientific Society* 69, 1-29.

ROMANO, A.H. (1966). Dimorphism. In *The fungi* ed. G.C. Ainsworth & A.S. Sussman, Vol.2, 181-209. New York: Academic Press.

ROPER, J.A. (1958). Nucleo-cytoplasmic interactions in *Aspergillus nidulans*. *Cold Spring Harbor Symposium Quantitative Biology* 23, 141-54.

SAXENA, R.K. & SINHA, U. (1973). Conidiation of *Aspergillus nidulans* in submerged liquid culture. *Journal of General and Applied Microbiology* 19, 141-6.

SCOTT, W.A. & SOLOMON, B. (1975). Adenosine 3',5'-cyclic monophosphate and morphology in *Neurospora crassa:* drug-induced alterations. *Journal of Bacteriology* 122, 454-63.

TERENZI, H.F., FLAWIÁ, M.M., TELLEZ-IÑÓN, M.T. & TORRES, H.N. (1976). Control of *Neurospora crassa* morphology by cyclic adenosine 3',5'-monophosphate and dibutyryl cyclic adenosine 3',5'-monophosphate. *Journal of Bacteriology* 126, 91-9.

VUJIČIĆ, R. & MUNTANJOLA-CVETKOVIĆ, M. (1973). A comparative ultrastructural study of conidium differentiation in the Cladosarum-like mutant 22B of *Aspergillus aureolatus*. *Journal of General Microbiology* 79, 45-51.

YUILL, E. & YUILL, J.L. (1938). *Cladosarum olivaceum*, a new Hyphomycete. *British Mycological Society Transactions* 22, 194-200.

ZONNEVELD, B.M.J. (1972). The significance of α-1,3 glucan of the cell wall and α-1,3, glucanase for cleistothecium development. *Biochimica et Biophysica Acta* 273, 174-84.

ZONNEVELD, B.M.J. (1974). α-1,3 glucan synthesis correlated with α-1,3 glucanase synthesis, conidiation and fructification in morphogenetic mutants of *Aspergillus nidulans*. *Journal of General Microbiology* 81, 445-51.

Mitochondrial Genetics of *Aspergillus nidulans*

G. TURNER and R. T. ROWLANDS

Department of Bacteriology, University of Bristol, Bristol, U.K.

INTRODUCTION

Mitochondrial genetics involves the study of two genetic systems, since mitochondrial biogenesis and function are dependent on nuclear and extranuclear genomes, the former being the greatest contributor of information, and the latter being associated with a unique DNA species located within the mitochondrion. For reviews of this subject, see Sager (1972) and Lloyd (1974). Since the discovery of mitochondrial DNA (Nass & Nass, 1963; Nass, Nass & Afzelius, 1965), extranuclear genetics has drawn the greater attention, and apart from cytochrome c in *Saccharomyces cerevisiae* (Sherman *et al.*,1966; Sherman *et al.*,1974) there has been little identification of nuclear genes responsible for coding mitochondrial proteins, even though many nuclear mutations affecting mitochondrial function have been reported. Most of this paper will refer to extranuclear mitochondrial genetics.

Recent years have witnessed rapid developments in the study of extranuclear mitochondrial genetics of the yeast *Saccharomyces cerevisiae*, and detailed mapping of the mitochondrial genome has been possible using a combination of physical and genetical techniques on the respiratory deficient cytoplasmic petite mutants (Kroon & Saccone, 1976). Progress with obligately aerobic eukaryote has been much slower for a number of reasons, one of which is the impossibility of deleting large sections of the mitochondrial genome, or of obtaining a wide range of mutations which impair mitochondrial function, without preventing growth of the organism. Amongst the filamentous fungi, extranuclear mitochondrial mutants have been isolated in *Neurospora crassa* (Mitchell & Mitchell,1952; Bertrand & Pittenger, 1972), *Aspergillus nidulans* (this paper), *Aspergillus amstelodami* (Handley, 1975), *Coprinus lagopus* (Casselton & Condit, 1972; Casselton & Kirkham, 1975) and *Podospora anserina* (Belcour, 1975), but only in *Aspergillus nidulans* has extranuclear recombination so far been reported.

Although it has not yet been formally demonstrated that extranuclear mutations affecting mitochondrial function are located in the mitochondrial DNA of Aspergillus, this is the most likely possibility. The mitochondrial DNA of *Aspergillus nidulans* has a molecular weight of approximately 21 x 10^6, and twisted circular molecules with a contour length of 10-10.6 have been observed by electron microscopy on lysed mitochondria (López-Perez & Turner, 1975). This is considerably smaller than the mitochondrial DNA of *Neurospora crassa*, which consists of covalently closed circular molecules of 20μ contour length (MW 40 x 10^6; Agsteribbe, Kroon & van Bruggen, 1972; Clayton & Brambl, 1972).

ISOLATION OF MITOCHONDRIAL MUTANTS

Resistance to inhibitors of mitochondrial function

A.nidulans has been screened for sensitivity to a large number of inhibitors of mitochondrial function (Rowlands, 1974; Gunatilleke, 1974), but most of those tested were found to have little or no inhibitory action against the organism *in vivo*. Of those which were effective, only chloramphenicol, oligomycin and carboxin (inhibitors of mitochondrial protein synthesis, mitochondrial ATP-ase and succinate dehydrogenase respectively) have been used successfully to isolate mitochondrial mutants. Of course, not all mutants isolated as resistant to inhibitors of mitochondrial function are necessarily affected at the mitochondrial level, and biochemical investigation is necessary to determine whether this is the case.

A large number of oligomycin-resistant mutants have now been isolated (Rowlands & Turner, 1973, 1974a; Turner & Watson, 1976) at 2 extranuclear sites (*oliA* and *oliB)*, and in addition, there are at least 5 different nuclear phenotypes, all those mutants so far examined being located on chromosome VII. *oli 2* has been mapped (Käfer, 1975) and is located between *phen B* and *sF* on the right hand arm of chromosome VII.

Chloramphenicol resistant mutants were first reported by Gunatilleke, Scazzocchio & Arst (1975). Two out of 123 mutants tested were extranuclear, and located at the same site (*camA)*.

The nuclear mutants mapped at a minimum of 3 loci on linkage groups I, II and V, though at least one of these loci, *camD*, may modify the cytoplasmic membrane rather than the mitochondrion, since it confers hypersensitivity to cycloheximide, an inhibitor of cytoplasmic protein synthesis. More recently, 31 extranuclear chloramphenicol-resistant mutants have been isolated from a chloramphenicol-sensitive strain (Lazarus & Turner, 1976). Though they all appear to map at or very close to (*camA*), they do show some variation in phenotype. This is most marked in (*camA51*), which grows very slowly in the absence of chloramphenicol, and is not inhibited by presence of the drug. All of the chloramphenicol resistant mutants were obtained following mutagenesis with N-methyl-N'-nitro-N-nitrosoguanidine (NTG).

Although no extranuclear mutants have been isolated which are resistant to carboxin, nuclear mutants with increased *in vitro* resistance of succinate dehydrogenase, succinate-cytochrome c reductase and succinate oxidase to the inhibitor have been reported (Gunatilleke, Arst & Scazzocchio, 1976). The mutations mapped at 3 loci on linkage groups III, VII, and VIII. One of these, *cbxA17*, mapped on linkage group III between *phenA* and *amdS*. Similar mutants have also been reported by van Tuyl (1975).

Cold sensitivity

During a search for cold-sensitive mutants with possible defects in robosome assembly, Waldron & Roberts (1973) isolated a cold-sensitive extranuclear mutant (*cs67*) following NTG mutagenesis. This mutant grows more slowly at 20°C than does the wild-type strain, but appears to be normal at 37°C. (*camA51*) is also cold sensitive.

Respiratory deficiency

Attempts to directly isolate leaky respiratory deficient mutants in *A.nidulans* have so far been unsuccessful. Houghton (1970) isolated nuclear mutants which were unable to grow on TCA cycle intermediates, and which grew poorly on glucose, but further investigation showed that the mutant mycelium was unable to take up the intermediates (Houghton, 1971). Unsuccessful attempts to identify respiratory deficient mutants have also been made using tetrazolium dyes as respiratory indicators (Rowlands, 1974; Gunatilleke, 1974).

Frequency of extranuclear mutants

Extranuclear mutants are rare in Aspergillus, comprising only about 2% of total spontaneous or NTG induced mutants. The reason for this low frequency is not clear, since the corresponding figure for yeast is approximately 50%. (Coen *et al.*,1970), and a method of increasing the extranuclear/nuclear mutant ratio would be very useful. In yeast, this has been achieved by a variety of methods, including the use of Mn++ as a fairly specific mutagen for mitochondrial DNA (Putrament *et al.*, 1975). No reliable method has yet been devised for *A.nidulans*, and those devised for yeast have proved ineffective.

GENETICAL BEHAVIOUR OF THE EXTRANUCLEAR MUTANTS

Heterokaryons

The heterokaryon test as proposed by Jinks (1963) has been used as a convenient test for extranuclearity in the mutants isolated. Since in general, nuclear recombination is not observed in heterokaryons, any reassortment of the marker in question with known nuclear markers after formation of a heterokaryon with another strain is good evidence for extranuclearity.

One interesting feature of heterokaryons containing mixtures of mitochondrial types (i.e. heteroplasmons) is their strong tendency to segregate into homoplasmic areas following continued growth of the forced heterokaryon on minimal medium (Fig. 1). This contrasts with the situation in *Neurospora crassa*, where it is possible to maintain heteroplasmons much longer (Pittenger, 1956; T.H. Pittenger, personal communication). However, although conidia taken from homoplasmic areas are of one extranuclear type only, it is sometimes possible to detect the other extranuclear type in hyphal samples if a suitable selection procedure is available. For instance, by plating hyphal blocks from wild-type homoplasmic areas of *(oliA1)/(+)* heterokaryons on to oligomycin-containing medium, *(oliA1)* can be re-isolated at a frequency higher than the spontaneous mutation rate (Rowlands & Turner, 1973).

While it is known that each conidium contains one nucleus only (Yuill, 1950), it is not clear how many mitochondria or mitochondrial genomes are present per conidium. However, conidia taken from heteroplasmons and spread to single colonies sometimes retain more than one extranuclear genotype per conidium (Rowlands & Turner, 1974b), though these soon segregate unless some suitable form of selection pressure is maintained.

Sexual crosses

In crosses between wild-type *(+)* and *(oliA1)* strains, cleistothecia, whether hybrid or selfed, always contain one extranuclear type only (Rowlands & Turner, 1973; Mason & Turner, 1975). The same behavious has also been reported in the other extranuclear mitochondrial mutants (Waldron & Roberts, 1973; Gunatilleke *et al.*, 1975). Although this behaviour suggested a type of "uniparental" or maternal inheritance, it is known that *Aspergillus nidulans*

Fig. 1. Sectoring of rutamycin resistance/sensitivity, i.e. *(oliAl)/(+)*, in 8 day old heterokaryons of *A.nidulans*. Each horizontal pair of plates consists of the original heterokaryon (left) and its replica on rutamycin-containing medium (right) obtained by replicating with damp velveteen. The sectors are independent of nuclear markers, since yellow and white conidiating areas can be seen to be well mixed all over the heterokaryons (left hand column) (from Rowlands & Turner, 1973, with permission).

---

is homothallic and there is no cytological evidence for differentiated male and female organs preceding cleistothecium formation. Nevertheless, studies with ascospore colour mutants (Apirion, 1963) showed that the ascospore wall colour was determined by nuclear genes in a non-autonomous fashion. Cleistothecia, whether selfed or hybrid, contained ascospores of one colour only, even though the genes involved in colour determination had segregated in a 1:1 ratio in hybrid cleistothecia. It appeared that the wall colour was thus determined by one of the pair of nuclei involved in the cross, which could be referred to as the "maternal" nucleus. This offers a way of examining the inheritance pattern of the extranuclear markers, but only when one other problem has been overcome. Since crosses in *Aspergillus nidulans* are normally carried out between heterokaryon compatible strains, there is an opportunity for reassortment of extranuclear/nuclear genomes to occur prior to formation of the sexual stage, as is observed in the heterokaryon test. Under such conditions, it is impossible to maintain association of a particular extranuclear marker with a particular nucleus. This problem had been solved in *Aspergillus amstelodami* to demonstrate maternal inheritance of the ragged (*rgd)* mutation, by using heterokaryon incompatible strains in the crosses, thereby preventing heterokaryon formation while still permitting sexual crosses (Handley, 1975). A number of heterokaryon incompatible strains of *Aspergillus nidulans* were obtained from Dr J.H. Croft, University of Birmingham, and it was confirmed that the extranuclear characters could not be transferred to these strains via heterokaryons. Crosses were set up between suitably marked strains, and a pattern of strict maternal inheritance was observed for the extranuclear markers (Rowlands & Turner, 1976b). Our data also revealed a further interesting feature. The crosses were carried out between our Glasgow derived strains and two strains from Birmingham which both differed from the Glasgow strains at heterokaryon incompatibility loci *hetA* and *hetB*. Instead of obtaining hybrid cleistothecia of two ascospore colour types within one cross, only one type was obtained, and the colour depended on the Birmingham strain used in the cross. One of the strains always behaved as the maternal parent, and the other as paternal relative to the Glasgow derived strains. It seems possible that the "sex" of these strains may be determined by nuclear genes, and this is being investigated at present.

EXTRANUCLEAR RECOMBINATION

Opportunity for recombination

Extranuclear recombination was first demonstrated between *(oliA1)* and *(csA67)* in conidia derived from heterokaryons between strains carrying these two mutations (Rowlands & Turner, 1974b), and both the double mutant and wild-type recombinant were recovered. It seems that the heterokaryon provides a suitable means for bringing extranuclear genomes into contact, and that recombination takes place, presumably following mitochondrial fusion. Unlike heteroplasmons, the double mutant recombinants have always proved stable, never having segregated out to give the single mutants after many generations, their phenotypic behaviour reflecting an additive effect of the two individual mutations (Turner & Rowlands, 1976). Since this time, recombinations has been demonstrated in heterokaryons between four extranuclear sites: *(oliA)*, *(csA)*, *(camA)* and *(oliB)* (Rowlands & Turner, 1975; Turner & Watson, 1976).

In crosses between heterokaryon compatible strains, extranuclear recombinants are also observed in cleistothecia, again following the rule that only one extranuclear genotype is observed in any one cleistothecium (Mason & Turner, 1975). Also for a given cross, some extranuclear markers are recovered in the cleistothecia more frequently than others. However, the question arose whether the recombination observed occured during the sexual stage or during a preceding heterokaryon phase. Crosses containing extranuclear markers were set up between heterokaryon incompatible strains, and to date, no extranuclear recombination has been observed in such crosses. This might be expected in view of the maternal inheritance pattern already demonstrated, which suggests that only one "mitochondriotype", that associated with the maternal nucleus, is allowed into the cleistothecium. In nature, mitochondrial genome recombination may therefore be severely limited by the presence of heterokaryon incompatibility genes. It has already been suggested that such genes may prevent the spread of fungal viruses and infectious cytoplasmic conditions (Caten, 1972). This may play an important role in the evolution of the mitochondrial genome.

Measurement of recombination

In order to map the mitochondrial genome, it is necessary to devise some way of quantifying the observed recombination between extranuclear markers. When a heterokaryon is formed by forcing on minimal medium, a variable amount of cytoplasmic mixing takes place.

Since recombination presumably depends on the opportunity for mitochondria to come into contact, the frequency of recombination will depend on the degree of cytoplasmic mixing. As an index of this, we have determined the percentage of reassorted conidia emerging from the heterokaryon. The recombinants obtained are then expressed as a ratio of the reassorted types, i.e. % recombination = (no. of recombinants/no. of reassorted types) x 100. It is important in these extranuclear crosses to use heterokaryons which are as young as possible (2 days on minimal medium), since homoplasmic sectors soon develop of both parental and recombinant phenotypes, and the "recombination frequency" would then really reflect the size of the recombinant sector (Fig. 1). In addition, since the frequency of reciprocal recombinants is variable, selective methods which identify only one of a pair of reciprocal recombinants are unsatisfactory.

An alternative method to assess recombination frequency involves scoring the recombinant sectors in a large number of heterokaryons, since sector formation appears to be a random process. Although the results obtained by this method have not contradicted the results from conidial analysis (Gunatilleke, 1974; Rowlands, 1974), the method has not been used extensively.

Extranuclear recombination is also observed during sexual crosses between heterokaryon compatible strains, but the bias against certain markers and the practical difficulties of examining large numbers of perithecia has ruled this out as a method of analysis (Mason & Turner, 1975).

The results which we have obtained to date indicate that recombination between *(oliA)*, *(camA)* and *(csA)* occurs with equal frequency, and we believe that this reflects a lack of tight linkage between these sites (Rowlands & Turner, 1975). Two recently isolated oligomycin-resistant mutants O^R322 and O^R332 appear to be at a different but closely linked site from *(oliA)*, since a low level of recombination can be detected between *(oliA)* and the new site *(oliB)* (Fig. 2; Turner & Watson, 1976). Although it is not yet known whether this represents intra- or inter-genic recombination, it may be possible to distinguish between these alternatives by examination of the changes brought about by these various mutations to oligomycin resistance in the sub-units of the mitochondrial ATP-ase.

MITOCHONDRIA OF ASPERGILLUS

Isolation of mitochondria

To date, there has been no comprehensive paper on the respiratory properties of the mitochondria of *A.nidulans*, though several have appeared on other species of Aspergillus, including *A.niger*

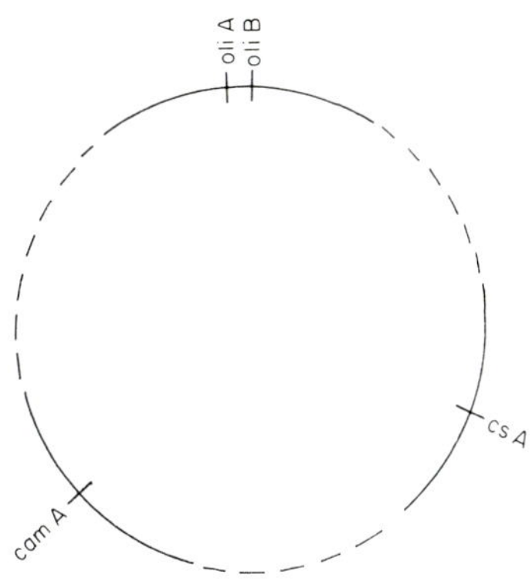

Fig. 2. Extranuclear loci in *A.nidulans*. Mutants of several different phenotypes are known to be located at or near *camA*. Mutants mapping at *oliA* all possess a similar phenotype, as do the two mutants mapping at *oliB*. Only one mutant is known for *csA*. Dotted lines between loci represent an absence of tight linkage.

(Watson & Smith, 1967; Higgins & Friend, 1968) and *A.oryzae* (Kawakita, 1970b). One of the problems has been the difficulty of breaking the mycelium without seriously damaging the mitochondria. A variety of methods has been applied to Aspergillus, including a motor-driven, all-glass homogenizer (Watson & Smith, 1967), a VirTis high speed homogenizer on frozen mycelial pellets (Higgins & Friend, 1968) and a specially constructed roller mill (Kawakita, 1970b). Using such methods, mitochondria capable of respiratory control have been isolated by the above authors, but in our experience, none of these methods have yielded mitochondria from *A. nidulans* sufficiently intact for detailed respiration studies. It is possible that the mitochondria of *A.nidulans* are more fragile than those of other species. In attempts to improve mitochondrial yield and quality, the grinding mill devised by Weiss, von Jagow, Klingenberg & Bücher (1970) for *Neurospora crassa* has been used routinely in our work and in that of Gunatilleke (1974) and Gunatilleke *et al.*(1976), but even this is not ideal. With rough, freshly-ground wheels, we have been able to demonstrate respiratory control with 2-oxoglutarate and pyruvate plus malate as substrates, though this ability is lost as the wheels become polished with further use (G.Turner, unpublished). Further improvements in mitochondrial preparation may have to wait until a method is devised for large scale preparation of protoplasts to avoid the high shear forces involved in all the other methods.

Respiratory properties of the mitochondria

Measurement of respiration with various TCA cycle intermediates, including succinate, 2-oxoglutarate, malate plus pyruvate and citrate and observations on the effects of the inhibitors rotenone, antimycin A and cyanide (G. Turner, unpublished; Gunatilleke, 1974) have suggested that the mitochondrial electron transport system of *A. nidulans* fairly closely resembles those of *A.oryzae* and *A.niger*, which seem to have 3 phosphorylation sites for NAD-linked substrates. In all these species, externally added NADH is rapidly oxidised by intact mitochondria, a characteristic feature of the mitochondria of many micro-organisms, but not of intact mammalian mitochondria (Lloyd, 1974). However, only a small proportion of this respiration with NADH is sensitive to rotenone, while most of it is sensitive to antimycin A and cyanide. This suggests that much of the externally added NADH is oxidised via a pathway which by-passes site I, but not sites II and III.

Cyanide (1μmM) and antimycin A (10 μg/mg protein) inhibit about 85-95% of the respiration observed with substrates such as succinate and 2-oxoglutarate, but a proportion of cyanide/antimycin-insensitive respiration usually remains. This forms an even larger proportion of total respiration in stationary phase mycelium, and Kawakita (1970a) showed that *A.oryzae* grown in the presence of antimycin A developed 100% antimycin-resistant respiration. Mitochondria isolated from such mycelium still showed respiratory control, though with lower P/O ratios, suggesting loss of a coupling site (Kawakita, 1971). His data suggested that an alternative pathway was present which involved site I, but not sites II and III. Although all the cytochromes were present, cytochrome b remained fully reduced even after aeration of the mitochondria, suggesting that the normal cytochrome pathway was not operating. Growth of *A.nidulans* in the presence of either cyanide or antimycin A does not lead to a marked increase in inhibitor-resistant respiration (G.Turner, unpublished), and studies on extranuclear mutants have suggested that this species is unable to grow using the alternative pathway (Turner & Rowlands, 1976).

Cytochromes

The room temperature reduced minus oxidized difference spectrum of *A.nidulans* mitochondria reveals α-bands at 553 nm, 560 nm and 608 nm for cytochromes c, b and a respectively, the absorption band for cytochrome a being at a slightly longer wavelength than those seen in mammalian and yeast mitochondria (605 nm). β-bands for cytochromes c and b are seen at 523 nm and 531 nm, and α-bands for b and ($a+a_3$) at 432 nm and 445 nm respectively. At the temperature of liquid nitrogen, these bands are shifted slightly towards the blue end of the spectrum, and additional bands are resolved (Turner

& Rowlands, 1976; Fig.3). Cytochrome b is resolved into 3 species with α-bands at 555 nm, 559.5 nm and 564.5 nm, and cytochrome c into $c_{548.5}$ and c_1 at 552.5 nm. Cytochrome c possesses a β-band at 521 nm, while β-bands associated with b-cytochromes are seen at 529 nm and 537 nm. Addition of antimycin A to succinate-reduced mitochondria, followed by re-aeration, results in the re-oxidation of cytochromes c, c_1 and $(a+a_3)$, leaving the b-cytochromes reduced (Fig. 3b). Very similar spectra have been reported for the mitochondria of *A. oryzae* (Wakiyama & Ogura, 1972).

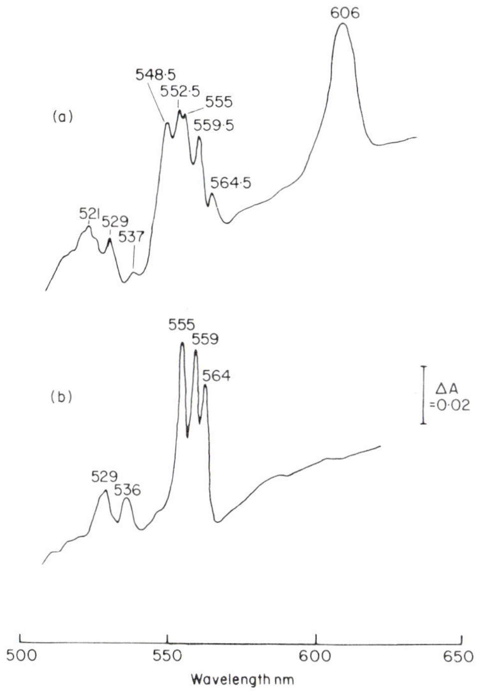

Fig. 3. Low temperature difference spectra of isolated mitochondria from *A.nidulans*. a) succinate-reduced anaerobic minus oxidised, 4.3 mg of protein/ml; b) succinate plus antimycin A-aerated minus oxidized, 5.1 mg of protein/ml. Path length 2 mm (Turner & Rowlands, 1976).

EFFECTS OF THE EXTRANUCLEAR MUTATIONS ON MITOCHONDRIAL FUNCTION

Oligomycin-resistant mutants

(oliA1) was originally observed to be slow growing on drug-free medium (Rowlands & Turner, 1974a) and subsequent genetic manipulation has failed to separate the slow growth character from oligomycin resistance, suggesting that it might be a result of the one mutation. Examination of the cytochrome spectrum of the mutant revealed that cytochrome c was present at approximately 1.5 times the concentration of the wild-type strain, with little change in the other cytochromes (Fig. 4b).

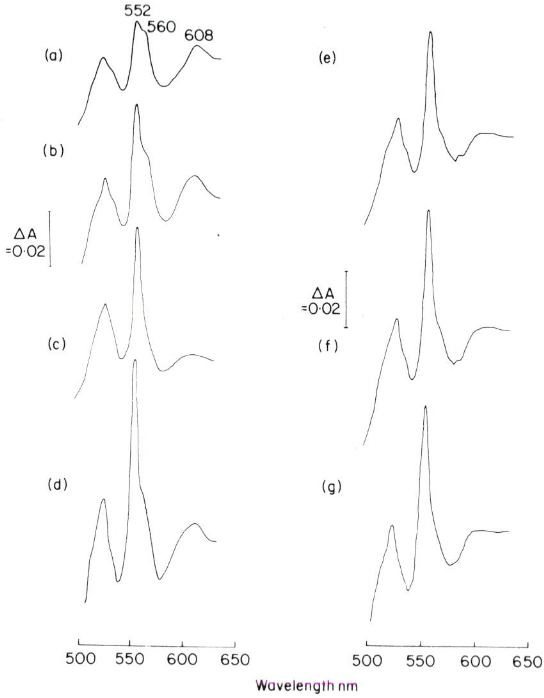

Fig. 4. Room temperature cytochomrome spectra (dithionite reduced minus oxidized) of the whole mycelia of the wild-type and extra-nuclear mutants of *A. nidulans*. a) wild-type; b) *(oliA1)*; c) *(csA67)* grown at 20°C; d) *(oliA1, csA67)* grown at 20°C; e) wild-type (chloramphenicol-sensitive) strain grown with chloramphenicol, 0.5 mg/ml; f) (*camA51*) grown in drug-free medium; g) *camA51)* grown with chloramphenicol, 0.5 mg/ml. Protein concentration approximate 4.5 mg/ml in each cuvette; path length 1 cm (Turner & Rowlands, 1976; Lazarus & Turner, 1976).

In addition, an increase in the proportion of cyanide-insensitive respiration was detected in this mutant (Turner & Rowlands, 1976). Recent experiments have demonstrated that the mitochondrial ATP-ase of *(oliA1)* shows an increased resistance to oligomycin. While 1.5 μg oligomycin/mg mitochondrial protein is required to achieve 50% inhibition of the wild-type ATP-ase activity, 8 μm/mg is the corresponding figure for *(oliA1)*. It is possible that an alteration in a mitochondrially-coded polypeptide of the ATP-ase complex leads to an increased resistance to oligomycin, and at the same time to lesions in the respiratory system, resulting in the slow growth observed. In this connection, it is interesting to note that O^R332, which can be distinguished genetically from *(oliA1)*, has a very similar, though not identical, phenotype.

Cold-sensitive mutant

When grown at 20°C, *(csA67)* exhibits marked changes in cytochrome content, and a respiration which is 100% cyanide-insensitive in young cultures (Turner & Rowlands, 1976). The spectrum (Fig.4c) resembles that observed in wild-type strains grown in the presence of chloramphenicol, and suggests a lesion in the mitochondrial protein synthesizing system, though no direct evidence has yet been obtained for this. The stable, double mutant recombinant *(oliA1, csA67)*, obtained from extranuclear crosses behaves like *(oliA1)* at 37°C, but shows additive effects of the two lesions when grown at 20°C.

The increase in cyanide-insensitive respiration seems to be a common feature of extranuclear mutants of *A.nidulans* in which slow growth and cytochrome alterations are apparent. However, there is as yet little information about the composition and significance of this pathway. In *Neurospora crassa*, where cyanide-insensitive respiration is also seen in both nuclear and extranuclear slow-growing mitochondrial mutants (Lambowitz & Slayman, 1971; Edwards & Kwiecinski, 1973), the pathway appears to branch off from the cytochrome-utilizing electron transport chain in the region of ubiquinone, thus by-passing phosphorylation sites II and III, and lowering P/O ratio for substrates which are normally oxidised via these sites (von Jagow, Weiss & Klingenberg, 1973). Although it has been suggested that this might be a way of reoxidizing reduced NAD when lesions are present in the usual cyanide-sensitive pathway (Lambowitz, Smith & Slayman, 1972), this does not seem to be the case for *A.nidulans*, since growth yield studies showed that growth of *(csA67)* at 20°C is just as efficient as at 37°C, even though the growth rate is lower. This suggests that unlike the case of *A. oryzae* grown on antimycin A, the pathway is not actually used during growth, and is only observed experimentally when the normal respiration is blocked by cyanide. Further support for this hypothesis comes from observations on the effect of the inhibitor salicyl hydroxamate (SHAM) on the respiration of *(csA67)* grown at 20°C. SHAM is a specific inhibitor of cyanide-insensitive respiration in higher plants (Schonbaum, Bonner, Storey & Bahr, 1971) and *Neurospora crassa* (Lambowitz & Slayman, 1971). In *A.nidulans*, however, SHAM will not inhibit the cyanide-insensitive respiration unless cyanide is also present, suggesting that electrons are diverted down the alternative pathway only when the cyanide-sensitive pathway is blocked (Turner & Rowlands, 1976). Although the concentration of cytochrome oxidase is rather low in *(csA67)* grown at 20°C, experiments with isolated mitochondria have shown that the cytochrome pathway is still functional.

Chloramphenicol-resistant mutants

The mutant *(camA112)* isolated by Gunatilleke *et al.*(1975) shows no abnormalities on drug-free medium, but grows on 5 mg/ml chloramphenicol, a level which severely inhibits growth of the wild-type strain. Since protein synthesis on the mitochondrial ribosomes is the site of action of chloramphenicol, it is possible that either the permeability of the mitochondrial inner membrane or the mitochondrial ribosomes themselves have been modified by the mutation, and work is in progress to try to distinguish between these possibilities. Evidence supporting the latter hypothesis has come from observations made on another extranuclear chloramphenicol-resistant mutant *(camA51)*. This mutant appears to be at the *(camA)* locus, since no recombination has been observed between it and *(camA112)*, but phenotypically, it is quite different from the latter. On drug-free medium, it gorws very slowly, and its cytochrome spectrum resembles that of wild-type strains grown in the presence of chloramphenicol (Fig. 4). It is possible that an alteration to the mitochondrial ribosome has resulted in a defective protein synthesizing system, while at the same time increasing the resistance to chloramphenicol.

NUCLEAR MUTANTS AND NUCLEAR/EXTRANUCLEAR INTERACTION

Oligomycin resistance

During the search for extranuclear mutants resistant to oligomycin, a large number of nuclear mutants were isolated, and some of these have been examined in more detail (Rowlands & Turner, 1974a). A varied pattern of cross resistance was observed to the ATP-ase inhibitor venturicidin, and more recently, hypersensitivity to triethyltin. These results suggested that some of the nuclear mutants might have altered ATP-ases, and tests have shown that this is indeed the case. The resistance increase varies from twofold to fivefold as observed in isolated mitochondria, indicating nuclear coding of some of the ATP-ase complex. Combinations of nuclear/ extranuclear oligomycin-resistant mutants have been made via heterokaryons, and all of these show enhanced resistance to oligomycin *in vivo* (Rowlands & Turner, 1976a). One of these combinations also shows enhanced resistance of the mitochondrial ATP-ase *in vitro*, indicating interaction of the nuclear and extranuclear products at the level of the ATP-ase. In other combinations, nuclear/extranuclear interaction has been observed at the level of the cytochromes, in that the altered cytochrome spectrum produced by *(oliA1)* has been corrected. Recently, progress has been made in the purification and characterization of the mitochondrial ATP-ase of *A.nidulans* (H.Küntzel, personal communication), and this should enable us to learn more about the coding of the enzyme complex.

Suppression of an extranuclear mutation

Following NTG mutagenesis, Waring & Scazzocchio (1976) have isolated both nuclear and extranuclear suppressors of the mutation *(csA67)*. The nuclear suppressors map at two loci in chromosomes III and VII respectively.

CONCLUSION

Its well-characterized nuclear genetics and the ease with which it can be grown and handled for biochemical work make *A. nidulans* an ideal organism for biochemical genetic studies. The results reported here, showing that extranuclear mutants can be isolated and mapped, confirm the potential of this organism for continued studies on the role of both nuclear and extranuclear genomes in mitochondrial biogenesis.

REFERENCES

AGSTERIBBE,E., KROON,A.M. & van BRUGGEN,E.G.J.(1972). Circular DNA from mitochondria of *Neurospora crassa*. *Biochimica et Bio-physica Acta* 269, 299-303.

APIRION,D. (1963). Formal and physiological genetics of ascospore colour in *Aspergillus nidulans*. *Genetical Research, Cambridge* 4, 276-83.

BELCOUR,L. (1975). Cytoplasmic mutations isolated from protoplasts of *Podospora anserina*. *Genetical Research, Cambridge*,25, 155-61.

BERTRAND,H. & PITTENGER,T.H. (1972). Isolation and classification of extranuclear mutants of *Neurospora crassa*. *Genetics* 71, 521-33.

CASSELTON,L.A. & CONDIT, A.(1972). A mitochondrial mutant of *Coprinus lagopus*. *Journal of General Microbiology* 72, 521-7.

CASSELTON,L.A. & KIRKHAM,J.B.(1975). Growth and ultrastructural studies on the mitochondrial mutant of *Coprinus lagopus*.

Archives of Microbiology 106, 215-20
CATEN,C.E.(1972). Vegetative incompatibility and cytoplasmic infection in fungi. *Journal of General Microbiology* 72, 221-9.
CLAYTON,D.A. & BRAMBL,R.M. (1972). Detection of circular DNA from mitochondria of *Neurospora crassa. Biochemical and Biophysical Research Communications* 46, 1477-82.
COEN,D., DEUTSCH, J., NETTER,P., PETROCHILO,E. & SLONIMSKI,P.P.(1970). Mitochondrial genetics I. Methodology and phenomenology. In *Control of Organelle Development: Symposia of the Society for Experimental Biology, No.XXIV* pp.449-496 Ed. P.L. Miller. Cambridge: University Press.
EDWARDS,D.L. & KWIECINSKI,F. (1973). Altered mitochondrial respiration in a chromosomal mutant of *Neurospora crassa. Journal of Bacteriology* 116, 610-8.
GUNATILLEKE,I.A.U.N. (1974). *Biochemistry and genetics of mitochondrial function in Aspergillus nidulans.* Ph.D.Thesis, University of Cambridge.
GUNATILLEKE,I.A.U.N., ARST,H.N. Jr. & SCAZZOCCHIO,C. (1976). Three genes determine the carboxin sensitivity of mitochondrial succinate oxidation in *Aspergillus nidulans. Genetical Research Cambridge* 26, 297-305.
GUNATILLEKE,I.A.U.N., SCAZZOCCHIO,C. & ARST,H.N. Jr. (1975). Cytoplasmic and nuclear mutations to chloramphenicol resistance in *Aspergillus nidulans. Molecular and General Genetics* 137,269-76.
HANDLEY,L. (1975). *Cytoplasmic inheritance in Aspergillus amstelodami.* Ph.D. Thesis, University of Birmingham.
HIGGINS,E.S. & FRIEND,W.H. (1968). Comparative mitochondrial function in two strains of *Aspergillus niger. Canadian Journal of Biochemistry* 46, 1515-21.
HOUGHTON,J.A. (1970). A new class of slow-growing non-perithecial mutants of *Aspergillus nidulans. Genetical Research, Cambridge* 16, 285-92.
HOUGHTON,J.A. (1971). Biochemical investigation of the slow-growing non-perithecial (*sgp*) mutants of *Aspergillus nidulans. Genetical Research* 17, 237-44.
von JAGOW,G., WEISS,H. & KLINGENBERG, M. (1973). Comparison of the respiratory chain of *Neurospora crassa* wild type and the *mi* mutants *mi-1* and *mi-3*. *European Journal of Biochemistry* 33,140-57
JINKS,J.L. (1963). Cytoplasmic inheritance in fungi. In *Methodology in Basic Genetics*, ed. W.J. Burdette, pp 325-354. San Francisco: Holden-day.
KÄFER,E. (1975). Reciprocal translocations and translocation disomics of Aspergillus and their use for genetic mapping. *Genetics* 79, 7-30.
KAWAKITA, M. (1970a). Studies on the respiratory system of *Aspergillus oryzae*. I. Development of respiratory activity

during germination in the presence and absence of Antimycin A. *Plant and Cell Physiology* 11, 377-84.

KAWAKITA, M. (1970b). Studies on the respiratory system of *Aspergillus oryzae*. II. Preparation and some properties of mitochondria from mycelia. *Journal of Biochemistry (Tokyo)* 68, 625-31.

KAWAKITA, M. (1971). Studies on the respiratory system of *Aspergillus oryzae*. III. Properties of mitochondria from mycelia grown in the presence of Antimycin A. *Journal of Biochemistry (Tokyo)* 69, 35-42.

KROON, A.M. & SACCONE, C. (1976). ed. The genetic function of mitochondrial DNA. *10th International Bari Conference on Mitochondria*, in the press.

LAMBOWITZ,A.M. & SLAYMAN,C.W. (1971). Cyanide-resistance respiration in *Neurospora crassa*. *Journal of Bacteriology* 108, 1087-96.

LAMBOWITZ, A.M., SMITH, E.W. & SLAYMAN, C.W. (1972). Oxidative phosphorylation in *Neurospora* mitochondria. Studies on wild-type, *poky*, and chloramphenicol-induced wild-type. *Journal of Biological Chemistry* 247, 4850-65.

LAZURUS, C.M. & TURNER,G. (1976). Novel extranuclear chloramphenicol-resistant phenotypes in *Aspergillus nidulans*. *Heredity*, in the press.

LÓPEZ-PÉREZ, M.J. & TURNER, G. (1975). Mitochondrial DNA from *Aspergillus nidulans*. *FEBS Letters* 58, 159-63.

LLOYD, D. (1974). *The Mitochondria of Micro-organisms*. London, New York & San Francisco: Academic Press.

MASON, J.R. & TURNER, G. (1975). Transmission and recombination of extranuclear genes during sexual crosses in *Aspergillus nidulans*. *Molecular and General Genetics* 143, 93-9.

MITCHELL, M.B. & MITCHELL, H.K. (1952). A case of "maternal inheritance" in *Neurospora crassa*. *Proceedings of the National Academy of Sciences of the United States of America* 38, 442-9.

NASS, M.M.K. & NASS, S. (1963). Intramitochondrial fibers with DNA characteristics. I. Fixation and electron staining reactions. *Journal of Cell Biology* 19, 593-611.

NASS, S. & NASS, M.M.K. (1963). Intramitochondrial fibers with DNA characteristics. II. Enzymatic and other hydrolytic treatments. *Journal of Cell Biology* 19, 613-29.

NASS, M.M.K., NASS, S. & AFZELIUS,B.A. (1965). The general occurrence of mitochondrial DNA. *Experimental Cell Research* 37, 516-39.

PITTENGER, T.H. (1956). Synergism of two cytoplasmically inherited mutants in *Neurospora crassa*. *Proceedings of the National Academy of the United States of America* 42, 747-52.

PUTRAMENT,A., BARANOWSKA,H., EJCHART,A. & PRAZMO, W.(1975). Manganese mutagenesis in yeast. A practical application of manganese for the induction of mitochondrial antibiotic-resistant mutations.

Journal of General Microbiology 90, 265-70.

ROWLANDS, R.T. (1974). *Extranuclear mutants of Aspergillus nidulans*. Ph.D. Thesis, University of Bristol.

ROWLANDS, R.T. & TURNER,G. (1973). Nuclear and extranuclear inheritance of oligomycin resistance in *Aspergillus nidulans*. *Molecular and General Genetics* 126, 201-16.

ROWLANDS,R.T. & TURNER,G. (1974a). Physiological and biochemical studies of nuclear and extranuclear oligomycin-resistant mutants of *Aspergillus nidulans*. *Molecular and General Genetics* 132, 73-88.

ROWLANDS,R.T. & TURNER, G. (1974b). Recombination between the extranuclear genes conferring oligomycin resistance and cold-sensitivity in *Aspergillus nidulans*. *Molecular and General Genetics* 133, 151-61.

ROWLANDS, R.T. & TURNER,G. (1975). Three-marker extranuclear mitochondrial crosses in *Aspergillus nidulans*. *Molecular and General Genetics* 141, 69-79.

ROWLANDS,R.T. & TURNER,G. (1976a). Interactions between nuclear and extranuclear oligomycin-resistant mutants of *Aspergillus nidulans*. *Heredity*, in the press.

ROWLANDS, R.T. & TURNER,G. (1976b). Maternal inheritance of extranuclear mitochondrial markers in *Aspergillus nidulans*. Submitted for publication.

SAGER,R. (1972). *Cytoplasmic genes and organelles*. New York & London: Academic Press.

SCHONBAUM,G.R., BONNER,W.D.Jr., STOREY,B.T. & BAHR, J.T. (1971). Specific inhibition of the cyanide-insensitive respiratory pathway in plant mitochondria by hydroxamic acids. *Plant Physiology* 47, 124-8.

SHERMAN,F., STEWART,J.W., MARGOLIASH, E., PARKER,J. & CAMPBELL,W., (1966). The structural gene for yeast cytochrome c. *Proceedings of the National Academy of Sciences of the United States of America* 55, 1498-504.

SHERMAN,F., STEWART,J.W., JACKSON,M., GILMORE, R.A. & PARKER,J.H. (1974). Mutants of yeast defective in iso-1-cytochrome c. *Genetics* 77, 255-84.

TURNER,G. & ROWLANDS,R.T. (1976). Cytochrome abnormalities and cyanide resistant respiration in extranuclear mutants of *Aspergillus nidulans*. *Journal of Bacteriology* 125, 389-97.

TURNER,G. & WATSON,C. (1976). A new extranuclear oligomycin-resistant locus in *Aspergillus nidulans*. *Heredity*, in the press.

van TUYL,J.M. (1975). Acquired resistance to carboxin in *Aspergillus nidulans*. *Netherlands Journal of Plant Pathology* 81,122-3.

WAKIYAMA,S. & OGURA,Y. (1972). Studies on the respiratory system of *Aspergillus oryzae*. IV. Low temperature difference spectra of mitochondria from mycelia grown in the presence and absence of chloramphenicol. *Journal of Biochemistry (Tokyo)* 71, 295-300.

WALDRON,C. & ROBERTS, C.F. (1973). Cytoplasmic inheritance of cold-sensitive mutant in *Aspergillus nidulans*. *Journal of General Microbiology* 78, 379-81.

WARING,R.B. & SCAZZOCCHIO,C. (1976). Suppressors of a mitochondrially inherited mutation in *Aspergillus nidulans*. *Heredity*, in the press.

WATSON,K. & SMITH,J.E. (1967). Oxidative phosphorylation and respiratory control in mitochondria from *Aspergillus niger*. *Biochemical Journal* 104, 332-9.

WEISS,H., von JAGOW,G., KLINGENBERG,M. & BÜCHER,T. (1970). Characterization of *Neurospora crassa* mitochondria prepared with a grind-mill. *European Journal of Biochemistry* 14, 75-82.

YUILL,E. (1950). The numbers of nuclei in conidia of Aspergilli. *Transactions of the British Mycological Society* 33, 324-31.

Aspects of the Population Genetics of *Aspergillus nidulans*

J. H. CROFT and J. L. JINKS

Department of Genetics, University of Birmingham, Birmingham, U.K.

INTRODUCTION

A main reason for the success of many genetical studies which have been made of *Aspergillus nidulans* has undoubtedly been the use of mutant strains derived from a single isolate. This isolate was originally designated A69 by Yuill (1939, 1950) and later NRRL 194 by Thom & Raper (1945). Many mutant strains were derived from this isolate in the Glasgow laboratory of Pontecorvo (Pontecorvo *et al.*, 1953) and more recently in many other laboratories in strains derived from these Glasgow held stocks (see Clutterbuck, 1974). Thus, the majority of strains of *A.nidulans* held in genetical laboratories can be considered to be isogenic other than at the known mutant loci carried by those strains. However, when a sample of independent isolates from the natural population is examined it is found that *A.nidulans* is a very variable species and more detailed studies have revealed a number of interesting features of the genetical structure of this population. In the present account the variation found in the natural population is described and related to the complex system of vegetative or heterokaryon incompatibility which is found in this species.

It is necessary to define the population which has been studied. *A.nidulans* is one of the twenty-seven varieties or subspecies which can be distinguished on taxonomic criteria and which have been described by Raper & Fennell (1965) in the *Aspergillus nidulans* group. The Birmingham collection of some one hundred and fifty independent wild isolates of *A.nidulans* was made mainly during the period from May 1962 to July 1963. A small number were isolated in 1954. The majority were isolated from soil samples collected from various regions in England and Wales (Grindle, 1963a,b). A few were isolated from other natural situations in these same regions. The collection is not a random sample of the population in the strict sense but it is probably representative of the population which existed over the collection period. The genetical studies of these isolates were carried out in the laboratory using the standard methods employed for the growth and analysis of *A. nidulans* (Pontecorvo *et al.*,1953; Clutterbuck, 1974) or slight modifications of these.

VARIATION IN THE NATURAL POPULATION

Inspection and measurement of independent isolates of *A. nidulans* reveals the existence of much variation between isolates for a number of different morphological, physiological and biochemical properties. The majority of characters vary in a continuous manner and are under the control of many genes. Variation in some of the characters falls into discrete classes and the genetical control of these can be assigned to individual loci (Table 1). Where the character is discretely variable conventional methods of studying its inheritance can be used, but for continuously variable characters it is necessary to employ the methodology of biometrical genetics (Mather & Jinks,1971).

Two general methods of examining the continuous variation found in the natural population have been used. The first involves the measurement of the particular character after the growth of wild isolates under one set of controlled environmental conditions. This method provides limited information about a large number of isolates since many isolates can be grown in one experiment. The sensitivity of this general method can be greatly increased if the wild isolates are compared, not in one, but in several environments. This enables measurement, not only of the character itself, but also of the sensitivity of the isolates to changes in the environment (i.e. genotype x environmental interaction). The second method involves the sexual crossing of a pair of wild isolates and the measurement of the character in a random sample of the progeny and in the parents themselves. These data provide a large amount of information about the genetical differences between that one pair of wild isolates and estimates can be made of the contribution of genotype and environment to the variation among the progeny, the number of effective factors and their association or dispersion in the parents, and the additive and non-additive components of the

Table 1. Characters for which variation has been examined in wild isolates of *Aspergillus nidulans*.

Character		Reference
(a) Continuously variable		
General colony morphology:	growth form	1,2
	pigmentation	1,2
	conidiation	1,2
	cleistothecial density	1,2,3
Radial or linear growth rate		2,3,4,5,6
Growth yield		6
Control of cleistothecial production		7
Crossing ability		8
Penicillin titre		9,10
(b) Discretely variable		
Penicillin non-production		11
Cordycepin production		12
Enzyme electrophoretic mobility		13,14
Heterokaryon or vegetative incompatibility		1,2,15

References: 1, Grindle, 1963a; 2, Grindle, 1963b; 3, Butcher *et al.*,1972; 4, Jinks *et al.*,1966; 5, Butcher, 1969; 6, Caten & Lawrence unpublished; 7, Croft, 1966; and unpublished; 8, Butcher, 1968; 9, Holt & Macdonald, 1968a; 10, Merrick & Caten, 1975; 11, Cole *et al.*, 1977; 12, Merrick, 1973; 13, Kurzeja & Garber, 1973; 14, Moorhouse unpublished; 15, Moorhouse Dales and Croft unpublished.

of the genetical variation. In a haploid organism these non-additive effects are due solely to epistasis (Caten & Jinks, 1976).

Comparisons of radial growth rates have been made on a number of occasions (e.g. Grindle, 1963b; Jinks *et al.*,1966; Caten and Lawrence unpublished) At 35°C and growing on Czapek medium at neutral pH radial growth rates of wild isolates vary from about 4.5 to about 7 mm. per day. When grown in a range of environments the responses to the different environments may vary greatly (Butcher *et al.*,1972). In Fig. 1 it can be seen that some conditions can exaggerate the differences in growth rate between isolates (e.g. Czapek medium, 35°C, pH 5), whereas other conditions may virtually eliminate all of these differences (e.g. malt extract, 25°C, pH 7).

In addition this figure illustrates the wide variation in response to changes in the environment which may be shown by different isolates. For example, as an extreme case, a change in pH from 7 to 5 of Czapek medium at 35°C results in a marked increase of the rate of growth of isolate 43 but a marked decrease in the rate of growth of isolate 28.

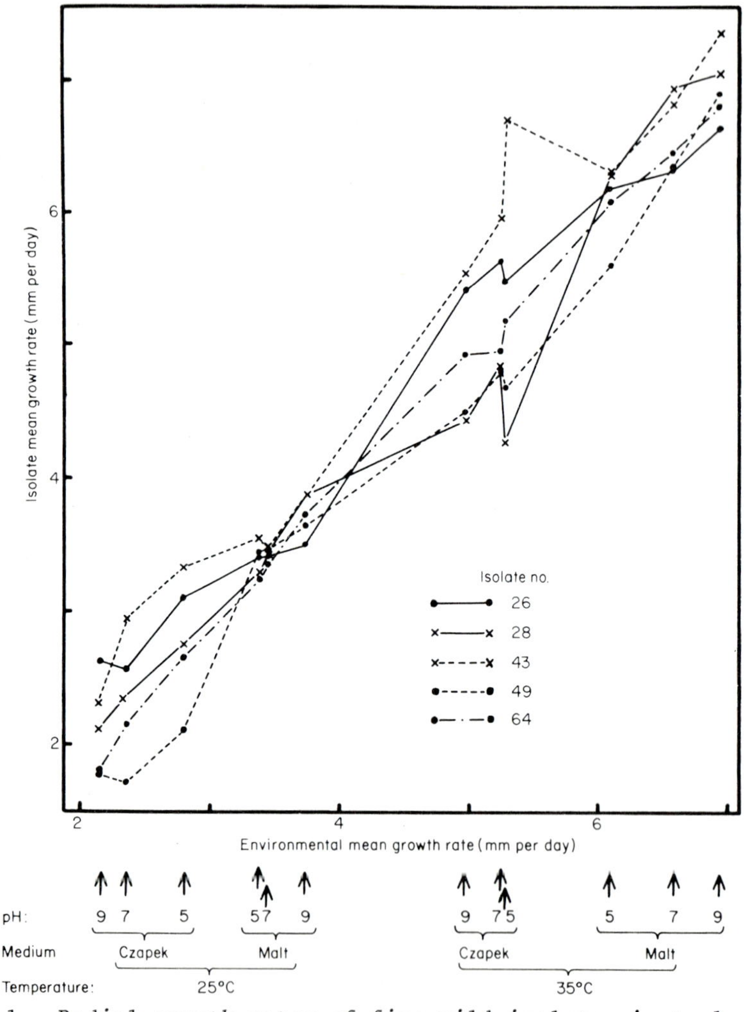

Fig. 1. Radial growth rates of five wild isolates in twelve laboratory environments. The environmental mean values were estimated from all twenty-five wild isolates which were grown in this experiment. (See Butcher *et al.*, 1972).

In order to investigate the genetical determination of growth rate the variation among the progenies of crosses between pairs of wild isolates has been analysed by Jinks *et al.*,(1966), Butcher (1969), and Caten and Lawrence (unpublished). These crosses show there to be a wide range of variability from a statistically not significant value for the genetical component (i.e. there is no detectable heritable variation present and the parental strains are near isogenic) to cases where the genetical component accounts for 85% of the total variation among the progeny. In one series of crosses (Caten and Lawrence unpublished results) the progeny means did not differ significantly from the mid-parental value, indicating the mode of action of the genes to be additive, but in another series of crosses (Jinks *et al.*,1966; Butcher, 1969) it was found that directional epistatic effects were present (Fig. 2).

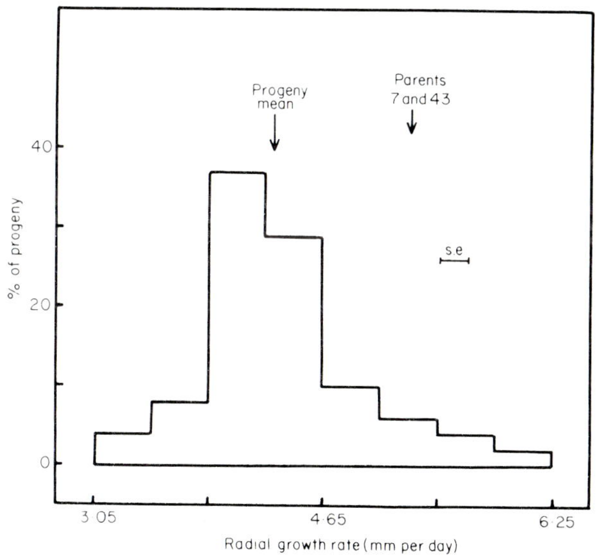

Fig.2. Example of the segregation of radial growth rate among the progeny of a cross between two wild isolates. In this case the elleles determing radial growth rate were dispersed in the two parents and the cross showed considerable epistasis. Heritability was 82.7% (From Jinks *et al.*,1966).

Progeny means corresponded to the mid-parental values only when the heritable variation in the cross was low. As the progeny variance increased the progeny mean became progressively more depressed below that of the mid-parental value thus providing evidence for genetical imbalance in these progenies.

The considerable genetical variation present in wild isolates has been clearly demonstrated in selection experiments. Selection for high and low rates of radial growth among the progeny of a cross between isolates 43 and 184 exhausted the genetical variation after only three generations and resulted in strains with growth rates of over 8 and under 5 mm per day (Fig.3). Caten and Lawrence also carried out a selection programme for a growth yield constant (measured as g dry weight per g substrate) in synthetic medium with glucose as the growth limiting substrate.

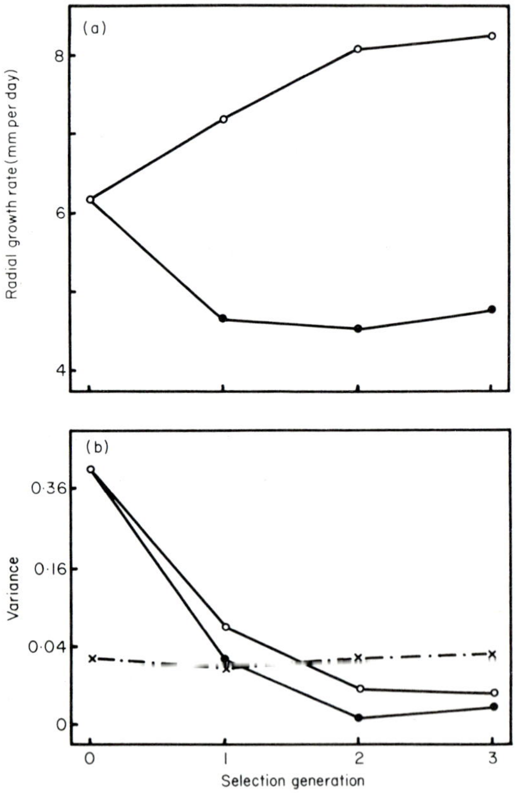

Fig. 3. (a) response of progeny mean, and (b) the values of the genetical and environmental variances during three generations of selection for high and low rates of radial growth.
0 - high slection line; ● - low selection line; x - environmental

variance. (From Caten and Jinks, 1976).

This parameter was estimated for twenty-five wild isolates and was found to vary from 0.371 to 0.559 with a mean value of 0.507 ± 0.008. Selection was carried out among the progeny of a cross which showed 65% heritability for this character and produced homogeneous populations with mean yield constants of 0.420 ± 0.005 and 0.482 ± 0.005 after two generations of selection. As is often the case in selection experiments correlated responses were found and low yield was associated with high radial growth rate, high conidiation and low cleistothecial density, with the reverse resulting from selection for high yield. These correlations suggest a physiological and genetical inter-relationship between these characters. However, it should be noted, when wild isolates are examined, that similar correlations are not necessarily present.

A second character which has been studied in detail is penicillin production. There is considerable variation among wild isolates for penicillin titre and its mode of inheritance follows a generally similar pattern to that for radial growth rate. Holt & Macdonald (1968a,b) first demonstrated the production of penicillin in this population of wild isolates and its segregation among the progeny of crosses. Merrick & Caten (1975) examined fifty wild isolates and found that forty-nine produced detectable amounts of penicillin which ranged from about 2 to 14.4 units per ml with a mean of 8.0 ± 0.4 units per ml. Crosses were made between a number of pairs of these isolates and estimates of the heritability of penicillin titre ranged from zero to 77% (Table 2).

Table 2. Inheritance of penicillin titre in seven crosses between wild isolates of *Aspergillus nidulans*. (From Merrick & Caten, 1975).

Cross	Parental Mean Titre P1	P2	Progeny Range	Progeny Mean	Heritability
109 x 65	11.8	10.4	7.4 - 21.5	14.6	0.47*
111 x 65	12.6	11.8	10.9 - 16.4	13.9	0.07
159 x 131	7.9	5.3	3.9 - 11.7	6.7	0.06
183 x 189	10.3	9.5	4.2 - 19.2	11.1	0.38*
139 x 82	9.6	4.2	0 - 18.9	6.1	0.58*
134 x 176	4.8	3.8	2.7 - 7.2	4.3	0
.54 x 176	5.1	4.9	0.1 - 17.4	3.9	0.77*

* Significantly different from 0.

'he mode of gene action was normally additive but there was evidence f epistatic effects in one cross. Programmes of selection for ncreased penicillin titre were successful and in four independent election lines increases from the initial progeny means of around

10 units per ml to values approaching 20 units per ml were made after three or four generations of selection (Merrick, 1975a,b, 1976).

Other continuously variable characters have not been investigated in as much detail as radial growth rate and penicillin titre. Butcher (1968) carried out an investigation of the genetical control of the frequency of hybrid cleistothecia in mixed cultures of wild isolates and found considerable genetical variability for this character in the population. The contribution of each of the pair of isolates to the production of hybrid fruit bodies was additive. Thus each wild isolate has a characteristic crossing potential and the frequency of hybrid cleistothecia as a proportion of the total fruit bodies present in a mixed culture can be predicted from the sum of the crossing potentials of each of the parental strains.

A number of aspects of the production of cleistothecia are under genetical control. The density of cleistothecia produced varies from one isolate to another (Grindle, 1963a,b; Butcher *et al.*, 1972), and many isolates segregate among their asexual or selfed sexual progenies for the ability or inability to produce cleistothecia (Croft, 1966). This segregation is reciprocal in the sense that colonies without cleistothecia can give rise to colonies with cleistothecia and *vice versa* and a cytoplasmic factor is involved in the difference between the two types of colony. In general, selfed ascospores produce the highest proportion of colonies with cleistothecia that is possible in the case of any particular isolate. This value is similar to the proportion produced by conidia found on cleistothecial colonies. However, conidia found on colonies without cleistothecia may give rise to a high, low or intermediate proportion of colonies with fruit bodies the actual value being characteristic for each isolate (Fig.4). Progeny from sexual crosses clearly demonstrate the genetical control of this instability. For example, the segregations among colonies grown from conidia taken from cleistothecial and non-cleistothecial colonies derived from single ascospores of the cross between isolates 37 and 114 show heritability values of 83% and 87% respectively (see Fig.5 for details). In the case of the non-cleistothecial colonies (Figure 5b) the bimodal distribution suggests that one of the factors segregating is having a large effect.

The characters which have been described above are all continuously variable and under the control of many genes. Though nothing can be inferred about the properties of the individual genes, estimates of their number, or more strictly of the number of effective factors, can be made (Chovnick & Fox, 1953; Croft & Simchen, 1965, Mather & Jinks, 1971). Mean estimates of nine (ranging from three to seventeen), eight (five to fifteen), & five have been made for radial growth rate, penicillin titre and growth yield respectively (Caten & Jinks, 1976). However, in the case of the characters which are discretely variable the individual loci

Fig. 4. Distributions of cleiostothecial cover on colonies grown from conidia taken from (a) colonies without cleistothecia and (b) colonies with cleistothecia in three wild isolates.

responsible for their control can either be inferred or have been demonstrated. These characters are cordycepin production (Merrick, 1973), penicillin non-production under the control of the *npeA* gene which has been located on linkage group IV (Cole *et al.*, 1977; Holt *et al.*, 1976) and electrophoretically detectable enzyme variation (Kurzeja & Garber, 1973; Moorhouse unpublished). Moorhouse studied eight enzyme systems and found polymorphism for six of these. Examination of progenies of crosses clearly demonstrated the genetical basis of some of the variation in electrophoretic mobility in the case of phosphatases and esterases. The final discretely variable character is heterokaryon incompatibility and this will be considered in detail.

THE HETEROKARYON INCOMPATIBILITY SYSTEM

Grindle (1963a,b) first detected the presence of the heterokaryon incompatibility system in *A.nidulans*. By growing one wild

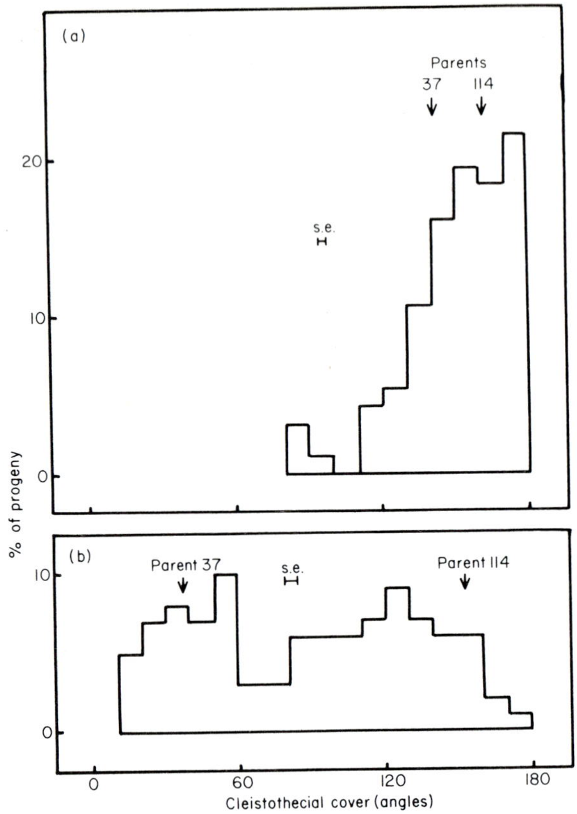

Fig. 5. Segregation of degree of cleistothecial cover among the progeny of a cross between two wild isolates. A sample of single ascospore colonies from the cross was first obtained. Each was then grown by spreading conidia so that colonies with cleistothecia and colonies without cleistothecia were obtained for each single ascospore strain. Colonies taken from (a) colonies with cleistothecia and (b) colonies without cleistothecia were then spread and the proportion of the surface area of the resultant colonies (in quarter colonies) was estimated. The data was transformed into angles for analysis.

isolate in a mixed culture with a conidial colour mutant of another wild isolate on a medium which promoted conidiation and searching the culture visually for the presence or absence of heterokaryotic conidial heads, pairs of isolates were scored for their ability or inability to form heterokaryons with each other. Both strains used were prototrophic and thus no forcing of heterokaryosis was involved

(Caten & Jinks, 1966). Using this test it was found that wild isolates could be divided into a number of heterokaryon compatibility (h-c) groups so that all members of any one group readily form heterokaryons with each other but not with members of any other group. A similar situation is found in other species of Aspergillus, such as *A.niger* (Gossop *et al.*,1940), *A.fonsecaeus* (Raper & Fennell, 1953), *A.heterothallicus* (Kwon & Raper,1967),*A.terreus* and *A. versicolor* (Caten, 1971), and members of the *A.glaucus* group (Jones, 1965), and heterokaryon incompatibility is known in other Ascomycetes, particularly Podospora and Neurospora (see Esser & Blaich, 1973 for review). In *A.nidulans* one hundred of the one hundred and fifty wild isolates currently available fall into 19 h-c groups, none of which are compatible with the Glasgow strains. Some of these groups have only a low number of independent members and little can be said about their geographical distribution. However, many of the groups have a larger number of members and the individuals within each of these groups were isolated from very widespread areas within England and Wales (Fig. 6). Members of more than one group are usually found in any one area, and often from a single soil sample. The distribution of the members of the small groups is often widespread. For example, the two wild isolates contained in h-c group G originated from Durham and Cornwall.

Despite the heterokaryon incompatibility system and the homothallic sexual breeding system of *A.nidulans* it has been found possible to carry out sexual crosses between most pairs of wild isolates (Butcher, 1968). There is considerable evidence available to suggest that in a cross, one of the parents acts as a 'protocleistothecial' parent (Apirion, 1963) and the demonstration of the maternal inheritance of a mitochondrially located marker (Rowlands & Turner, 1977) strongly suggests that sexual crossing is not dependent upon vegetative heterokaryon formation, even though the formation of a balanced heterokaryon between compatible pairs of strains may increase the frequency of hybrid fruit bodies to the extent that the phenomenon of 'relative heterothallism' has been reported (Pontecorvo *et al.*,1953).

Jinks & Grindle (1963) demonstrated that heterokaryon incompatibility was under nuclear control and the segregation of incompatibility among the progenies of between h-c group crosses has shown that the control is heterogenic, that is incompatibility results from heterozygosity at one or more of a series of incompatibility or *het* loci. Statistical estimates of the number of *het* genes segregating in some crosses between h-c groups have indicated that a large number of genes are involved (Jinks *et al.*,1966) and detailed analysis of crosses, including complex backcrossing schemes, have confirmed this. It has now been demonstrated that members of h-c groups R and B, and R and Q differ at three and eight *het* loci (Caten *et al.*,1972) and groups A and B, A and Glasgow, and B

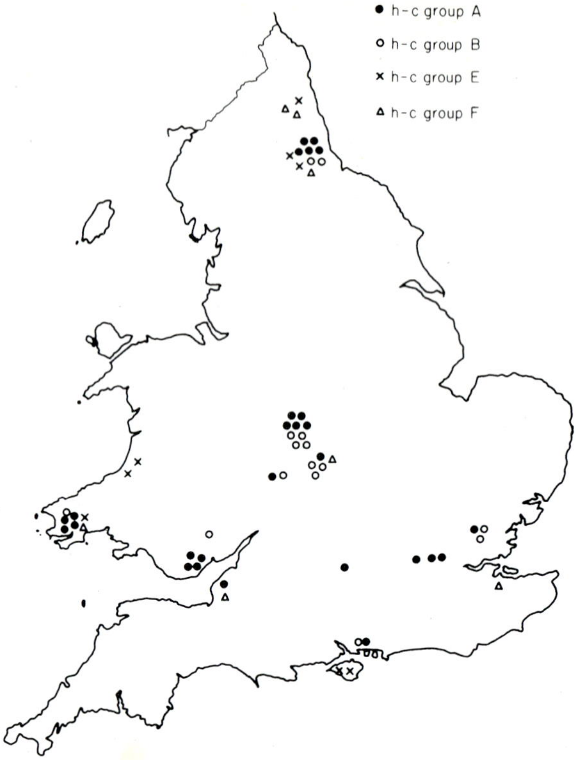

Fig. 6. Localities of origin of members of four h-c groups.

and Glasgow differ at seven, six and two *het* loci respectively (Moorhouse unpublished results). Consideration of the latter three combinations leads to the conclusion that one of the genes involved in this triangle of crosses must be heterozygous in all three crosses with a different allele in each isolate though these have not yet been fully identified. Thus there is evidence for the existence of multiple allelic series at *het* loci.

The only detailed study of *het* genes in *A.nidulans* at present is of the two genes which segregate in crosses between Glasgow strains and h-c group B isolates. These have been designated *het*A and *het*B, with Glasgow strains being *het*A1 *het*B1 and group B strains *het*A2 *het*B2. The recombinant classes, *het*A1 *het*B2 and *het*A2 *het*B1, segregate in equal frequencies with the parental classes in crosses, showing the two loci to be unlinked. When the forcing of heterokaryons between pairs of strains carrying the above *het* gene combinations and complementing auxotrophic markers on minimal medium is attempted normal heterokaryons are formed only when both strains carry identical alleles at both *het* loci. Combinations heterozygous at one or both loci produce only slow growing red-brown pigmented cultures. Repeated transfers from such cultures have in some cases led to the isolation of colonies with a sectoring mode of growth similar to that of normal heterokaryons but with a red-brown pigmentation in the thin mycelium and with very few conidial heads. However, the occasional observation of heterokaryotic conidial heads and the isolation of parental strains from these colonies leave little doubt that they are vegetative heterokaryons (Moorhouse and Croft unpublished). Recently using protoplast fusion techniques (Ferenczy *et al.*,1975; Anné & Peberdy, 1976) it has proven possible regularly to produce heterokaryons heterozygous at a range of *het*loci (Dales and Croft unpublished results). The morphology of heterokaryons obtained in this way has varied from almost normal to extremely poorly growing dependent on the combinations of *het* genes involved. For example, a heterokaryon heterozygous for *het*B is severely restricted whereas one heterozygous for *het*A is more nearly normal. In addition, variation in the genetical background may be involved in determining the expression of *het* genes. No specific effect of any individual *het* gene upon crossing ability has been detected in *A.nidulans* but in *A.amstelodami* one gene which reduces the frequency of hybrid fruit bodies has been noted (Caten, 1973).

Soft overlay platings of the few conidial heads formed by heterokaryons heterozygous at *het*A or *het*B separately, and at both loci together, have yielded diploid colonies at about the normal frequency These heterozygous diploids are also abnormal in morphology, tending to have a thin, red pigmented mycelium with few, but large diameter conidia. When grown on a nutritionally supplemented medium containing benlate (Hastie, 1970) haploids with normal morphologies can be isolated. Analysis of a sample of these haploids derived from a diploid heterozygous at both *het*A and *het*B, and one of the parents of which was a multiply marked translocation free tester strain, have clearly located *het*A and *het*B on linkage groups V and VI respectively (Table 3).

NATURAL VARIATION AND HETEROKARYON INCOMPATIBILITY

When the variation between wild isolates discussed in the first section is examined in relation to the heterokaryon incompatibility system it is clear that, for all characters, there is more variation between than within h-c groups. Grindle (1963b) noted that members of one h-c group tended to have a similar general morphology and he also demonstrated intra-group similarity and inter-group differences in radial growth rate. The analysis of progeny variances of intra- and inter-group crosses illustrates this clearly (Jinks *et al.*, 1966). All inter-group crosses produced highly significant variation among their progenies with a genetical component accounting for over 70% of the progeny variance, but two intra-group crosses gave no significant genetical variation in one case and a significant genetical component accounting for only 45% of the total variation in the other. It is of interest that both parents of the first of these intra-group crosses were isolated from the same locality and the parents of the second cross came from widely separated localities. However, this choice of parents may have been no more than a chance occurrence.

Table 3. Segregation of heterokaryon incompatibility genes, *het*A and *het*B, among haploid segregants from the diploid between Glasgow tester strain MSD and a h-c group B strain carrying the markers *paba*A1.3 and *Acr*A1. The diploid was homozygous for *Acr*A1 so there is no test for location on linkage group II. The allele, *paba*A1.3 was induced in wild isolate 1. It showed complete linkage to *y*A.

Linkage Group	Comparison	Haploid Segregants Parental		Recombinant		% Recombinants
I	yA : hetA	23	22	24	28	53.6
	ya : hetB	28	28	19	22	42.3
	pabaA : hetA	22	23	28	24	53.6
	pabaA : hetB	28	28	22	19	42.3
III	phenA : hetA	37	12	34	14	49.5
	phenA : hetB	33	9	38	17	56.7
IV	pyroA : hetA	25	22	24	26	51.5
	pyroA : hetB	27	25	22	23	46.4
V	lysB : hetA	51	45	1	0	1.0
	lysB : hetB	25	20	27	25	53.6
VI	sB : hetA	24	21	25	27	53.6
	sB : hetB	49	47	0	1	1.0
VII	nicB : hetA	21	15	31	30	62.9
	nicB : hetB	29	24	23	21	45.4
VIII	riboB : hetA	24	18	28	27	56.7
	riboB : hetB	27	22	25	23	49.5
	hetA : hetB	21	25	25	26	52.6

Butcher (1968) showed there to be much more variation between than within h-c groups for crossing ability, and Merrick & Caten (1975) and Merrick (1976) came to a similar conclusion for penicillin titre. Thirty-one isolates which belonged to known h-c groups were assayed, and of the genetical variation between isolates 67% was due to differences between h-c groups. Of the crosses which were carried out only those between parents belonging to different h-c groups gave significant genetical variation among their progenies. (Table 2).

Other characters which show this relationship to h-c groups are cordycepin production which is known only in members of h-c group B (Merrick, 1973), and penicillin non-production in members of h-c groups F and G (Holt & Macdonald, 1968a; Merrick & Caten, 1975). The latter character is due to mutation at the *npe*A locus (Cole *et al.*,1977) and, although it has not yet been demonstrated that all members of either of these groups carry the same allele, there is evidence that the mutant alleles involved in the two groups are different. The allele carried by a group G isolate is leaky, and leakiness segregates as a single gene in a cross between this isolate and a member of h-c group F. Analysis of the variation in electrophoretic mobility found for certain enzymes illustrates the tendency for members of an h-c group to have similar banding patterns but it also reveals that there is more variation within groups than might have been expected from the examination of the other characters (Moorhouse unpublished results).

The most striking demonstration that members of any one h-c group are more closely related to each other than to members of different h-c groups has come from the analysis of the response of isolates to different environments (Butcher *et al.*,1972). A biometrical model was fitted to data obtained for radial growth rate and cleistothecial density in twelve laboratory environments for twenty-five isolates belonging to five h-c groups. Variation for these characters was partitioned into environmental, additive genetical, and linear and non-linear genotype x environmental interaction components. The variation due to the additive genetical and the genotype x environmental interaction effects was mainly due to differences between h-c groups rather than to variation within groups, but the method was also very sensitive at detecting individual isolates which varied within a group. In Fig.7 the additive genetical effects and the linearly related genotype x environmental interactions for cleistothecial density are compared and the clustering of members of each h-c group is apparent. In addition, the comparison clearly shows that isolate 29 in h-c group A and isolates 133 and 147 in h-c group H deviate considerably from other members of their respective groups. This may mean that these isolates are unrelated to the other members of their groups, but although isolate 29 deviates for the second character, growth rate, examination of the plots of mean response of each isolate against each overall environmental mean indicates that the close relationship of all members of any one h-c group is maintained (Fig.8).

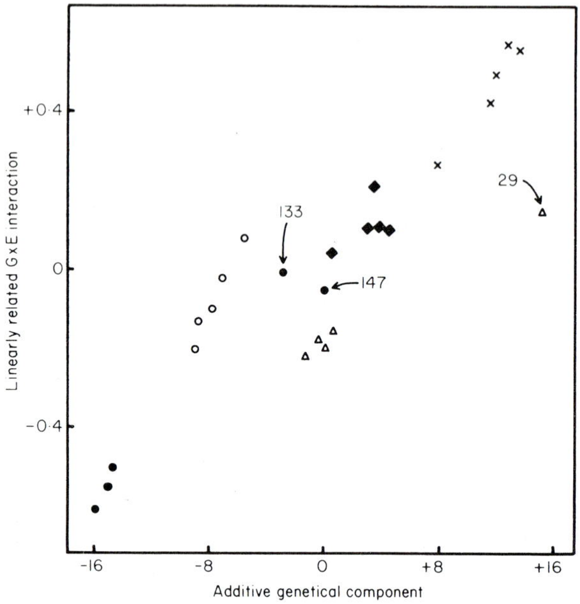

Fig. 7. Plot of linearly related genotype x environmental interaction and additive genetical component for twenty-five isolates belonging to five h-c groups grown in twelve laboratory environments. (From Butcher *et al.*,1972).

Despite the various evidence for some variation within h-c groups the only conclusion which can be drawn is that all members of any one h-c group so far isolated from nature are clonally related to each other. Jinks *et al.*,(1966) concluded that the ability of pairs of isolates to form heterokaryons is determined by the degree of overall similarity of their respective genomes. Now that specific *het* genes are known to be involved in heterokaryon incompatibility it is important to be sure that the variation in the other characters is not due to pleiotropic effects of these same genes. This has been shown clearly for radial growth rate where a pair of compatible strains, both with the genotype *het*A1 *het*B1, were derived from a cross between a Glasgow strain and an h-c group B wild isolate. These were then crossed and their progeny displayed significant

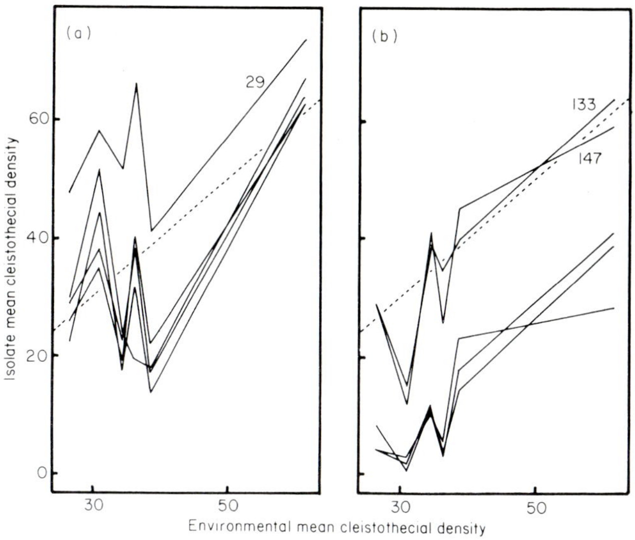

Fig. 8. Cleistothecial densities (number of cleistothecin per 7 mm diameter circle) of ten wild isolates grown in six laboratory environments. The environmental mean values were estimated from all twenty-five wild isolates which were grown in this experiment. (a) h-c group A. (b) h-c Group H. The diagonal broken lines represent the overall isolate means plotted against the environmental means. (From Butcher *et al.*,1972).

genetical variation. Selection among these progeny rapidly produced lines with high and low rates of growth respectively (Fig. 9) showing the *het* genes to be independent of those controlling radial growth rate.

DISCUSSION

In common with all other species which have been adequately examined (see, for example, Lewontin, 1973; Mather, 1973), much variation and polymorphism is found in the natural population of *A.nidulans*. Although other species of Aspergillus have not been studied in as much detail it is clear that they also show much variation, not only in other sexual species such as *A.amstelodami*, but also in asexual species such as *A.versicolor* or *A.terreus*. Only a proportion of the potential genetical variation is visible in the wild isolates themselves. Much is hidden away in balanced and dispersed combinations and is only displayed when crosses are carried out between wild isolates in the laboratory. Indeed, the progeny

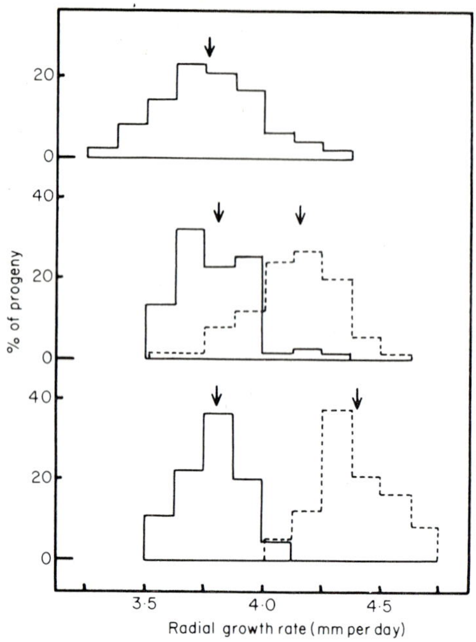

Fig. 9. Selection for high and low rates of radial growth within a single heterokaryon compatibility genotype. The arrows represent the progeny means. Each generation was grown at a separate time, so only within generation comparisons can be made.

of a single cross often display more variation than is actually seen among wild isolates, and selection among such a progeny will maximise the genetical differences which are present by responding to selection for both extremes. In general it may be concluded that, in nature, balanced intermediate phenotypes are favoured.

The evidence presented in this account leads to the conclusion that the wild population of *A.nidulans* is divided into a number of independent sub-populations and that each sub-population is a clonally related group of strains in which evolution may be proceeding independently with little gene exchange between them. Thus each h-c group may be evolving into one of a group of sibling species (Mather, 1965) and it is of interest to note that the taxonomic classification of Aspergillus presently orders the various species into groups of sibling species (Raper & Fennell, 1965).

The importance of heterokaryon incompatibility in maintaining this division is not clear, but it will be determined by the relative importance of sexual, parasexual and asexual reproduction in the natural population. At the present time there is not much information available concerning the ecology and natural reproductive biology of of moulds such as *A.nidulans*. However, the fact that *A.nidulans* is closely related to many successful and variable imperfect species suggests that asexual reproduction may be the norm and that the sexual system may be a relict feature. This is further supported by the observation that members of any one h-c group are dispersed widely throughout England and Wales and also there is some evidence that the genomes of different h-c groups are unbalanced when recombined. The presence of the heterokaryon incompatibility system and its mode of action would appear to rule out the possibility that parasexuality is of importance as a mechanism for recombination between h-c groups, though it could act as a means for the recombination of new genetical variation which is accumulating within an h-c group.

Little is known about the selective forces which maintain the variation and polymorphism found in the population. It is possible to speculate that the different forms occupy different ecological niches, or that cyclical advantages accompanied by balancing disadvantages may occur due, say, to seasonal fluctuations or the sequence of saprophytic decay, but no firm statement can be made at present. One explanation often put forward in the case of diploid organisms, namely that of heterozygous advantage, clearly cannot be operating in a haploid organism. The extent to which enzyme polymorphisms have been found within h-c groups appears, at this stage, to be large in comparison with the other characters. This general problem has been discussed at length elsewhere (for example, Kimura & Ohta, 1971) but it is tempting to speculate that these enzyme polymorphisms constitute a class of variation which may be selectively neutral or at least which come under selective forces different to those operating on the other characters. Finally, there is evidence from *A.amstelodami* that the heterokaryon incompatibility system is acting as a barrier to infection by the invasive cytoplasmic character 'vegetative death' (Caten, 1972). It is thus possible that incompatibility may be important as a cellular defence mechanism by preventing the spread of viruses and other suppressive cytoplasmic determinants from cell to cell in nature.

The model of the *A.nidulans* population outlined above was derived from the examination of a sample of the population taken at one point in time. From the information available it is not possible to decide whether the proposed structure of the British population arose at a point now distant in time and that all evolution since then has taken place by mutation and selection and possibly by parasexual recombination *within* each isolated sub-population, or whether a 'steady state' exists where the generation of new variation may also be occurring continually by occasional sexual crossing or by the introduction of new material from the world population. There is clearly a need to resample the British population and also to look at the structure of the world population in the terms described in this account.

Acknowledgements:

The authors wish to thank C.E.Caten, R.B.G.Dales, Jean Lawrence, and J.Moorhouse for permission to use their unpublished results.

REFERENCES

ANNE,J. & PEBERDY,J.F. (1976). Induced fusion of fungal protoplasts following treatment with polyethylene glycol. *Journal of General Microbiology* 92, 413-7.

APIRION,D. (1963). Formal and physiological genetics of ascospore colour in *Aspergillus nidulans*. *Genetical Research* 4, 276-83.

BUTCHER,A.C. (1968). The relationship between sexual outcrossing and heterokaryon incompatibility in *Aspergillus nidulans*. *Heredity, London* 23, 443-52.

BUTCHER,A.C. (1969). Non-allelic interactions and genetic isolation in wild populations of *Aspergillus nidulans*. *Heredity, London* 24, 621-31.

BUTCHER,A.C., CROFT,J.H. & GRINDLE,M. (1972). Use of genotype-environmental interaction analysis in the study of natural populations of *Aspergillus nidulans*. *Heredity, London* 29, 263-83.

CATEN,C.E. (1971). Heterokaryon incompatibility in imperfect species of *Aspergillus*. *Heredity, London* 26, 299-312.

CATEN,C.E. (1972). Vegetative incompatibility and cytoplasmic infection in fungi. *Journal of General Microbiology* 72, 221-9.

CATEN,C.E. (1973). Genetic control of heterokaryon incompatibility and its effects on crossing and cytoplasmic exchange in *Aspergillus amstelodami*. *Genetics, Princeton* 74, s40.

CATEN,C.E. & JINKS,J.L. (1966). Heterokaryosis: its significance in wild homothallic Ascomycetes and Fungi Imperfecti. *Transactions of the British Mycological Society* 49, 81-93.

CATEN,C.E. & JINKS,J.L. (1976). Quantitative genetics. In *Second International Symposium on the Genetics of Industrial Microorganisms*, ed. K.D. Macdonald. pp. 93-111. Academic Press: London.

CATEN,C.E., BUTCHER,A.C. & CROFT,J.H. (1972). Genetical control of heterokaryon formation in *Aspergillus*. *Microbial Genetics Bulletin* 34, 16.

CHOVNICK,A. & FOX,A.S. (1953). The problem of estimating the number of loci determining quantitative variation in haploid organisms. *American Naturalist* 87, 263-7.

CLUTTERBUCK,A.J. (1974). *Aspergillus nidulans*. In *Handbook of Genetics*. ed. R.C. King. Vol.1, pp.447-510. Plenum Press:New York.

COLE,D.S., HOLT,G. & MACDONALD,K.D.(1977). Relationship of the genetic determination of impaired penicillin production in naturally occurring strains to that in induced mutants of *Aspergillus nidulans*. *Journal of General Microbiology*. in press.

CROFT,J.H. (1966). A reciprocal phenotypic instability affecting development in *Aspergillus nidulans*. *Heredity, London* 21, 565-79.

CROFT,J.H. & SIMCHEN, G. (1965). Natural variation among monokaryons of *Collybia velutipes*. *American Naturalist* 99, 451-62.

ESSER,K. & BLAICH,R. (1973). Heterogenic incompatibility in plants and animals. *Advances in Genetics* 17, 107-52.

FERENCZY,L., KEVEI,F. & SZEGEDI,M. (1975). High frequency fusion of fungal protoplasts. *Experientia* 31, 1028-30.

GOSSOP,G.H., YUILL,E. & YUILL,J.L. (1940). Heterogeneous fructifications in species of *Aspergillus*. *Transactions of the British Mycological Society* 24, 337-44.

GRINDLE,M. (1963a). Heterokaryon compatibility of unrelated strains in the *Aspergillus nidulans* group. *Heredity, London* 18, 191-204.

GRINDLE,M. (1963b). Heterokaryon compatibility of closely related wild isolates of *Aspergillus nidulans*. *Heredity, London* 18,397-405.

HASTIE,A.C. (1970). Benlate induced instability of *Aspergillus* diploids. *Nature, London* 226, 771.

HOLT,G. & MACDONALD,K.D. (1968a). Penicillin production and its mode of inheritance in *Aspergillus nidulans*. *Antonie van Leeuwenhoek* 34, 409-16.

HOLT,G. & MACDONALD,K.D. (1968b). Isolation of strains with increased penicillin yield after hybridization in *Aspergillus nidulans*. *Nature, London* 219, 636-7.

HOLT,G., EDWARDS,G.F.St.L. & MACDONALD,K.D. (1976). The genetics of mutants impaired in the biosynthesis of penicillin. In *Second International Symposium on the Genetics of Industrial Microorganisms*. ed. K.D. Macdonald, pp. 199-211.

JINKS,J.L. & GRINDLE,M. (1963). The genetical basis of heterokaryon incompatibility in *Aspergillus nidulans*. *Heredity, London* 18, 407-11.

JINKS, J.L.,CATEN,C.E., SIMCHEN,G. & CROFT,H.J. (1966). Heterokaryon incompatibility and variation in wild populations of *Aspergillus nidulans*. *Heredity, London* 21, 227-39.

JONES, D.A. (1965). Heterokaryon compatibility in the *Aspergillus glaucus* Link group. *Heredity, London* 20, 49-56.

KIMURA,M. & OHTA,T. (1971). Protein polymorphism as a phase of molecular evolution. *Nature, London* 229, 467-9.

KURZEJA,K.C. & GARBER,E.D. (1973). A genetic study of electrophoretically variant extracellular amylolytic enzymes of wild-type strains of *Aspergillus nidulans*. *Canadian Journal of Genetics and Cytology* 15, 275-87.

KWON,K.J. & RAPER,K.B. (1967). Heterokaryon formation and genetic analyses of color mutants in *Aspergillus heterothallicus*. *American Journal of Botany* 54, 49-60.

LEWONTIN,R.C. (1973). Population genetics. *Annual Review of Genetics* 7, 1-17.

MATHER,K. (1965). The genetical interest of incompatibility in fungi. In *Incompatibility in Fungi* ed. K. Esser and J.R. Raper, pp. 113-117. Springer Verlag, Berlin.

MATHER, K. (1973). *Genetical structure of populations.* Chapman and Hall: London.

MATHER,K. & JINKS,J.L. (1971). *Biometrical genetics.* 2nd Edn. Chapman and Hall: London.

MERRICK,M.J. (1973). Quantitative genetics of antibiotic production in *Aspergillus nidulans. Ph.D Thesis. University of Birmingham.*

MERRICK, M.J. (1975a). Hybridization and selection for increased penicillin titre in wild type isolates of *Aspergillus nidulans. Journal of General Microbiology* 91, 278-86.

MERRICK,M.J. (1975b). The inheritance of penicillin titre in crosses between lines of *Aspergillis nidulans* selected for increased productivity. *Journal of General Microbiology* 91, 287-94.

MERRICK,M.J. (1976). Hybridization and selection for penicillin production in *Aspergillus nidulans* - a biometrical approach to strain improvement. In *Second International Symposium on the Genetics of Industrial Micro-organisms.* ed. K.D. Macdonald, pp. 229-242.

MERRICK,M.J. & CATEN,C.E. (1975). The inheritance of penicillin titre in wild-type isolates of *Aspergillus nidulans. Journal of General Microbiology* 86, 283-93

PONTECORVO,G., ROPER,J.A., HEMMONS,L.M., MACDONALD,K.D. & BUFTON,A.W.J (1953). The genetics of *Aspergillus nidulans. Advances in Genetics* 5, 141-238.

RAPER,K.B. & FENNELL,D.L. (1953). Heterokaryosis in *Aspergillus. Journal of Elisha Mitchell Scientific Society* 69, 1-29.

RAPER,K.B. & FENNELL,D.L. (1965). *The genus Aspergillus.* Williams and Wilkins: Baltimore.

ROWLANDS,R.T. & TURNER,G. (1977). Maternal inheritance of extra-nuclear mitochondrial markers in *Aspergillus nidulans.* Submitted for publication.

THOM,C. & RAPER,K.B. (1945). *A Manual of the Aspergilli.* Williams and Wilkins, Baltimore.

YUILL,E. (1939). Two new *Aspergillus* mutants. *Journal of Botany, London* 77, 174-5.

YUILL,E. (1950). The numbers of nuclei in conidia of Aspergilli. *Transactions of the British Mycological Society* 33, 324-31.

Fermenter Cultivation of *Aspergillus*

R. COCKER* and R. N. GREENSHIELDS

Department of Applied Microbiology, University of Strathclyde, Glasgow, U.K.

Department of Biological Sciences, University of Aston in Birmingham, Birmingham, U.K.

INTRODUCTION

Considerations of the cultivation of fungi have often given rise to some surprising misconceptions. One of these, recently highlighted by Miall (1975) was the idea that the mould fermentation industry arrived with the advent of antibiotics, whereas the cultivation of moulds in general and *Aspergillus* spp. in particular have been part of man's industry for centuries. The Japanese 'Koji' process for example, which provides the basis for the manufacture of saké, is recorded to have been carried out since at least the eighth century. Here *Aspergillus oryzae* is cultivated to convert cereal preparations prior to their alcoholic fermentation. Likewise, another early mould fermentation is that of the preparation of gallic acid for tanning and printing, utilising *Aspergillus niger* to hydrolyse the ester form found in gall nuts.

Charles Wiezmann is credited with the concept of the deliberate use of microorganisms for the production of useful biochemicals rather than their use following accidental discovery of their end-product properties. However, one might question this by examining Currie's work in 1917 (Miall, 1975) when he showed that Wehmer was wrong in assuming that the natural product of *Aspergillus niger* was oxalic acid and that citric acid was only characteristic of *Citromyces* species. The result was the basis of today's citric acid industry

* *Now at Glaxo Laboratories Limited, Ulverston, Cumbria, U.K.*

where specialised strains of *Aspergillus spp.* are exhorted to prodigious feats of mono-end-product conversion in a most deliberate fashion, almost to their destruction.

Defining the title of this chapter, "The fermenter cultivation of *Aspergillus*"can also give rise to misconceptions. To many mycologists, a fungus is a virtually two-dimensional subject particularly when grown on agar plates but to the microbial biochemist or the fermentation scientist who is interested in the mould for what it is or does, it is three-dimensional since he grows it in submerged culture, constrained by a container or fermenter.

It is only recently that cultivation of fungi has been of importance in the production of biomass, particularly for its protein content. Cultivation in the fermentation industry has often meant some if not a serious limitation of biomass potential by physical, genetic or chemical manipulation.

The catholic definition of the term fermenter is the convenient vessel or containment system limiting the environment of the micro-organism for physical, chemical and biological reasons; cultivation is the growth of the micro-organism within these constraints and fermentation is the biochemical activity of the micro-organism produced by the constraints on the growth within the fermenter. The fermenter cultivation of *Aspergillus* therefore includes all fermentations for all purposes of study and of industry, although the true cultivation of biomass is becoming industrially important. A wide variety of factors can be considered but can conveniently be grouped by considering the following scheme.

INOCULUM

In industry fungal spores have until recently been considered as a dormant stage in the life cycle of fungi but they have been shown to contain enzymes necessary for a variety of metabolic activities (Vezina & Singh, 1975). Not only can they synthezise proteins, but they are able to transform fatty acids and steroids, carbohydrates and flavanoids and possibly even hydrocarbons and alkaloids. It is not surprising to find that spore germination is dependent on the fungal genetic make-up and is sensitive to the formenter conditions, both physical and chemical. Spore inocula are usually the best way of starting fermentations for experimental or industrial processes but run into problems on scale-up. These are the problems of their production in quantity, clumping when in excess, the apparent dormancy (the lag phase) due to the biochemical change from nutrient exhaustion (particularly nitrogen-limitation) to full growth medium, and purity and genetic changes. Inoculation with spores is by suspension in liquids containing detergents but this may also provide problems, changing the chemical environment and in some fermenter configurations, causing froth-flotation problems. Scale-up often relies on mycelial inocula and here it is important to maintain adequate growth to avoid biochemical and physiological effects attributable to senescence.

BIOLOGICAL FACTORS

Growth

To a large extent the processes of growth involve non-reversable, time-dependent processes. Once spores have swollen, polarity ensues and one or two wall areas grow out to form hyphal filaments. In surface culture most filamentous fungi including *Aspergillus* species develop aerial, surface and submerged hyphae, whereas in submerged culture three-dimensional hyphalgrowth results in a variety of morphological forms. Factors which influence stages in this development are, therefore, irrevocable once the moment of effectiveness has passed.

Colony formation

Under the conditions present in bubble-column fermenters (Cocker, 1975) colonies have been observed to form in a variety of ways. The aggregation of swollen spores or of newly-germinated spores gave rise to relatively large colony nuclei. Adhesion of ungerminated spores to germ-tubes also took place and the result was a tendency to reduce the overall concentration of potential colonies in the medium. However, spore-agglutination was not a prerequisite for colony formation under these conditions and colonies were observed to form from single spores and from short filaments. Filaments which had fragmented from larger colonies also had the ability to form new colonies, for instance, in continuous culture. Less frequently observed was the tendency for larger aggregates to entrain hyphae at the germ-tube stage. The above processes are represented diagramatically in Fig.1.

These results appear to be at variance with the observations of other workers (Burkholder & Sinnot, 1945; Takahashi & Yamada 1960; Takahashi, Hidaka & Yamada, 1965) where colony formation was apparently encouraged by the initial clumping of spores. There is evidence that the tendency of spores to clump together at the stage where they are swollen is a function of the frequency of collision between spores since the superficial gas velocity in bubble-column fermenters can be used to modify the degree of clumping (Daunter, 1971). Increasing the inoculum size also encourages clumping (Morris, 1973), and thus it seems that the more random agglomeration-patterns found in these fermenters are the result of conditions which minimise spore-collisions. The observation that single spores or hyphal fragments can form colonies demonstrates that clumping is not necessary for colony-formation. The introduction of spores into a static environment, for example a broth-culture in a test-tube allows the formation of perfectly spherical colonies, and it would seem that the contrast between the above behaviour and that found for cultures in stirred fermenters depends on the absence or presence of agitation.

It has been found (Yanagita, 1957; Bhatnagar & Krishnan, 1960a, b,) that the processes of spore-swelling and germination are quite complex, and depend on both physical and biochemical reactions. Factors which inhibit metabolism at the germination stage allow a greater time for mixed patterns of agglutination to take place, resulting in a more random distribution of colony-size, colony-shape, and in a smaller overall number of colonies. An inverse relationship between colony-size and colony-concentration is easily demonstrated by replacing aliquots of culture with fresh medium under otherwise standard conditions. Colonies up to 25mm in diameter were produced using this technique, but a reduction in the size of the spore-inoculum was not successful in increasing colony-size, apparently because the concentration of embryo colonies is governed more by the extent of agglutination than by the concentration of spores in the inoculum (Cocker, 1975).

Aggregation to form colonies may also be seen to occur when organisms other than filamentous fungi are cultured in fermenters, but these do not possess the distinctive radial symmetry of the fungal types (Cocker, 1975). The mechanism by which this symmetry is maintained is elusive, but it probably depends on interactions between hyphae which are designed to prevent undue competition for nutrients between hyphae. The detailed study of fungal morphology during submerged culture is subject to considerable difficulties of description and quantitation. Mainly because of a lack of adequate methods, many workers have contented themselves with subjective verbal descriptions such as 'flocculent', 'pelleted', 'granular', and 'loose pellets' (Whitaker & Long, 1973).

For certain morphologies, notably the purely filamentous form on the one hand, and the essentially spherical form on the other, measurements of various hyphal and colonial dimensions are possible, including mean internode length (Trinci, 1970), hyphal growth unit (Trinci, 1973), and radial growth-rate and peripheral hyphal density (Trinci, 1969). An alternative approach to the problem of describing morphology is that of recording pictorially the colonial forms which arise. This approach may have advantages in that it is indiscriminate and records details of the morphological development which may not be considered important until later. Photography, autoradiography or even line-drawings can provide faithful methods of description and communication. It is interesting to note that workers using such methods (e.g. Burkholder & Sinnott, 1945; Duckworth & Harris, 1949; Clark, 1962; Bent & Morton, 1963;

Yanagita & Kogane, 1963a) often avoid the descriptive and conceptual limitations of verbal descriptions and give equal emphasis to forms of growth other than 'filamentous' and 'pelleted'. The problem with pictorial representations is that they are not in themselves quantitative, although they may be used to form ranking-systems. Cocker & Greenshields (1975) have argued that the use of verbal descriptions is ambiguous, and preferred to use such terms as 'floc' or 'colony', using a ranking-system to qualify this description. Thus, seven 'types' were used as reference-points in a continuous range of morphological forms and were defined by reference to photographs (Cocker, 1975).

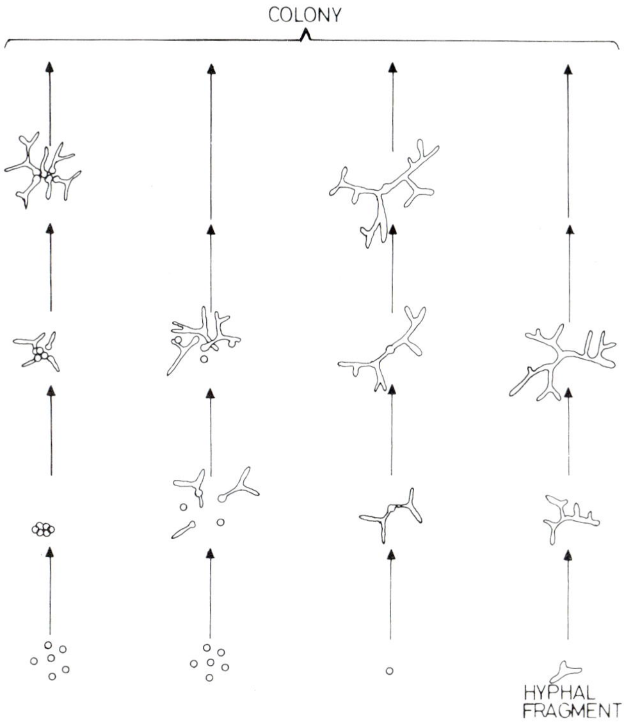

Fig. 1. Diagrammatic representation of the agglutination and growth mechanisms which participate in the formation of colonies.

The influence of the genotype

The intrinsic suitability of any particular mould strain to the fermentation is determined ultimately by its genotype. It is well-known that the morphology of fungal hyphae varies from class to class and from species to species; indeed, this has often been the basis for their taxonomic distinction. Even between distinct strains and mutants of the same species, differences in hyphal structure or dimensions may occur, and therefore one would expect that differences will also occur with respect to the morphological development of various species and strains in submerged culture.

One of the earliest and most comprehensive impressions of species variation with respect to colonial morphology in submerged cultures is given by the series of photographs by Burkholder & Sinnott (1945). It is important to note that the morphological forms they obtained were all the result of shake-flask culture under very similar conditions and therefore indicate the varied response of a range of species to a particular environment. A similarly broadbased study using simple bubble-column fermenters was carried out by Daunter (1971) as part of a biomass screening survey.

Hyphal branching

Butler (1966) has speculated that the gross features of colonies formed during submerged culture of filamentous fungi are a direct consequence of the branching characteristics of the constituent hyphae. It seems reasonable to expect that the degree of branching is also largely responsible for the packing density of hyphae in colonies. The interaction of such features as agitation, mass-transfer and mixing with various morphologies is also likely to differ in important ways, as suggested by Greenshields *et al.* (1972). Methods of controlling and characterising hyphal branching are therefore of interest. Robertson (1959, 1965) reports success in controlling branching of *Hyphomyces* species by varying the concentrations of trace-metals, griseofulvin and succinic acid in the culture medium, and many other workers have reported factors which have been found to control branching. A particularly well-documented phenomenon is the production of abnormally short, highly-branched, swollen hyphae under adverse conditions such as lack of oxygen, high acidity, high ferricyanide activities, and toxic metals etc. (e.g. Foster, 1949).

Hyphal interactions

Any adhesive interaction between hyphae is likely to affect the response of the mycelium to agitation effects. Works devoted to the flocculation of yeast cells has indicated the importance of a phosphate-mannan-protein complex, the formation of which depends on the protein and carbohydrate constituents of the cell-wall (Eddy & Rudin, 1958; Rainbow, 1966). It may be that mould cell walls have an analogous adhesive mechanism, or that they are able to entwine without the help of such adhesive properties. On the other hand, fungal hyphae of many species are known to 'repel' each other, as in the case of well-established colonies on agar plates, which form straight edges at boundaries between colonies with a gap between. It is not known what the effect of this interaction is during submerged culture. It is not known to what extent individual hyphae are sensitive to the presence of other hyphae, but the dominant radial symmetry of fungal colonies which are allowed to develop freely is remarkable. On solid media, hyphae only rarely grow back toward the colony centre, and the distance between hyphae is controlled by some mechanism, forming a regular, neat, fringe to the colony. Interesting exceptions to such rules are provided by the spirally-developing colonies of certain strains of *A. niger* (Yuill & Yuill, 1954), by the ability of 'young' colonies to grow together, and by the tendency of certain combinations of species to grow into each other.

Mechanical interactions between hyphae can be expected to influence the response of colonies to shear forces or the rheological properties of the suspension as a whole.

Mechanical properties of hyphae

Much of the behaviour of a mycelial suspension will depend on the mechanical properties of the hyphae; to what extent are they able to flex? How strong are they? Biological variation between strains of *A. niger* may be a significant source of variation in engineering terms.

Growth kinetics

Most workers have found that the growth of various fungi in submerged culture may obey either a cube-root relationship:-

$$M = K_c . t . M_o^{1/3}$$

where M = mass of mycelium per unit volume t minutes after the mass was M_o, per unit volume and K_c is an empirical constant (Emerson, 1950; Machlis, 1957; Marshall & Alexander, 1960), or that it may follow an exponential relationship:-

$$M = M_o . e^{-\alpha t}$$

where α is an experimental constant (Zalokar, 1959; Choudhary & Pirt, 1965a). It is important to remember, however, that these results are empirical results based on total dry weight or cell nitrogen, and therefore include non-viable cells in the estimation. Pirt (1966) argued that exponential growth was associated mainly with filamentous growth and cube-root growth with 'pellets', where mass transfer into cells in the core of the colonies was restricted. He also envisaged a more loosely-packed type of colony in order to explain his earlier observation that the dry weight values for *A. niger* colonies could follow an exponential law (Choudhary & Pirt, 1965b).

Several workers (e.g. Emerson, 1950; Trinci & Banbury, 1967) have reported that on solid medium, individual hyphae extend according to linear kinetics, and Katz, Goldstein & Rosenberger (1972) have proposed that for an exponentially growing mycelium, the rate of extension reaches a maximum characteristic of the strain. In order to explain the observed exponential growth of fungi in submerged cultures, it has been proposed that linear hyphal extension, coupled with the formation of new branch points, can account for exponential growth (Plomley, 1959; Kabitschek, 1970; Caldwell & Trinci, 1973). However, under certain conditions, individual hyphae may elongate according to exponential kinetics. Trinci (1969, 1971a) and Plomley (1959) have both reported that germ-tubes of a total of nine different species (including an *A. niger* strain) elongate exponentially. It has also been shown (Trinci, 1969, 1971a; Trinci & Gull, 1970) that the estimated specific growth rate in such cases was higher than that of submerged cultures. It has been pointed out by Mitchison (1973) that an exponential increase in length or volume of cells does not necessarily mean that the dry weight will increase at the same rate, but nevertheless, the increase in dry weight is probably exponential. Trinci (1971a) reported that for the strains he tested, the initial high specific growth rate entered a deceleration phase when the germ-tube reached a certain length. He does not show whether the deceleration resulted in a lower doubling time or in linear kinetics. Machek & Fencl (1973) used a special growth cell to observe the elongation of hyphae in a liquid medium flowing past the hypha under investigation. They found that hyphae could elongate according to either linear or exponential rules, but that there were many transient, intermediate types of growth. From their graph (Fig. 2) it can be calculated that the doubling-time is approximately 1.7 h, whereas Trinci (1971a) calculated a mean doubling time of 2.54 h, standard deviation 0.42 h (sample-size > 10). The apparent contradiction may be a result of strain differences or of differences in the conditions, and it is interesting to note that in Machek & Fencl's system (Fig. 2, curves I and II) the initial growth within 15 min of the start must be at higher logarithmic rate than subsequently, and that (curve II) the decceleration phase certainly starts after about 20-30 μm of hypha have been formed. Fig. 3 shows that the sum of elongation of all the branches forming from a single length of hypha probably follows a logarithmic relationship. Smith (1924) also reported that on solid media individual hyphae longer than about 200 μm were able to extend exponentially. The model for exponential growth based solely on the influence of branching on linearly-extending apices may therefore be too simple. The comparison of a fungal hypha with a series of single cells (Kubitschek, 1970) neglects the functional differentiation within a hypha which allows several cells to contribute to the growth of an apical cell.

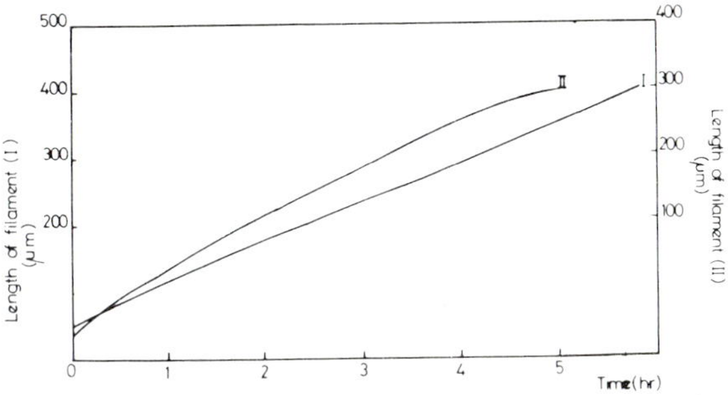

Fig. 2. *Aspergillus niger* filament elongation (Machek & Fencl, 1973).

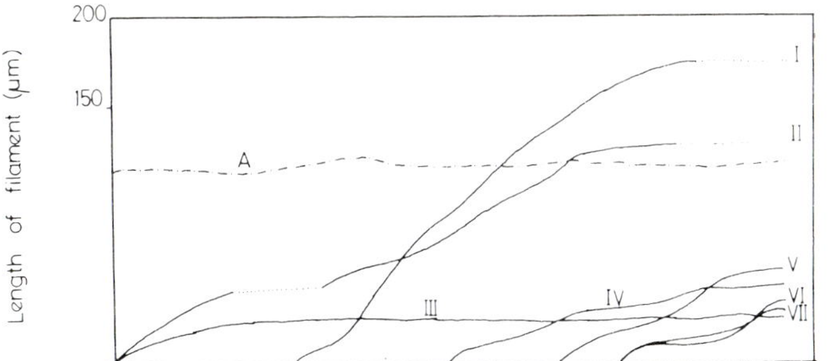

Fig. 3. Length of nongrowing portion of filament and of branches. Curves I-VII indicate the length of branches; Curve A the length of the nongrowing portion of the filament (Machek & Fencl, 1973).

Functional aspects of growth

It has been established that the growth of each hyphal apex is physiologically dependent on a length of subapical hypha which has ceased to extend, but which contributes its biosynthetic activities to the extending apical region (e.g. Zalokar, 1959; Yanagita & Kogane, 1962; McClure, Park & Robinson, 1968;

Nagasaki, 1968; Fencl & Machek, 1970; Megae *et al.*, 1970; Gooday, 1971; Trinci, 1971b). The approximate length of this functionally active zone may be determined in the case of colonies growing on solid medium by determining the minimum distance from the apex at which the hypha may be severed without reducing the extension rate (Ryan, Beadle & Tatum, 1943; Clutterbuck & Roper, 1966; Trinci, 1971b). In the case of side branches from such hyphae, there is no such information available, but we are led to consider what happens when branches are too short to have their own 'support zone'. Plomley (1959) has proposed the concept of a 'growth unit', implying that each branch is supported by a specific portion of the main hypha. Measurements for *A. niger* strains by the peripheral growth zone method indicate that at the periphery, the approximate length of hypha involved is 1200 μm (Lhoas, 1968; Trinci, 1971b). Branches are usually formed much closer together than 1200 μm, and therefore probably share the same 'growth unit'. This might restrict the rate of growth of branches until they reach a certain length, as observed by Machek & Fencl (1973).

In attempting any kind of extrapolation of the available data to the study of submerged cultures, there is a need to consider the effect of the extra degrees of freedom present in this type of system. It is possible to speculate, for instance, that if the peripheral growth zone of an approximately spherical colony 5 mm in diameter were 1200 μm, then this zone would occupy 13 - 14 mm^3 of a total colony volume of 66 mm^3, i.e. approximately one-fifth of the colony volume. However, it may be erroneous to assume that the peripheral growth zone will not change in character and dimensions in such a different environment. Pirt (1966) envisaged that some very compact spherical colony-types have a peripheral zone to which growth is restricted because it is the only part of the colony in which mass transfer is adequate, resulting in 'cube root' growth. However, his observation that a less compact type of colony could grow exponentially (in terms of dry weight) suggests that radical differences in functional organization may occur between different colony types. The extra spatial dimension in submerged culture places a much greater importance on the role played by hyphal branches in mycelial proliferation. Even the concept of a peripheral growth zone may become redundant if mass transfer occurs freely between all parts of the mycelium and the medium. This is the situation envisaged by the many authors for filamentous growth in submerged cultures, but the intermediate types of growth between the filamentous form and the compact spherical form may also come into this category. *A. niger* strains grow in stirred fermenters as the filamentous form often produce filaments considerably less than 1200 μm in length (the 'growth zone' length) and the question arises of whether the growth zones are reduced *pro rata* or the filaments are undergoing physiological damage caused by rupture within 'growth zones'. There is an obvious need to understand more fully the interaction of the inherent physiological organisation of fungi with the environments provided by the various forms of fermenter culture, but in acquiring this knowledge, much information may come to light about the physiology of *Aspergillus*.

CHEMICAL ENVIRONMENT

This is basically predetermined by the medium composition, although metabolite production and nutrient uptake also affect growth at later stages of batch culture. A great number of medium ingredients have been shown to influence the growth of *Aspergillus* species in fermenters. Interpretation of the conflicting and apparently contradictory published data is complicated by the existence of genetic variation between species and strains, and the alterations in response to individual ingredients caused by the effects of other ingredients (see later).

Trace metals

Trace metals are known to have a profound effect on the growth of *Aspergillus* species, particularly in the production of organic acids. The use of *A.niger* strains for the bioassay of potassium, nitrogen, phosphorus, molybdenum, magnesium and manganese in agricultural soils indicates the extent to which the genus may be affected (Mehlich *et al.*,1933; Zodrow *et al.*1966; Walczyna, 1966; Nicholas & Fielding, 1950; Sulochana & Lakshmanan, 1968). In addition, calcium, copper, iron, mercury, zinc, lead, lithium, sodium, rubidium and aluminium have all been shown to affect growth of *Aspergillus* species. In liquid culture, the roles played by these cations is complicated by the fact that several are flocculating and chelating agents and alter the physico-chemical balance of the medium. In general, a deficiency or excess of most of these cations results in sub-optimal fermentation, morphological abberations and poor product-accretion.

MAGNESIUM AND CALCIUM Salts of magnesium and calcium are very often used as precipitating or flocculating agents in chemical processes and are also known to assist flocculation of antibody/antigen mixtures and of suspended yeast-cells. Nowosielski (1960) used the *A.niger* method of Nicholas & Fielding (1950) to measure the magnesium content of soil extracts made in 10% ammonium acetate or hydrochloric acid and found good agreement with results from flame-photometry. This co-linearity held until a threshold point is reached at around 0.1% magnesium when abnormal growth was found

(Zodrow *et al.*,1966). At higher concentrations, particularly above 0.306% magnesium, the medium probably becomes toxic to *A. niger* (Kauffmann-Cosla & Tudor, 1946). Measurement of biomass yields should be carefully conceived, since magnesium content has been shown to alter the water content of *A.niger* tissues (Kauffmann-Cosla & Tudor, 1946). A triple interaction has been found between the action of different concentrations of zinc, cadmium, and magnesium salts, with magnesium acting as an antagonist to the toxic effects of the other two (Laborey & Lavollay, 1967). The importance of genetic differences is highlighted by the findings of Zodrow *et al.*(1966), who tested 17 *A.niger* strains over the range 0 - 1000 μg/10 ml. Six strains did not respond to magnesium, eight responded at about the same rate, and three responded very markedly.

Calcium seems to have little effect on *Aspergillus* species except when other essential trace-elements are missing (Garnier & Kientizler, 1958).

MOLYBDENUM *Aspergillus niger* strains have been used as indicators of molybdenum content in soil (Nicholas & Fielding, 1950) and in chemical solutions using molybdenum-starved spores (Büssler, 1970). Molybdenum-deficient cultures were found to have shortened, irregularly thickened, malformed hyphae. More specifically, Raman *et al.* (1962) found that molybdate was inhibitory to growth of *A.niger* in the range 2 - 10 mg/10 ml, but that methionine prevented both inhibition by, and uptake of, molybdenum ions.

MANGANESE Sulochana & Lakshmanan (1968) tested five strains of *A.niger* for their quantitative responses (mycelial dry weight) to graded concentrations of manganese in otherwise optimal media as regards trace elements. No significant increase in weight occurred above 0.006 mg/ml. This corresponds with earlier results by Steinberg (1939).

COPPER, IRON AND ZINC The essential importance of these three heavy metals to the nutrition of fungi in general has been recognised since the very first studies. Copper has been determined biologically using *A.niger* (Upitis *et al.*,1966) and whilst all three are essential in trace amounts, they are very easily found in inhibitory concentrations (Fiedler, 1959; Laborey & Lavollay, 1967; Upitis *et al.*,1966).

LITHIUM, SODIUM, POTASSIUM AND RUBIDIUM The presence of these ions stimulates sporulation in the order Li > Na > K > Rb, and potassium has a marked growth promoting effect which none of the others completely replaces (Fiedler, 1959; Garnier & Kientizler, 1968). A sodium/potassium relationship similar to that found in most flora and fauna has been demonstrated. The growth response to potassium is particularly marked, and this is another metal whose bioassay in soils has been carried out using *A.niger* (Mehlich *et al.*,1933;

Walczyna, 1966).

ALUMINIUM Bertrand (1963) cites aluminium as a "Dynamic trace element for *Aspergillus niger*". There may be other trace elements of nutritional importance to *A.niger* but it seems highly unlikely that any of these would be lacking.

HEAVY METALS The toxicity of zinc, copper and cadmium has already been discussed, but several other heavy metals are likely to be present in fermentation media. Zlochevskaya & Rabotnova (1966) showed that a concentration of 296 mg/free lead ions per litre only slightly affected mycelium production and did not suppress sporulation. Mercurial compounds of various kinds may also accidentally occur in culture fluids; from agricultural fungicides, for example. *A.niger* will tolerate relatively high concentrations of mercury and organomercurials, probably by a mechanism involving the protective over-production of non-proteinaceous thiol compounds (Ashworth & Amin, 1964). The toxicity of silver and mercury does, however, double for each threefold increase in concentration exposure time (McCallan & Miller, 1957). It seems that the chief concern with lead, mercury and silver may be their toxicity to animals in the end-product of biomass fermentations. The rare-earth elements, whilst inhibitory, can be detoxified by the formation of rare earth oxalates (Talburt & Johnson, 1967). Borax has been found toxic to *A.niger* at levels around 1.3 g/1 (Bowen & Gauch, 1966).

Chelating Agents

The presence of chelating agents has a profound effect on both nutritious and toxic concentrations of free trace-metal ions and hence on the metabolism, morphology and sporulation of *Aspergillus* species. The production of organic acids by *Aspergillus* species is heavily dependent on the nutritional control of metabolism using chelating agents to prevent the functioning of enzymes which need metallic cofactors. Specific blocking of metabolic pathways allows the accumulation of kojic, itaconic, citric, gluconic or oxalic acids. Examples of chelating treatments for citric acid production using ferrocyanide salts are given by Gold & Keiber (1967; Schoedler (1968); Perlman (1943); Gerhardt *et al.* (1966); Bernhauer *et al.* (1949) and Martin & Waters (1952).

Ferrocyanide salts are also potent inhibitors of respiration in their own right, and obviously metabolic inhibition such as this seriously retards growth rates, but work using EDTA has shown that the final maximum weight under otherwise constant conditions was the same with or without EDTA (Bertrand & de Wolf, 1960). Many naturally occurring media constituents, e.g. organic acids such as citric, gluconic and humic acid, are chelate-formers and it is interesting to note that these may act as buffering agents for the metals in a manner analagous to that of pH buffers (Choudhary & Pirt, 1965a,b).

Nitrogen and Phosphorus

A.niger is heavily dependent on phosphorus compounds (Marchesini, 1966) which are known to be incorporated into the RNA of the highly active peripheral extending zone of *A.niger* colonies and pellets (Yanagita & Kogane, 1962, 1963a,b). A common inorganic source of nitrogen for moulds is as the nitrate or ammonium salts. Nitrate is reduced via nitrite to ammonium ion both intra and extra-cellularly (Kostychev & Tsvetkova, 1920; Iteryott, 1936; Ivanov *et al.*,1936) and converted to amino acids. Whilst both nitrate and ammonium ion may be absorbed by the hyphae, fumarate, succinate and malate assist the absorption of ammonium more than that of nitrate; citrate, α -ketoglutarate and oxaloacetate vice-versa (Peter, 1956). Amino acids, purine or pyrimidine compounds may be utilised instead (see later) and it has also been shown that *A. niger* can utilise betaine, nicotine, quinine, pyridine and even cyanides by adaptive enzyme formation when starved of nitrogen sources (Ivanov *et al.*,1936).

Amino Acids

The presence of individual amino acids in addition to inorganic nitrogen sources has a stimulatory effect on growth rate (Delic-Koch, 1957; Kasatkina, 1961; Liczuk & Galeziewska, 1968). Mixtures of amino acids have an even greater stimulatory effect. Low nitrogen-to-carbon ratios in the medium increase the likelihood of sporulation taking place (Galbraith & Smith, 1969b) and synthesis of fats becomes a prominent metabolic function (Prescott & Dunn, 1959

Purine and Pyrimidine concentrations

Most pyrimidines and purines other than vitamins seem to help growth, since they play an important part in metabolism. However, derivatives and analogues of nucleic acid bases, for example, caffeine, theobromine, 5-bromouracil etc. can be strongly inhibitory by prevention of nucleotide base synthesis, or by nucleotide base-substitution and consequent mutagenic effects.

Lipid content

Of the various lipid classes the fatty acids are probably the most likely class to be found in fermentation media of vegetable origin. They are important for *Aspergillus* species as inhibitors of growth, spore swelling and spore germination, particularly the short chain fatty acids, including formic, acetic, butyric, propionic, capric, caprylic, valeric, isovaleric, levulenic, undecyclic,

2,3-nonenoic, 2,3-decenoic, 10-undecenoic, 2-3-dodecenoic and 2,3-tridecenoic acids (Fencl & Leopold, 1956; Philip *et al.*1963; Ozaki & Baba, 1943; Millis *et al.*,1963; Romano & Kornberg, 1969).

The acids are thought to act at the cell surface by preventing uptake of nutrients and they are known to prevent phosphate and sulphate absorption in *Aspergillus* species. This effect can be partly reversed by excess phosphate but not by excess potassium, sodium or ammonium ions. The dissociated and associated forms show very different levels of inhibitory effect (Fencl & Leopold, 1956). Higher fatty acids, including oleic and linoleic, but not palmitic acid, found in plants, can be utilised as sole carbon source by *A.niger* (Waheed-Hhan & Chughtai, 1960). Peanut oil and tributyrin havealso been found to inhibit spore germination for one *A.niger* strain (Millis *et al.*,1963). However, although it was found that peanut oil and maize oil could be utilised by *A.niger* as sole carbon source, they could not replace sucrose for citric acid production and thus they did not form an alternative supply of acetyl coenzyme A in this case (Millis *et al.*,1963). Lewis (1970) reports the formation of methylketones from fatty acids and his paper is a source of references on fatty acid utilisation by *Aspergillus* species and other fungi. He concludes that fatty acids may play an important part in the regulation of metabolism in *Aspergillus* spp. The effect of sterols on growth of *Aspergillus* species is interesting. In certain cases, 0.5 to 1.0% cholesterol additions were found to increase the dry weight of shake cultures of *A.niger* by 122 to 160% (Matkovics, 1957; Matkovic & Pulay, 1958) and Hanifa Moursi & Refai (1968) found the following effects of hormones on *A.niger* (Table 1). The results are not sufficiently quantitative but they do merit further investigation. It is well known that moulds can transform steroids, (as a means of detoxification probably) and *A.niger* is also known to produce ergosterol independently of a steroid source (Izuka *et al.*,1962; 1965) in amounts ranging from 1.10% to 0.31% of dry weight (Data from 60 standard strains).

The effect of lipids as surfactants may also be significant, depending on the amount in the medium.

Alcohols

The low molecular weight alcohols seem to have a similar inhibitory effect on *Aspergillus* species. Methanol in particular inhibits sporulation, sugar-consumption, growth-rate and stimulates citric acid production. The most effective concentrations were 3% (Uchi *et al.*,1965; Qadeer *et al.*,1968) and 49% (Echevarria & Varela, 1959). Ethanol and propan-2-ol were found to have a similar but less marked effect at an optimum concentration of 2% (Qadeer *et al.*, 1968) and methanol, ethanol, propan-2-ol, butan-1-ol and 2-methylpropan-1-ol were all found to affect the growth and morphology of

Table 1. Effect of plant growth regulators on growth rate of *Aspergillus niger*

Growth regulators	Concentrations in ppm 0.0	0.5	1.0	5.0	25.0	50.0
CONTROL (mean of five)	185	-	-	-	-	-
1-Naphthylacetic acid (NAA)	-	232*	202	196	179	166
2-Naphthoxyacetic (NOAA)	-	233	264	237	208	197
Indole-3-yl-propionic acetic acid (IPA)	-	203	225	212	176	134
2,4,5-trichloro-phenoxyacetic acid	-	202	226		187	126
Colchicine	-	281	309	287	282	227

(dry wt. in mg after 6 days at $28^{\circ}C$)

Adapted from Sankhla & Sharma (1969)

* Each weight is the mean of three

A.niger (Qadeer *et al.*,1968; Puente & Regueiro, 1962).

Alcohols can have a strong antifoam effect, especially at chain lengths of about eight carbon atoms, and they can be expected to be detrimental to aeration efficiency in submerged cultures. However, at higher chain lengths colloidal properties take effect and seem to promote an overall beneficial action on growth of *Aspergillus* species.

Available sulphur

Sulphur is yet another element for which *Aspergillus niger* has been used as a bioassay agent to indicate the amount actually available to plants (Katta & Lynd, 1965; Naik & Das, 1964). Maximum growth (final dry weight) was obtained with three different strains at 2 mg sulphur as Na_2SO_4/50 ml culture at $34^{\circ}C$. Sulphate, sulphite thiosulphate or preferably organic sulphur compounds such as cystine, cysteine and methionine can be utilised by *A. niger* but thiocyanate and sulphide ions cannot. Growth rate is depressed by sulphide ion concentration. The concentration of this ion may be increased by bacterial activity in stored media.

Tracer experiments (Robson, 1957, 1958) have shown that radioactive sulphur is accumulated largely in the spores. Choline sulphate seems to act as a reservoir for energy and sulphur, since spores can germinate on sulphur free media and the compound is broken down during germination.

Carbohydrates

A knowledge of both the types of sugar and the amounts of each is necessary for fermentation using *Aspergillus*. Not very much information is available, but that which is available indicates interactions between utilisation of sugars: D-xylose stimulates galactose utilisation (Lindberg, 1967), fructose is utilised in preference to glucose (Brannon, 1923) and a galactose/glucose mixture is utilised far better than either sugar singly (Horr, 1936).

With a variety of single sugars as sole carbon source, Hasija & Wolf (1970) found that the growth rate and final mycelial dry weight were not the same in all cases. Hexoses were generally as good as pentoses, with certain exceptions which seem to depend on the strain. Disaccharides and starch were also satisfactory with the exception of lactose on which growth was poor. Lactose has been found in washed mycelium and this has been taken to show that it is used directly (Lilly & Barnett, 1953). Takami (1968) found that vitamin production by *A. niger* was affected by the type of sugar and that maltose, sucrose and glucose gave better yields of vitamins than other sugars tested.

Aspergillus species also produce polysaccharides, some of which may escape into the medium and alter its characteristics, for example, polyols and glucosans such as nigeran (Kawamura & Baba, 1964; Barker *et al.*, 1953a,b, 1954, 1955, 1956, 1957, 1958). *A. niger* is not as tolerant of high osmotic pressure as *Aspergillus glaucus* and 50% glucose has been cited as being 'toxic' (Kaufmann-Cosla & Tudor, 1946). The speed with which *A. niger* uses up or excretes sugars at high sugar concentrations could therefore have a bearing on its growth rate through an osmotic effect.

Another very important carbohydrate parameter is the level of various sugar acids and TCA cycle acids in the medium. *Aspergillus niger* is famous as an industrial provider of citric and gluconic acids, and it has been noted that besides having a drastic depressant effect on pH, these acids may act as energy-storage compounds in a manner analogous to that in higher plants (Ranson, 1965). In the citric acid fermentation, the concentration of citric acid begins to decline after a peak value because it is used as a substrate. It is interesting to note the competitive advantages of acid production; both in an industrial fermenter and in a natural environment, carbohydrates of value to other organisms are removed and the pH lowered to a level inhibitory to many microorganisms.

Surfactants

Antifoam and detergent compounds both have the effect of reducing oxygen-transfer in submerged fermenters (Murphy *et al.*, 1959; Jones, 1972). The addition of antifoam increases the bubble size and detergents can cause serious foaming problems. Some antifoams may inhibit the mould, or be used as a carbon source. An interesting feature of detergents is that they may affect the morphology. Takahashi *et al.*, (1960, 1965) found that 0.05 - 0.1% non-ionic detergents encouraged filamentous growth of *A. niger*. For example, sucrose monosteareate, sorbitan monolaurate, sorbitan trioleate and Span 20 increased dry weight and prevented pellet formation. Tween 20 was inhibitory; it encouraged formation of large pellets. This effect was also measured using mycelial inocula, so the effect was not that of controlling spore clumping.

Plant growth factors and alkaloids

Plant growth factors are undoubtedly present in most industrial media because of the predominance of vegetable extracts. Plant hormones increase dry weight yields of *A. niger* at low concentrations but are inhibitory at high concentrations, e.g. 2,4-D is an effective stimulant at concentrations as low as 0.5-1 ppm or 10^{-6}M (Smith & Shennan, 1966; Ito & Surano, 1952; Arnold *et al.*, 1966; Sankhla & Sharma, 1969). Other plant substances which are known to affect *A. niger* strains include coumarins, which are effective inhibitors at 100-1000 ppm (Knypl, 1963). Certain halogen and alkyl derivatives of coumarins have been proposed as fungicides (Stauffer Chemical Co., 1966). Tannins at 15-20% w/v stimulate 'luxuriant' growth (Neithammer, 1956).

Polymers

Synthetic polymers including polyvinyl acetate, polyacrylate, polymethacrylate and carboxypolymethylene have been observed to stimulate growth of *A. niger* strains (Botri, Cieri & Giodani,

1964; Elmayergi & Moo-Young, 1973; Elmayergi, Scharer & Moo-Young, 1973). It may be that certain detergent-types may have a boundary-layer effect at the cell-wall/medium interface. Similar effects have been reported when synthetic polymers such as polyvinyl acetate, polyacrylate, polymethacrylate, and carboxypolymethylene were incorporated in the medium for deep culture of *A. niger* strains (Botri, Cieri & Giodani, 1964; Elmayergi & Moo-Young, 1973; Elmayergi, Scharer & Moo-Young, 1973). In view of the similarity in the physical behaviour and the structure of the polymers and the detergents which have the above effect, it seems that they either enhance hydrodynamic slip at the hyphal surface (Oldroyd, 1949; Oldroyd & Toms-Davies, 1949; Shaver & Merrill, 1959; Hoyt & Fabula, 1965; Moo-Young, Hirose & Alt, 1970), or they facilitate more rapid diffusion across the cell-wall/medium interface (Elmayergi & Moo-Young, 1973). Either effect would be expected to improve mass-transfer and therefore increase the growth-rate.

FERMENTER CONFIGURATION

Before dealing with the physical environment it is useful to consider fermenter designs, since these can predetermine physical parameters.

The simplest man-made fermenter configuration suitable for mould-culture is the open tray or dish, as used in the Koji process and still used for citric acid production in Britain and Eastern Europe. Various adaptations have been used including flat bottles and even flexible pillow-tanks (Kybal, 1977). The stirred tank reactor (S.T.R.) is at present predominant amongst possible reactor configurations. Ever since the rapid conscription of stirred tank reactors for the production of penicillin and food yeast during World War II, manufacturers of other enzymes antibiotics, etc. have for pragmatic reasons adopted the established technology. A large proporation of laboratory fermenters have followed suit in pursuit of the established goals of ideal mixing, homogeneity, and consequent ease of measurement and control. However, by far the largest-scale application of aerobic fermentation, that of sewage treatment, employs reactors which depend deliberately on heterogeneity and differ radically from the S.T.R. concept, and there is at the present time an upsurge of interest in alternative types of reactor, among them various types of overtly tubular reactor such as the pressure cycle fermenter, various types of bubble-column and loop fermenters with either mechanical or pneumatic recycling. These developments were recently reviewed by Katinger (1977).

In view of the widespread role of bubble-columns in chemical process industries, it is perhaps surprising that they have been only slowly adopted by the fermentation industry. Patents and publications describing various designs of bubble-column in beer production (Smith & Greenshields, 1972), vinegar (Shimwell, 1955; Simonin & Bernard, 1958), activated sludge treatment (Chipperfield, 1967) and citric acid production (Martin & Waters, 1952; Clark & Lentz, 1963) did not stimulate much parallel activity by the fermentation industry in general.

In recent years, and with the benefit of lengthy experience in actual fermentations, there has been a realization that stirred fermenters have drawbacks. Large stirred fermenters require expensive seals and moving parts and absorb considerable quantities of power (Finn, 1967). They also require very large quantities of cooling water in consequence of the need to dissipate frictional heat. Where filamentous fungi are concerned, the problems are compounded by the typical suspension rheology. Not only does the viscosity become very high (Solomons, 1962, reports up to 20,000 cp. for *A. niger*) increasing power requirements, but the rheology becomes non-newtonian, with associated mixing-problems. Thus, large and small scale fermentations with all types of filamentous fungi may be subject to problems of poor mixing, dead zones, 'stalled' impellers, blockages and wall growth.

Filamentous fungi such as *Aspergillus* species are capable of profound interactions with any fermenter, such that fermenter cultivation always involves a dynamic reciprocal reaction between fermenter and organism. Some recent designs of fermenter are intended to exploit this reciprocal interaction, for example, biodisc fermenters adapted from designs used in effluent treatment plants have been used to cultivate *Aspergillus* species as films growing on rotating discs which are part-submerged in the growth-medium (Divers, Anderson & Blain, personal communication). The tendency of the fungus to grow as a film on the discs allows the medium to be changed whilst retaining the biomass, permits physiological and biochemical observation of this mode of growth, and accomplishes gas-transfer in a simple fashion.

PHYSICAL ENVIRONMENT

The physical environment of organisms in a fermenter is the outcome of the interaction of fermenter design and the physical behaviour of the medium and of the organism itself.

Aeration

In submerged culture air or oxygen are usually introduced. The concentration of oxygen and of carbon dioxide dissolved in the medium require to be controlled within critical limits, a surfeit or deficit being injurious.

In stirred fermenters aeration may be induced by vortex conditions but may be assisted by introducing compressed air. Baffles are usually included to induce turbulent flow patterns thereby improving heat transfer and mass transfer. High impeller speeds result in mycelial fragmentation, with consequent increase in viscosity and loss of desired mass transfer characteristics.

In air-lift or bubble-column designs column geometry, combined with judicious control of superficial gas velocity and/or liquid velocity are used to control the degree of turbulence induced by the friction at the wall. Sheer rates are comparatively low and, therefore, fragmentation considerably less frequent than in stirred fermenters. This permits studies of colonial physiology and morphology in submerged culture.

Rheology

At high biomass levels mycelial fermentations are prone to develop high viscosities with non-newtonian rheology. External measurements of broth rheology do not determine the problem precisely and the most satisfactory solution for stirred fermenters is to characterise broth rheology directly via torque measurement at the impeller shaft.

Heterogeneity

Stirred fermenters are often considered homogeneous but at high biomass concentrations or in very large fermenters broth circulation rates are reduced and dead spots may occur. Air lift fermenters and bubble-column fermenters involve heterogeneity which is dependent on colony sedimentation rates. This feature has been deliberately used in continuous versions to achieve biomass retention (Pannell & Greenshields, 1977). Even more deliberate is the heterogeneity found in tray and biodisc fermenters where the mycelium forms a pellicle which is differentiated, and mass-transfer is presumably slow. The deliberate acceptance of heterogeneities of this sort promises novel insights into the physiology of filamentous fungi under such conditions. Pannell & Greenshields (1975, 1977) have claimed that where bubble-column fermenters are used for continuous fermentation, biomass retention allows the use of low nutrient concentrations, high hydraulic loadings and even non-aseptic cultures making the fermentation suitable for effluent treatment from processes such as food-manufacture.

In summary it can be seen that fermenter cultivation of *Aspergillus* spp. is possible in a variety of environments and allows the mycologist to broaden the dimensions of his study of the biochemical and physiological behaviour of this genus. Its commercial application has merely touched on the surface of possibilities, and the founding work of Raistrick is still limitless in its potential for exploitation.

REFERENCES

ARNOLD, W.R., SANTELMANN, P. & LYND, J.W. (1966). Picloram and 2, 4 -D. Effects with *Aspergillus niger* proliferation. *Weeds* 14, (1), 89-90.

ASHWORTH, L.J. & AMIN, J.V. (1964). A mechanism for mercury tolerance in fungi. *Phytopathology* 54, 1459-63.

BARKER, S.A. & CARRINGTON, T. (1953). Studies of *Aspergillus niger* II. Transglycosidation by *A. niger*. *Journal of the Chemical Society* 3588-93.

BARKER, S.A., BOURNE, E.J. & CARRINGTON, T. (1954). Studies of *Aspergillus niger*. III. The structure of a trisaccharide synthesised from glucose. *Journal of the Chemical Society* 2125-9.

BARKER, S.A., BOURNE, E.J. & STACEY, M. (1953). Studies of *Aspergillus niger*. The structure of the polyglucosan synthesised by *Aspergillus niger* 152. *Journal of the Chemical Society* 3084-90.

BARKER, S.A., BOURNE, E.J. & THEANDER, O. (1956). Studies of *Aspergillus niger*. The enzymic synthesis of a new trisaccharide. *Journal of the Chemical Society* 2064-7.

BARKER, S.A., GOMEZ-SANCHEZ, A. & STACEY, M. (1958). Studies of *Aspergillus niger*. X. Polyol and disaccharide production from acetate. *Journal of the Chemical Society* 2583-6.

BARKER, S.A., BOURNE, E.J., HEWITT, G.C. & STACEY, M. (1955). Studies of *Aspergillus niger*. IV. The synthesis of linked glucosaccharides. *Journal of the Chemical Society* 3734-40.

BARKER, S.A., BOURNE, E.J., O'MANT, D.M. & STACEY, M. (1957). Studies of *Aspergillus niger*. VI. The separation and structures of oligosaccharides from nigeran. *Journal of the Chemical Society* 2448-54.

BENT, K. & MORTON, A.G. (1963). Formation and nature of swollen hyphae in *Penicillium* and related species. *Transactions of the British Mycological Society* 46, 401-8.

BERNHAUER, K., RAUCH, J. & GROSS, G. (1949). Uber die Sauerbildung aus Zucker durch *Aspergillus niger*. XIII. Mitteilung: Die Citronsauerbildung aus Zuckerubenme Hasse in der Oberflaschenkultur.

BERTRAND, D. (1963). Aluminium : a dynamic trace element for *Aspergillus niger*. *Comptes Rendus* 257, 3057-9.

BERTRAND,D. & De WOLF, A. (1960). *A.niger* growth in the presence of chelating agents. *Comptes Rendus* 250, 1543-5.

BHATNAGAR, L. & KRISHNAN, B.(1960a). *Archiv für Mikrobiologie* 36, 131-8.

BHATNAGAR,L., & KRISHNAN,B. (1960b). *Archiv für Mikrobiologie* 37, 211-4.

BOTRI,C., CIERI,L. & GIODANI,F. (1964). Influenza di Alcuni Polimeri Fermentaxione del *Penicillium chrysogenum. Farmaco Ed. Prat.* 17,507-11.

BOWEN,J.E. & GAUCH,H.G. (1966). Nonessentiality of boron in fungi and the nature of its toxicity. *Plant Physiology* 41, 319-24.

BRANNON,J.M. (1923). Influence of glucose and fructose on growth of fungi. *Botanical Gazette* 76, 257-73.

BURKHOLDER,P.B. & SINNOT,E.W. (1945). Morphogenesis of fungal colonies in submerged fermentation. *American Journal of Botany* 32, 424-31.

BUSSLER,W. (1970). Detection of molybdenum in a drop of nutrient solution by the deficiency symptoms shown by *Aspergillus niger. Zeilschrift Pflanzenlernähr* 125, 16-23.

BUTLER,G.M. (1966). Vegetative structures. In *The fungi*,Vol.2 pp.83-112. Edited by G.C. Ainsworth and A.S.Sussman. New York and London: Academic Press.

CALDWELL,I.Y. & TRINCI, A.P.J. (1973). The growth unit of the mould *Geotrichum candidum. Archiv für Mikrobiologie* 88, 1-10.

CHIPPERFIELD,P.N.J. (1967). *Effluent and water treatment conventions.* London: Academic Press.

CHOUDHARY,A.Q. & PIRT,S.J. (1965a). Metal-complexing agents as metal buffers in media for the growth of *Aspergillus niger. Journal of General Microbiology* 44, 99-107.

CHOUDHARY,A.Q. & PIRT,S.J. (1965b). The influence of metal-complexing agents on citric acid production by *Aspergillus niger. Journal of General Microbiology* 41, 99-107.

CLARK,D.S. (1962). Submerged citric acid fermentation of ferricyanide-treated beet molasses. Morphology of pellets of *Aspergillus niger. Canadian Journal of Microbiology* 8, 133-6.

CLARK,D.S. & LENTZ, C.P. (1963). Submerged citric acid fermentation of sugar beet molasses: effect of pressure and recirculation of oxygen. *Biotechnology and Bioengineering* 5, 193-206.

CLUTTERBUCK,A.J. & ROPER,J.A. (1966). A direct determination of nuclear distribution in heterokaryons of *Aspergillus nidulans. Genetical Research* 7, 185-94.

COCKER,R. (1975). *Ph.D.Thesis. University of Aston in Birmingham.*

COCKER,R. & GREENSHIELDS, R.N. (1975). Annual Research Meeting. Institute Chemical Engineers, University of Bradford, West Yorks.

DAUNTER,B. (1971). *M.Sc.Thesis. University of Aston in Birmingham.*

DELIC-KOCH,G.A. (1957). The influence of essential amino-acids on moulds. *Phyton (Buenos Aires)* 9, 69-79.

DUCKWORTH, E.B. & HARRIS,G.C.M. (1949). The morphogenesis of *Penicillium chrysogenum* in submerged culture. *Transactions of the British Mycological Society* 32, 224-35.
ECHEVARRIA,J.L.M. & VARELA,B.R. (1959). The production of citric acid by fermentation. III. The influence of alcohols on surface fermentation. *Microbiologie Espanol* 12, 139-60.
EDDY,A.A. & RUDIN,A.D. (1958). Comparison of the respective electrophoretic and flocculation characteristics of different strains of *Saccharomyces*. *Journal of the Institute of Brewing* 64, 139-43.
ELMAYERGI,H. & MOO-YOUNG, M. (1973). Effect of a polymer additive on mass-transfer into mould pellets.*Biotechnology and Bioengineering Symposium* 4, 507-12.
ELMAYERGI,H., SCHARER,J.M. & MOO-YOUNG, M. (1973). Effects of polymer additives on fermentation parameters in a culture of *Aspergillus niger*. *Biotechnology & Engineering* 15, 845-59.
EMERSON,S. (1950). *Journal of Bacteriology* 60, 221-30.
FENCL, Z. & LEOPOLD,J. (1956). *Aspergillus niger:* inhibition of spore-germination in acidified molasses. *Coll.Czech. Chem. Comm.* 21, 1028-34.
FENCL, Z. & MACHEK,F. (1970). Differentiation in the filament of *Aspergillus niger*. *Folia Mikrobiologia* 15, 211-8.
FIEDLER,H.J. (1959). The action of alkalis on *Aspergillus niger*. *Zentr. Bakteriol. Parasitenk.* Abt. II 112, 666-707.
FINN,R.K. (1969). Energy costs of oxygen-transfer. *Process Biochemistry* June, pp.17-20.
FOSTER,J.W. (1949). *Chemical activities of fungi*. New York: Academic Press.
GALBRAITH,J.C. & SMITH,J.E. (1969a). Filamentous growth of *Aspergillus niger* in submerged shake-flask culture. *Transactions of the British Mycological Society* 52, 237-46.
GALBRAITH,J.C. & SMITH,J.E. (1969b). Sporulation of *Aspergillus niger* in submerged liquid culture. *Journal of General Microbiology* 59, 31-5.
GARNIER,F.J. & KIENTIZLER,L. (1958). Etude de l'influence des ions potassium, sodium et calcium sur certains propertes du mycelium d'un souche *d'Aspergillus niger*. *Bull, Ec. Nat Sup. Agron.*, Nancy 10, 3-6.
GERHARDT,P., DORRELL, W.W. & BLADWIN,K.L. (1966). Citric acid fermentation of beet molasses. *Journal of Bacteriology* 52, 555-61.
GOLD,W. & KEIPER,R.J. (1967). Production of citric acid by fermentation. U.S. Pat. No. 3,326,774.
GOODAY,G.W. (1971). An autoradiographic study of the hyphal growth of some fungi. *Journal of General Microbiology* 67, 125-33.
GREENSHIELDS,R.N., ALAGARATNAM,R., COOTE,S.D.J., MORRIS,G.G., IMRIE, F.K.E. & SMITH,E.L. (1972). The growth and morphology of microorganisms in tower fermenters. *Proceedings IVth International*

Fermentation Symposium (Tokyo).
HANIFA MOURSI,S.A. & REFAI,M. (1968). The effects of hormones on some fungi *in vitro*. *Mykosen* II(1), 747-54.
HASIJA,S.K. & WOLF,F.T. (1970). The effect of different carbon sources on the sugar composition of the mycelium of *Aspergillus niger*. *Mycopathology et Mycologica Applicata* 39, 337-43.
HORR,W.H. (1936). *Plant Physiology* 81-99.
HOYT,J.W. & FABULA,A.G. (1965). The effects of additives on fluid friction. *ONR-Skipsmodelltanken 5th Symposium. Naval Hydrodynam. (Bergen)*.
ITERYOTT,D. (1936). The conditions for nitrogen, especially nitrate, absorption by *Aspergillus niger*. *Flora* 31, 60-86.
ITO,K. & SURANO,S. (1952). Effects of vegetable hormones against pathogenic fungi. I. The effects of sodium 2,4-dichlorophenoxyacetate on the growth of some pathogenic fungi. *Bulletin Pharmaceutical Research Institute Osaka Medical College* No.3, 29-35.
IVANOV,N.N., IVANOV,V.E. & LA RIONOVA,N.N. (1936). The nitrogen fixation of fungi. *Arch. Sci.Biol. U.S.S.R.* 43, 378-81.
IZUKA,H., YAMAMOTO,K. & NAITO,A. (1962). Steroids and microorganisms. V. Ergosterol produced by strains of *Aspergillus*. *Nippon Nogei Kagaku Kaishi* 36, 640-4.
IZUKA,H., YAMAMOTO,K. & NAITO,A. (1965). In *Chemical Abstracts* 62, 2214n.
JONES,D.D. (1972). *M.Sc.Thesis, University of Aston in Birmingham.*
KASATKINA,D. (1961). The effect of some amino-acids on the growth and acid-production by the *Aspergillus niger* T.1. mutant. *Mikrobiologyia* 30, 447-52.
KATTA,J.I. & LYND,J.Q. (1965). Sulfur bioassay investigations with *Aspergillus niger*. *Botanical Gazette* 126, 120-3.
KATINGER,H. (1977). *IV FEMS Symposium, University of Vienna.*
KATZ,D., GOLDSTEIN,D. & ROSENBERGER,R.F. (1972). Model for branch initiation in *Aspergillus nidulans* based on measurements of growth parameters. *Journal of Bacteriology* 109, 1097-110.
KAUFFMANN-COSLA,O. & TUDOR,O. (1946). *Annual Acad. Romane. Mun Sect. Stintif.* 21, Mem. 6 pp.73.
KAWAMURA,M. & BABA,A. (1964). Starch-like material produced by *Aspergillus niger*. *Shigakenritsu Tanki Daigaka Gakujutsu Zasshi* 5, 22-9.
KNYPL, J.S. (1963). *Nature* 200, 4908-9.
KOSTYCHEV,S. & TSVETKOVA,E. (1920). The utilisation of nitrates by moulds for the production of nitrogenous compounds. *Zeitschrift Physiologie Chemica*. III. 171-200.
KUBITSCHEK,H.E. (1970). Evidence for the generality of linear cell-growth. *Journal of Theoretical Biology* 28, 15-9.
KYBAL,J. (1977). *IVth FEMS Symposium, University of Vienna.*
LABOREY,F. & LAVOLLEY,J. (1967). Toxicity of zinc and cadmium

during *Aspergillus niger* growth. Antagonism of these ions, and magnesium, zinc, cadmium interaction. *C.R. Acad. Sci. Paris Ser D*. 264, 2937-40.

LEWIS,H.L. (1970). Caproic acid metabolism and the production of 2-pentanone and gluconic acid by *Aspergillus niger*. *Journal of General Microbiology* 63, 203-10.

LHOAS,P. (1968). Growth rate and haploidisation of *Aspergillus niger* on a medium containing p-fluorophenylanine. *Genetical Research* 12, 305-15.

LICZUK,Z. & GALEZIEWSKA,E. (1969). Amino acids of beet molasses and their usefulness for citric acid fermentation. *Przem. Fermant. Rolny*. 12, 7-9.

LILLY,V.G. & BARNETT,H.L. (1953). The utilisation of sugars by fungi. *Bulletin West Virginia Agricultural Experimental Station.*

LINDBERG,M. (1967). Stimulating effect of xylose on the utilisation of galactose in some fungi. *Nature* 213, 395-6.

MACHEK,F. & FENCL. Z. (1973). Differentiation as a basis for the understanding of product-formation. *Biotechnology and Bioengineering Symposium* No.4, 129-42.

MARCHESINI,A. (1966). Effects of chelating agents on the solubility of soil-phosphorus. IV: relationship between phosphorus availability and the growth of *Aspergillus niger*. *A. Microbiology Enzymology* 16, 147-51.

MARSHALL,K.C. & ALEXANDER,M. (1960). Growth characteristics of fungi and actinomycetes. *Journal of Bacteriology* 80, 412-8.

MARTIN,S.M. & WATERS,W.R. (1952). *Industrial and Engineering Chemistry* 44, 228-11.

MATKOVICS,B. (1957). The effects of sterols on micro-organisms. *Naturwissenschaften* 44, 400-1.

MATKOVICS,B. & PULAY,G. (1958). The action of steroids on micro-organisms. I. The effect of cholesterol on the dry weight of *Rhizopus arrhizus, Penicillium chrysogenum, Aspergillus niger* and *Mucor mucedo*. *Zentr. Bakteriol. Parasitenk. Abt. II* III 440-2.

McCALLAN,S.E.A. & MILLER,L.P. (1957). Effects of fungicides on oxygen consumption and viability of mycelial pellets. *Contribution of Boyce-Thomson Institute* 18, 4482-95.

McCLURE,W.K., PARK, D. & ROBINSON,P.M (1968). Apical organisation in somatic hyphae of fungi. *Journal of General Microbiology* 50, 177-82.

MEGEE,R.D., KINOSHITA,S., FREDRICKSON,A.G. & TSUCHIYA,H.M. (1970). Differentiation and product-formation in moulds. *Biotechnology and Bioengineering* 12, 771-8.

MEHLICH,A., TRUOG,E., & FRED,E.B. (1933). The *Aspergillus niger* method of measuring available potassium in soils. *Soil Science* 35, 259-60.

MIALL,L.M. (1975). Historical development of the fungal

fermentation industry. In *The filamentous fungi* Vol.1, pp. 104-121. Edited by J.E.Smith and D.R.Berry. London: W Edward Arnold Publishers.

MILLIS,N.D., TRUMPY,B.H. & PALMER,B.M. (1963). The effects of lipids on citric acid production by an *Aspergillus niger* mutant. *Journal of General Microbiology* 30, 365-79.

MITCHISON,J.M. (1963). The cell cycle of a fission yeast. *Symposium Society Cell Biology* 2, 151-67.

MOO-YOUNG, M., HIROSE,T. & ALT,S. (1970). *Proceedings Vth International Congress Rheology. University Park Press, Maryland* pp.233.

MORRIS, G.G. (1973). *Ph.D. Thesis, University of Aston in Birmingham.*

MURPHY, D., CLARK,D.S. & LENTZ,C.P. (1959). Aeration in tower-type fermenters. *Canadian Journal of Chemistry and Engineering* 157-61.

NAGASAKI,S. (1968). Physiological aspects of various enzyme activities in relation to the culture age of *Aspergillus niger* mycelium. *Journal of General and Applied Microbiology* 14, 147-61.

NAIK, M.S. & DAS, N.B. (1964). Available sulphur status of Indian soils by the *Aspergillus niger* method. *Journal of Indian Soil Science* 12, 151-5.

NEITHAMMER,A. (1956). Distribution and growth of *Aspergillus niger. Gartenbauwissenschaft* 3, 98-101.

NICHOLAS,D.J.D. & FIELDING,A.H. (1950). Use of *Aspergillus niger* as a test organism for determining molybdenum available in soil to plants. *Nature* 166, 342-3.

NOWOSIELSKI,O. (1960). Determination of magnesium by a simplified *Aspergillus niger* method. *Roczniki Gleboznawcze* 9, 89-102.

OLDROYD,J.S. (1949). A suggested method of detecting wall effects in turbulent flow through tubes. *Proceedings 1st International Congress of Rheology, Scheveningen* pp.130.

OLDROYD,J.S. & TOMS-DAVIES,E. (1949). Discussion of papers. *Proceedings 1st International Congress Rheology, Scheveningen.*

OZAKI,A. & BABA,S. (1943). Substances contained in technical starch inhibiting growth of and acid formation by *Aspergillus niger. Journal of Agricultural Chemical Society.*

PANNEL, S.D. & GREENSHIELDS,R.N. (1975). *Annual Research Meeting, Institution of Chemical Engineers, University of Bradford, West Yorkshire.*

PANNELL,S.D. & GREENSHIELDS,R.N. *IVth FEMS Symposium Vienna University.*

PERLMAN,D. (1943). Factors affecting the production of citric acid by *Aspergillus niger. M.Sc.Thesis, University of Wisconsin.*

PETER,M. (1956). Absorption and assimilation of nitrate by *Aspergillus niger. Ber. Schweiz. Botan. Ges.* 66, 277-317.

PHILIP,E.T., HALL,A.N. & WALKER,T.K. (1963). *Archives of Biochemistry and Biophysics* 102, 238-41.

PIRT, S.J. (1966). A theory of the mode of growth of fungi in the form of pellets in submerged culture. *Proceedings of the Royal Society B* 166, 369-73.

PLOMLEY, N.B.J. (1959). Formation of the colony in the fungus *Chaetomium*. *Australian Journal of Biological Science* 12, 53-64.

PRESCOTT, S.C. & DUNN, P. (1959). *Industrial Microbiology (Third Edition)* New York: McGraw-Hill.

PUENTE, A. & REGUERO, B. (1962). Production of citric acid by fermentation. VIII. Effect of the addition of alcohols in the submerged fermentation of molasses. *Microbiologie Espanol* 15, 171-86.

QADEER, M.A., KHAN, K.M., CHUGHTAI, M.I.D. & ASHRAF, M. (1968). Effects of alcohols in the biosynthesis of citric acid by *Aspergillus niger*. *Pakistan Journal of Biochemistry* 1, 41-8.

RAINBOW, C. (1966). Flocculation of brewer's yeast. *Process Biochemistry* 1, 489-92.

RAMAN, N., SASTRY, S.K. & SARMA, P.S. (1962). *Biochemica et Biophysica Acta* 195-7.

RANSON, S.L. (1965). In *Plant Biochemistry* pp 493-532. Edited by J. Bonner and J.E. Varner. New York: Academic Press.

ROBERTSON, N.F. (1959). Experimental control of hyphal branching and branch forms in hyphomycetous fungi. *Journal of the Linnaean Society, London* 56, 207-14.

ROBERTSON, N.F. (1965). The fungal hypha. *Transactions of the British Mycological Society* 48, 1-8.

ROBSON, J.E. (1957). Sulfur metabolism in fungi. *Radioisotopes Science Research Proceedings* 4, 29-43.

ROBSON, J.E. (1958). Sulfur metabolism in fungi. *Radioisotopes Science Research Proceedings International Conference, Paris.*

ROMANO, A.H. & KORNBERG, H.L. (1969). Regulation of sugar uptake by *Aspergillus nidulans*. *Proceedings of the Royal Society B* 173, 475-90.

RYAN, F.J., BEADLE, G.W. & TATUM, E.L. (1943). The tube method of measuring the growth-rate of *Neueospora*. *American Journal of Botany* 30, 789-99.

SANKHLA, H.C. & SHARMA, L.C. (1969). Utilisation of growth-regulators by *Aspergillus niger* van Teighem. *Labdevelopment Journal of Science and Technology* 7-B (4), 334-50.

SCHOEDLER, K. (1968). Surface and submerged culture fermentation of citric acid from blackstrap molasses. *Zucker 5* (March) 121-2.

SIMONIN, R.F. & BERNARD, M. (1958). *British Patent No. 805,698.*

SHAVER, R.G. & MERILL, E.W. (159). Polymer solutions in straight cylindrical tubes. *A.I. Chemical Engineering* J 5, 181-9.

SHIMWEIL, J.L. (1955). *British Patent No. 727,039.*

SMITH, E.L. & GREENSHIELDS, R. (1973). *Biotechnology and Bioengineering* 4, 519-26.

SMITH, J.E. & SHENNAN, J.L. (1966). The effects of substituted phenoxyacetic acids and phenoxybutyric acids on the growth and respiration of *Aspergillus niger*. *Journal of General Microbiology* 42, 293-300.

SMITH, J.H. (1924). On the early growth-rate of individual hyphae. *New Phytologist* 23, 67-78.

SOLOMONS, G.L. (1962). In *Continuous culture of microorganisms*. Edited by Malek *et al.* *Czech. Acad. Sci. Prague*.

STAUFFER CHEMICAL CO. (1966). *Netherlands Patent No. 6,608,382*.

STEINBERG, R.A. (1939). Relation of carbon-nutrition to trace-elements and accessory requirements of *Aspergillus niger*. *Journal of Agricultural Research* 53, 749-54.

SULOCHANA, C.B. & LAKSHMANAN, M. (1968). *Aspergillus niger* technique for the bioassay of manganese. *Journal of General Microbiology* 50, 285-93.

TAKAHASHI, J. & YAMADA, K. (1960). Effects of some physical conditions on submerged mould culture. VI. Effects of non-ionic surface-active agents on the mycelial forms of *Aspergillus niger*. *Journal of Agricultural Chemical Society*, Japan 32, 310-21.

TAKASHI, J., HIDAKA, H. & YAMADA, K. (1965). Effects of mycelial forms on citric acid fermentation in submerged culture. *Agriculture & Biological Chemistry* (Tokyo) 29, 331-6.

TAKAMI, W. (1968). Vitamin by-products from citric acid fermentation by *Aspergillus niger*. *Hakko Kyokaishi* 26, 550-4.

TALBURT, D.F. & JOHNSON, G.T. (1967). Some effects of rare earth elements and yttrium on microbial growth. *Mycologia* 59, 492-503.

TRINCI, A.P.J. (1970). *Archiv für Mikrobiologie* 73, 353-601.

TRINCI, A.P.J. (1971a). Exponential growth of the germ-tube of fungal spores. *Journal of General Microbiology* 67, 345-8.

TRINCI, A.P.J. (1971b). Influence of the peripheral growth zone on the radial growth-rate of fungal colonies on solid media. *Journal of General Microbiology* 67, 325-44.

TRINCI, A.P.J. & BANBURY, G.H. (1967). A study of the growth of the tall conidiophores of *Aspergillus giganteous*. *Transactions of the British Mycological Society* 50, 525-33.

TRINCI, A.P.J. & GULL, K. (1970). Effect of actidione, griseofulvin, and triphenyltin acetate on the kinetics of fungal growth. *Journal of General Microbiology* 60, 287-92.

UPITIS, V., RINKIS, G.& PAKALNE, D. (1966). Comparative determination of available copper in soils by chemical and microbiological methods. *D. Sb. Latv. Filial. Vses. Obschchest. Pochvoved.* 4, 54-67.
VEZINA, C. & SINGH, K. (1975). Transformation of organic compounds by fungal spores. In *The Filamentous Fungi* Vol. 1, pp 158-192. Edited by J.E. Smith and D.R. Berry, London: Edward Arnold Publishers.
WAHEED-KHAN, A.& CHUGHTAI, I.D. (1960). *Current Science (India)* 29, 100-101.
WALCZYNA, J. (1966). Suitability of the *Aspergillus niger* method for the determination of available nitrogen, phosphorus, and potassium in grassland soils. (in Polish). *Rocz. Glebozn. 16 (1) 275-296 (in Chem. Abst. 1967 67 42914.*
WHITAKER, A. & LONG, P.A. (1973). Fungal pelleting. *Process Biochemistry* 8, 11-19 (Nov.).
YANAGITA, T. (1957). Biochemical aspects of the germination of *Aspergillus niger* conidiophores. *Archiv für Mikrobiologie* 26, 329-44.
TANAGITA, T. & KOGANE, F. (1962). Growth and cytochemical differentiation of mould colonies. *Journal of General and Applied Microbiology* 8, 201-13.
YANAGITA, T. & KOGANE, F. (1963a). *Journal of General and Applied Microbiology* 9 (2) 179-187.
YANAGITA, T. & KOGANE, F. (1963b). Cellular differentiation of growing mould colonies with special reference to phosphorus metabolism. *Journal of General and Applied Microbiology* 9, 313-30.
YUILL, J.L. & YUILL, E. (1954). Spiral development of mycelium in *Aspergillus niger* colonies. *Nature* 203,643.
ZALOKAR, M. (1959). Enzyme activity and cell-differentiation in *Neurospora. American Journal of Botany* 46, 555-60.
ZLOCHEBSKAYA, I.V. & ROBOTNOVA, I.L. (1966). Toxicity of lead to *Aspergillus niger. Mikrobiologyia* 35, 1044-52.
ZODROW, K., STEFNAIAK, O. & JANUCHOWSKI, J. (1966). Suitability of some *Aspergillus niger* strains for the quantitative determination of magnesium. (in Polish). *Rocz. Wysz. Szk. Rolm. Pozn. 30 608-619.* In *Chem. Abs.* 1970, 72, 75564.

Industrial Enzymes Produced by Members of the Genus *Aspergillus*

P. BARBESGAARD

Novo Industri A/S, DK-2880 Bagsvaard, Novo Alle 1, Denmark

INTRODUCTION

The year 1896 marked the start of modern microbial enzyme technology. In that year the Japanese Takamine marketed in the West his famous preparation Takadiastase. The preparation was a rather crude mixture of hydrolytic enzymes which were prepared by growing *Aspergillus oryzae* on wheat bran. The method used was very close to the methods which have been used for thousands of years in the preparation of Koji in the Orient. So what happened in 1896 was really a transfer of technology from East to West.

The enzyme industry has since grown to a substantial size and many new enzyme sources have been found, but the Aspergillus enzymes still play an important part. In Table 1 are listed the most important microbial enzymes produced today. Production on a world-wide basis is given in terms of tons of active enzyme protein produced per year, together with the approximate time of availability. Koji is probably the greatest product, but the exact amount produced is difficult to evaluate.

Table 1. The most important industrial enzyme preparations from micro-organisms.

Name	Commercially available before			Current production (t.enzyme protein/year)
	1900	1950	1976	
Koji	X			?
Bacillus protease		X		500
Amyloglucosidase			X	300
Bacillus amylase		X		300
Glucose isomerase			X	50
Microbial rennet			X	10
Fungal amylase	X			10
Pectinase		X		10
Fungal protease		X		<10

AMYLOGLUCOSIDASE, FUNGAL AMYLASE,PECTINASE, AND FUNGAL PROTEASE

Amyloglucosidase

Amyloglucosidase is by far the most important of these enzymes It is produced by several members of the *Aspergillus niger* group: *A.niger, A.awamori, A.phoenicis, and A.foetidus*. The enzyme attacks both the alpha-1.4 and the alpha-1.6-bonds in amylose and amylopectin (Fig.1), and it is primarily used to hydrolyse starch completely to glucose.

Amyglucosidase as we know it today is a relatively new product achieving commercialisation in the fifties as a result of development work initiated in the U.S. during the Second World War in order to economize on malt in the manufacture of alcohol.

The initial products were made by cultivation of the fungus on moistened bran, or in submerged culture by using crushed grain, and the whole culture was used in the saccharification process. Good, complete saccharification was obtained, although laboratory results were often confusing, and it was impossible to obtain a complete hydrolysis of the starch without using a concomitant yeast fermentation. The problems were soon solved when it was realised that *A.niger*, apart from producing the well-known alpha-amylase and amyloglucosidase, also produces a transglucosidase which catalyses the transfer of glucose from 1.4-bound oligosaccharides to 1.6-boun oligosaccharides (Fig.2). In the starch hydrolysis during the alcohol fermentation this was of no importance, because the maltose was constantly removed by the yeast so that the equilib rium would be pushed to the left, but in starch hydrolysis where a rather high maltose concentration prevails in part of the process, this equilibrium would be pushed to the right with the result that the final

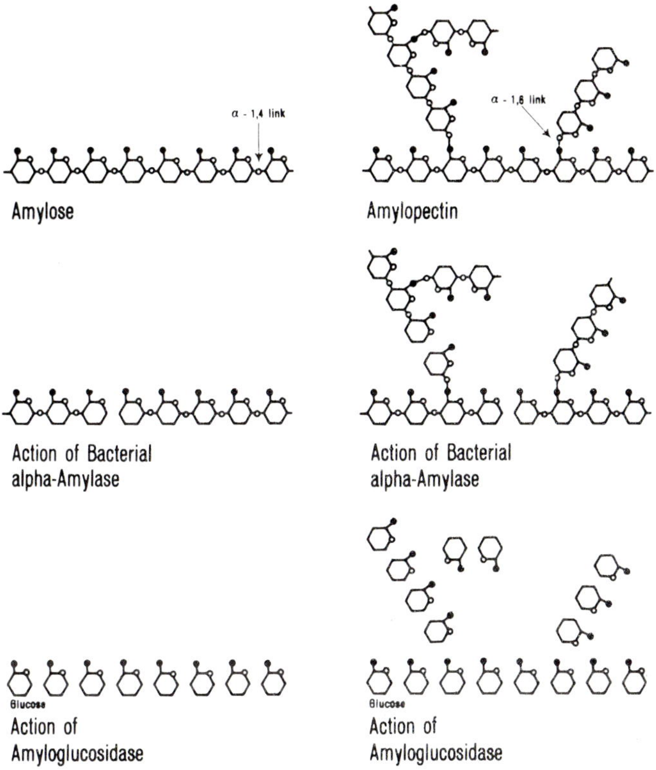

Fig. 1. Action of bacterial a-amylase and amyloglucosidase on amylose and amylopectin.

hydrolysate would contain a mixture of oligosaccharides and glucose. The solution to the problem was to remove the transglucosidase and fortunately this could be done by way of absorption, e.g. on bentonite or by mutation of the producing fungus. When this transglucosidase was removed it was possible to use the *A.niger* to obtain an almost complete hydrolysis of starch to glucose.

The process was soon adopted by the starch industry which previously had had to use acid hydrolysis in the manufacture of dextrose, a process which resulted in a relatively low yield of dextrose. By using the new process, yields of about 95% dextrose were made possible. A modern starch hydrolysis process is shown in Fig.3. Two processes are involved; first the starch must be liquefied, by the use of acid hydrolysis, or preferably by bacterial alpha-amylase. Then the hydrolysis is completed by the use of amyloglucosidase. The hydrolysate will contain a number

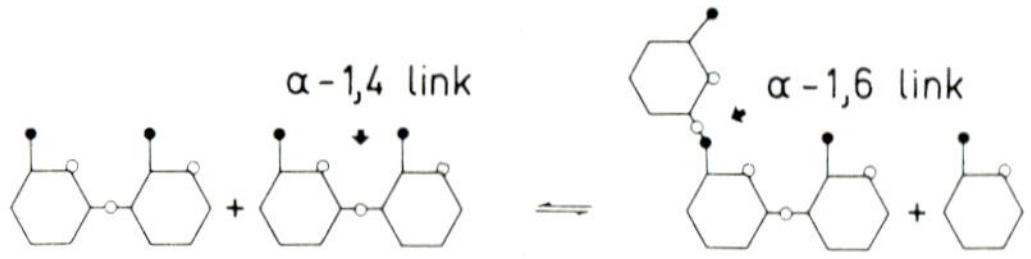

Action of transglucosidase

Fig. 2. Schematic representation of the transfer of glucose from 1.4-bound to 1.6-bound oligosaccharides by the action of transglucosidase.

of impurities which will have to be removed by ion-exchange resins or activated carbon, so that a sparkling clear and colourless product remains. Finally, the dextrose is crystallized out and the mother liquor containing the oligosaccharides is returned to the process.

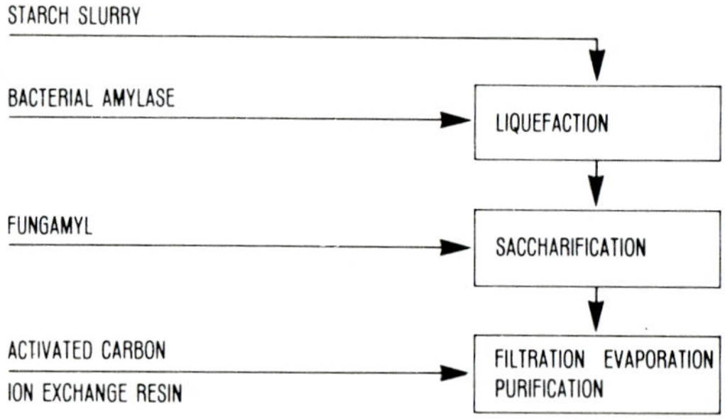

Fig. 3. Process diagram for the production of dextrose from starch.

The starch manufacturer has the following expectations of the amyloglucosidase product he buys: low cost; standardised activity; chemical and microbial stability; freedom from disturbing side-activities; and guaranteed absence of toxic substances and harmful micro-organisms. The activities of the enzyme manufacturer aim at meeting these requirements and hopefully doing so a little better than the competitors. This requires good technology and above all careful and extensive control at every stage of the manufacturing process. A flow sheet of the manufacturing process as shown in Fig.4 reveals nothing extraordinary. A standard fermentation batch process in standard fermenters is followed by a simple recovery process. This consists of removal of the mycelium by filtration or centrifugation, if necessary followed by a removal of the transglucosidase by absorption on bentonite or similar materials followed by a concentration. This may be done conventionally by evaporation or by reverse osmosis. Finally, a preservative is added and the preparation standardized to the declared activity. The problems lie in the details. Amyloglucosidase is a very competitive product and the manufacturer can only make a profit if he is able to produce the enzyme in very high yield. This is obtained by using super-producing mutants and by optimizing the fermentation medium and the fermentation conditions. In this way it will often be possible to obtain yields better than 1 gram of active enzyme protein per litre.

The 'in production' quality and purity of the product is ensured by a number of control measures. The microbiological control consists of macroscopic and microscopic evaluation of the fermentation at frequent intervals and furthermore streaking of agar plates. In this way it can be established that the culture is pure and that the producing organism has its normal appearance. The fermentation has to be run under completely aseptic conditions, but usually this does not create too many problems because the fermentation takes place at a low pH so that growth of bacteria is inhibited, but if a batch is infected it will be discarded even if it has good enzyme activity. The recovery process is not aseptic, but the degrees of contamination are kept extremely low partly due to the low pH and partly because the liquid is cooled to a temperature below $10^{\circ}C$ as soon as possible after it has left the fermenter. The non-toxicity of the product is secured by using good and non-toxic raw materials throughout the process, for instance the main ingredient in the fermentation medium is usually good-quality maize. Furthermore, equipment of stainless steel or other inert material is used throughout the process. All this would help little if the fungus produced toxic substances during its growth, such as mycotoxins or other toxic metabolites. The absence of such materials is ensured by using strains of species which are not known as toxin-producers, and supported furthermore by testing the product by extensive tests for known mycotoxins as for instance aflatoxin, and by feeding studies on animals which show the general safety of the product. Table 2 gives an example of a specification for a food grade enzyme.

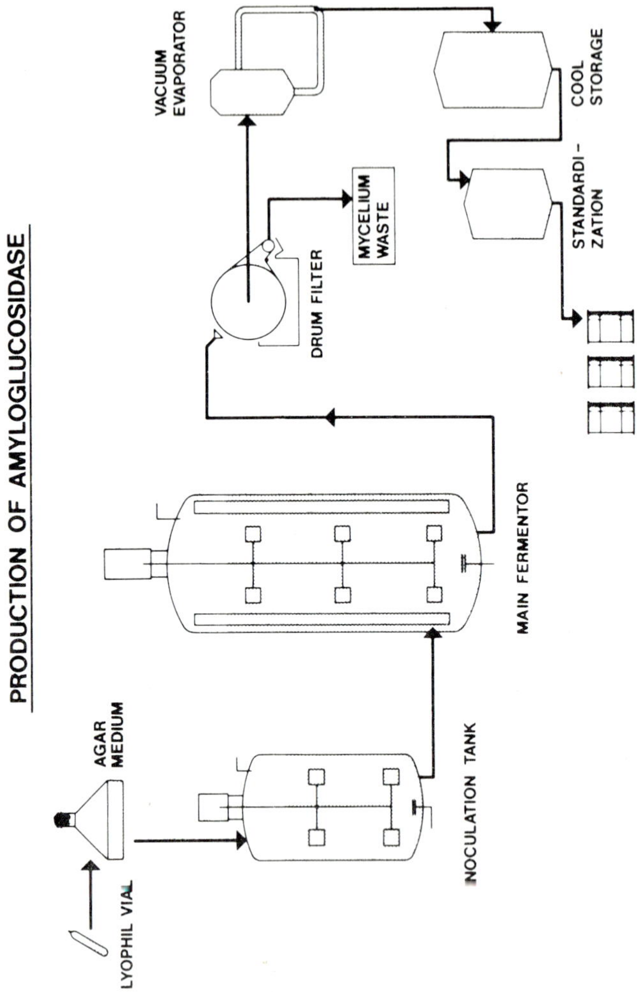

Fig. 4. Flow sheet of fermentation and purification process in the production of transglucosidase.

Table 2. NOVO food grade specifications.

	Method	Limit	
Heavy metals PPM	FCC k972 p 92o II	4o	*)
Lead PPM	Novo Atomic Absorption	4o	*)
Arsenic PPM	FCC 1972 p 865	3	*)
Aflatoxin	AOAC 197o 26.o18	Not detected	*)
Total viable count/g	Novo B 5/3	5×10^4	
Coliform/g	Novo B 6/3	30	*)
E. coli	Novo B 16/3	Not detected	
Moulds/g	Novo B 7/3	100	
P.aeruginosa/10 g	USP XVIII p 85o	Not detected	*)
Salmonella/ 50 g	Novo B 1o/2	Not detected	*)

*) WHO or FCC Recommendation

Fungal Amylase

The next Aspergillus contribution to the list of important enzyme preparations is fungal amylase. While amyloglucosidase is a relatively new product, fungal amylase is the oldest microbial enzyme marketed in the West, being the main active component in the famous Takadiastase preparation. In modern fermentations the enzyme is produced by strains of *Aspergillus oryzae*. It was made originally by Takamine by cultivating the fungus on moist wheat bran, and this process is still in use in Japan, but it is also possible to produce the enzyme in submerged culture, the method preferred in the West. It is characteristic that the fungus grown on moist bran will form several enzymes such as protease and lactase in good yield together with the amylase, whereas only amylase is produced in the submerged culture. Because of the close relationship of *A.oryzae* with aflatoxin-producing Aspergilli, these products have been suspected of aflatoxin content. Fortunately, the amylase-producing strains do not form aflatoxins but as a safety measure the products are always tested for aflatoxin content.

Fungal amylase is an alpha-amylase, so that it breaks down starch and dextrins by hydrolyzing the alpha-1.4-bonds. Unlike amyloglucosidase it does not attack alpha-1.6-bonds and the reaction products are mainly maltose and maltotriose. A major use is in the baking industry. The content of alpha-amylase in wheat flour varies considerably and wheat deficient in amylase will, as a standard procedure, be fortified with amylase. For this purpose amylase is superior to malt amylase because it is free from protease activity. Its low inactivation temperature is an additional advantage to both malt amylase and bacterial alpha-amylase. In the starch industry fungal amylase is used in the production of maltose syrups. The process is outlined in Fig.5. The first step is a liquefaction of the starch which is partly hydrolysed to dextrins. This may be done by acid treatment, but in this case where it is important that the minimum amount of dextrose is formed, a bacterial alpha-amylase should always be used. Fungal amylase is used in the saccharification step for the degradation of dextrins, mainly to

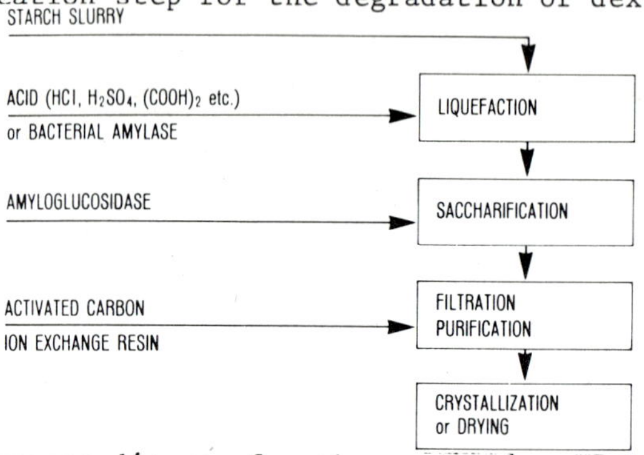

Fig. 5. Process diagram for the production of high maltose syrups by use of fungal amylase (Fungamyl).

to maltose. The maltose syrup is then filtered, concentrated by evaporation, and purified by use of activated carbon and ion-exchange resins. The characteristics of high-maltose syrups are given in Table 3. They are characterised not only by a concentration of maltose up to 60% but also by a very small amount of dextrose. They are used inter alia in the canning industry and in the production of hard candy, where it is important that these syrups are rather non-hygroscopic and have a low tendency to crystallize.

Table 3. Characteristics of high maltose syrups.

DE	40-50%
Maltose ($C_{12}H_{22}O_{11}$)	45-60%
Dextrose ($C_6H_{12}O_6$)	2- 7%

Fungal amylase is also used in the production of the so-called high-conversion syrups (Fig.6). Here the process starts from an acid-treated syrup and in the saccharification step a mixture of fungal amylase and amyloglucosidase is used. The result is shown in Table 4. These syrups contain maltose and dextrose in in approximately equal amounts. They are used in the production of e.g. soft candy, where it is of importance that they are stable to crystallization at room temperature above 4°C, even at very high concentrations. They also find some application in brewing and baking and in the production of soft drinks. The relative economic importance of microbial amylases is given in Table 5. It is interesting to note the importance of amyloglucosidase, whereas fungal amylase, which initially was the dominating enzyme, now plays a very modest role.

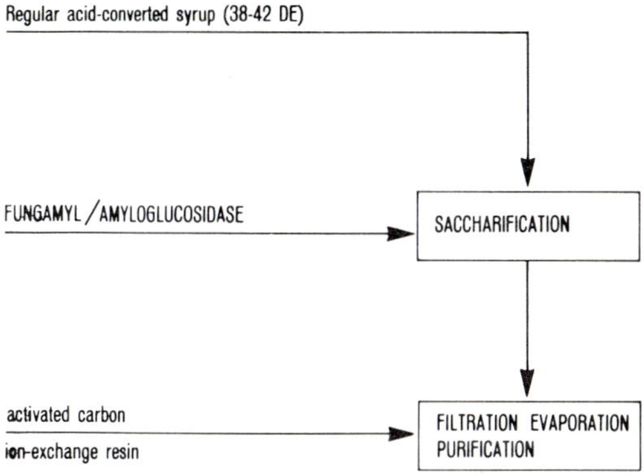

Fig. 6. Process diagram for the production of high conversion syrups by use of fungal amylase and amyloglucosidase.

Table 4. Characteristics of high conversion syrups.

DE	63-67%
Dextrose ($C_6H_{12}O_6$)	34-40%
Maltose ($C_{12}H_{22}O_{11}$)	33-37%

Table 5. Total sales of microbial amylases

AMG	50%
Bacterial amylases	40%
Fungal amylase	10%

Pectinases

Another enzyme product of longstanding importance is pectinase. Pectin is universally distributed throughout the plant kingdom and likewise pectinase is formed by a larger number of micro-organisms. Commercial production has been attempted with several micro-organisms, but at present only *Aspergillus niger* enzymes are of commercial interest. Pectic substances are built of galacturonic acid units linked together in alpha-1.4-bonds. The carboxyl groups may be free as in pectic acid, or esterified by methyl groups. The term pectins refers to substances with a varying degree of methoxylation. All commercial pectinases are mixtures of several pectic enzymes. The classical scheme includes two types: pectin methyl esterase, which demethylates pectin to pectic acid, and polygalacturonases, which degrade pectic acid further by hydrolyzing the 1.4-glucosidic links, thus finally splitting the molecule into units of galacturonic acid. There is little doubt that most or all commercial preparations contain these two enzymes, but as a matter of fact the whole story is considerably more complicated.

The traditional method of production of pectinase is by cultivation on moist bran. This is still the preferred method for commercial production. The main reason is that by this method it is possible to produce all the pectic enzymes in good yields. In submerged cultivation some of the enzymes are formed in very good yields, for example polygalacturonase, whereas others are produced in negligible amounts. If a good pectinase were to be made by submerged cultivation one would have to mix several products. This makes the process more expensive and so far non-competitive with the cultivation on moist bran. The pH-optimum of most commercial preparations is about 5. But when talking about such an optimum one must remember not only that it reflects the optima of different enzymes, but also that the optimum is different for different substrates.

Pectinases are used almost exclusively in the wine and juice industries and for two main purposes:

1. They are used as so-called filtration enzymes in the pressing process. The high content of pectic substances in grapes and fruits retains water and thus decreases the yield of juice. After degradation of the pectins by use of pectinases a much higher yield is obtained. Furthermore, a better extraction of flavour and colour substances is obtained.

2. They are also used as clarification enzymes in 'must' treatment. In production of 'must' a very cloudy product is normally obtained, because the pectic substances retain small particles in suspension. After a more or less complete degradation of the pectins it is rather easy to clear the

'must' by filtration or sedimentation.

Use of pectinase in the juice industry started in 1930, while use in the wine industry was initiated in the forties. Total sales of pectinases are now little lower than the total sales of bacterial amylases, and estimated as a commercial product, pectinase holds an intermediate position between amyloglucosidase and fungal amylase.

Fungal Protease

The last of the four Aspergillus enzymes is of considerably smaller economic importance up to the present.

The most significant protease is the one produced by *A.oryzae*. This product contains a range of enzymes apart from protease, such as amylase, pectinase, cellulase, and lactase. The product is not very different from the fungal amylase described earlier, and in fact the old Takadiastase could be used equally well as amylase or protease. Today fungal proteases will usually be optimized with respect to protease content, but they may still be produced by the same strains as the amylases. *A.oryzae* produces at least three proteases, an acid, a neutral, and an alkaline, which make the preparation active over a very broad pH-range.

By far the most important use is in the baking industry, and for this it is essential that the enzyme is active at a low pH, contrary to bacterial proteases. The rather low inactivation temperature is also of importance. It acts by modifying the gluten, thereby improving the handling properties of the dough. Its use is highly dependent on the properties of wheat and on baking traditions. It is therefore used extensively in some countries and not at all in others.

The proteases from *A.saitoi* and *A.niger macrosporus* have an acid pH-optimum. They find limited application, the most important being as digestive aids where the *A.oryzae* enzyme is also used. These preparations are very popular in some countries for example Japan. The therapeutic value of such preparations is debatable, mainly because most of the preparations available contain too low enzyme activity to be of any importance.

MISCELLANEOUS ENZYMES

The unique capacity of the Aspergilli for hydrolysis of various substrates has attracted scientists for many years and innumerable publications dealing with new enzymes from Aspergilli can be found. Most of these have not found any commercial application and will not be mentioned here. Some, which are of actual or potential importance, are listed in Table 6.

Table 6. Miscellaneous enzymes produced by *Aspergillus* species.

Glucose oxidase	Lipase
Catalase	α-galactosidase
β-glucanase	β-galactosidase
Tannase	Pentosanase (Xylanase)
Naringinase	Cellulase

Glucose oxidase

Glucose oxidase oxidises glucose to gluconic acid and hydrogen peroxide. It is an intracellular enzyme produced by *A.niger*. The enzyme has two main uses: it is used in the food industry to remove residual glucose or residual oxygen from certain products, and it is used in clinical analysis for determination of glucose. Its use in the food industry is declining while its use as an analytical reagent is increasing. The enzyme may be formed as a by-product of the gluconic acid fermentation.

Catalase

Catalase is generally incorporated in glucose oxidase preparations to catalyze the decomposition of hydrogen peroxide. In cold pasteurization of milk excess hydrogen peroxide is likewise removed by catalase. A further application is in the conversion of latex to foam rubber. The enzyme can be obtained from *A.niger*.

Beta-Glucanase

The presence of beta-glucan in barley and malt may give filtration difficulties in beer production. The problem can be solved by addition to the mash of beta-glucanase, which hydrolyses the beta-1.4 and the beta-1.3-bonds of beta-glucan. Compared with bacterial beta-glucanase this enzyme has certain advantages. Due to a lower pH-optimum it can be used for glucan break-down in fermentation or lagering, if this has not been accomplished in the mashing process. The enzyme is produced by *A.niger* or *A.phoenicis*.

Tannase

In certain foods tannins give astringent and undesirable flavou: Tannase has been suggested for use in such cases. The enzyme is produced by strains of *A.niger* and *A.oryzae*. It has only been used to a very limited extent.

Naringinase

In citrus peel and juice a bitter tasting glucoside, naringin, can be hydrolysed to non-bitter components by use of naringinase. Naringin is a rhamnose-containing glucoside, and the enzyme preparation contains rhamnosidase as well as beta-glucosidase activity. The enzyme is produced by *A.niger* and was detected as an impurity in commercial pectinase preparations.

Lipases

Lipases are produced by many micro-organisms, and *A.niger* is no exception. Potential markets are in the dairy industry, in digestive aids, and in the detergent industry. Apart from this, their uses have been suggested in many areas, e.g. in the degreasing of bones for production of gelatin.

Alpha-Galactosidase

Alpha-galactosidase hydrolyses sugars like melibiose and raffinose. Its main use would be in the production of sugar from beet molasses and in the purification of soy protein. The enzyme can be produced by several Aspergillus species, as for instance *A.niger*.

Lactase

Lactase can be used in the dairy industry for the production of new or improved products, and in the baking industry, and in the pharmaceutical industry as a digestive aid. There are three well-known sources of the enzyme: *Saccharomyces fragilis*, *E.coli* and *A.niger*. In the dairy industry the *fragilis* lactase is at present the preferred enzyme. For use in the baking industry the fungal enzymes seem to have an advantage due to their lower pH-optimum. There is at present a growing interest in lactase for the treatment of waste whey. The demonstration of lactose intolerance in an unexpectedly high number of people has excited interest in an enzyme preparation which would allow these people to consume milk.

Pentosanases

At present pentosanase preparations are used in the baking industry to hydrolyse the insoluble pentosans of wheat flour. This is a market of minor importance, but through the xylanases the pentosanases have a link to the field of hemicellulase and cellulase, and thereby to a field of much wider and much more controversial aspects of application.

Cellulase

It is not likely to be an exaggeration to say that among enzymes of possible practical importance cellulase is at present the one which is most extensively explored. Cellulases are produced by a very large number of micro-organisms. Preparations based on enzymes from *A.niger* are available, but for most applications the strain of choice will be *Trichoderma viride*. The reason for this is that the Aspergillus enzyme has very little activity towards native cellulose. It acts preferably on modified cellulose or cellulose derivatives.

It is customary to conclude a discussion of this nature with some predictions of the future. Will the Aspergilli which have served Mankind as enzyme sources for thousands of years continue to give new important products?

There is a strong possibility that they will since the Aspergilli can hydrolyse almost every compound and are found in varying ecological niches over a wide range of temperatures. They grow quickly and sporulate profusely and from a nutritional viewpoint are inexpensive. These features make them very popular for bulk fermentation. The large number of species in the genus and the wide variety of enzymes they produce as a group make them readily adopted as stepping stones to new applications. An already effective example is the recent development of glucose isomerase, where Aspergillus produced amyloglucosidase is incorporated in the process leading from starch to fructose syrups.

New and exciting developments in which the Aspergilli will be expected to play their part include the use of immobilised enzyme systems for producing specific metabolites. The potential for molecular transformations and the synthesis of defined polymers will take us out of the hydrolytic era and into the next.

Citric Acid Production by *Aspergillus niger*

D. R. BERRY,* A. CHMIEL** and Z. Al OBAIDI*

* *Dept. of Applied Microbiology, University of Strathclyde, Glasgow, U.K.*
** *Technical University of Łodz, Łodz, Poland*

INTRODUCTION

Although the production of citric acid by fermentation using *Aspergillus niger* has been a commercial process since 1923 (Prescott & Dunn, 1959), there still remains many gaps in our knowledge and understanding of the fermentation. Recent reviews indicate that we do not know with any certainty which biochemical pathways are involved in the formation of citric acid in the *A. niger* fermentation and that there is still no unanimity concerning the optimum physiological conditions required to obtain a high yield (Smith, Nowakowska-Waszczuk & Anderson, 1974; Lockwood, 1975). Resolution of these problems is not facilitated by the fact that much of the published work is derived from academic institutions which have little knowledge of the commercial process, and that most of the information obtained by commercial companies remains unavailable. This lack of information is conspicuous in the field of genetics and strain selection (see Pontecorvo, 1976). This review attempts to give an adequate appreciation of the general state of knowledge of the fermentation so that any new material and the problems still remaining can be appreciated in context.

DESCRIPTION OF FERMENTATION PROCESS

The process of citric acid production can be divided into three phases, viz. the preparation of the substrates and inoculum, the fermentation stage, and product recovery. Traditionally, the fermentation is inoculated with conidia which have been produced in surface culture. This technique is, as far as we are aware, still used for the inoculation of production scale fermentations using both surface and submerged liquid culture. Molasses has been most widely used as the carbon substrate, while substrates such as starch hydrolysates have also been used. When crude substrates are used it is necessary to reduce the level of iron and other trace metals which interfere with citric acid production. This is normally achieved either by precipitation with ferricyanide or by the use of a cation exchange resin. Recently, a successful fermentation has been obtained using date syrup which had been deionised using a cation exchange column (Al Obaidi & Berry, unpublished results). The critical nature of the medium composition will be discussed below.

Two processes of fermentation are in widespread use for the commercial production of citric acid. In the traditional shallow pan process, pans made from aluminium or stainless steel are filled with medium, inoculated by blowing spores across the surface, and then incubated in racks in a controlled environment. As the fermentation develops a thick mat of mycelium forms over the surface of the medium and this is then responsible for the conversion of the carbohydrate in the medium into citric acid. In the submerged process medium containing up to 16% sugars is inoculated. The fermentation then continues in two distinct phases, an initial growth phase and a subsequent production phase in which growth is restricted.

In the final stage of the process, product recovery, the mycelium is filtered off and the citric acid precipitated out as calcium citrate by the addition of calcium hydroxide. The citric acid is regenerated using sulphuric acid then cleaned up before being finally purified by crystallisation.

STRAIN SELECTION

Strains of *Aspergillus niger* vary considerably in their capacity to produce citric acid and excrete it into the growth medium at a high concentration. Unfortunately, there have been few reports in the Western literature of experiments in strain improvement. No doubt more has been carried out within the confines of the industrial laboratory.

The amount of citric acid produced by *A. niger* strains can be increased by induced mutation (Gardner, James & Rubbo, 1956;

Imshenecky, Solnceva & Kuranova, 1960; Millis, Trumpy & Palmer, 1962; Scherbakova, 1963, 1964; Ilczuk, 1968; Gradel, 1969; Seichertova & Leopold, 1969a; Scherbakova & Lanskaya, 1971). Two- or more-stage mutagenesis and combined treatment by physical and chemical mutagenic agents has been found to be more advantageous than the use of a single mutagen to improve citric acid production in *A. niger* (Imshenecky & Kuziurina, 1966; Scherbakova, Karadzhova & Eremakova, 1974; Nowakowska-Waszczuk & Zakowska, 1976.

Das & Nandi (1969) and Das (1972) compared the effectiveness of nitrogen mustard, colchicine, ultraviolet and γ radiation for producing improved yields of citric acid and found that γ irradiation was the most effective.

Reports in the literature suggest that marked differences in the sensitivity to metals occurs between strains. Strains having a reduced sensitivity to metals have been reported by Hisanaga & Nakamura (1966).

The occurrence of parasexual hybridization in *A. niger* may provide an alternative technique for obtaining active strains for citric acid production (Ciegler & Raper, 1957; Seichertova & Leopold, 1969b; Ilczuk, 1971a,b,c).

MORPHOLOGY OF THE ORGANISM

Aspergillus niger can grow in liquid culture either as loose filamentous mycelium or in pellet form depending upon the growth conditions. The characteristics of fungal pellets and the factors influencing their development have been reviewed by Whitaker & Long (1973). Most studies on citric acid fermentation indicate that pellets with a loose structure and a diameter not exceeding 1 mm provide the best mycelial form for producing high yields of citric acid (Snell & Schweiger, 1949; Cimerman, Johanides & Skafar, 1976; Zhuravsky, 1955; Martin, 1957; Clark, 1962; Chmiel, 1973). Pellet formation is influenced by many environmental factors; pH, aeration and agitation, and inoculum size appear to be particularly critical in producing a suitable inoculum for the citric acid fermentation. The relationship between growth conditions, the morphology of the mycelium and citric acid production have been discussed by Martin & Demain (1977).

In surface culture the mycelium grows as a folded mat, 2.3 cm thick on the surface of the medium. Different layers of the mat can be differentiated and citric acid biosynthesis appears to be associated with the bottom layer of cells which have direct contact with the medium.

KINETICS OF THE FERMENTATION

The kinetics of citric acid production of a sucrose mineral salts medium have been studied by Chmiel (1975a) who has distinguished five successive phases in the fermentation (Fig. 1).

The period of spore germination was followed by an exponential growth phase. The maximum value of specific growth rate in this phase was dependant on temperature: μ max = 0.13 h^{-1} at 26° and 0.28 h^{-1} at 32°C. After exponential growth there was a period of growth disturbance which can be explained by perturbation in the metabolism of *A. niger* caused by the exhaustion of ammonia and phosphate from the medium.

This was followed by the period of citric acid production which continued until the available carbohydrate was exhausted when production ceased. Analysis of the kinetics of the fermentation (Fig. 1) clearly shows the separation of the periods of maximum mycelial growth and of product formation.

A major factor in the efficient production of citric acid, particularly in submerged culture, is the amount of biomass formed. Usually good citric acid producing strains form 10-15 g dry weight of mycelium per litre of medium. Chmiel (1977) has studied the effect of biomass density on citric acid production rate in shake flask culture (Fig. 2). The results indicated that at low mycelium concentrations (up to 4 g/1) the rate of citric acid formation was proportional to the biomass level:

$$\frac{\Delta P}{\Delta t} = \beta X,$$

where $\Delta P/\Delta t$ is the rate of citric acid production (g/1/h); X is the mycelium concentration (g/1); and β is a metabolic activity of the cells, i.e. the maximum rate of citric acid production per unit of biomass dry weight depended on the physiological state of the cells (g/g/h). The metabolic activity of three-day old mycelium was 0.15 g/g/h.

The rate of product formation reached a maximum level approximately 1 g/1/h at a mycelium concentration of about 9-12 g/ and decreased at higher mycelial density (Fig. 2). The changes in the specific rate of product formation, i.e. the citric acid production rate per unit of biomass, with the different mycelium concentrations resulted from an inhibition of citric acid synthesis by decrease in the level of dissolved oxygen in the high mycelium densities, (Fig. 2). It is well known that in submerged cultivation conditions the oxygen absorption rate decreases with increasing amounts of mycelium (Brierley & Steel, 1959; Chain, Gualandi & Morisi, 1966).

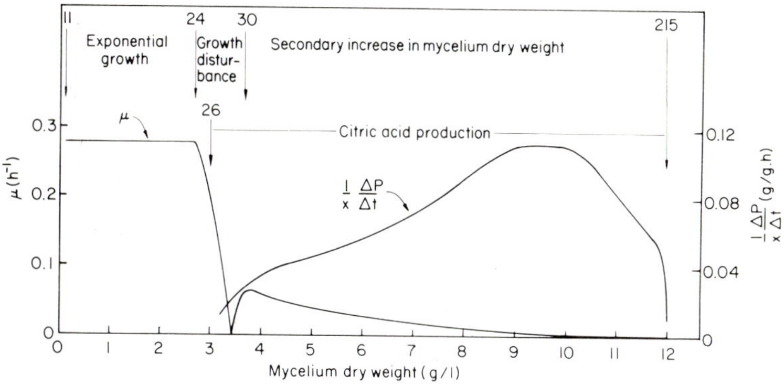

Fig. 1. Relationship between the biomass proliferation and the specific growth rate (μ) and specific rate of citric acid production ($\frac{1}{X}\frac{dP}{dt}$) during the time course of shaken culture of *A. niger* B-64-5. Temperature 32° for the first 48 h and 26° during the following, citric acid accumulation stage. The arrows point out the time of changes in the process kinetics (Chmiel, 1975a).

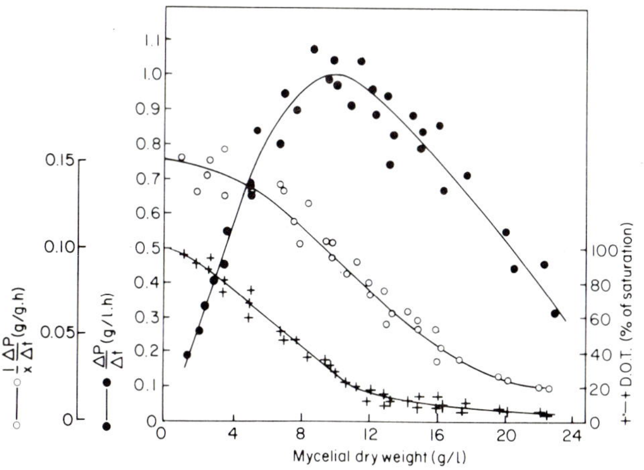

Fig. 2. Effect of mycelium concentration on the rate ($\frac{\Delta P}{\Delta t}$) and specific rate ($\frac{1}{X}\frac{\Delta P}{\Delta t}$) of citric acid production and the dissolved oxygen tension (D.O.T.)

Legend to Fig. 2 cont'd....

Conditions: 1 litre shake flasks containing 100 ml of second-stage medium and different amounts of 72 h-old mycelium; temperature 26°, process duration 24 h (Chmiel, 1977).

PHYSIOLOGY OF THE FERMENTATION

The citric acid fermentation is normally carried out by a batch process in which the conditions undergo a continuous change as a result of consumption of nutrients, multiplication of cells, and accumulation of products, etc. Although some of these changes may be in a direction which is favourable to citric acid production, e.g. the reduction in pH as citric acid is produced, this is not necessarily so. It has been reported that the accumulation of high concentrations of citric acid may inhibit the rate of citric acid formation. Addition of 10 g/l citric acid to the fermentation resulted in such a reduction. This appears to be associated with a reduction in the pH to below 2.0 and to involve specific effects on the metabolism of *A. niger* (Chmiel, 1977).

In this section an attempt will be made to summarise the effects of certain critical physiological factors on the rate of citric acid production.

Carbohydrates

The carbohydrate sources used in the citric acid fermentation are usually crude substrates such as molasses and starch hydrolysates. If these substrates contain high levels of metal ions they must be removed using one of the techniques described previously. The formation of citric acid is clearly not subject to any form of catabolite repression since up to 16% sugars can be added at the beginning of the fermentation. Although some of this is used during the growth phase, high levels remain at the beginning of the production phase. Using a two-stage process it has been shown that 16% sugars do not inhibit citric acid production (Chmiel, 1975b, 1977). Shu & Johnson (1948) reported that higher yields could be obtained on sucrose than on molasses and suggested that molasses may contain inhibitory substances other than metals which had been removed. However, the presence of stimulatory substances in molasses has also been reported (Leopold & Valtr, 1961; Fencl & Vintrova, 1963; Ilczuk, 1967; Janota-Bassalik, Olczyk & Chodorawska, 1970).

Phosphate

The influence of phosphate levels on the citric acid fermentation is somewhat confusing. It has been reported that low levels of phosphate favour citric acid production (Scücs, 1944; Karow & Waksman, 1947) and that the ratio of alkali metals to phosphate is critical (Bhatti, 1967). The addition of 0.1 g/l of KH_2PO_4 to the production phase has been reported to cause a significant reduction in the rate of citric acid production (Chmiel, 1975b). A similar inhibition was found to be associated with an increase in 5-ketogluconic acid, oxalic acid and malic acid by Martin & Steele (1955). However, Shu & Johnson (1948a) have reported a strain (Wis 72-4) in which citric acid production was favoured by up to 5 g/l phosphate. They repeated the experiments of Scücs (1944) and found that low levels of phosphate were only essential if iron and zinc were present in excess. They suggested that the results of Scücs (1944) could be attributed to the use of impure medium components; however, it seems more likely that strain differences were more critical (Table 1).

Table 1. Regulatory action of the mineral constituents of medium on the growth and citric acid accumulation of *A. niger*

Constituent	Concentration	Stimulation
NH_4^- , NO_3^-	Less than 1.0 g/l	Citric acid production
	High, above 1.0 g/l	Mycelium growth
PO_4^{--}	Deficiency	Citric acid production
	High, above 0.1 g/l	Mycelium growth
Trace metals		
FE^{++}	Less than 1.0 mg/l	Citric acid production
Zn^{++}	Less than 1.0 mg/l	
Mn^{++}	Less than 1.0 μg/l	
Cu^{++}	0.1 - 1.0 mg/l	
Hydrogen ions	High, pH 2.0-3.0	Citric acid production
	Above pH 3.0	Mycelium growth

Nitrogen

Ammonium nitrate, ammonium sulphate, ammonia, sodium nitrate and potassium nitrate have all been used as nitrogen source for the citric acid fermentation. The use of ammonium carbonate has been shown to minimise the formation of pigment and slime and also facilitates the extraction procedure (Snell & Schweiger, 1949). It was further reported that ammonium carbonate suppressed oxalic acid production.

Although citric acid fermentation appears to be nitrogen limited, the addition of small quantities of nitrogen to the production phase does stimulate acid production (Shepherd, 1963). Addition of 0.5 g/l ammonium nitrate to the production stage of the two stage process described by Chmiel (1975b) also stimulated citric acid production. Higher levels of ammonium nitrate have, however, been reported to stimulate oxalic acid production.

pH

The pH of the medium is critical for citric acid production. The most favourable hydrogen ion concentration being between 2.0 and 3.0. This low pH is not only optimal for citric acid production but also minimises the formation of undesirable organic acids such as gluconic and oxalic acids. It also reduces the risk of contamination by other organisms. A higher pH is favoured during the initial stages of the fermentation since this facilitates spore germination; however, pH drops during the fermentation as the citric acid accumulates in the weakly buffered medium. The optimal starting pH varies depending upon the nature of the substrate; whereas pH 2.4 - 4.0 is optimal for defined media an initial pH of 6.0 - 7.5 is required in molasses medium. It has been reported that molasses contains small quantities of acetic acid which inhibits germination at low pH values.

Temperature

Temperature is an important factor in the control of the physiology of *Aspergillus niger*. Temperatures of 28-30°C have been proposed to obtain high yields and rapid rates of accumulation of citric acid (Prescott & Dunn, 1959; Karklinsh & Probok, 1972). At higher temperatures the fermentation process is very rapid and abundant mycelial growth occurs causing large amounts of sugar to be oxidized to CO_2 thus lowering the yield of citric acid. Higher levels of oxalic and gluconic acids are also produced. The lower temperature, on the other hand, permits higher yield of citric acid and minimal amounts of other acids. However, this prolongs the process (Kitos, Campbell & Tomlinson, 1953; Martin, 1957; Chmiel, 1975a). A correct pattern of temperature control during the citric acid fermentation should be differentiated according to the changes in physiologica state of the organism and with the economics of the fermentation process (Kovats, 1946; Jakubowska & Florianowicz, 1965; Karklinsh & Probok, 1972; Chmiel, 1975a). The temperature course applied in the Soviet Union for the surface citric acid fermentation can be quoted as an example: inoculation with conidia is carried out at 36-42°, mycelium growth at 34-36° and citric acid production at 32-34°C (Karklinsh & Probok, 1972).

Aeration and agitation

It is well known that the oxygen absorption rate decreases with increasing amounts of mycelium in submerged culture (Brierley & Steele, 1959; Chain, Gualandi & Morisi, 1966). This is important since oxygen has been shown to stimulate the formation of citric acid by *Aspergillus niger* (Shu, 1953; Clark & Lentz, 1961). The changes in the specific rate of citric acid formation observed by Chmiel (1977) (Fig. 2) resulted from an inhibition of its synthesis caused by a decrease in the level of dissolved oxygen at high mycelial densities. Kwapisz (1976) maintained the dissolved oxygen level at 60% of saturation by increasing the air flow rate and the agitation speed. In these experiments the respiration rates of the cells reached a maximum at the period of maximum mycelial growth rate then was considerably reduced during the period of citric acid production.

It is known, however, that the cells are more sensitive to oxygen deficiency just into the citric acid production phase. If air supply to the medium was stopped for some time in this fermentation phase, product formation was limited or fully stopped in spite of the resumption of aeration (Batti, 1966; Kovats & Gackowska, 1976). The later the air stoppage occurred in the fermentation, and also the longer time it was off, the greater was this effect. However, it has been reported that the substitution of air with oxygen causes a reduction in the growth of the mycelium and a stimulation of citric acid production.

Metals and other medium components

The concentration of metals in the medium appears to play an important role in the control of the citric acid fermentation. The addition of excess iron has been shown to lower the yield of citric acid (Schweiger, 1961). However, small quantities of iron accompanied by a limited quantity of zinc are essential to obtain a high citric acid yield (Shu & Johnson, 1948b; Trumpy & Millis, 1963). Wold & Suzuki (1976a) have reported that the onset of the acid accumulation phase is controlled by the concentration of zinc. At low zinc concentrations (below 1 μM) growth became limited and citric acid formation began. If additional zinc was added during the production phase (up to 2 μM) then it was found that growth could be resumed. Copper has been used as an antagonist to iron (Schweiger, 1961). The addition of copper also offers the additional advantage that it selectively inhibits the growth of contaminating penicillia (Schweiger, 1959). Manganese also has an inhibitory effect on the citric acid fermentation (Shu & Johnson, 1947). However, Tomlinson, Campbell & Trussell (1951) have reported that manganese and copper were essential for citric acid production in surface culture.

Many variations in the requirements of the citric acid fermentation for metals have been reported in the literature. There can be little doubt that if high yields are to be obtained these need to be investigated each time a new strain or substrate is used. In our own studies, different metal requirements were observed when the experiments were carried out in flasks and in a 7 l. Chemap fermenter, even when the same strain and the same carbon substrate were used. Two published media which have been found to be satisfactory are shown in Table 2.

TABLE 2. Comparison of media used in shake flask culture and in stirred fermenter culture

Fermenter System (Snell & Schweiger, 1949)		Shake Flask System (Shu & Johnson 1948a)	
Sucrose	140 g/1	Sucrose	140 g/1
$(NH_4)_2CO_3$	2.0 g/1	$(NH_4)_2CO_3$	2.5 g/1
KH_2PO_4	0.2 g/1	KH_2PO_4	2.5 g/1
$MgSO_4 \cdot 7H_2O$	0.1 g/1	$MgSO_4 \cdot 7H_2O$	0.25 g/1
Zn^{++}	5.0 mg/1	ZN^{++}	0.25 mg/1
		Cu^{++}	0.06 mg/1
		Fe^{+++}	1.3 mg/1
pH 2.6		pH 3.8	

BIOCHEMISTRY

It is generally considered by most workers that the excess citric acid produced in the citric acid fermentation is synthesised by the normal mitochondrial biosynthetic pathway, namely the condensation of acetyl Co A with oxaloacetate derived from the tricarboxylic acid cycle (Smith, Nowakowska-Waszczuk & Anderson, 1974) (Fig. 3). The evidence for this is rather sparse and can be considered in two parts; (1) whether the TCA cycle operates in *Aspergillus niger* during the period of citric acid biosynthesis and (2) whether the citric acid produced extracellularly is formed by the TCA cycle. Ahmed, Smith & Anderson (1972) have presented evidence that certain TCA cycle enzymes, viz. the citrate condensing enzyme, aconitase, NAD and NADP dependent dehydrogenasess, and isocitrate dehydrogenase are as high during the production phase as during the growth phase of the fermentation. These results support those of La Nauze (1966) and disprove the theory of Ramakrishnan, Steel & Lent (1955) that overproduction of citric acid occurred because of the disappearance of aconitase and isocitrate dehydrogenase. Furthermore, Ahmed *et al.*, (1972) demonstrated that the TCA cycle was active in mitochondria isolated during the production phase. Mitochondria isolated from the production and growth phases were tightly coupled and gave good RC and ADP/O ratios when different TCA cycle intermediates were used as substrates. Studies on the level of intermediates in the TCA cycle gave no support to the concept that the TCA cycle is blocked during citric acid production (Ahmed, 1970) (Table 3).

Apart from the report by Shu, Funk & Neish (1954) that 40% of the citric acid was formed from recycled dicarboxylic acid, there is little direct evidence that extracellular citrate is derived from the TCA cycle. If it is produced by this pathway then the necessary question which arises is, which biochemical mechanism operates to replenish the supply of organic acids, i.e. one of the CO_2 fixation reactions from phosphoenol pyruvate or pyruvate, or the glyoxylate cycle. The glyoxylate cycle enzymes appear to be present (Ahmed *et al.*, 1972) and both pyruvate carboxylase and phosphoenol pyruvate carboxykinase have been reported in *A. niger* (Woronik & Johnson, 1960). Verdoff & Spradlin (1976) reported that pyruvate carboxylase, isocitrate lyase and oxalacetic hydrolase are more active in *A. niger* during oxalic acid and citric acid production.

The energy balance of the citric acid fermentation has been studied by Meyrath (1967) who concluded that the theoretical maximum yield was 85% w/w from hexose but that this could only be achieved in conditions in which citric acid was the sole metabolic product of the culture. Verhoff & Spradlin

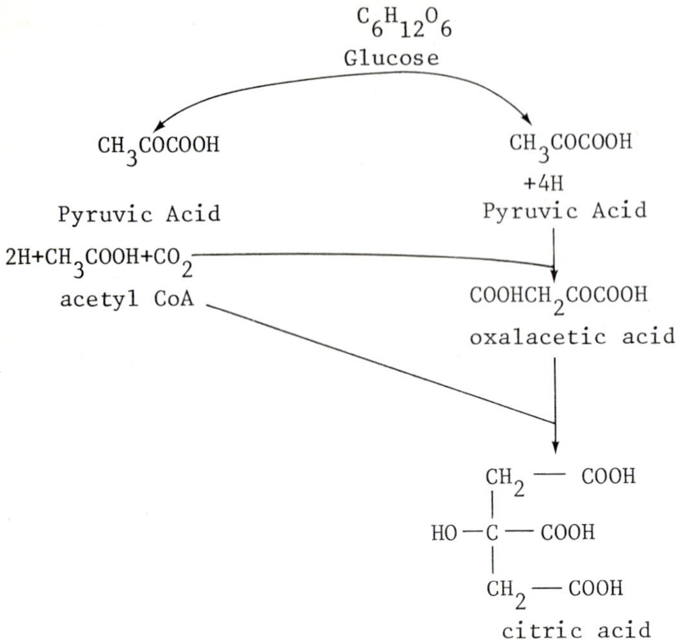

Fig. 3. A simplified and idealised scheme of citric acid synthesis in *A. niger* (Meyrath, 1967)

(1976) have studied the application of mass and energy balances to the citric acid fermentation using the assumption that after the termination of growth all the available carbon is converted into citric acid. It is also considered in the study that oxalic acid is often produced as an undesirable by-product and also that there is little net carbon dioxide formation during the production phase. Their analysis supports the metabolic pathway shown in Fig. 4.

They offer the following speculation on the control of citric acid formation:- "Initially, consider cells growing normally during exponential growth phase. As they enter the declining phase of growth, the media carbon to nitrogen ratio becomes high and the cells produce lipids. As the lipid concentration gets high, the glyoxylate cycle enzymes are induced in an attempt to reconvert the lipids to sugars. However, there is plenty of sugar and the net result is an increase in oxaloacetate which induces the oxaloacetate hydrolase to produce oxalic acid and acetate. Oxalic acid is an inhibitor in the citric acid cycle from succinate to fumarate generating an excess of succinate. With the accumulation of both succinate and oxalic

Table 3. Levels of intermediates in *Aspergillus niger* at different stages of the citric acid fermentation

Intermediates	Duration of fermentation		
	48h	120h	168h
Pyruvic	5.5	5.5	6.0
Fumaric	5.0	5.7	5.1
Succinic	5.7	6.2	6.0
α -ketoglutaric	4.6	5.5	5.2
cis Aconitic	3.5	4.0	4.0
Malic	3.0	4.2	3.5
Citric	6.0	12.0	14.0
d Isocitric	2.5	3.0	2.5

(Values represent moles/20 g wet weight mycelium)

acid which is converted to glyoxylate, the isocitrate lyase enzyme reaction is reversed fromthat normally associated with the glyoxylate cycle. Before the citric acid cycle is stopped and this reaction is completely reversed a certain amount of oxalic acid production is possible. If the fermentation coniditions from this point forward are correctly maintained, *A. niger* can continue to produce citric acid according to the metabolic scheme shown in Fig. 4."

The critcal influence of trace metals on the citric acid fermentation is not clearly understood. Iron has been reported to cause a stimulation of aconitase activity and a concurrent decrease in the rate of citric acid biosynthesis (La Nauze, 1966). The effect can be reversed by the addition of copper which acts as an antagonist of iron. Zinc has been reported to influence several enzyme systems associated with sugar metabolism in *Aspergillus niger*. Bertrand & de Wolfe (1959) considered that zinc was necessary for the induction of phosphofructose dehydrogenase and triose phosphate dehydrogenase in *A. niger* and also

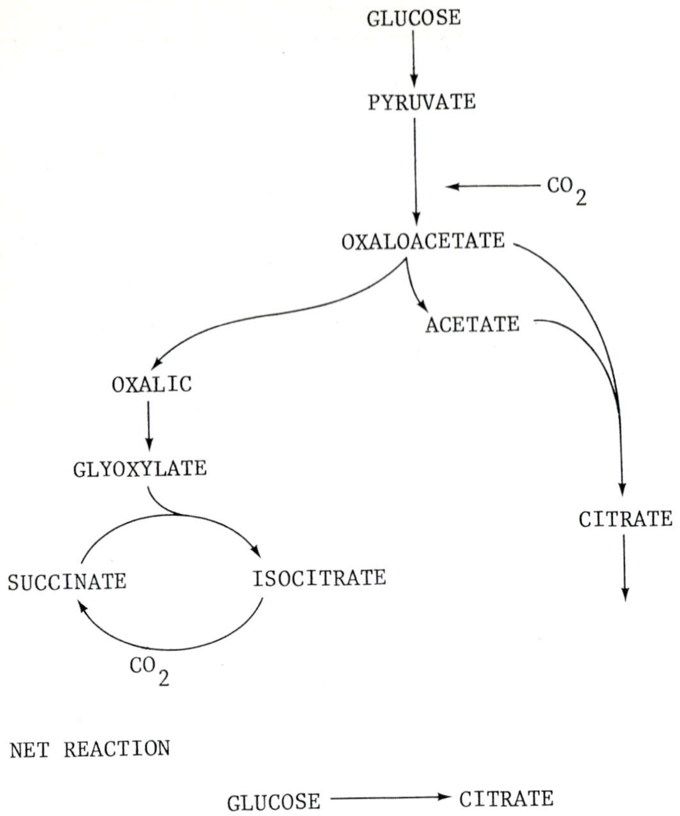

Fig. 4. Metabolic pathways for citric acid production (Verhoff & Spradlin, 1976).

influenced the activities of aldolase, glucose 6-phosphate dehydrogenase and 6-phosphogluconate dehydrogenase. The levels of aldolase and glucose 6-phosphate dehydrogenase were found to be reduced in zinc deficient cultures by Geser (1962) but the level of 6-phosphogluconate dehydrogenase was increased.

The addition of cAMP (10^{-6} M) to the citric acid fermentation stimulated the production of citric acid and the rate of sucrose utilisation (Wold & Suzuki, 1973). This appears to be a specific effect since adenosine, ATP and cAMP did not have any effect below 10^{-3} M; however, cAMP was slightly inhibitory and the ADP inhibited by 42%. The addition of theophylline at 10^{-2} M also stimulated citric acid production, but not to the same extent as cAMP. Subsequent work indicated that the effect of cAMP could only be obtained when *Aspergillus niger* was growing on a medium containing low levels of zinc, indicating that cAMP was not primarily concerned with the induction of citric acid biosynthesis, which was still dependent upon a deficiency of zinc (Wold & Suzuki, 1976 a,b). The rate of growth of the fungus was influenced by the level of cAMP. During the growth phase the rate of growth was augmented by cAMP whereas during the production phase it was depressed. The mechanism of action of cAMP is not known although evidence has been presented that it may be involved in the development of the correct pellet form, for citric acid production. Addition of cAMP was found to increase the adhesiveness of conidia and facilitate their adhesion into clumps (Wold & Suzuki, 1973b). Such clumps have been implicated in the formation pf pellets. More recently, the relationship between pellet size and intracellular and extracellular levels of cAMP at different stages of the citric acid fermentation has been studied (Al Obaidi & Berry, unpublished results) (Fig. 5). It was found that treatment which gave a good yield of citric acid had low levels of intracellular and extracellular cAMP in the early stages of growth but had high levels at the beginning of the production phase.

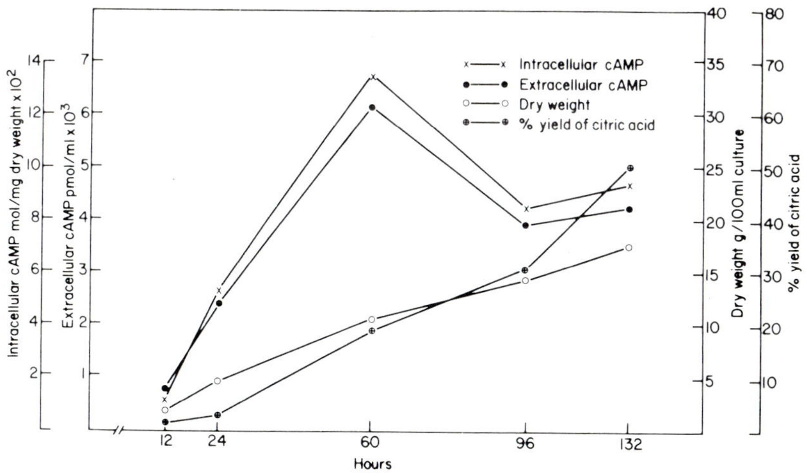

Fig. 5. Levels of intracellular and extracellular cAMP during a typical citric acid fermentation in shake flasks (Al-Obaidi & Berry, unpublished results).

FERMENTATION TECHNIQUES

Many different techniques have been employed to carry out the citric acid fermentation. Traditionally, industry still use the surface culture technique. However, at the present time a much greater emphasis has been placed on the use of submerged culture. Batch techniques in stirred tank reactors are in general use (Lockwood, 1975) and there is considerable interest in the potential use of continuous culture.

Many difficulties arise in the continuous culture of filamentous fungi connected with the morphology of mycelial growth, the physiological differentiation of mycelium cells and the technical problems of mould cultivation, i.e. wall-growth, over growth on pipes, etc. There are, however, a few data concerning continuous cultivation of *Aspergillus niger* in citric acid fermentation processes.

Zagrodzki, Wysocki & Szwajcowska (1969) have used a continuous culture system for surface citric acid fermentation. The process was carried out for 20-21 days giving the citric acid production rate of about 0.7-1.0 kg/day at the flow rate of medium of 8-9 l/day. Citric acid yield calculated from sucrose used was 65%.

Zhuravsky, Terenteva, Aglish & Fishkova (1971) have developed a semi-continuous method for proliferation of vegetative inoculum for submerged citric acid fermentation. Lesniak (1974) has obtained very interesting results in studies with semi-continuous and continuous submerged processes in 10 and 60 l fermenters. The rate of citric acid production for the semi-continuous fermentation was 20-26 g/l day, and for the continuous process 25-29 g/l day. Product yield, calculated from sugar consumption, was 75-85 and 80-88% respectively. The fermentation was kept running for 21 days. During this time some morphological and physiological changes of mycelium were found which caused the reduction in the citric acid-producing activity of *A. niger* mycelium.

Tower fermenters have also been used for the continuous production of citric acid (Greenshields & Smith, 1971; Martin & Waters, 1952). In tower fermenters the morphology of the organism has been shown to be critical in its influence on the rheological properties of the broth and to influence the growth rate of the organism (Cocker, 1975).

A different technique which has been used for the production of citric acid is solid state fermentation (Hisanaga & Nakamura 1966; Yamada, 1965). A strain of *Aspergillus niger* which was tolerant to trace metals was used to carry out a fermentation in which sweet potato or wheat bran were used as the carbon substrate. Another possible technique which is being investigated in Strathclyde University is the use of the biodisc fermenter (Divers, Anderson & Blain, unpublished results).

PRODUCTION OF OTHER ORGANIC ACIDS BY *ASPERGILLUS* SPECIES

The organisms of the genus *Aspergillus* produce a rich variety of organic acids. In this review only citric acid has been referred to with the occasional reference to oxalic acid which can occur as an undesirable by-product of the fermentation. Other organic acid fermentations which have been studied in some detail are the production of itaconic acid by *A. terreus* and *A. itaconicus* (Smith *et al.* 1974; Jakubowska, this volume) and gluconic acid production by *A. niger* (Lockwood, 1975). However, a long list of organic acids has been reported as being produced by different *Aspergillus* sp (Laskin & Lechevalier, 1973). These include succinic acid, glutaric acid, fumaric acid, glycolic acid, malic acid, itatartaric acid and glyoxylic acid. A list of the major companies in the world producing organic acids by the fermentation route is given in the Handbook for Microbiology (Laskin & Lechevalier, 1973).

REFERENCES

AHMED, S.A. (1970). The effect of different systems of cultivation and of various factors on the production of citric acid and the role of mitochondria in this process. *Ph.D. Thesis, University of Strathclyde, Glasgow, Scotland.*

AHMED, S.A., SMITH, J.E., ANDERSON, J.G. (1972). Mitochondrial activity during citric acid production by *Aspergillus niger*. *Transactions of the British Mycological Society* 59, 51-60.

BATTI, M.A. (1967). Process for producing citric acid. *United States Patent* 3, 355, 067.

BERTRAND, D. & DE WOLF, A. (1959). Premieres, recherches sur le role de l'oligoelement zinc chez. *L'Aspergillus niger*. *Bulletin Societe de Chimie Biologique* 41, 545-54.

BRIERLEY, N.R. & STEEL, R. (1959). Agitation-aeration in submerged fermentation. Part II. Effect of disperse phase on oxygen absorption in fermenter. *Applied Microbiology* 7 57-61.

CHAIN, E.B., GUALANDI, P. & MORISI, G. (1966). Aeration studies. Part IV. *Biotechnology and Bioengineering* 8, 595-613.

CHMIEL, A. (1973). Biosynthesis of citric acid by the submerged method. Part I. *Aspergillus niger* as a 'producer' of citric acid. *Przemysl Fermentacyjny i Rolny* 17, 11-12, 29-34 (in Polish).

CHMIEL, A. (1975a). Kinetic studies on citric acid production by *Aspergillus niger*. Part I. Phases of mycelium growth and product formation. *Acta Microbiologica Polonica Ser B7(24)* 185-93.

CHMIEL, A. (1975b). Kinetic studies on citric acid production by *Aspergillus niger*. Part II. Two-stage processes. *Acta Microbiologica Polonica Ser. B* 7(27), 237-42.
CHMIEL, A. (1977). Kinetics of citric acid production by pre-cultivated mycelium of *Aspergillus niger*. *Transactions of the British Mycological Society* (in press).
CIEGLER, A. & RAPER, K.B. (1957). Application of heterokaryons of *Aspergillus niger* to commercial type fermentations. *Applied Microbiology* 5, 106-10.
CIMERMAN, A., JOHANIDES, V. & SKAFAR, S. (1976). Effect of native polymer additives on pellet differentiation in submerged citric acid fermentation. In *Abstracts of Fifth International Fermentation Symposium*. Ed. H. Dellweg, p. 405, Berlin.
CLARK, D.S. (1962). Submerged citric acid fermentation of ferri-cyanide-treated beet molasses: morphology of pellets of *Aspergillus niger*. *Canadian Journal of Microbiology* 8, 133-6.
CLARK, D.S. & LENTZ, C.P. (1961). Submerged citric acid fermentation of sugar beet molasses: effect of pressure and recirculation of oxygen. *Canadian Journal of Microbiology* 7, 447-53.
COCKER, R. (1975). The morphological development of *Aspergillus niger* in tower fermenter cultures. *Ph.D. Thesis, University of Aston, Birmingham.*
DAS, A. (1972). Strain selection in citric acid fermentation. A review. *Current Science* 41, 593-6.
DAS, A. & NANDI, P. (1969). Improved production of citric acid from *Aspergillus niger* by mutagen treatment. *Indian Journal of Experimental Biology* 7, 275-6.
FENCL, Z. & VINTROVA, E. (1963). A new metabolic regulator of citric acid production. *Nature, London* 199, 905-8.
GARDNER, J.F., JAMES, L.V. & RUBBO, S.P. (1956). Production of citric acid by mutants of *Aspergillus niger*. *Journal of General Microbiology* 14, 228-37.
GESER, G. (1962). Untersuchungen an *Aspergillus niger* van Tiegh. Über Zinkaufnahme und Aktivitätsveränderunge einiger Enzyme bei Zinkmangel. *Archiv für Mikrobiologie* 41, 408-40.
GRADEL, A. (1969). Züchtung von *Aspergillus niger* -Stammen für die technische Citronensäuregärung durch Kombinationen einer Adaptationsmethode nach Leopold und UV-Bestrahlung. *Die Nahrung* 13, 265-9.
GREENSHIELDS, R.N. & SMITH, E.L. (1971). Tower fermentation systems and their applications. *Chemical Engineer* May, No. 249, 182-90.
HISANAGA, W. & NAKAMURA, S. (1966). *Japan Patent 16555/66*.
ILCZUK, A. (1967). Effect of best molasses medium on the yield of citric acid fermentation. *Gazeta Cukrownicza* 75, 11-13.
ILCZUK, Z. (1968). Genetics of citric acid producing strains of *Aspergillus niger*. I. Citric acid synthesis by morphological mutants induced with UV. *Microbiologica Polonica* 17, 331-6.

ILCZUK, Z. (1971a). Genetik der Citronensaure erzeugendedn Stamme *Aspergillus niger*. 3 Mitt Citronensauresynthese durch erzaungene Heterokaryonen swischen auxotrophen Mutanten von *A. niger*. *Die Nahrung* 15, 251-62.
ILCZUK, Z. (1971b). Genetik der Citronensaure erzeugendan Stamme *Aspergillus niger*. 4 Mitt. Citronensaure synthese durch heterokaryotische Diploide von *A. niger*. *Die Nahrung* 15, 381-8.
ILCZUK, Z. (1971c). Genetic aspects of selection of *Aspergillus niger* strains useful for citric acid fermentation. Part II. *Przemysl Fermentacyjny i Rolny* 15, No. 8, 13-8 (in Polish).
IMSHENECKY, A.A. & KUZUIRINA, L.A. (1966). *Aspergillus niger* variability produced by combined effect of ethylenimine and UV-irradiation. *Mikrobiologiya* 35, 813-6 (in Russian).
IMSHENECKY, A.A., SOLNCEVA, L.I., & EURANOVA, N.F. (1960). Experimental variability of *Aspergillus niger*. Part 2. Formation of citric acid by mutants of *A. niger* produced by UV-irradiation. *Mikrobiologiya* 29, 351-7 (in Russian).
JAKUBOWSKA, J. & FLORIANOWICZ, T. (1965). The effect of temperature changes on growth and citric acid formation by *Aspergillus niger*. *Zeszyty Naukowe Politechniki Lodzkiej, 67 Chemia Apozywcza* 9, 21-33 (in Polish).
JANOTA-BASSALIK, L., OLCZYK, C. & CHODOROWSKA, M. (1970). Searching, in beet molasses, activators or inhibitors for citric acid fermentation. *Acta Microbiologica Polonica Ser B.* 2, 137-40.
KARKLINSH, R. YA. & PROBOK, A.D. (1972). Biosintez organicheskich kislot. *Riga, Zinatne* (in Russian)
KAROW, E.O. & WAKSMAN, S.A. (1947). Production of citric acid in submerged culture. *Industrial Engineering Chemistry* 39, 821-5.
KITOS, P.A., CAMPBELL, J.J.R. & TOMLINSON, N. (1953). Influence of temperature on the trace elements requirements for citric acid production by *Aspergillus niger*. *Applied Microbiology* 1, 156-9.
KOVATS, J. (1946). The influence of temperature on the rate of citric acid fermentation by *Aspergillus niger*. *Przemyzl Chemiczny* 2, 122-8 (in Polish).
KOVATS, J. & SACKOWSKA, L. (1976). Effect of interruption in aeration of medium in deep citric acid fermentation on molasses solution. *Przemysl Fermentacyjny i Rolny* 20, 22-5 (in Polish).
KWAPISZ, E. (1976). *Dr. Thesis, Technical University of Lodz, Poland.*
LA NAUZE, I.M. (1966). Aconitase and isocitric dehydrogenase of *Aspergillus niger* in relation to citric acid production. *Journal of General Microbiology* 44, 73-81.

LASKIN, A.I. & LECHEVALIER, H.A. (1973). *CRC Handbook of Microbiology*, Vol. 3. Ohio: CRC Press.
LEOPOLD, H. & WALTER, Z. (1961). Über einen die Citronensäuer production durch *Aspergillus niger* Stimulierenden Faktor in Melasse. *Zentralblatt für Bakteriologie* II Abteilung 114, 45-6.
LESNIAK, W. (1974). Introductory studies on continuous citric acid fermentation. Presented in Symposium *Intensification of citric acid production in Poland* Lodz, Poland.
LOCKWOOD, L.B. (175). Organic acid production. In *The Filamentous Fungi* Vol. 1, pp 140-57. Edited by Smith, J.E. & Berry, D.R. London: Edward Arnold.
LOCKWOOD, L.B. & SCHWEIGER, L.B. (1967). Citric acid itaconic acid fermentations. In *Microbial Technology*; pp 183-99 Edited by Peppler, H.J.; New York, Amsterdam, London: Reinhold Publishing Corporation.
MARTIN, J.F. & DEMAIN, A.L. (1977). In *The Filamentous Fungi* Vol. III. Edited by Smith, J.E. & Berry, D.R. London: Edward Arnold. In press.
MARTIN, S.M. (1957). Citric acid production by submerged fermentation. *Industrial and Engineering Chemistry* 49, 1231-2.
MARTIN, S.M. & STEEL, R. (1955). Effect of phosphate on production of organic acid by *Aspergillus niger*. *Canadian Journal of Microbiology* 1, 470-2.
MARTIN, S.M. & WATERS, W.R. (1952). Production of citric acid by submerged fermentation. *Industrial & Engineering Chemistry* 44, 2229-33.
MILLIS, N.F., TRUMPY, B.H. & PALMER, B.M. (1963). The effect of lipids on citric acid production by an *Aspergillus niger* mutant. *Journal of General Microbiology* 30, 365-79.
MEYRATH, J. (1967). Citric acid production. *Process Biochemistry*. 2, 25-27, 56.
NOWAKOWSKA-WASZCZUK, A. & ZAKOWSKA, Z. (1976). The use of N-methyl-N'-nitro-N-nitrozoguanidine, N-methyl-N'-nitrozourea and UV rays to improve the citric acid production in *Aspergillus niger*. Presented in this Symposium as a short paper.
PRESCOTT, S.D. & DUNN, C.G.(1959). *Industrial Microbiology* New York: McGraw-Hill Inc.
PONTECORVO, G. (1976). Presidential Address at 2nd International Symposium on the genetics of Industrial Microorganisms, pp 1-4 Edited by MacDonald, K.D. London: Academic Press.
RAMAKRISNAN, C.V., STEEL, R. & LENTZ, C.P. (1955). Mechanism of citric acid formation and accumulation in *Aspergillus niger*. *Archives of Biochemistry and Biophysics* 55, 270-3.
SCHWEIGER, L.B.(1959). Method of producing citric acid by fermentation. *United States Patent 2, 916, 420.*

SZUCS, J. (1977). Method of producing citric acid by fermentation. *United States Patent 2, 353, 771.*
SCHWEIGER, L.B. (1961). Production of citric acid by fermentation. *United States Patent 2, 970, 084.*
SEICHERTOVA, O. & LEOPOLD, H. (1969a). Die Aktivierung von Stammen des *Aspergillus niger.* I. Die Bildung von Mutanten durch wiederholte Bestrahlung der Stamme mittels VU-licht. *Zentralblatt für Bacteriology II Abteilung* 123, 557-63.
SEICHERTOVA, O. & LEOPOLD, H. (1969b). Die Benutzung der para-sexuallen Hybridisation. *Zentralblatt für Bacteriology II Abteilung* 123, 564-9.
SHEPARD, M.W. (1963). Method of producing citric acid by fermentation. *United States Patent 3,083, 1217.*
SCHERBAKOVA, E. YA. (1963). UV induced variability of *Aspergillus niger.* *Mikrobiologiya* 32, 43-49. (in Russian).
SCHERBAKOVA, E. YA. (1964). Characteristics of biochemically active *Aspergillus niger* variants produced through UV-irradiation. *Mikrobiologiya* 33, 49-55. (in Russian).
SCHERBAKOVA, E. YA. (1970). Variability of *Aspergillus niger* induced by chemical mutagens. Chlekopekarnaya (in Russian) and *Konditerskaya Promyshlennost* No. 9 7.
SCHERBAKOVA, E. YA., KARADZHOVA, Z.S. & ERMAKOVA, V.P. (1974). Study of variability of *Aspergillus niger* producing citric acid induced by separate and combined effect of nitrozomethylurea and UV light. *Mikrobiologiya* 43, 508-513 (in Russian).
SCHERBAKOVA, E. YA. & LANSKAYA, L.N. (1971). A new active strain of *Aspergillus niger* No. 288/9 for submerged citric acid production. *Trudy Leningradskogo Nauchno-Issledovatelskogo instituta Pishchevoy Promyshlennosti* 1, 16-26 (in Russian).
SHU, P., FUNK, A. & NEISH, A.C. (1954). Mechanism of citric formation from glucose by *Aspergillus niger.* *Canadian Journal of Biochemistry and Physiology* 32, 68-80.
SHU, P. & JOHNSON, M.J. (1948a). The interdependence of medium constituents in citric acid production by submerged fermentation. *Journal of Bacteriology* 56, 577-85.
SHU, P. & JOHNSON, M.J. (1948b). Citric acid production by submerged fermentation with *Aspergillus niger.* *Industrial and Engineering Chemistry* 40, 1202-5.
SNELL, R.L. & SCHWEIGER, L.B. (1949). Production of citric acid by fermentation. *United States Patent 2,492,667.*
SMITH, J.E., NOWAKOWSKA-WASZCZUK, A. & ANDERSON, J.G. (1974). Organic acid production by mycelial fungi. In *Industrial aspects of biochemistry.* Edited by Spencer, E. Vol. 1, pp. 297-317. Amsterdam: Elsevier.

TOMLINSON, N., CAMPBELL, J.J.R. & RUSSELL, P.C. (1951). Influence of zinc, iron, copper, manganese on the production of citric acid by *Aspergillus niger*. *Journal of Bacteriology* 61, 17-25.

TRUMPY, B.H. & MILLIS, N.F. (1963). Nutritional requirements of an *Aspergillus niger* mutant for citric acid production. *Journal of General Microbiology* 30, 381-93.

VERHOFF, F.H. & SPRADLIN, J.E.(1976). Mass and energy balance analysis of metabolic pathways applied to citric acid production by *Aspergillus niger*. *Biotechnology and Bioengineering* 18, 425-33.

WHITAKER, A. & LONG, P.A. (1973). Fungal pelleting. *Process Biochemistry* 8, November, 27-31.

WOLD, W.S.M. & SUZUKI, I. (1973a). Cyclic AMP and citric acid accumulation by *Aspergillus niger*. *Biochemical and Biophysical Research Communications* 50, 237-41.

WOLD, W.S.M. & SUZUKI, I. (1973b). Promotion of conidial aggregation in *Aspergillus niger* by cyclic AMP and 5' GMP. *Biochemical and Biophysical Research Communications* 55, 824-30.

WOLD, W.S.M. & SUZUKI, I. (1976). Regulation by zinc and adenosine 3'5'-cyclic monophosphate of growth and citric acid accumulation in *Aspergillus niger*. *Canadian Journal of Microbiology* 22, 1093-101.

WORONICK, C.L. & JOHNSON, M.J. (1960). Carbon dioxide fixation by cell-free extracts of *Aspergillus niger*. *Journal of Biological Chemistry* 235, 9-15.

YAMADA, K. (1965). Science in Japan p 401 in American Association for the Advancement of Science, Washington, D.C.

ZAGRODZKI, S., WYSOCKI, B. & SZWAJCOWSKA, K. (1969). A continuous method of surface citric acid fermentation. *Reczniki Technologii i Chomii Sywnosci* 15, 115-26 (in Polish).

ZHURAVSKY, G.I. (1955). A submerged method of citric acid production by *Aspergillus niger*. *Mikrobiologiya* 24, 332-340.

ZHURAVSKY, G.I., TERENTOVA, O.F., AGLISH, I.B. & FISHKOVA, E.S. (1971). Continuous cultivation of *Aspergillus niger* in submerged citric acid production. *Trudy Leningradskogo Nauchne-Issledovatelskoge Instituta Pishchevoy Promyshlennosti i* 27-40 (in Russian).

Itaconic and Itatartaric Acid Biosynthesis

J. JAKUBOWSKA

Institute of Technology of Fermentation and Microbiology, Technical University, Łodz, Poland

INTRODUCTION

Itaconic acid production by *Aspergillus terreus* is an important industrial process providing, from cheap raw materials, a high yield of valuable product for commercial use.

Development of itaconic acid fermentation is connected with a broad application of itaconate and its various derivatives, mainly in the textile and chemical industries. It is used as an intermediate in the preparation of pharmaceuticals, surfactants, in the production of synthetic fibres and plastics, surface coatings, alkyl resins, and in obtaining polymeric materials (Prescott & Dunn, 1959; Lockwood, 1954; Knuth, 1967).

Itaconic acid was first isolated as a fungal metabolite from a culture of *A.itaconicus* (Kinoshita, 1931) while later it was shown to be formed by certain strains of *A.terreus* (Calam, Oxford & Raistrick, 1939).

Extensive studies were carried out in 1945 to 1953 in the Fermentation Division of the Northern Regional Research Laboratories at Peoria to obtain the best itaconic acid producers by induced mutations (Hollaender, Raper & Coghill, 1945; Raper, Coghill & Hollaender, 1945; Lockwood *et al.*,1945). Research on optimal conditions for itaconic acid production in surface and submerged fermentations were carried out at laboratory and semi pilot scale

(Moyer & Coghill, 1945; Lockwood & Nelson, 1946; Lockwood & Ward, 1945; Lockwood & Reeves, 1945; Pfeifer, Vojnowich & Heger, 1952). For the high acidifying isolated strain *A.terreus* NRRL 1960 (Lockwood & Reeves, 1945) the yield of itaconic acid was about 60% from glucose (Pfeifer *et al.*, 1953). The effect of external influences such as source of carbon, pH and several inhibitors on growth and itaconic acid production in *A.terreus* NRRL 1960, was intensively studied by Larsen and co-workers (Larsen & Hovden, 1954; Eimhjellen & Larsen, 1955; Larsen & Eimhjellen, 1955; Jenssen, Larsen & Ormerod, 1956).

Studies developed by Bentley & Thiessen (1957a,b,c) were focussed on elucidating the metabolic pathway of itaconic acid formation, in particular, the role of TCA cycle intermediates.

A.itaconicus was recommended by Japanese investigators for fermentation of beet and cane molasses (Kyowa Hakko, 1959), for black strap molasses and refined wood hydrolysates (Tabuchi, 1961, 1963; Kobayashi, 1967). Differences of some physiological features of itaconic acid producing strains representing both species were pointed out by Tabuchi (1963)(Table 1).

Table 1. Physiological characteristics of *A.terreus* and *A.itaconicus* (Tabuchi, 1963).

Some parameters	*A.terreus* K-26	*A.itaconicus* 34
Optimal pH	2.0	4.0
Optimal temperature ^{o}C	37	28
Limiting nutrient	phosphorus supplied	S compounds to be less than 10 ppm
Sporulating medium	after Moyer & Coghill	addition of 5% NaCl needed
Average rate of batch Fermentation per day in%	2.7 sugar consumed	1.9 sugar consumed
	1.6 itaconic acid formed	0.7 itaconic acid formed
By-product acid formed from sugar consumed	less than 3% of itatartaric acid	about 15% of itatartaric acid

Kobayashi and co-workers from 1963 to 1972 have made significant progress in improving the process of itaconic acid production and to reduce the cost. This has taken into consideration some bioengineering problems, viz. regulation of mycelium concentration, research on heat transfer, modelling of batch and continuous fermentation systems and application of modern techniques for effective separation of itaconic acid from fermentation fluids (Kobayashi & Nakamura, 1966; Nakamura & Kobayashi, 1967; Nakagawa & Kobayashi, 1968; Kobayashi, Nakamura & Nakagawa, 1972).

Table 2. References on itatartaric acid biosynthesis

Author	Strains used	Acids estimated in fermented liquors g/100 ml: itaconic	itatartaric	β hydroxyparaconic	Remarks
Stodola *et al.* (1945)	*A. terreus* NRRL 265 UV-mutant	3.6	0.3	not determined	Surface fermentation in flasks
Arpai (1958)	*A. terreus* UV-mutant	not determined	about	not determined	surface fermentation in flasks
Tabuchi (1963)	*A. itaconicus* IAM-35	5.0	0.2	2.0*	surface fermentation
Kobayashi & Nakamura 1964	*A. terreus* K-26	4.2 total	0.65	not determined	submerged fermentation in 200 l Waldhof fermenters
Jakubowska *et al.* (1967b, 1968)	*A. terreus* NRRL 1960 its UV - and gamma mutants	4.2-5.7 4.2-5.7	0.8-1.2	0.65-0.8	surface fermentation in flasks

* determined by paper chromatography under conditions where β-hydroxyparaconic acid has not been separated from malic acid.

From 1964 to 1975 investigations of the metabolic pathway for fermentative production of itatartaric acid by *A.terreus*, undertaken by Jakubowska and co-workers, were concentrated on five-carbon dicarboxylic branched acid metabolism, in particular, to elucidate the relationship between itaconic and itatartaric acid formation (Jakubowska & Nowakowska-Waszcuk, 1968; Jakubowska & Metodiewa, 1974b, 1975; Zakowska, 1973).

Other acids formed in itaconic acid fermentation. Itatartaric acid was shown to be accumulated in fermented liquors by *A.terreus* and *A.itaconicus*. This unusual C_5 branched dicarboxylic acid was first discovered by Stodola *et al.* (1945) with *A.terreus* resulting from genetic changes induced by mutation. As far as it is known itatartaric acid is not found in nature nor in any other living organism. References in the literature on itatartaric acid occurrence are summarized in Table 2. Since it is not available as a commercial product its usefulness in this sense is unknown.

The origin of itatartaric acid accumulation in itaconic acid fermentation is not understood. According to Arpai (1958a,b) itatartaric acid is formed by oxidation of itaconic acid. However, this supposition is not satisfactorily supported by experiments, viz.(a) it was based on measurements of oxygen uptake by crude enzyme extracts from mycelium; (b) the reaction mixture consisted of itaconic acid in buffered citrate, which might be a substrate of the reaction; (c) the product of oxidation has not been determined; and (d) itaconate oxidase has not been isolated.

Studies on the metabolic pathway for itatartaric acid formation has provided additional information on the acidifying abilities of *A.terreus*. Several isolated biochemical mutants of *A. terreus* NRRL 1960, resulting from UV-and-gamma ray irradiation, were surveyed for their stability and acidifying capabilities (Jakubowska *et al.*,1967a,b, 1974). In 1975 6 strains described in aforenamed papers were deposited in the International Collections with the following accession numbers: UV_{17} as NRRL-A-16-108; 19/IV as NRRL-A-16-109; 14/II as ATCC 32587; 21/I/β/ as ATCC 32588; 25/III as ATCC 32589; K as ATCC 32590.

Characteristics of some physiological properties of active itaconic acid producers is presented in Table 3. High acidifying mutant strains were shown not to be altered in morphological and essential physiological features when compared with parental strains. However, moderate growth of mycelium, which was thin and delicate, together with delayed and reduced sporulation, was characteristic for active itaconic acid producers. Specific media were needed to stimulate conidial formation (Sobocka, Lasota-Oulbowska & Makiedonska, 1970). Itaconic acid content was greater than for the parent strain (up to 4.5 to 5.0 g/100 ml). Itatartaric acid was shown to be formed as a primary product undergoing conversion into the gamma-lactone in acid medium. Equilibrium of both

Table 3. Some physiological properties of *A.terreus* NRRL-1960 and its mutants.

Characteristics	Strain	NRRL 1960	UV-17	21*I (=)	25/III	K
On modified Larsen-Hovden medium with 10% glucose	Itaconic acid g/100 ml (Friedkin,1945)	4,20	5,11	5,70	5,12	5,65
	Itatartaric acid+ β -hydroxyparaconic ac.* g/100 ml	0,3	1,3-1,7	1,5-1,7	1,5	1,0-1,5
	Other acids detected on chromatograms	citric, malic, 2-oxo-glutaric, succinic citramalic	2-oxogluta-ric,malic, succinic, citramalic citric	2-oxoglu-taric	citric, malic	citramalic, malic citric
	Mycelium d.wt.g/100 ml	2.06	0.32-0.45	0.36-0.40	0.48-0.60	0.33-0.45
	Utilization of itaconate	+	–	+	+	+
Nitrogen source:						
NH_4^+	– growth	+	+	+	+	+
	– acid production	+	+	+	+	+
NO_3^-	– growth	+	+	+	+	+
	– acid production	–	–	+	–	–

* determined approximately by paper chromatography (Jakubowska *et al.* 1967b).

compounds in solution in pure water was established at a ratio near 1:1 at pH 2.1 and at 35°C (Jakubowska *et al.*,1970). In fermented liquors after 10 days the ratio of lactone to itatartaric acid amounted to 0.86 (Jakubowska *et al.*,1967b). For some strains itaconic acid represented 90% of total acidity.

Traces of several other acids were detected in fermented liquors, viz. succinic, malic, α-oxoglutaric, citraconic, citric and citramalic acids (Jakubowska & Metodiewa, 1974a; Jakubowska & Zakowska,, 1975). Besides itaconic acid only citraconic acid reacted with bromine solution (Friedkin, 1945). It was formed in amount not exceeding 0.1 g/100 ml (Jakubowska & Nowakowska-Waszczuk, 1968). Of the volatile acids only acetic and propionic were found, the latter being formed in the early phase of growth and was quickly consumed.

Techniques for separation and identification of pure citramalic, itatartaric acid and its gamma-lactone form have been developed by Zakowska (1973).

Laevorotatory citramalic acid was first described as a metabolite of mycelial fungi in the itaconic acid fermentation (Jakubowska *et al.*,1974; Jakubowska & Zakowska, 1975). A sample of (-)-CM was isolated pure from the exchange medium supplied with 0.5% citraconic acid, instead of glucose. The conversion yield of citraconate into (-)- citramalate after 5 days incubation of pre-grown mycelium (strain 23/III) was about 90% (Zakowska, 1973).

Pure citramalic acid was identified as laevorotatory citramalic acid with the following characteristics: white crystals, melting point 108-110°C, optical rotation : $(\alpha)_D^{20^\circ} = -29^\circ$,

elemental analysis: C = 40,8%; H = 5.8%.
Similar data were given by Hulme (1954) for (-)- CM isolated from apple tissue.
β - hydroxyparaconic acid was soluble in water as Ca-salt,

optical rotation: $(\alpha)_D^{20^\circ} = -30^\circ$. Elemental analysis :

C = 38.34%, H= 3.05%. Itatartaric acid, as a biological product, was shown to be optically active. It appears in water solution as a mixture with the gamma-lactone. Ca and Pb salts of itatartarate were insoluble in water. Synthetic itatartaric acid obtained as a standard by chemical oxidation of itaconic acid (Fittig & Kohl,1899), was a racemate, being easily separated from the lactone, due to the solubility of its Ca and Pb salts in water (Zakowska 1973, 1976).

INFLUENCE OF NITROGEN NUTRIENTS ON ACID FORMATION

Acidifying activity of *A.terreus* was shown to be dependent on nitrogen source, in particular, when supplied in the form of

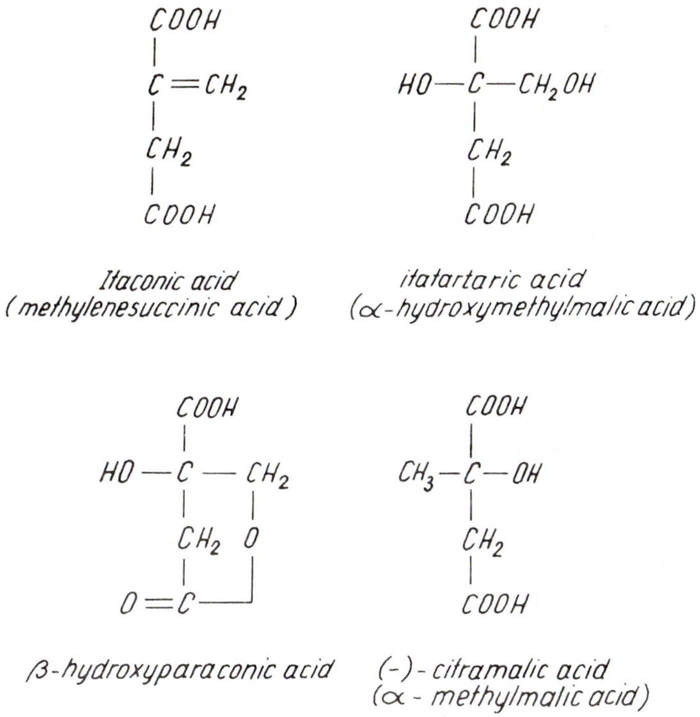

Fig. 1. C_5 branched dicarboxylic acids formed by *Aspergillus terreus*.

ammonium or nitrate ions (Nowakowska-Waszczuk *et al.*,1969). Comparison of ammonium sulphate, ammonium nitrate, magnesium nitrate or glutamic acid, each being supplied as the sole source of nitrogen and at the most appropriate dose corresponding to 0.5-0.6 g N/litre of mineral glucose medium (Lockwood & Ward,1945; Pfeifer *et al.*,1952), revealed that ammonium ions were the most favourable for strain NRRL 1960 as well as for its mutants. Nitrate supported abundant growth up to 4 g dry wt. mycelium/100 ml without acid formation. Similar observations have been reported by others (Moyer & Coghill, 1945). In the same conditions only the mutant 21/I produced a normal amount of itaconic acid in nitrate medium. Ammonium sulphate gave moderate growth (0.9-1.0 g. mycelial dry wt), forming up to 4-5 g of itaconic acid and about 0.54 g of itatartaric acid. Ammonium nitrate stimulated growth (about 2 g dry wt), itaconic acid was produced at the same level, but only traces of itatartaric acid were present.

A response of certain mutants to nitrate present in culture

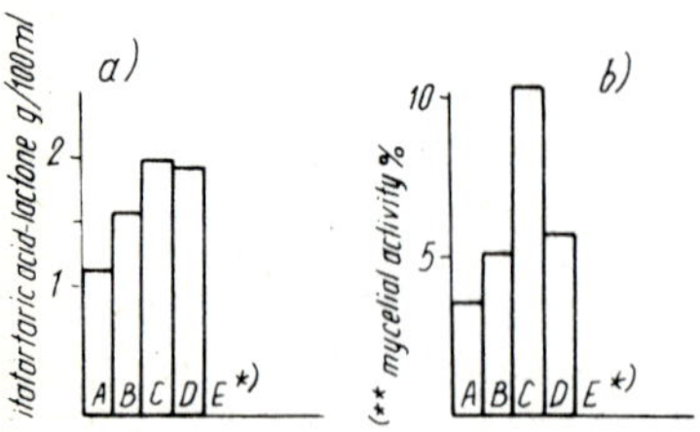

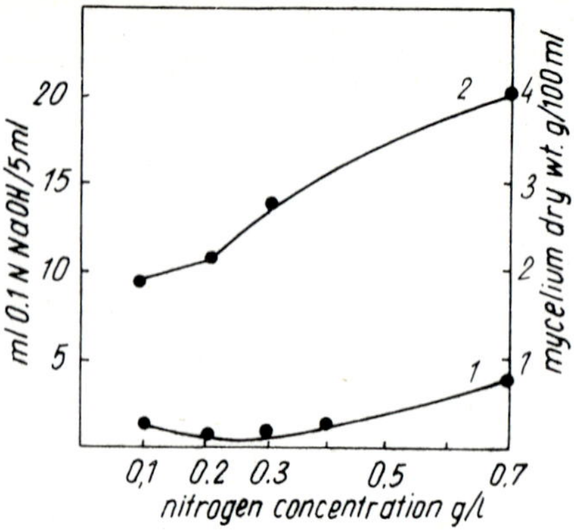

Fig. 2. Itartartaric acid production (a) and mycelial activity (b) in relation to various nitrogen nutrients.
Nitrogen given in A = NH_4^+, NO_3^-, B = NH_4^+, (half NO_3^- dose), C = NH_4^+, D = glutamic acid, E = NO_3^-

* heavy growth, no acids, ** mycelial activity =

$$\frac{\text{itac. a + itartart.a - lact.g}}{\text{mycelium dry wt.g}}$$

Fig. 3. Effect of nitrate on the growth and acid production (after 7 days of cultivation, strains 14/II, 25/III)

1 - titratable acidity, 2 - mycelium

is often recommended as an addition to incomplete mineral media in small amounts, viz. 4.0 ml/litre or 0.18% (Lockwood & Ward, 1945; Pfeifer *et al.*,1952. In complete mineral media containing magnesium, potassium, calcium, phosphorus, sulphur and microelements supplied with tap water ommission of CSL does not cause any decrease of itaconic acid yield (Larsen & Hovden, 1954). The value of CSL for itaconic acid production has been established to be related to its mineral content, mainly potassium, phosphorus and calcium. When the aforenamed media were supplied with KH_2PO_4, 0.01% and Ca 1 mg/100 ml, CSL was not required for growth of mycelium as well as for the fermentation process (Nowakowska-Waszczuk *et al.*,1969).

The requirement for amino acids by *A.terreus* is related to the main source of nitrogen in the cultivation medium. In the presence of nitrate all 13 amino acids introduced with CSL to mineral medium were completely exhausted after 11 days of fermentation (strain 14/II). However, in the presence of N-NH_4 ions or glutamic acid all amino acids remained unused. It is worth pointing out that with glutamic acid no growth was obtained unless potassium phosphate was added. Some gamma-mutants (14/II, 25/III) were shown to be able to produce amino acids from inorganic nitrogen. Without CSL in the medium, containing ammonium sulphate and potassium phosphate, the following amino acids appeared on the 5th day of cultivation: glutamic and aspartic acids, serine, L-cystine, glycine and L-glutamine. On the 14th day all these amino acids were consumed (Jakubowska & Nowakowska-Waszczuk, 1968).

Two mineral media are to be recommended for *A.terreus:* (1) according to Larsen (1957) the best yield of itaconic acid is assured in mineral medium with: 10% glucose or sucrose; $(NH_4)_2SO_4$, 0.3; $CaSO_4$ (anhydrous), 0.3; $MgSO_4.7H_2O$,0.05; KH_2PO_4,0.01 all in percent in tap water, without regulated pH; (2) after Jakubowska *et al.* (1974) for research of metabolism the aforenamed medium (Larsen, 1957) slightly modified, viz. instead of tap water and $CaSO_4$ include distilled water saturated with $CaSO_4$ and enriched in microelements viz. $FeSO_4$ 0.01; $CuSO_4$ 0.02; $ZnSO_4$ 0.01 all in per cent. Initial pH 5.6-5.7.

INFLUENCE OF SEVERAL CARBOHYDRATES

Glucose and sucrose are known to be easily converted into itaconic acid by *A.terreus* (Lockwood & Ward, 1945; Pfeifer *et al.*, 1952; Larsen & Hovden, 1954); however, glucose media were definitely superior for itatartaric acid formation than sucrose containing media. *A.terreus* was able to ferment 16.5% glucose (Lockwood & Ward, 1945) giving about 5.0% itaconic acid, i.e. about 30% yield on the initial sugar. Some mutants were active in the concentration 10 to 22% glucose, however, the residual sugar gradually increased (Jakubowska & Nowakowska-Waszczuk, 1968).

The effect of several substrates on itaconic acid production has been greatly clarified by Larsen and co-workers (Larsen & Hovden, 1954; Eimhjellen & Larsen, 1955). It was shown that growth and acidifying ability of *A.terreus* NRRL 1960 were dependent upon the carbon source and on the degree of acidity of the culture medium. Only sucrose, D-glucose and cellobiose gave the best yields of itaconic acid, viz. 57,52 and 41% w/w of substrate added, respectively.

Intensive growth but decreased itaconic acid production (about 50%) occurred in the presence ofD-mannose, D-fructose, D-xylose, L-arabinose or glycerol. Maltose, L-rhamnose, D-galactose, raffinose, starch and glycogen supported abundant growth without itaconic acid formation. However, maltose was readily converted into itaconate regardless of pH, between 2.1 to 6.0, by non-proliferating mycelium, originating from acid-glucose medium. This phenomenon lead to the conclusion that the formation of the enzyme systems responsible for itaconic acid synthesis occurs only in acid medium, as is observed in fermentation by rapidly lowering the pH (2.1 to 2.2).

Using a non-proliferating mycelial mat, incubated in glucose solution, pH 2.1, the repartition of carbon-atoms of glucose has been recovered in the following products: itaconic acid 54.2; L-malic acid 1.0; CO_2 24.0; and mycelium 15.0. At pH 6.0, regulated with $CaCO_3$ or phosphates, only traces of L-malic, succinic and fumaric acids were detected (Larsen & Eimhjellen, 1955).

MECHANISM OF ITACONATE AND ITATARTARATE FORMATION

The metabolic pathway for fermentative itaconic acid production is not completely recognized. Itaconate is considered to be a secondary metabolite originating from the condensation of an acetate-derived chain with a TCA cycle intermediate (Turner, 1971). The usually proposed sequence of reactions for itaconic acid formation from glucose leads through pyruvate as a primary metabolite generated via the EMP pathway.

The establishment of the pathway was experimentally derived by the isolation of cell-free extracts with aldolase activity from mycelial pads of *A.terreus* (Bentley & Thiessen, 1957b). Pyruvate is believed to be utilized providing the TCA cycle intermediates, viz. acetic, oxaloacetic, citric and cisaconitic acids (Bentley & Thiessen, 1957a,b,c). The operation of the TCA cycle in *A.terreus* is considered to be confirmed by the occurrence of α-oxo glutaric, malic, succinic and citric acids in the culture media and by the presence of an active cis-aconitate decarboxylase (EC 4.1.1.6) in cell free extracts of *A.terreus* NRRL which is responsible for transformation of cis-aconitic acid into itaconic acid (Bentley & Thiessen, 1957b,c). In tracer studies 1,2-^{14}C acetate and labelled 2,3-^{14}C succinate were shown to be incorporated into itaconate (Bentley & Thiessen, 1957a).

From the biochemical point of view it is acceptable that in *A.terreus* an enzyme system associated with citric acid metabolism is involved in itaconic acid formation (Florkin & Stotz, 1968; Lockwood, 1975).

Recently, studies on the mechanism of biosynthesis of itaconic acid revealed no evidence of active TCA cycle enzymes in *A.terreus* (Nowakowska-Waszczuk, 1973). The mitochondria extracted from mycelium of active itaconic acid producing mutants were not able to oxidize TCA cycle intermediates in the presence of ADP, cytochrome c or after disruption of mitochondrial membranes. Reduced pyridine nucleotides, in particular NADH, were the only substrates respired. The high RQ quotients, up to 24.3 in citrate and 26.3 in cis-aconitate, excluded oxidation of citrate in the TCA cycle, which for citrate gives RQ equal to 1.33 (Table 4) (Jakubowska & Metodiewa, 1974b).

The role of citrate and cis-aconitate has been greatly

medium was considered to be an important property indicating that nitrate can be recommended for those strains that do not dissimilate nitrate (Nowakowska-Waszczuk, 1971). Characteristic behaviour of strains in relation to nitrogen source is illustrated in Figs.2-5 (Nowakowska-Waszczuk *et al.*,1969).

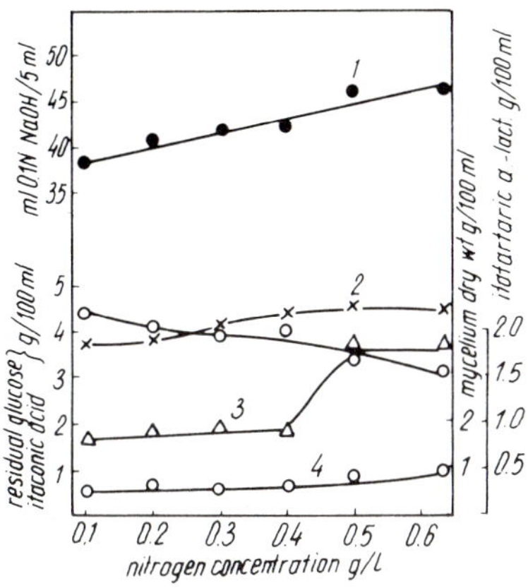

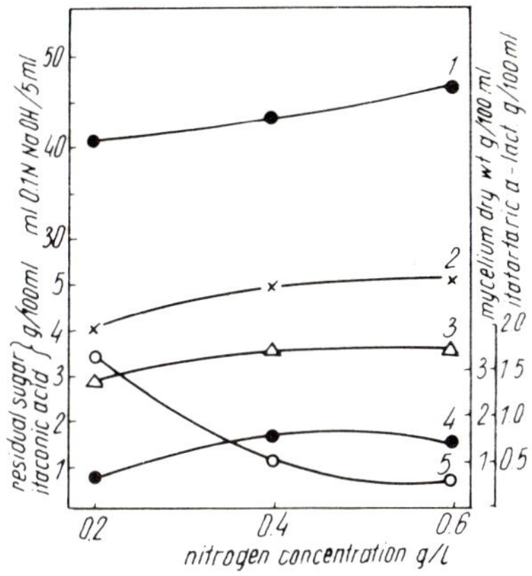

Fig. 4 Effect of ammonium ion on growth and acid production (after 11 days of cultivation, strain 14/II, 25/III)
1-titratable acidity, 2-itaconic acid, 3-itatartaric acid-lactone, 4-mycelium, 5-residual glucose.

Fig. 5 Effect of glutamicacid on growth and acid production (after 11 days of cultivation, strain 14/II)
1-titratable acidity, 2-itaconic acid, 3-itatartaric acid-lactone, 4-mycelium, 5-residual glucose.

Glutamic acid supported moderate growth with distinct stimulatory effect on acid production, viz. mycelial dry wt. did not exceed 1.5 g dry wt., itaconic acid 5.0g, and about 1.8 g of itatartaric acid-lactone was accumulated in 100 ml of the medium. It is obvious that glutamic acid provided both amino-nitrogen and the carbon skeleton, α-oxoglutaric acid. The latter has been considered as a repressor of several enzymes of the TCA cycle in other organisms and is a possible regulator of acid metabolism in *A.terreus* (Hanson & Cox, 1967; Leal, Gomez-Miranda & Nicolas, 1966).

Corn steep liquor, rich in amino acids and in mineral salts, clarified by Larsen and co-workers in experiments carried out with non-proliferating mycelia of *A.terreus* NRRL 1960 suspended in glucose solution, pH 2.1. Both acids were shown to be involved in itaconic acid production (Jenssen, Larsen & Ormerod, 1956).

Table 4. Metabolism of citrate and cis-aconitate crude cell-free extracts, strain 25/III, Warburg technique.

Substrate	μ moles/h/mg protein				RQ
	citrate	O_2 absorbed	CO_2 released	Carbonyl compounds	
Citrate	utilized 1.725	0.145	1.217	0.09	24.3
Cis-aconitate	produced 0.508	0.155	3,28	-0.0195	26.3

Reaction system in 1 ml: phosphate buffer 80 μ moles, Mg^{++}, pH 6.5, protein 1.4 mg, citrate or cis-aconitate 5 μ moles, - decrease of carbonyl compounds.

Citric acid was estimated to be a more immediate precursor than cis-aconitic acid. In the presence of fluoroacetate (5×10^{-3}M) there was a pronounced increase in the yield of itaconic acid both from glucose and in particular from citric acid. Fluoroacetate when converted into fluorocitrate markedly inhibits the activity of aconitase. It was concluded that "cis-aconitic acid inhibits the TCA-cycle at the stage of cis-aconitic acid itself, thus diverting a greater proportion of the metabolic intermediates to the pathway leading to the formation of itaconic acid. It seems possible that the presence of cis-aconitic acid promotes the conversion of activated citrate into itaconic acid" (Eimhjellen & Larsen, 1955). In Fig. 6 two mechanisms responsible for itaconic acid formation from activated citrate are suggested, viz. (1) conversion of activated citric acid into citramalic acid by decarboxylation and dehydration of citramalic acid to form itaconic acid and (2) conversion of activated citrate into cis- aconitate and then decarboxylation to itaconate. However it was not confirmed for D,L-citramalate, which was not used by non-proliferating mycelia.

Recently, it was confirmed that (-)-citramalic acid is formed and used by *A.terreus* NRRL 1960 and its mutants (Jakubowska & Metodiewa, 1974a,b; Jakubowska *et al.*,1974) in fermentation systems, by pregrown mycelium and in reactions with crude oell-free

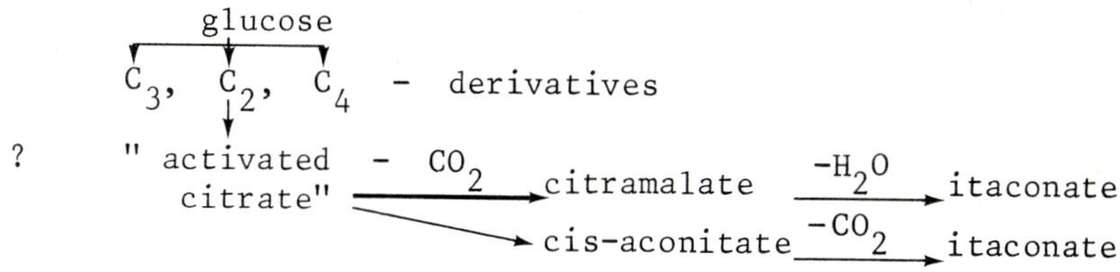

Fig. 6 Supposed metabolic pathway for itaconic acid formation (Eimhjellen & Larsen, 1955).

extracts, (-)-CM was converted into itaconic and itatartaric acids (Fig. 7, Table 5). All tested strains behaved alike. From labelled glucose 1-^{14}C radioactivities corresponding to the positions of itaconic, itatartaric and β-hydroxyparaconic acids, as end-products, were found, while citramalic, citric and gluconic acids present in the 2nd and 4th day of fermentation, were exhausted. It is worth pointing out that at pH 6.4 only itaconic acid was formed from (-)-CM, while at pH 2.2. besides itaconic acid, itatartaric and malic acids were formed.

Table 5. Products formed from itaconate and citramalate by cell-free extracts of *Aspergillus terreus*

Strain	Itaconic acid anaer. aerob.	Citramalic acid anaer.	aerob.
NRRL 1960	MA	IA, ITA.	MA
K	MA	IA	ITA
UV-17	non metabolized	IA	ITA
		MA, CA	IA, ITA
19/IV	non met. MA, SA	ITA	MA

Abbreviations: MA-malic a.; CM-citramalic a.; SA-succinic a.; IA-ita-conic a.; ITA-itatartaric a.; CA-citric a.

From 1,5-^{14}C citrate radioactive (-) citramalic and malic acids were formed (Fig.8). Iodoacetate (10^{-3}M) added to replacement medium containing citric acid, as a source of carbon caused distinct modification of acid formation in comparison with the control, viz. itatartaric and its lactone were not present, while succinic and 2-oxoglutaric acid, absent in the control, accumulated. It was shown that iodoacetate inhibits the metabolism of citric acid in the direction of cleavage as well as the conversion of cis-aconitate to itaconate.

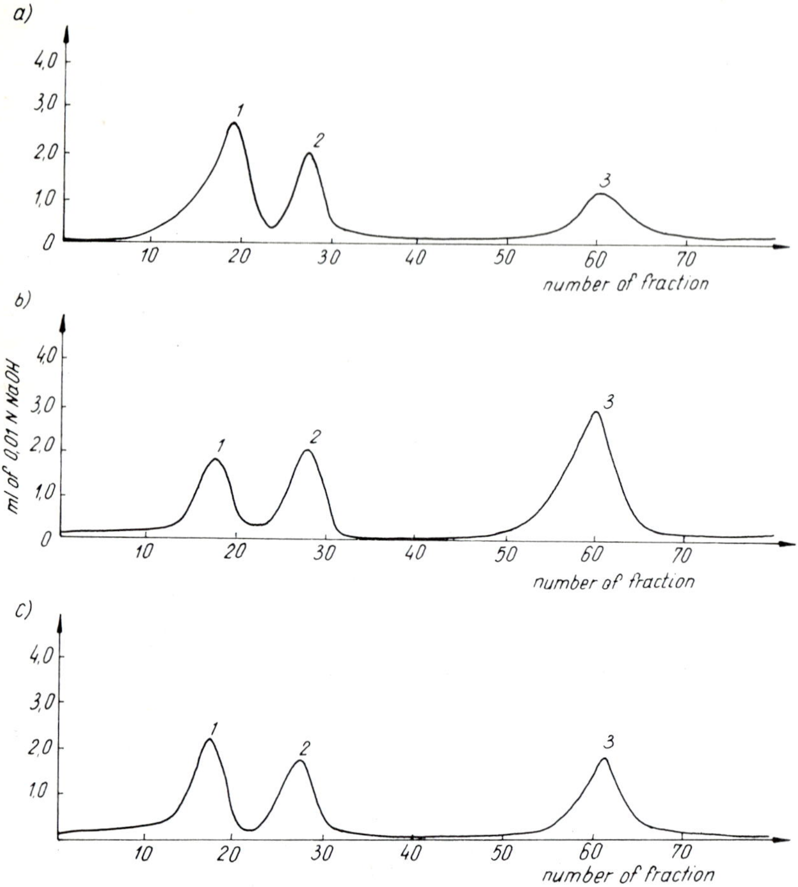

Fig. 7. Anaerobic metabolism of citramalate by cell-free extracts of *A. terreus:* a) extract from strain NRRL-1960, b) extract from strain K, c) extract from strain UV-17.

1) citramalic a., 2) itaconic a.; 3) itatartaric a. Reaction mixture (final volume 4 ml) : crude cell-free extract (9,5-10 mg protein, about 2 ml), K-phosphate buffer, (pH 6,5)80 μm; $MgCl_2$ - 80 μm, sodium citramalate 80 μm.

The key-position of (-)- citramalic acid in the metabolic shunt leading to itatartaric acid formation, is presented on the simplified pattern of reactions, assuming that the enzymatic condensation between pyruvate and acetylcoenzyme A (EC 4.1.3.25) is operating in

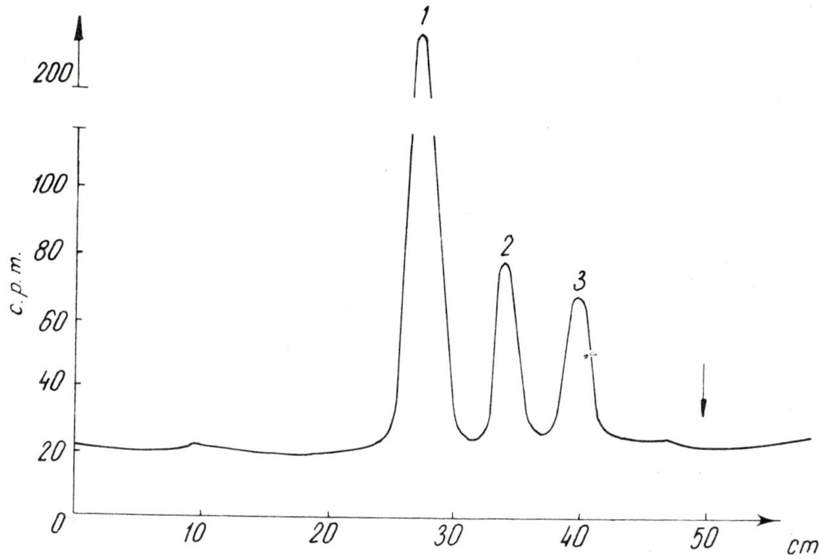

Fig. 8. Radioactivity of metabolic products formed from citrate 1,5 - ^{14}C by crude cell-free extract of *A.terreus* strain UV-17. 1-citric a.; 2-malic a.; 3-citramalic a.

A.terreus. (Fig.9). These enzymatic reactions have been established for several micro-organisms (Gray & Kornberg, 1960; Losada *et al.*, 1964; Sai *et al.*, 1969; Sai, 1968), in particular, when the TCA cycle was found not to be operating (Greenfield & Claus, 1972). Using radioactive 1-^{14}C acetylcoenzyme A and pyruvate as substrates and pure enzyme, obtained from RD-mutants of yeast, radioactivity has been recovered in (-) CM (Amaha & Sai, 1969; Sai *et al.*, 1969). According to the mechanism operating in other living organisms citraconic and mesaconic acids, derived from decarboxylation of cis-aconitic acid are converted by enzymatic hydration into laevo-and dextrorotatory citramalic acids, respectively, as a penultimate precursor for itaconic acid (Wang *et al.*, 1961, Nagai, 1963; Subramanian *et al.*, 1966; Raghavendra *et al.*,1963; Losada *et al.*, 1964).

A similar sequence of reactions was discussed for *A.terreus* by Smith *et al.* (1974). Citramalic acid is considered to be an active intermediate formed from glucose, and is universally observed in living cells, ie. plant-and animal tissues, in aerobic and anaerobic bacteria and in *Saccharomyces* (Dunbus, 1970). Among several organic acids only citraconic acid provided an effective increase of (-) CM accumulation (Sai & Amaha, 1965), due to an active citraconase being responsible for enzymatic hydration of

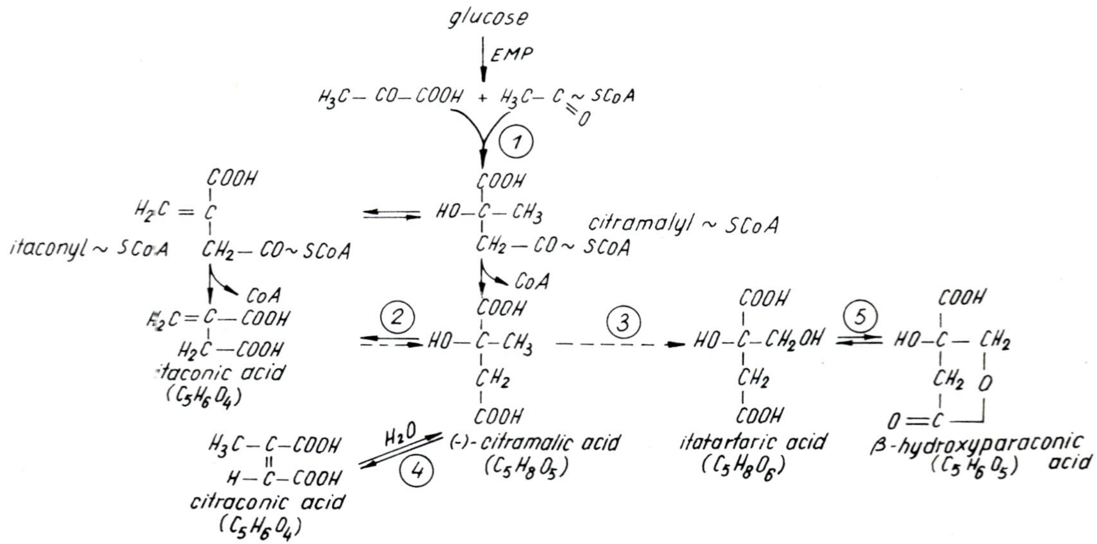

Fig. 9. Simplifiec pattern of reactions supposed for itatartaric acid formation.
Explanation of sequence of reactions*
1 - citramalate lyase (EC 4.1.3.22, EC 4.1.3.25); 2 - citramalate hydratase; 3 - ? unknown but experimentally confirmed; 4 - citraconate hydratase (EC 4.2.1.35); and 5 - in pure itatartaric acid sclution.
* aforenamed steps were confirmed experimentally for *Aspergillus terreus*

citraconate to citramalate (Raghavendra *et al.*,1963). The same effect was found in *A.terreus* (Jakubowska & Metodiewa, 1974b; Zakowska, 1973).

The role of pyruvate and citramalate has been experimentally confirmed for several strains of *A.terreus*, viz. (1) pyruvic acid as a source of carbon was shown to support growth and formation of all acids present in glucose medium (Nowakowska-Waszczuk, 1973; Jakubowska & Metodiewa, 1974 a; (2) pyruvate has been converted (in cell-free extracts) initially into itaconate and citramalate (Fig.10,11), later when pyruvate and citramalate were consumed, a distinct increase of itaconate and formation of itatartarate and

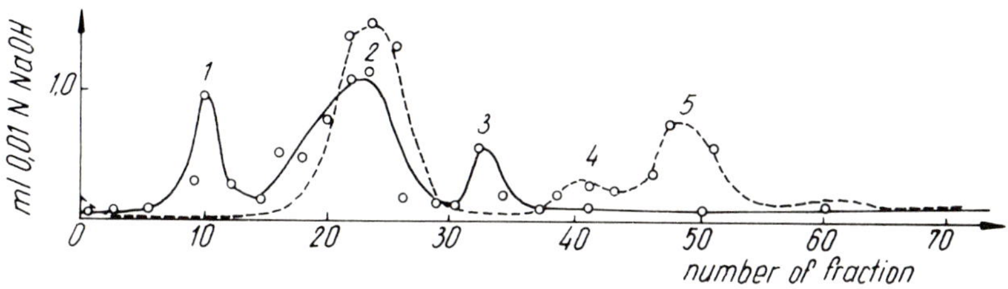

Fig. 10. Conversion of pyruvate in cell-free extract of *A.terreus* strain K.
1-citramalic a.; 2-itaconic a.; 3-pyruvic a.; 4- hydroxyparaconic a.; 5-itatartaric a.; ——— -30 min incubation of reaction mixture at 30°C, - - - - 2 hr incubation of reaction mixture at 30°C. Reacti mixture contained: 2 ml of crude cell-free enzyme extract (equivalent to 8-10 mg of protein), K-phosphate buffer (pH 6.5), 80 μM, $MgCl_2$ 80 μM, sodium pyruvate 80 μM.

Table 6. Influence of L-leucine on acid formation from pyruvate in extracts of *A.terreus*

Strains	Control	5 x 10^{-3}M L-leu/sample
NRRL-1960	IA,CA, 2-oxogl. ITA	MA, IA, 2-oxogl.
K	IA, ITA	IA, MA, CA
UV-17 (2P)	IA,MA, CA, ITA	IA, MA
24/I	IA,MP, SA, ITA	IA
19/I	SA, MA, ITA	IA
25/III	IA,CM (traces),ITA	IA, CA
21/I	IA,CA, CM, CM,ITA	IA

- in anaerobic incubation of reaction mixture.
Abbreviations: MA-malic a.; CM-citramalic a.; SA-succinic a.; IA-itaconic a.; ITA-itatartaric a.; CA-citric a.

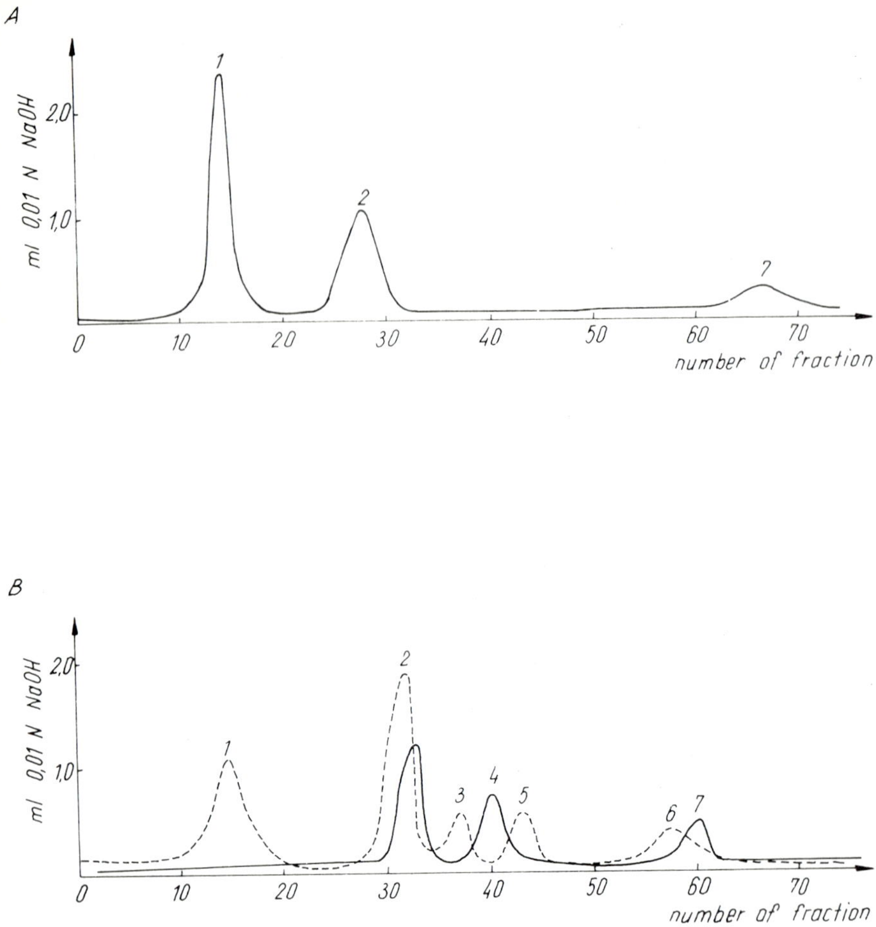

Fig. 11A. Aerobic metabolism of pyruvate* by cell-free extract of *A.terreus* strain NRRL-1960**
Fig. 11B. Anaerobic metabolism of pyruvate by cell-free extracts of *A.terreus* strain NRRL-1960**
1-malic a,; 2-itaconic a. 3-citric a.; 4-pyruvic a.; 5-2-oxoglutaric a.; 6,7-itartaric and β-hydroxyparaconic acid: ——— - after 2 hr of incubation at 30°C --- after 6 hr of incubation at 30°C

* Incubation of reaction mixture 2 hr at 30°C
** Strain grown 3 days on Larsen-Hovden medium (1954)

its lactone occurred (Jakubowska & Metodiewa, 1974a); (3) in the same conditions in presence of L-leucine (5.10^{-3}M) only citrate and itaconate were formed from pyruvate (Table 6).

L-leucine is known as a specific inhibitor for citramalate lyase (Sai *et al.*,1968, 1969). Its carbon skeleton, isovaleric acid, was shown to be harmful for some filamentous fungi (Coll & Leal 1972) and for several mutants of *A.terreus* (Jakubowska & Metodiewa, 1975, 1976).

Some differences with respect to the conditions favourable for itaconic and itatartaric acid biosynthesis indicate that the latter is not originating from itaconate conversion, viz: the consumption of itaconic acid by growing mycelium stimulated growth without distinct increase of itatartaric acid (Jakubowska & Nowakowska-Waszczuk, 1968), while in prolonged cultivation of strain 25/III at pH 2.2 there was no decrease of itaconic acid but a distinct increase of itatartaric acid occurred (Fig.12).

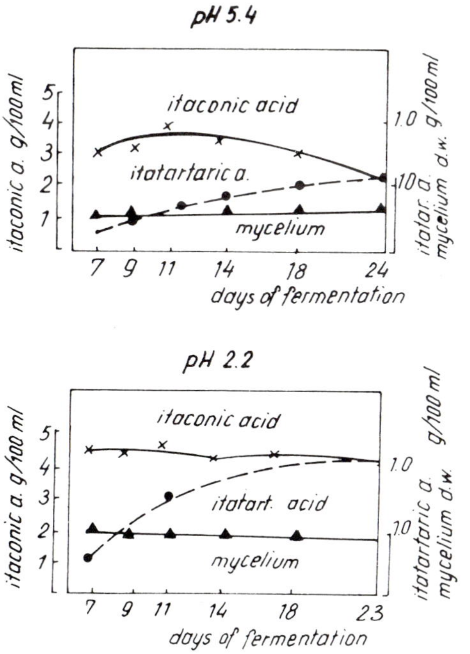

Fig. 12. Surface prolonged fermentation on Larsen and Hovden medium. Strain 25/III.

Using uniformly labelled itaconate - ^{14}C and by extraction of mitochondrial fraction obtained from strain K radioactivity was detected in aconitic and mesaconic acids (Fig. 13) (Jakubowska & Metodiewa 1974a). In two-step cultivation with pre-grown mycelium mats, supplied with itaconic acid as source of carbon in mineral medium, different behaviour characterized the investigated strains, viz: strain UV_{17} (being unable to consume itaconic acid in cultivation medium) was able to utilize the itaconic acid with an increase of mycelium and formation of malic, citric and citraconic acids; gamma-mutant 19/IV had little ability to use itaconic acid for growth and metabolism; and strain 25/III and 14/II consumed itaconate resulting in an increase of mycelium without acid formation.

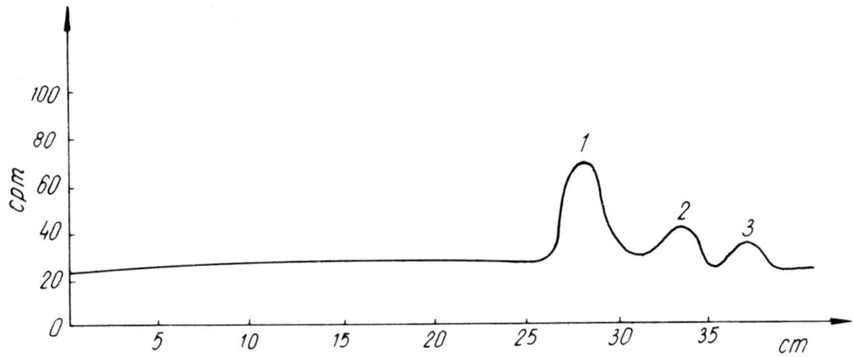

Fig. 13. Radioactivity of metabolic products formed from labelled itaconate - ^{14}C* by extract of mitochondrial acetone powder of *A.terreus* strain K
1-aconitic a.; 2-itaconic a.; 3-mesaconic acid

* reaction mixture: enzyme extract (0,95 - 1.00 mg of protein, itaconate - ^{14}C - 25 μm, $MgCl_2$-5μm, phosphate buffer (pH-7.4)- 20μm, final volume - 2 ml, incubation 1 hr at 30°C.

It is evident that the mechanism of itaconic and itatartaric acid formation needs further elucidation. In particular, the presence of citramalate-lyase (EC4.1.3.22 and/or EC 4.1.3.25) and regulation of the phase of growth and acid production in *A.terreus* warrants further study.

REFERENCES

AMAHA,M & SAI,T. (1969). Some aspects of (-)-citramalic acid accumulation by respiration deficient mutants of yeasts. *Antonie van Leeuwenhoek, 5* Supplement: Yeast Symposium G 15,16 .

ARPAI,J. (1958a). Enzymatická oxydácia kyseliny itakonovej na kyselinu itavinuu. *Chemicke Zvesti* 12, 532-9.

ARPAI,J. (1958b). Itaconic oxidase: an enzyme from an ultraviolet induced mutant of *Aspergillus terreus*. *Nature, London* 182,661-2.

BARKER,H.A. (1967). Citramalate lyase of *Clostridium tetanomorphum*. *Archiv für Mikrobiologie* 59, 4-12.

BENTLEY,R. & THIESSEN,C.P. (1957a). Biosynthesis of itaconic acid in *Aspergillus terreus*. I. Tracer studies with C^{14}-labelled substrates. *Journal of Biological Chemistry* 226. 673-87.

BENTLEY,R. & THIESSEN,C.P. (1957b). Biosynthesis of itaconic acid in *Aspergillus terreus*. II. Early stages in glucose dissimilation and the role of citrate. *Journal of Biological Chemistry* 226, 689-701.

BENTLEY,R. & THIESSEN,C.P. (1957c). Biosynthesis of itaconic acid in *Aspergillus terreus*. III. The properties and reaction mechanism of cis-aconitic acid decarboxylase. *Journal of Biological Chemistry* 226, 703-20.

CALAM,C.T., OXFORD,A.E. & RAISTRICK,H. (1939). The biochemistry of micro-organisms. LXIII. Itaconic acid a metabolic product of a strain of *Aspergillus terreus* Thom.*Biochemical Journal* 23 1488-95.

COLL,J. & LEAL,J.A. (1972). Utilization of L-leucine as nitrogen source by fungi. *Transactions of the British Mycological Society* 59, 107-14.

DUBUS,G. (1970). Contribution a l'étude de l'acide citramalique dans les vins. I. Identification. *Connassance de la Vigne et du Vin* 4, 399-430.

EIMHJELLEN,K.E. & LARSEN,H. (1955). The mechanism of itaconic acid formation by *Aspergillus terreus*. The effect of substrates and inhibitors. *Biochemical Journal* 60, 139-47.

FITTIG, R & KOHL,W. (1899). Oxidation zweibasischer ungesattiger Sauren. *Justus Liebig's Annales der Chemie* 305, 41-7.

FLORKIN,M. & STOTZ,E. (1968). Enzymes associated with citric acid cycle metabolites. In *Comprehensive Biochemistry,* Vol.20 471-4. Elsevier Pub. Co: Amsterdam.

FRIEDKIN,M. (1945). Determination of itaconic acid in fermentation liquors. *Industrial Engineering Chemistry* 17, 637-8.

GRAY,C.T. & KORNBERG,H.L. (1960). Enzymatic formation of citramalate from acetylcoenzyme A and pyruvate in *Pseudomonas ovalis* Chester catalyzed by "pyruvate transacetase" *Biochimica et Biophysica Acta* 42, 379-82.

GREENFIELD,S. & CLAUS,G.W. (1972). Nonfunctional tricarboxylic acid cycle and the mechanism of glutamate biosynthesis in *Acetobacter suboxidans*. *Journal of Bacteriology* 112, 1295-301.

HANSON,R.S. & COX, D.F. (1967). Effect of different nutritional conditions on the synthesis of TCA cycle enzymes. *Journal of Bacteriology* 93, 1171-87.

HOLLAENDER,A., RAPER,K.B. & COGHILL,R.D. (1945). The production and characterization of ultraviolet-induced mutations in *Aspergillus terreus*.I. Production of the mutations. *American Journal Botany* 32, 160-5.

HULME,A.C. (1954). The isolation of L-citramalic acid from the peel of the apple fruit. *Biochimica et Biophysica Acta* 14,6-43.

JAKUBOWSKA,J. & METODIEWA,D. (1974a). Investigations of the metabolic pathway for fermentative production of itatartaric acid from glucose or sucrose, to provide new industrial outlets for these agricultural commodities. Final Report 1-115, Research for U.S. Dept. of Agriculture, Printed Lodz, Poland.

JAKUBOWSKA,J. & METODIEWA,D. (1974b). Studies on the metabolic pathway for itatartaric acid formation by *Aspergillus terreus*. II. Use of (-)-citramalate, citraconate and itaconate by cell-free extracts. *Acta Microbiologica Polonica* Ser.B 6(23),51-61.

JAKUBOWSKA,J. & METODIEWA,D. (1975). Investigations on metabolic regulation of itatartaric acid and its lactone formation in mutants of *Aspergillus terreus*. Final Report, 1-10, Research for Institute of Biochemistry and Biophysics, Polish Academy of Sciences (Polish).

JAKUBOWSKA,J. & NOWAKOWSKA-WASZCZUK,A. (1968). Investigation on the fermentative production of itatartaric acid from glucose, sucrose or molasses. Final Report 1-106. Research for U.S. Department of Agriculture, Printed Lodz, Poland.

JAKUBOWSKA,J. & ZAKOWSKA,Z. (1975). On some organic acids formed by *Aspergillus terreus* in itaconic acid fermentation. I. Spectrum of non-volatile acids. *Zeszyty Naukowe Politechniki Lodzkiej, Chemia Spozywoza*, z. 25, 29-52 (Polish).

JAKUBOWSKA,J., METODIEWA,D. & ZAKOWSKA,Z. (1974). Studies on the metabolic pathway for itatartaric acid formation by *Aspergillus terreus*. I. Metabolism of glucose and some C_5 and C_6-carboxylic acids. *Acta Microbiologica Polonica* 6(23) Ser.B., 43-50.

JAKUBOWSKA,J., ZAKOWSKA,Z. & JESZKA,J. (1970). Identification of organic acids and some properties of itatartaric acid and its lactone formed in itaconic acid fermentation. *Zeszyty Naukowe Politechniki Lodzkiej, Chemia Spozywcza*, z. 17, 21-32 (Polish).

JAKUBOWSKA,J.,OBERMAN,H.,MAKIEDONSKA,A. & FLORIANOWICZ,T. (1967). The itaconic and itatartaric acid formation by UV and gamma irradiated isolates of *Aspergillus terreus* NRRL 1960. *Acta Microbiologica Polonica* 16, 53-68.

JAKUBOWSKA,J., NOWAKOWSKA-WASZCZUK,A., LASOTA-DULKOWSKA,E., ZAKOWSKA,Z. & SOBOCKA,B. (1967b). *Acta Microbiologica Polonica* 16, 227-36.

JENSSEN,E.B., LARSEN,H. & ORMEROD,J.G. (1956). Formation of itaconic acid from the Krebs cycle tricarboxylic acids by extracts of *Aspergillus terreus*. *Acta Chemica Scaninavica Proceedings* 10, (6) 1047-54.

KINOSHITA,K. (1931). Uber eine neue *Aspergillus* Art, *A.itaconicus*. *Botanical Magazine* (Tokyo), 45, 45-61.

KNUTH, Ch.,J. (1967). Itaconic acid. In Kirk-Othmer Encyclopedia of Chemical Technology, 12, 83-86. John Wiley Pub. New York.

KOBAYASHI,T. (1967). Itaconic acid fermentation. *Process Biochemistry* 2(9), 1-5.

KOBAYASHI,T. & NAKAMURA, I. (1966). Dynamics in mycelial concentration of *Aspergillus terreus* k-26 in steady state of continuous culture.*Journal of Fermentation Technology* 44 (6), 264-74.

KOBAYASHI,T., NAKAMURA,I. & NAKAGAWA, M. (1972). Process design for itaconic acid fermentation, 215-221. Fermentation Technology Today, Proceedings of the IV International Fermentation Symposium, Kyoto 1972, ed. Gyozo Terui, Society of Fermentation Technology, Japan.

KYOWA HAKKO KOGYO KABUSHIKI HAISHA (1959). Procede de fabrication d'acide itaconique par fermentation. Brevet d'Invention N°1, 1,89667 C 12d, Paris.

LARSEN,H. (1957). Itaconic acid and related compounds. In *Methods of enzymology* Vol.3 pp 437-443. Edited by S.P. Colowick, N.Kaplan. Academic Press: New York.

LARSEN,H. & EIMHJELLEN, K.E. (1955). The mechanism of itaconic acid formation by *Aspergillus terreus*. I. Effect of acidity. *Biochemical Journal* 60, 135-9.

LARSEN,H. & HOVDEN,H. (1954). Over teknisk viktige faktorer for itaconsyregjaeringen. *Norges Teknisk-Naturvitenskapelige Forskningsrad, Rapport til*, 1-28.

LEAL,J.A., GOMEZ-MIRANDA,B. & NICOLAS,G. (1966). Formation of toxic substances by fungi in the degradation of some amino acids. *Canadian Journal Microbiology* 12, 1073-6.

LOSADA,M., CANOVAS,J.L. & RUIZ-AMIL,M. (1964). Oxaloacetate, citramalate and glutamate formation from pyruvate in baker's yeast. *Biochemische Zeitschrift* 340, 60-74.

LOCKWOOD,L.B. (1954). Itaconic acid. In *Industrial fermentations*, Edited by Underkofler, L.A. & Hickey,R.J. I. 488-497. Chemical Publishing Co.Inc: New York.

LOCKWOOD,L.B. (1975). Organic acid production. In *The filamentous fungi*. Vol.I, pp.151-7. Edited by Smith,J.E. & Berry,D.R. Edward Arnold: London.

LOCKWOOD,L.B. & NELSON,G.E.N. (1946). Some factors affecting the production of itaconic acid by *Aspergillus terreus* in agitated cultures. *Archives of Biochemistry* 10(5), 365-73.
LOCKWOOD,L.B. & REEVES,M.D. (1945). Some factors affecting the production of itaconic acid by *Aspergillus terreus*. *Archives of Biochemistry* 6, 455-69.
LOCKWOOD,L.B. & WARD,C.E. (1945). Fermentation process for itaconic acid. *Industrial Engineering Chemistry* 37, 405-6.
LOCKWOOD,L.B., RAPER,K.B., MOYER,A.J. & COGHILL,R.D.(1945). The production and characterization of UV-induced mutations of *Aspergillus terreus*. III. Biochemical characteristics of the mutations.*American Journal of Botany* 32,(4), 214-7.
MOYER,A.J. & COGHILL,R.D. (1945). The laboratory scale production of itaconic acid by *Aspergillus terreus*. *Archives of Biochemistry* 7, 167-83.
NAGAI,J. (1963). Studies on itaconate metabolism. II. Citramalate metabolism in *Pseudomonas fluorescens* grown on itaconate. *Journal of Biochemistry (Tokyo)* 54, 34-40.
NAKAGAWA,M. & KOBAYASHI,T. (1968). Studies on product separation from fermented liquor. II. Concentrating itaconic acid from fermented liquor with ion exchangers. *Journal of Fermentation Technology* 46(3), 158-68.
NAKAMURA,I. & KOBAYASHI,T. (1967). Studies on itaconic acid fermentation. (VII). Heat transfer of fermentation broths in stirred fermentor. *Journal of Fermentation Technology* 45(12,1139-46.
NOWAKOWSKA-WASZCZUK,A. (1971). A response of *Aspergillus terreus* mutants to nitrite. *Acta Microbiologica Polonica,* Ser.B,3,13-22.
NOWAKOWSKA-WASZCZUK,A. (1973). Utilization of some TCA-acid-cycle intermediates by mitochondria and growing mycelium of *Aspergillus terreus*. *Journal of General Microbiology* 79, 19-29.
NOWAKOWSKA-WASZCZUK,A., ZAKOWSKA,Z. & SOBOCKA,B. (1969). The effect of nitrogen sources on acid production by *Aspergillus terreus*. *Acta Microbiologica Polonica* Ser.B, 1,105-10.
PFEIFER,V., F.,VOJNOWICH,Ch. & HEGER,E.N. (1952). Itaconic acid by fermentation with *Aspergillus terreus*. *Industrial and Bio-Engineering Chemistry* 44, 2975-80.
PFEIFER,V.F., NELSON,G.E.N., VOJNOWICH,Ch. & LOCKWOOD,L.B. (1953). U.S. Patent. Production of itaconic acid. 2,657,173 Oct. 27.
PRESCOTT,S.C. & DUNN,C.G. (1959). *Industrial microbiology*. McGraw Hill Book Co Inc: New York.
RAGHAVENDRA RAO, M.R., SUBRAMANIAN,S.S., RAHATEKAR,H.J. & PARANJAPE, S.V. (1963). Enzymatic hydration of citraconate to (-)citramalate. *Biochemical and Biophysical Research Communications* 12, 78-82.
RAPER,B.B., COGHILL,R.D. & HOLLAENDER,A. (1945). The production and characterization of ultraviolet, induced mutations in *Aspergillus terreus*. II. Cultural and morphological characteristics of the

mutations. *American Journal Botany* 32, 165-76.

SAI,T. (1968). Studies on (-)citramalic acid formation by respiration-deficient yeasts mutants. IV. The inhibition of CM formation by L-leucine and the induced formation of (-)CM from citraconic acid. *Bulletin of Brewing Science* 14, 61-72.

SAI,T. & AMAHA,M. (1965). Accumulation of (-)citramalic acid by respiration-deficient mutants of brewers' yeasts. *Journal of General and Applied Microbiology* 11, 355-65.

SAI,T., AIDA,K. & UEMURA,T. (1968). Enzymatic formation of (-)-citramalate from acetyl coenzyme A and pyruvate in respiration-deficient *Saccharomyces carlsbergensis*. *Agricultural and Biological Chemistry* 32, 1398-400.

SAI,T., AIDA,K. & UEMURA T. (1969). Studies on (-)-citramalic acid formation by respiration-deficient yeast mutants. V. Purification and some properties of citramalate condensing enzyme. *Journal of General and Applied Microbiology* 15, 345-63.

SMITH,J.E., NOWAKOWSKA-WASZCZUK, A. & ANDERSON,J.G. (1974). Organic acid production by mycelial fungi. In *Industrial Aspects of Biochemistry*, edited by B.Spencer, pp.311-314.

SOBOCKA,B., LASOTA-DULKOWSKA,E. & MAKIEDONSKA,A. (1970). Selection of suitable sporulation media for scanty sporulating mutants of *Aspergillus terreus*. *Zeszyty Naukowe Politechniki Lodzkiej, Chemia Spozywcza* 17, 107-15 (Polish).

STODOLA,F.H., FRIEDKIN,M., MOYER,A.J. & COGHILL,R.D. (1945). Itatartaric acid, a metabolic product of an ultraviolet-induced mutant of *Aspergillus terreus*. *The Journal of Biological Chemistry* 161, 739-42.

SUBRAMANIAN,S.S., RAGHAVENDRA RAO, M.R. & BHATTACHARYYA,P.K. (1966). Stereochemistry of enzymatic hydration of citraconate acid mesaconate. *Indian Journal of Biochemistry* 3, 260-2.

TABUCHI,T. (1961). Formation of itatartaric acid by *Aspergillus itaconicus*. *Journal Agricultural Chemical Society, Japan* 35,541-4.

TABUCHI,T. (1963). Studies on the cultural characteristics of *Aspergillus itaconicus* and *Aspergillus terreus*, itaconic acid producers in submerged fermentation, Bulletin, March IV, 245-300, Laboratory of Industrial Microbiology, Faculty of Agriculture, Tokyo University of Education.

TURNER,W. (1971). *Fungal metabolites*. Academic Press: London.

WANG,S.F., ADLER,J. & LARDY,H.A. (1961). The pathway of itaconate metabolism by liver mitochondria. *Journal of Biological Chemistry* 236, 26-30.

ZAKOWSKA,Z. (1973). Organic acids formed by strains of *Aspergillus terreus*. Doctor thesis, Technical University, Lodz,Poland.

ZAKOWSKA,Z.(1976). Isolation and properties of itatartaric acid formed by *Aspergillus terreus*. *Zeszyty Naukowe Politechniki Lodzkiej, Chemia Spozywcza*, (Polish) (In Press).

The Genus *Aspergillus* and Biodeterioration

ANGELA R. THOMAS

Department of Applied Microbiology, University of Strathclyde, Glasgow, U.K.

INTRODUCTION

Biodeterioration is deterioration caused by biological agents - or biodeteriogens. A wide range of organisms including insects, birds, rodents, crustaceans, micro-organisms and even higher plants act as biodeteriogens. The activities of these organisms deteriorate not only organic materials or materials of organic origin such as stored produce and textiles, but many also affect inorganic substrates; for example, the biological corrosion of metals. Apart from causing spoilage problems by attacking various substrates in the form of raw products, manufactured or in-service materials, the growth and presence of biodeteriogens may also interfere with manufacturing processes by causing the breakdown of industrial emulsions, decreasing efficiency of machinery or even being responsible for malfunctions and blockages of equipment.

Biodeterioration has always existed due to the very nature of biological processes. Life is a cyclical process involving elemental cycles in which living organisms play their various roles, but when these roles interfere to the detriment of man's economy then they become biodeteriorative. Contrarily, the process of biodegradation involves these same activities but then they are exploited to man's advantage to solve waste disposal problems or to recycle raw materials from waste (Eggins & Allsopp, 1975).

Thus biodeterioration is of major importance in economic terms. Various estimates involving millions of pounds have been proposed to account for losses due to biodeterioration, but in fact it is a difficult sum to determine for a number of reasons. These have been summarised as:- lack of statistics on known biodeterioration because of disinterest in this aspect by biologists, and reluctance on the part of manufacturers particularly those in the food, cosmetic and pharmaceutical industries, to admit to experiencing biodeterioration problems; and failure to recognise biodeterioration either due to ignorance or complacency (Eggins, 1967). The Food and Agriculture Organisation (F.A.O.) of the United Nations have estimated that 5% of all food grains harvested are lost before consumption and this figure may be as high as 30% in some countries especially developing ones which are often least able to afford such losses (Christensen & Kaufmann, 1969), whilst post-harvest losses in tropical horticultural produce may exceed 50% in some cases (Coursey, 1972). This only accounts for one area of biodeterioration. Further, to the losses due to various forms of spoilage must be added the cost of the intangible aspects of biodeterioration - decreased efficiency of machinery, the expense incurred to service malfunctioning or blocked equipment to mention but a few instances. It has been stated that prevention of biodeterioration could be beneficial to the growth of developing countries and increase the economic efficiency of the developed countries (Eggins, 1968), but to-day with inflation, decreases in the world's resources and world food shortage it takes on an even greater significance.

The manifestations of biodeterioration may be classified into the following categories (Hueck, 1965):-

1. Mechanical
2. Chemical
    a. assimilatory
    b. dissimilatory
3. Soiling

Mechanical damage is synonymous with physical attack, for example, the gnawing by rodents or penetration by insects of various substrates. Chemical damage by assimilation occurs when organisms utilise a substrate or constituents thereof as nutrients and thereby weaken or change the properties of that substrate. The dissimilated metabolic products of organisms similarly may cause chemical damage - for example, corrosive organic acids, disfiguring pigments, or mycotoxins. Finally, soiling results from the mere presence or growth of organisms, which utilise an inert surface as a substrate for attachment thus disfiguring that substrate and making it an unacceptable commodity, or causing the fouling of machinery or even ships' hulls - in these instances the biodeteriogens utilise dirt contaminants or surrounding liquid media as a source of nutrients.

For further general information about biodeterioration, reference should be made to the publications listed in the bibliography.

The recognition and diagnosis of biodeterioration problems is often complex and requires skilled expertise and judgement. Depending on the nature of the biodeteriogen different techniques may be involved. Fungi are frequently implicated in biodeterioration problems probably due to their common saprophytic role and ability to utilise a wide range of substrates in a variety of ecological niches, and due to the nature of their growth which enables them to exploit a substrate (Eggins & Allsopp, 1975). Biodeterioration by selected organisms may occur whenever the nutrients and the conditions for growth - temperature, moisture, pH, aeration - are conducive. The diagnosis of a problem caused by a fungal biodeteriogen requires an approach based on mycological studies (Walsh, 1968; Eggins & Allsopp, 1975). The conditions under which a biodeteriorated substrate is discovered may not be those conditions under which biodeterioration actually occurred. Further, not only does the macroclimate play a role but also the microclimate of the substrate. Sometimes the biodeteriogens may not be in direct association with the substrate or may have ceased to exist on the substrate. Since the organisms isolated depend on the conditions and techniques of isolation, selective isolation techniques are important such as when xerophilic moulds are implicated in the deterioration of stored produce. To establish whether the isolates are active or passive on a particular substrate, or whether they are merely utilising dirt contaminants and not the substrate, or whether they are only secondary colonisers in a microbial succession, the application of Koch's postulates has been recommended (Hueck, 1965). That is to reproduce the biodeterioration symptoms of the original material by introducing pure cultures of the isolates on to fresh samples of the material.

Members of the *Aspergillus* genus are often identified as biodeteriogens probably due to a combination of reasons. The large number of species comprising the genus represent a broad spectrum of various ecological types capable of utilising a wide variety of substrates (Raper & Fennell, 1965). Many of the *Aspergilli* are amongst the most commonly occurring fungal species - various species are frequently isolated from soil, vegetation, air and certain substrates undergoing deterioration (Raper & Fennell, 1965; Hudson, 1969). The isolation of the *Aspergilli* from soil, vegetation and the air has been related to their growth on various materials including cotton fabrics, paints, plastics, leather and glass (Upsher, 1972), but as the species were not identified in each instance no direct relationships could be concluded.

The biodeteriorative roles of *Aspergillus* species on various substrates are described below.

WOOD

Aspergillus species are included amongst those fungi which grow and sporulate on wood surfaces producing green and black stains (Scheffer, 1973). These fungi can be classified as the 'moulds' which utilise dead cell contents of the wood and may produce effects similar to the early stages of attack by the 'soft-rot' fungi - indeed some fungi classified as 'moulds' cause soft rot under certain conditions (Kaarick, 1974). Similarly, species of Fungi Imperfecti and Ascomycetes have been found to occur frequently on wood fibreboard and were thought to cause discolouration only, until it was shown that they also caused decay (Merrill, French & Hossfeld, 1965).

An unidentified *Aspergillus* species was included in the list of fungi causing soft rot when this form of wood decay was first described (Savory, 1954). Soft rot fungi attack wood cellulose, but only modify or slightly degrade lignin, and occur on wood experiencing continual wetting, for example, the timber slats of cooling towers; as opposed to the active wood-decomposing 'brown' and 'white rot' fungi consisting mainly of the Basidiomycetes which occur under less wet conditions (Savory, 1954; Kaarik, 1974; Eaton, 1976). Soft rot fungi cause discolouration and softening of the surface layers of the wood making it more vulnerable to erosion in some situations - hardwoods are more susceptible than softwoods (Savory, 1954; Eaton, 1976).

Aspergillus ustus along with other fungal species comprised 92% of the isolates found colonising pine wood stakes over a twelve week period; *A.niger* and *A.ustus* being the prevalent species during the second and third week respectively of colonisation (Merrill & French, 1966). The fungal isolates occurred with equal frequency from heartwood and sapwood, but, although soft rot developed the fungi causing it were not determined. *Aspergillus* species, including *A.niger*, have been reported amongst the isolates from similar experiments even in the presence of wood preservatives (Gersonde & Kerner-Gang, 1968). *A.niger* was one of the fungal colonists on pieces of beech wood veneer when perfused with nutrients at pH values 3.7 and 4.5, but was not isolated at higher pH values (Sharp & Eggins, 1970), and has been shown to have considerable amylolytic activity and to cause some strength loss of cellulose strips (King & Eggins, 1973). More recently, *A.niger* was found to colonise beech wood stakes in the presence and absence of preservatives whether additional nutrients and water were supplied or not (Sharp, 1975). Also in this latter study, *A.fumigatus* was isolated occasionally from preservative treated stakes. *A.fumigatus* when grown for 3 months at 40°C on birch, pine and spruce sapwood blocks caused weight losses of 13.8%, 3.5% and 2.9% respectively (Kaarick, 1974). In a review of fungi causing soft rot of cooling tower timbers *A.ochraceus* and *A.niger* were reported amongst the causal organisms - the latter species also being quoted as causing weight losses in beech and Scots pine woods (Eaton, 1976). Finally, when wood fibreboard made from semichemically pulped aspen was exposed to a pure culture of *A.niger* over an 8 month period, weight and strength losses occurred, though when the weight loss was only small, strength loss was not measurable (Merrill *et al.*, 1965). These workers also found that *A.niger* consumed hemicelluloses first but as these were depleted increasing amounts of χcellulose were consumed - and it also caused a slight decrease in lignin content.

PAPER

The paper industry experiences biodeterioration not only on its finished products but also during manufacture. As previously stated, fibreboards prepared from wood pulp are subject to spoilage; thus it is therefore not surprising that the organisms responsible for the spoilage also may occur on stored wood pulp awaiting processing into paper. *Aspergillus fumigatus* and *A.niger* are among the microbial isolates obtained from certain moist wood pulps and have been implicated in 'slime' formation (Hughes, 1968). Water used in paper processing provides a plentiful nutrient supply consisting of sugar and minerals extracted from the wood pulp plus starch, size or other additives required for finishing the paper. The additives may be added to the paper at a later stage thus avoiding adding to the nutrient status of the processing water, but additives on paper when recycled will still enter the processing water. Since the processing water is recycled the various nutrients are maintained at a fairly constant high level. The temperature of the water varies between 5° and $40^{\circ}C$, and the pH ranges from pH 4.5 to 9, whilst aeration occurs in the vicinity of splashing and anaerobic conditions prevail in dead flow areas. Thus depending on the conditions, different micro-organisms including bacteria, yeasts, filamentous fungi and algae are encountered forming various slimes (Turner, 1967; Hughes, 1968). These slimes cause a number of problems. They decrease flow rate and mechanical efficiency, but may also affect the quality of the paper produced. If there is an intensive build up of slime it may be torn free from its surface of attachment and become mixed with the paper slurry. Slime, thus incorporated into the wet paper web, causes areas of weakness which can cause breakage of the paper whilst it is being rolled - resulting in machinery stoppage, spoiled paper requiring recycling and lost production time as the machines then have to be rethreaded. Even if breakage does not occur, spots of slime in the paper, although killed during drying, are the cause of thin 'window' patches or charred areas in the finished product. *Aspergillus* species have also been cited as being among the microflora which cause slime formation in cooling towers, but bacteria and algae are considered the main causes of the problem in this situation (Kelly, 1975; Eaton, 1976).

The two *Aspergillus* species which contribute to slime formation in paper processing are included with *Aspergillus flavus*, *A. terreus*, *A.versicolor* and the *A.glaucus* group amongst the organisms which cause spoilage due to their growth on paper products, books and paper board - some of which grow at relatively low water activities(see section on Stored Produce) (Hughes, 1968). *Aspergillus* species are also among the organisms reported to penetrate the paper and muslin wrappings of sterile hospital packs, which apart from causing wasteful spoilage could also present a health hazard (Standard, Mackel & Mallison, 1971; Standard, Mallison & Mackel, 1973).

TEXTILES

Raw cotton besides consisting of cellulose provides other nutrients for microbial growth. *Aspergillus niger*, *A.nidulans*, *A.fumigatus*, *A.versicolor* and *A.glaucus* are included in the fungal species which have been found causing damage to raw cotton fibres (Turner, 1967), which may occur before harvesting. In which case this is not strictly biodeterioration, but such damage may cause uneven dyeing during processing of the fabric (Siu, 1951).

Cotton textiles depending on type of finish, and temperature and humidity conditions, are subject to spoilage by a range of micro-organisms (Siu, 1951; Turner, 1967). Some cause spoilage by merely growing on the textiles utilising water soluble nutrients of the cotton and dressings, such as starch; while other organisms attack the cellulose causing strength and weight losses of the fabric. Spoilage, apart from visible microbial growth, may produce musty odours and stains. *Aspergillus terreus* has been cited as one of the organisms known to cause 'diamond-spot' stains, and *A.roseus* produces a pigment causing red stains - though stains are not always due to microbial pigments, but may be caused by interaction between microbial metabolites and finishing materials, particularly dyes (Siu, 1951).

The occurrence of numerous micro-organisms, including many *Aspergillus* species, on cotton textiles has been extensively reviewed (Siu, 1951), and those *Aspergillus* species which were reported as cellulolytic agree for the most part with other reports (Reese & Downing, 1951; Turner, 1967). Species of any particular *Aspergillus* group have been found generally to exhibit similar responses when tested for cellulolytic activity - members of the *A.terreus* *A.fumigatus* and *A.flavipes* groups being the most active (Reese & Downing, 1951). However these workers did find that occasionally isolates of the same species did not respond alike;

and it is reported that *A.niger* is unable to utilise cellulose in the absence of other carbon sources at neutral pH but is able to do so under acid conditions (Turner, 1967).

Aspergillus species have also been implicated with other micro-organisms in the deterioration and rotting of cotton carcass conveyor belting in coal mines (Chamberlain, Meek, Kewett & Finlay, 1966).

Jute, like cotton, is subject to microbial spoilage and to strength loss due to microbial cellulolytic activity, but jute consists of substantially greater quantities of hemicelluloses and lignin (Ghose & Bose, 1973). In their review of jute biodeterioration, these workers report that *Aspergillus* species are among the damaging fungal species when the jute is stored in shade, but in sunlit weather-exposed jute, damage is due to a different fungi. Also, fungal growth occurs on jute when its moisture content is above 15% - *Aspergillus glaucus* being predominant at lower moisture contents, whilst other *Aspergillus* species occur at higher moisture contents (see later). *Aspergillus terreus* was found to be the most active cellulase producer of the isolates from jute.

Aspergillus species have also been reported to grow on sterilised wool (Reese & Downing, 1951; Shaposhnikov, Kozlova & Azova, 1964).

LEATHER

Biodeterioration problems arise at all stages during the processing of leather, from the raw hides to the finished leather products (Turner, 1967; Orlita, 1968), but only those problems which involve *Aspergillus* species will be mentioned here.

Although vegetable tanning liquors contain a range of nutrients, the tannin inhibits microbial growth. However, initial tannage of hides is usually carried out in weaker 'suspender' liquors which are readily infected by micro-organisms. *Aspergillus flavus* and *A.niger* which have been grown in 10% tannic acid, are included with those fungi which cause deterioration of tanning liquors by destroying the tannins (Turner, 1967).

Aspergillus ochraceus and *A.wentii* have been found to be responsible for causing damage to drying leather by producing coloured spots on the leather (Orlita, 1968).

However the worst damage caused by *Aspergillus* species results from their growth especially in tropical conditions on finished leather products, in particular those containing a rich carbohydrate supply - involving *A.flavus-oryzae*, *A.fumigatus*, *A.glaucus*, *A.niger*, *A.ochraceus*, *A.repens*, *A.versicolor* and *A.wentii* (Turner, 1967; Orlita, 1968). Moulds grow on leather with moisture contents above 14%. Glycerol, sugars and the nature of the tanning material all affect the uptake of moisture by leather, whilst vegetable tanned leather is more conducive to microbial growth than chrome or oil tanned leathers. Different oils used to grease leather also affect microbial growth. Fungal growth on leather may cause discolouration - *A.niger* has been known to be responsible for producing grey spots. Although fungal growth on leather is thought not to cause strength loss of the leather, it can result in chemical changes by hydrolysing oils and grease to produce free fatty acids. Breakdown of the oils and greases alters the physical properties of leather by making it hard and less pliable.

PAINT

Water-thinned paints are subject to spoilage by bacteria in the can, but most paint biodeterioration problems are due to fungal growth causing visual spoilage and physical damage to water-thinned and solvent-thinned paint films (Ross, 1963; Ross, Sladen & Wiener, 1968). The biodeterioration of paint films depends not only on the paint constituents - which may be nutritive, antimicrobial or affect moisture absorption - but also on the type of substrate to which they are applied, and dirt accumulation on their surface (Ross, 1963; Turner, 1967; Ross *et al.*, 1968). *Aspergillus* species including *A.flavus* and *A.niger* have been isolated from paint films, but do not constitute the predominant fungal isolates (Ross, 1963; Ross *et al.*, 1968; O'Neill & Skinner, 1966; Turner, 1967) and are considered to be chance contaminants and opportunists on dirt deposits (Drescher, 1958; Ross *et al.*, 1968), although *A. niger* utilises linseed oil which has been used as a constituent of some solvent-thinned paints (Ross, 1963).

RUBBER AND PLASTICS

The biodeterioration of rubber and plastics is very complicated due to the variety and complexity of these materials which include natural and synthetic rubbers, and different plastics consisting of a range of compounds and additives.

A review of the literature on the biodeterioration of rubber (Cundell & Mulcock, 1972) "suggests that under certain environmental conditions natural rubber vulcanizates are susceptibe to microbioligical deterioration whereas most synthetic rubber vulcanizates appear resistant under the same conditions" and that Streptomycetes are the organisms usually implicated in the deterioration. Although *Aspergillus* species have been reported to grow and cause staining in some instances on diluted latex, crepe, smoked sheet and other materials used in rubber formulations, and rubber products (MacLachlan, Heap & Pacitti, 1966; Turner, 1967; Cundell & Mulcock, 1972; Zyska, Rytych, Zankowicz & Fundalej, 1972), it has been commented that the fungi utilise non-rubber constituents of the rubber or are opportunists on dirt contamination of the rubber (Rook, 1955; Heap & Morrell, 1968; Cundell & Mulcock, 1972). Certainly it is also reported that *A.niger* and *A.fumigatus* found growing on rubber were unable to decompose pure rubber hydrocarbon (Cundell & Mulcock, 1972).

The fungal biodeterioration of plastics may be due to two causes - disfigurement due to superficial microbial growth using the plastic as an attachment surface and/or utilising dirt as nutrient, or damage due to assimilatory and disssimilatory microbial action (Mills, 1974). Dissimilatory damage may be due to staining pigments produced by micro-organisms actively growing on the plastic or its dirt contaminants. *Aspergillus awamori* and *A.oryzae* have been reported to cause stains (Kestel'man, Yarovenko & Mel'nikova, 1972). Assimilation of plastic constituents causes mechanical changes in the plastic apart from facilitating disfiguring microbial growth. *A.niger, A.flavus, A.versicolor, A. amstelodami, A.oryzae* and *A.awamori* either in mixed or pure culture have been grown on or shown to have caused mechanical damage to various plastics (Hazeu, 1967; Turner, 1967; Darby & Kaplan, 1968; Evans & Levisohn, 1968; Heap & Morrell, 1968; Kestel'man *et al.*, 1972; Klausmeier, 1972; Lazar, 1975).

Plastic formulations consist of polymers as well as fillers, plasticisers, pigments and stabilisers (Klausmeier & Jones, 1961). Most polymers are microbially resistant, but melamine formaldehyde, cellulose nitrate, polyvinyl acetate and polyester type polyurethanes are susceptible (Mills, 1974). *Aspergillus niger* and *A. flavus* were amongst test organisms used to demonstrate the susceptibility of the former three compounds (Turner, 1967). *A.niger* besides other organisms, has been reported to attack polyester-based polyurethanes but not polyether-based polyurethanes (Evans & Levisohn, 1968; Heap & Morrell, 1968). *A.niger* has been used also in mixed culture with *A.flavus* and *A.versicolor* and other fungal species to test the susceptibility and resistance of polyester and polyether polyurethanes respectively (Darby & Kaplan, 1968).

The amount of pigment and stabiliser in a plastic is usually small and of little significance in microbial breakdown (Klausmeier & Jones, 1961; Mill, 1974), but the nature of the stabiliser may exert some influence as shown by the response of *Aspergillus flavus* and other fungal species to unplasticised PVC stabilised with calcium stearate or lead stearate (Lazar, 1975).

Fillers may consist of textiles, wood flow or other organic materials, but can be replaced with fibreglass or other inorganic materials (Klausmeier & Jones, 1961). It is probably the organic fillers that support the reported growth of *A.flavus* and *A.amstelodami* and other fungi on plastics in the absence of other assimilable substrates containing these compounds (Lazar, 1975).

It is the microbial utilization of plasticisers however that is the major cause of biodeterioration of plastics. Plasticisers are mostly used in the vinyl industry - they modify the secondary forces between polymer chains and make the plastic less rigid. Utilisation and removal of the plasticiser results in decreased elasticity and flexibility, and increased tensile strength and brittleness of the plastic (Klausmeier & Jones, 1961; Mills, 1974). Those plasticisers consisting of esters of fatty acids - lauric, oleic, ricinoleic and stearic acid derivatives, and oils containing glyceryl esters of these acids - were originally considered to be the susceptible plasticisers (Klausmeier & Jones, 1961). A number of *Aspergillus* species have been reported to grow on such plasticisers (Berk, Ebert & Teitell, 1957; Bomar, 1968; Reese, Cravetz & Mandels, 1955). However, such studies utilised selected pure cultures or mixed cultures, but by employing an enrichment technique 'funginert' plasticisers were shown to be susceptible, including tributyl citrate by an *A.terreus* isolate (Klausmeier & Jones, 1961). These same workers hypothesised that since it appeared that plasticiser esters were hydrolysed by extracellular microbial esterases, then opportunists growing on superficial debris might be induced to produce esterase of low specificity by the plasticiser ester and cause the degradation of the plasticiser, although the micro-organisms would not be able to utilise the plasticiser for growth. Subsequently it has been shown that micro-organisms in the presence of extraneous complex organic nutrients can cause the deterioration of plasticisers previously considered resistant (Klausmeier, 1966). Further, plasticisers shown to be resistant to mesophilic organisms have been found susceptible to attack by thermophilic fungi, including the thermotolerant *A.fumigatus* (Mills & Eggins, 1974).

The biodeterioration of plastics may be encountered in situations wherever plastics are utilised in conditions conducive to microbial growth including that of *Aspergillus* species, for example - litho-printer rollers (Evans & Levisohn, 1968); microbial enzyme production (Kesterl'man *et al.*, 1972); protective materials for pipelines (Pankhurst, Davies & Blake, 1972); coated textiles and leathercloth (Hazeu, 1967; Turner, 1967), though it may be the cotton textile which is the component degraded (Mills, 1974). It has even been shown that *A.fumigatus* can penetrate the hydrophilic polymer of contact lenses (Filppi, Pfister & Hill, 1973). One of the most serious effects due to the biodeterioration of plastic in which *Aspergillus* species including *A.niger* and *A.flavus* have been involved, is the effect on electrical instruments where microbial growth may cause increased conductance of insulating materials and affect the performance of the instruments (Teitell, Berk & Kravitz, 1955; Gauger *et al.*, 1966) though fungal growth in these instances may not always be due to assimilation of the plastic or plasticiser but due to adventitious growth on dirt contamination.

AVIATION KEROSINE

Aspergillus species are often reported among the isolates from fuel storage tanks, airfield fuelling systems and subsonic aircraft fuel tanks (Thomas, 1973). Such isolates of *A.flavus, A.fumigatus, A.fischeri, A.flavipes, A.niger* and the *A.notus* group have been reported to produce moderate to limited growth in mineral salts medium with different fuels (Darby, Simmons & Wiley, 1968; Thomas, 1973; Sheridan & Soteros, 1974). Fungal growth which occurs in storage and aircraft fuel tanks when water is present, can cause blockage of fuel filters in fuel lines and malfunctions of fuel gauge probes. *Cladosporium resinae* is the predominant fungal contaminant in these situations, however, and is capable of corroding the aluminium alloy of aircraft fuel tanks (Hendey, 1964).

With the emergence of supersonic aircraft, like Concorde, a new environment has been presented for fuel-utilising organisms (Thomas, 1973; Hill & Thomas, 1976; Thomas & Hill, 1976a,b,c,d). The fuel tanks of both subsonic and supersonic aircraft provide similar conditions for microbial growth except for temperature - the fuel in subsonic aircraft is cooled during subsonic flight whereas that in supersonic aircraft is heated during supersonic flight, as it is used as a heat sink for the dissipation of heat created by friction due to the high speeds of the aircraft. Depending on the rate at which the fuel is utilised it may heat up to between 40° and $50^{o}C$ or as high as 90° to $100^{o}C$ during supersonic flight lasting several hours. *Cladiosporium resinae* when screened with other selected fungal species for their ability to grow in the simulated conditions of such a fuel tank heating up to 40° and $45^{o}C$, failed to grow; but an albino strain of *Aspergillus fumigatus* produced vigourous growth under these conditions. Thus, the albino *A.fumigatus* which grew in mineral salts medium and aviation kerosine over the temperature range 10° to $50^{o}C$ and at pH values between 3.65 and 7.95 of the aqueous phase, was shown to be capable of growth in the fuel tanks of Concorde aircraft under some of the temperature conditions which occur during supersonic flight. Further studies showed that the spores of the fungus were able to survive higher temperatures in this environment and therefore could remain viable until conditions were favourable for growth. Growth of the fungus, besides being able to cause malfunctions and blockages, was found to corrode the aluminium alloy of which Concorde aircraft fuel tanks are constructed - probably due to the action of organic acid metabolites and galvanic cells formed as a result of the fungal growth. Other albino strains and a pale green pigmented strain of the fungus were isolated which grew well and had similar growth characteristics in aviation kerosine and mineral salts medium. Dark green pigmented strains produced either very scant or no growth under these conditions, although one such strain grew well on the individual alkanes which constitute aviation kerosine.

CORK

Several fungal species including *Aspergillus* species have been reported to grow on bottle corks - such growth possibly being responsible for 'mouldy' tastes and 'off' odours produced occasionally in wines (Turner, 1967; Marais & Kruger, 1975).

GLASS

Although glass itself does not provide nutrient for fungal growth, the optical lenses in various instruments including cameras, binoculars, microscopes, gunsights and surveying instruments are often found to be optically impaired due to fungal growth especially in humid conditions (Nagamuttu, 1967; Turner, 1967; Eggins & Allsopp, 1975). The fungi utilise dirt or leather, wax or the vegetable pigments of black lacquer, on other areas of the optical instruments in close proximity, as nutrient sources to grow adventitiously on the glass lenses. Superficial fungal growth can easily be cleaned off; however if such growth is allowed to remain it can cause permanent damage by etching the glass surface due to corrosive metabolic products. Amongst fungal species found growing in such situations, species of *Aspergillus* including *A.fumigatus*, *A.versicolor* and the *A.glaucus* group have been encountered.

BUILDING MATERIALS

Aspergillus species besides other fungal species are found growing and producing stains on plaster walls especially in humid factories where dirt may help to sustain their growth (Turner, 1967; Hueck-van der Plas, 1968).

COSMETICS and PHARMACEUTICAL PRODUCTS

Information about the biodeterioration of cosmetics and pharmaceutical products is difficult to obtain since "spoilage is handled with the minimum of publicity by manufacturers who often develop satisfactory solutions without proper identification" (Manowitz, 1961). The very nature of use of these preparations requires them to be free of microbial contamination, since passive microbial contaminants may be inherent or opportunist pathogens - *Aspergillus* species contaminating various preparations including talcs and ointments have been cited in this context (Savin, 1967; Butler, 1968; Bowman, White & Lyles, 1971; Bowman, Knoll, White & Mislivec, 1972), although aflatoxins have not been implicated (Beveridge, 1975). Some products have to be sterile, but it is not economically feasible to sterilise all products. Contamination of these nonsterile products may occur from various sources during manufacture even though precautions are taken to maintain their quality, or during consumer usage - thus quality maintenance of these products may depend on low water activity or pH, controlled storage conditions or preservatives (Manowitz, 1961; Butler, 1968; Beveridge, 1975). Oil-in-water emulsions are particularly vulnerable to microbial spoilage (Manowitz, 1961) and present particular preservation problems (Beveridge, 1975). *Aspergillus* species including *A.niger*, *A.tamarii* and *A.flavus* are amongst micro-organisms which cause spoilage of pharmaceutical emulsions by causing emulsion breakdown or production of acid, odours, unpleasant flavours or discolouration, apart from utilising the oil or emulgents as nutrients and producing unsightly growths (Manowitz, 1961; Eggins & Walters, 1963; Turner, 1967; Beveridge, 1975). *A.niger* has also been cited as attacking a large variety of cosmetics and pharmaceutical products over a wide pH range and having a high resistance to antimicrobial agents (Manowitz, 1961; Turner, 1967) and has been proposed as a test organism for preservatives (Manowitz, 1961; Olson, 1967; Tenebaum, 1973), though its response to inhibitory processes has been shown to be affected by its growth temperature (McRobbie & Parker, 1973).

STORED PRODUCE

Aspergillus species are involved in the spoilage of a wide variety of stored produce (Table 1). It must be noted that in many of these cases, other fungal species may be involved in causing spoilage apart from other micro-organisms as well as insects and rodents. However, in many instances the *Aspergillus* species are among the major spoilage organisms and in others they are the predominant cause of spoilage.

The species of *Aspergillus* found actively causing biodeterioration in any particular stored produce depends on the nature of the substrate of that produce - but even on the same substrate the *Aspergillus* species vary depending on physical factors, especially moisture and temperature. A number of *Aspergillus* species germinate and grow at low water activities - the limits of which vary with temperature (Ayerst, 1966, 1969; Sinha, 1973; Pitt, 1975). This property is a major factor in the predominance of *Aspergillus* species as 'storage fungi' on dry stored produce - including members of the *A.restrictus*, *A.glaucus*, *A.candidus*, *A.ochraceus* and *A.flavus* groups (Christensen & Kaufmann, 1974). The available moisture for fungal growth is dependent on the moisture content of a particular substrate which changes with variations in the relative humidity of the atmosphere (Pixton, 1967). Above the 'safe' minimum moisture content of a particular substrate below which no fungal growth occurs, different species which grow at different water activities become active as the moisture content of the substrate increases.

Fungal infestation is a major cause of spoilage of stored cereals and seeds including corn, rice and sorghum. Such infestation causes discolouration and loss of germinability, heating, mustiness, caking and decay. Loss of germinability reduces the worth of the product for use as viable seed; heating may result in combustion; while the other consequences of spoilage reduce the quality of the produce in terms of food value.

On other stored products, fungal infestation apart from causing general spoilage due to soiling, heating and decay, can produce other detrimental effects. Coffee processed from mouldy coffee beans may be musty and produce a liquor lacking in quality, aroma and flavour. Similarly, mould growth in cocoa beans affects their lipid content and increases their free fatty acid content which may result in reduced flavour besides producing offensive odours, taints and discolouration in the processed products. Also, many of the micro-organisms - mostly fungi of which *Aspergillus* species are predominant - occurring on stored seeds, fruits and nuts from which vegetable oils are extracted are lipolytic. Thus the micro-organisms reduce the lipid content of such produce while increasing the free fatty acid content. The extracted oils likewise may be affected during storage besides developing noxious tastes.

Of the fungal species growing on raw sugar, the *Aspergillus* species in particular invert the sugar thus reducing the sucrose content and increasing the invert content - *A.glaucus* and *A.repens* being the most active. The main consequence of microbial growth on stored tobacco - the *Aspergillus* species being predominantly implicated - is the loss of material due to rotting and breakdown of the tobacco constituents. Other consequences are 'off' odours produced in the tobacco which may affect the aroma of the smoked product; while the elasticity and strength of cigar binder and wrapper leaves may be affected.

Apart from the damage described above, contaminated and infested stored produce can present health hazards. Some *Aspergillus* species are known to be pathogenic and thus their presence and growth on stored produce increases the risk to the people handling the produce, of becoming infected (Austwick, 1965; Dennis, 1973). More in the context of true biodeterioration is the production of mycotoxins on stored food produce which renders the food unfit for consumption by humans or animals (Coveney, Peck & Townsend, 1966; Scott, 1965, 1973; Davis *et al.*, 1975) are found growing on stored produce.

Table 1. Stored products on which *Aspergillus* spp. play a deteriorative role.

Stored product	Reference
Bread	Eskin *et al.* (1971)
Butter, cheese and margarine	Eskin *et al.* (1971); Reese *et al.* (1955)
Cereal grains (barley, corn, oats, rice, sorghum & wheat)	Apinis (1972); Christensen & Kaufman (1974); Eskin *et al.* (1971); Flannigan (1969); Oblisam *et al.* (1972); Parvathappa *et al.* (1970); Pitt (1975); Wallace (1972)
Cocoa beans	Broadbent & Oyeniran (1968); Hansen & Welty (1970); Hansen *et al.* (1973); Maravalhas (1966); Raymond (1966)
Coffee beans	Butt (1966); Raymond (1966); Vincents Nissen (1967)
Copra	Eggins & Coursey (1968)
Cotton seed	Christensen *et al.* (1949); Eggins & Coursey (1968)
Dried fruits (prunes, raisins)	Follstad (1966); Pitt (1975); Pitt & Christian (1968)
Dried peas	Eskin *et al.* (1971)
Jam	Eskin *et al.* (1971); Pitt (1975)
Meat	Bullerman & Ayres (1968); Sutic *et al.* (1972); Wu (1974)
Molasses	Eskin *et al.* (1971)
Nuts	Austwick & Ayerst (1963); Chipley & Heaton (1971); Diener (1960); Doupnik & Bell (1971); Eskin *et al.* (1971); Esuruoso (1974); Lillard *et al.* (1970); Schindler *et al.* (1974); Taber & Schroeder (1967)
Olives	Eggins & Coursey (1968)
Palm fruit	Raymond (1966)
Pears	Eskin *et al.* (1971)
Peppercorns	Christensen *et al.* (1967)
Rapeseed and rapeseed cakes	Eskin *et al.* (1971); Poisson (1972)
Raw sugar	Eskin *et al.* (1971); Tilbury (1966)
Soybeans	Christensen & Kaufmann (1974)
Sunflower seeds	Christensen & Kaufmann (1974)

Table 1. (cont'd)

Stored product	Reference
Tobacco	Mitchell & Stauber (1975); Welty & Lucas (1968, 1969)
Vegetable oils	Coursey (1966); Eggins & Coursey (1968); Eskin *et al.*(1971); Raymond (1966); Tomlins & Townsend (1968)
Vegetables (cabbage, carrots, yams, etc.)	Adeniji (1970); Eskin *et al.* (1971); Ogundana (1972); Ogundana *et al.*(1970, 1975)

CONCLUSIONS

The information available about any *Aspergillus* species in any particular biodeterioration problem varies from its noted presence merely as an '*Aspergillus* sp.' to thorough accounts giving correct and verified identification of the species with details of the conditions and consequences of its growth in a particular situation.

The extent to which the part played by any biodeteriogen in any biodeterioration problem is investigated often depends on the relative importance of that biodeteriogen in the particular problem and the means by which it is controlled.

The prevention of biodeterioration may be effected by controlling environmental conditions; by replacing susceptible components with more resistant ones or by modifying them to make them more resistant; by coating susceptible materials with protective resistant materials; or with biocides including fungicides.

Refrigeration, drying, preservation with sugar or salt, storage in low levels of oxygen and high levels of carbon dioxide are examples of environmental control which provide unfavourable temperature, moisture, or atmospheric conditions for microbial growth. For environmental control to be effective full details with regard to the identification and growth requirements of the biodeteriogens involved must be investigated and established. Thus, for example, by knowing these details about the *Aspergillus* species which cause deterioration in stored produce, measures can be taken to provide unfavourable conditions for their growth (Ayerst, 1968; Christensen & Kaufmann, 1974).

The formulation and use of biocides, including fungicides, are complicated procedures requiring expertise to evaluate the requirements of different situations (Kaplan, 1968). Many of the references given in this paper provide information about control methods used in the various areas of biodeterioration, but reference to the publications listed in the bibliography will provide further information.

Where biocides, or fungicides, are used to control a common biodeterioration problem, the identification of the biodeteriogen during the initial diagnosis merely as an '*Aspergillus* sp.' for example, or even as 'a fungus' has often been thought to be sufficient. However in order to test the effectiveness of control methods employing biocides or resistant materials when they are being forumulated, known biodeteriogens which predominantly occur in particular biodeterioration problems are employed - in which case the identity and growth requirements of the biodeteriogens commonly occurring in each particular situation are important.

Since the testing of materials for biodeterioration resistance is somewhat complex (Jones, 1968; Wälchi, 1968) standard procedures exist for this purpose, for which specific test organisms are supplied by the Commonwealth Mycological Institute (Onions, 1975). The validity of these test methods is dependent on the detailed information obtained from investigations of biodeterioration problems as mentioned above. It is hardly surprising that *Aspergillus* species are often included as test organisms in these resistance tests for materials considering the substantial role members of the *Aspergillus* genus play in biodeterioration.

REFERENCES

ADENIFI,M.O. (1970). Fungi associated with storage decay of yam in Nigeria. *Phytopathology* 60, 590-2.

APINIS,A.E. (1972). Mycological aspects of stored grain. In *Biodeterioration of materials*, ed. A.H.Walters & E.H. Hueck-van der Plas, Vol.2, pp 493-8. Applied Science Publishers Ltd.: London.

AUSTWICK,P.K.C. (1965). Pathogenicity. In *The genus Aspergillus*, K.B.Raper & D.I. Fennell, pp.82-126. The Williams & Wilkins Company: Baltimore.

AUSTWICK,P.K.C. & AYERST,G. (1963). Groundnut mycoflora and toxicity. *Chemistry and Industry* Part 1, 55-61.

AYERST,G. (1966). The influence of physical factors on deterioration by moulds. In *Microbiological deterioration in the tropics, Society of Chemical Industry Monograph No.23*, 14-20.

AYERST,G. (1968). Prevention of biodeterioration by control of environmental conditions. In *Biodeterioration of materials*, ed. A.H.Walters & J.J.Elphick, pp.223-41. Elsevier Publishing Co.

Ltd.: Amsterdam, London, New York.

AYERST,G. (1969). The effects of moisture and temperature on growth and spore germination in some fungi. *Journal of Stored Products Research* 5, 127-41.

BERK,S., EBERT,H. & TEITELL,L. (1957). Utilization of plasticizers and related organic compounds by fungi. *Industrial and Engineering Chemistry* 49, 1115-25.

BEVERIDGE,E.G. (1975). The microbial spoilage of pharmaceutical products. In *Microbial aspects of the deterioration of materials*, ed. R.J. Gilbert & D.W. Lovelock, pp.213-35. *The Society for Applied Bacteriology Technical Series No.9*, Academic Press: London, New York, San Francisco.

BOMAR,M. (1968). Some results of antifungal tests on combinations of plasticisers and fungicides. *International Biodeterioration Bulletin* 4, 43-8.

BOWMAN,F.W., WHITE,M. & LYLES,R.L. (1971). Microbial contamination of unsterile antibiotic market samples: a survey. *Journal of Pharmaceutical Sciences* 60, 1099-101.

BOWMAN,F.W., KNOLL,E.W., WHITE, M. & MISLIVEC,P. (1972). Survey of microbial contamination of opthalmic ointments. *Journal of Pharmaceutical Sciences* 61, 532-5.

BROADBENT,J.A. & OYENIRAN,J.O. (1968). A new look at mouldy cocoa. In *Biodeterioration of materials*, ed. A.H. Walters & J.J. Elphick, pp.693-702. Elsevier Publishing Co.Ltd.: Amsterdam, London, New York.

BULLERMAN,L.B. & AYRES,J.C. (1968). Aflatoxin producing potential of fungi isolated from cured and aged meats. *Applied Microbiology* 16, 1945-6.

BUTLER,N.J. (1968). The microbiological deterioration of cosmetics and pharmaceutical products. In *Biodeterioration of materials*, ed. A.H. Walters & J.J. Elphick, pp.269-80. Elsevier Publishing Co.Ltd.: Amsterdam, London, New York.

BUTT,D.J. (1966). The microbiological deterioration of Uganda dry processed Robusta coffee. In *Microbiological deterioration in the tropics, Society of Chemical Industry Monograph* No.23,80-97.

CHAMBERLAIN,E.A.C., MEEK,I.R., KEWELL,C.B. & FINLAY,A.J. (1966). Microbiological breakdown of conveyor belting in coal mines. *Journal of Mines, Metals and Fuels* 14, 131-9.

CHIPLEY, J.R. & HEATON, E.K. (1971). Microbial flora of pecan meat. *Applied Microbiology* 22, 252-3.

CHRISTENSEN,C.M. & KAUFMANN,H.H. (1969). *Grain storage: the role of fungi in quality loss.* University of Minnesota Press: Minneapolis.

CHRISTENSEN,C.M. & KAUFMANN,H.H. (1974). Microflora. In *Storage of cereal grains and their products*, ed. C.M. Christensen, Monograph Series Vol. V - Revised, pp.158-92. American Association of Cereal Chemists Inc.: St. Paul, Minnesota.

CHRISTENSEN,C.M., OLAFSON,J.H. & GEDDES, W.F. (1949). Grain storage studies. VIII. Relation of molds in moist stored cottonseed to increased production of carbon dioxide, fatty acids and heat. *Cereal Chemistry* 26, 109-28.
CHRISTENSEN,C.M., FANSE, H.A., NELSON, G.H., BATES, F. & MIROCHA, C.J. (1967). Microflora of black and red pepper. *Applied Microbiology* 15, 622-6.
COURSEY,D.G. (1966). Biodeteriorative processes in palm oil in West Africa. In *Microbiological deterioration in the tropics, Society of Chemical Industry Monograph* No.23, 44-56.
COURSEY,D.G. (1972). Biodeteriorative losses in tropical horticultural produce. In *Biodeterioration of materials,* ed. A.H. Walters & E.H. Hueck-van der Plas, Vol.2. pp.464-71. Applied Science Publishers Ltd.: London
COVENEY,R.D., PECK,H.M. & TOWNSEND,R.J. (1966). Recent advances in mycotoxicoses. In *Microbiological deterioration in the tropics, Society of Chemical Industry Monograph* No.23, 31-43.
CUNDELL,A.M. & MULCOCK,A.P. (1972). Microbiological deterioration of vulcanized rubber. *International Biodeterioration Bulletin* 8, 119-25.
DARBY,R.T. & KAPLAN,A.M. (1968). Fungal susceptibility of polyurethanes. *Applied Microbiology* 16, 900-5.
DARBY,R.T., SIMMONS, E.G. & WILEY, B.J. (1968). A survey of fungi in a military aircraft fuel supply system. *International Biodeterioration Bulletin* 4, 39-41.
DAVIS,N.D., WAGNER,R.E., DALBY,D.K., MORGAN-JONES,G. & DIENER,U.L. (1975). Toxigenic fungi in food. *Applied Microbiology* 30,159-61.
DENNIS,C.A.R. (1973). Health hazards in grain storage. In *Grain storage: part of a system,* ed. R.N. Sinha & W.E. Muir, pp.367-87. Avi Publishing Co. Inc.: Westport, Connecticut.
DIENER,U.L. (1960). The mycoflora of peanuts in storage. *Phytopathology* 50, 220-3.
DOUPNIK,B. & BELL,D.K. (1971). Toxicity to chicks of *Aspergillus* and *Penicillium* species isolated from mouldy pecans. *Applied Microbiology* 21, 1104-6.
DRESCHER,R.F. (1958). Microbiology of paint films. IV. Isolation and identification of microflora on exterior emulsion paints. *American Paint Journal* 42, 80-102.
EATON,R.A. (1976). Cooling tower fungi. In *Recent advances in aquatic mycology,* ed. E.B. Gareth Jones, pp.359-87. Elek Science: London.
EGGINS,H.O.W. (1967). The economics of biodeterioration. *Environmental Engineering* 29, 15-6.
EGGINS, H.O.W. (1968). Ecological aspects of biodeterioration. In *Biodeterioration of materials,* ed. A.H. Walters & J.J. Elphick, pp.22-7. Elsevier Publishing Co.Ltd.: Amsterdam, London, New York.

EGGINS,H.O.W. & ALLSOPP,D. (1975). Biodeterioration and biodegradation by fungi. In *The filamentous fungi*, ed. J.E.Smith & D.R.Berry, Vol.1 pp.301-19. Edward Arnold: London.

EGGINS,H.O.W. & COURSEY,D.G. (1968). The industrial significance of biodeterioration of oilseeds. *International Biodeterioration Bulletin* 4, 29-38.

EGGINS,H.O.W. & WALTERS,V. (1963). The decomposition of pharmaceutical emulsions by fungi. *West African Journal of Biological and Applied Chemistry* 7, 2-8.

ESKIN,N.A.M., HENDERSON,H.M. & TOWNSEND,R.J. (1971). The biodeterioration of foods. In *Biochemistry of foods* pp.153-217. Academic Press: New York and London.

ESURUOSO, O.F. (1974). Fungi associated with kernel rot disease of cashew (*Anacardium occidentale* L.) in Nigeria. *International Biodeterioration Bulletin* 10, 57-9.

EVANS,D.M. & LEVISOHN, I. (1968). Biodeterioration of polyester-based polyurethane. *International Biodeterioration Bulletin* 4, 89-92.

FILPPI,J.A., PFISTER,R.M. & HILL,R.M. (1973). Penetration of hydrophilic contact lenses by *Aspergillus fumigatus*. *American Journal of Optometry and Archives of American Academy of Optometry* 50, 553-7.

FLANNIGAN,B. (1969). Microflora of dried barley grain. *Transactions of the British Mycological Society* 53,371-9.

FOLLSTAD,M.N. (1966). Microflora of nonprocessed raisins. *Phytopathology* 56, 1413.

GAUGER,G.W., SPRADLIN,B.C., EASTERDAY,J.L., KNUTH,D.T., DAVIDSON, R.S. & LITCHFIELD, J.H. (1966). Microbial deterioration of electronic components. I. Selected components and materials of construction. *Developments in Industrial Microbiology* 8,372-94.

GERSONDE,V.M. & KERNER-GANG,W. (1968). Untersuchungen an Moderfäule-Pilzen aus Holzstäben nach Freilandversuchen. *Material Organismen* 3, 199-212.

GHOSE,S.N. & BOSE,R.G. (1973). Biodeterioration of jute; part 1. Nature of microbiological growth. *Journal of Textile Association, March 1973* 11-6.

HANSEN,A.P. & WELTY,R.E. (1970). Microflora of raw cacao beans. *Mycopathologia et Mycologia Applicata* 44, 309-16.

HANSEN,A.P., WELTY,R.E. & SHEN,R. (1973). Free fatty acid content of cacao beans infested with storage fungi. *Journal of Agricultural and Food Chemistry* 21, 665-70.

HASEU,W. (1967). Results of the first inter-laboratory experiment in biodeterioration of plastics. *International Biodeterioration Bulletin* 3, 15-9.

HEAP,W.M. & MORRELL,S.H. (1968). Microbiological deterioration of rubbers and plastics. *Journal of Applied Chemistry* 18, 189-94.

HENDEY,N.I. (1964). Some observations on *Cladosporium resinae* as a fuel contaminant and its possible role in the corrosion of aluminium alloy fuel tanks. *Transactions of the British Mycological Society* 47, 467-75.

HILL,E.C. &THOMAS,A.R. (1976). Microbial aspects of supersonic aircraft fuel. In *Proceedings of the Third International Biodegradation Symposium*, ed. U.M. Sharpley & A.M. Kaplan, pp. 157-74. Applied Science Publishers Ltd.: England.

HUDSON,H.J. (1969). *Aspergilli* in the air-spora at Cambridge. *Transactions of the British Mycological Society* 52, 153-9.

HUECK,H.J. (1965). The biodeterioration of materials as a part of hylobiology. *Material Organismen* 1, 15-34.

HUECK-VAN DER PLAS, E.H. (1968). The microbiological deterioration of porous building materials. *International Biodeterioration Bulletin* 4, 11-28.

HUGHES,R.L. (1968). Microbiological degradation in the paper, printing and packaging industries. In *Biodeterioration of materials*, ed. A.H. Walters & J.J. Elphick, pp.281-90. Elsevier Publishing Co.Ltd.: Amsterdam, London, New York.

JONES,E.S.L. (1968). Some problems posed by quality screening for biodeterioration. In *Biodeterioration of materials*, ed. A.H. Walters & J.J. Elphick, pp.188-95. Elsevier Publishing Co.Ltd.: Amsterdam, London, New York.

KAARIK, A.A. (1974). Decomposition of wood. In *Biology of plant litter decomposition, ed.* C.H. Dickinson & G.J.F. Pugh, Vol. 1 pp.129-74. Academic Press: London and New York.

KAPLAN,A.M.(19 68). The control of biodeterioration by fungicides - philosophy. In *Biodeterioration of materials* ed. A.H. Walters & J.J. Elphick, pp.196-204. Elsevier Publishing Co.Ltd.: Amsterdam, London, New York.

KELLY, B.J. (1975). New chemical formulation will control microorganisms in hydrocarbon contaminated cooling water systems. *Materials Protection* 4, 62-7.

KESTEL'MAN,V.N., YAROVENKO,V.L. & MEL'NIKOVA,E.I. (1972). The corrosion of polymeric materials under the conditions of the microbiological synthesis of enzymes. *International Biodeterioration Bulletin* 8, 15-9.

KING, B. & EGGINS,H.O.W. (1973). Decay mechanisms of microfungi which might produce an enhanced permeability in wood. *International Biodeterioration Bulletin* 9, 35-43.

KLAUSMEIER, R.E. (1966). The effect of extraneous nutrients on the biodeterioration of plastics. In *Microbiological deterioration in the tropics, Society of Chemical Industry Monograph* No.23,232-43.

KLAUSMEIER, R.E. (1972). Results of the second inter-laboratory experiment on biodeterioration of plastics. *International Biodeterioration Bulletin* 8. 3-7.

KLAUSMEIER,R.E. & JONES,W.A. (1961). Microbial degradation of plasticizers. *Developments in Industrial Microbiology* 2, 47-53.

LAZAR,V. (1975). The study of microbial corrosion of plastics in Romania. *International Biodeterioration Bulletin* 11, 16-23.

LILLARD,H.S., HANLIN,R.T. & LILLARD,D.A. (1970). Aflatoxigenic isolates of *Aspergillus flavus* from pecans. *Applied Microbiology* 19, 128-30.

MacLACHLAN,J., HEAP, W.M. & PACITTI,J. (1966). Attack of bacteria and fungi on rubbers and plastics in the tropics. In *Microbiological deterioration in the tropics, Society of Chemical Industry Monograph* No.23, 185-200.

McROBBIE,D.I. & PARKER,M.S. (1973). The influence of growth temperature upon the resistance of a spoilage mould to inhibitory processes. *International Biodeterioration Bulletin* 9, 19-22.

MANOWITZ, M. (1961). Preservation of cosmetic emulsions. *Developments in Industrial Microbiology* 2, 65-71.

MARAIS,P.G. & KRUGER,M.M. (1975). Fungus contamination of corks responsible for unpleasant odours in wine. *Phytophylactica* I, 115-6.

MARAVALHAS,N. (1966). Mycological deterioration of cocoa beans during fermentation and storage. In *Microbiological deterioration in the tropics, Society of Chemical Industry Monograph* No.23, 98-104.

MERRILL,W. & FRENCH,D.W. (1966). Colonization of wood by soil fungi. *Phytopathology* 56, 301-3.

MERRILL,W., FRENCH,D.W. & HOSSFELD,R.L. (1965). Effects of common moulds on physical and chemical properties of wood fibreboard. Part II of a series of wood fibreboard studies. *Tappi* 48, 470-4.

MILLS, J. (1974). The biodeterioration of synthetic polymers and plasticizers. *CRC Critical Reviews in Environmental Control* 4, 341-51.

MILLS,J. & EGGINS,H.O.W. (1974). The biodeterioration of certain plasticisers by thermophilic fungi. *International Biodeterioration Bulletin* 10, 39-44.

MITCHELL,T.G. & STAUBER,P.C. (1975). Biodeterioration of tobacco. In *Microbial aspects of the deterioration of materials*, ed. R.J. Gilbert & D.W. Lovelock, *The Society for Applied Bacteriology Technical Series No.9*, pp.203-11. Academic Press: London, New York, San Francisco.

NAGAMUTTU,E. (1967). Moulds on optical glass and control measures. *International Biodeterioration Bulletin* 3, 25-7.

OBLISAM,G., SRINIVASA,H.P. & RANGASWAMI,G. (1972). Microbial deterioration of stored rice seeds. *Oryza* 9, 47-55.

OGUNDANA,S.K. (1972). The post-harvest decay of yam tubers and its preliminary control in Nigeria. In *Biodeterioration of materials*, ed. A.H. Walters & E.H. Hueck-van der Plas, Vol.2, pp.481-92 Applied Science Publishers Ltd.,: London.

OGUNDANA,S.K., NAQVI, S.H.Z. & EKUNDAYO,J.A. (1970). Fungi associated with soft rot of yams (*Dioscorea* spp) in storage in Nigeria. *Transactions of the British Mycological Society* 54, 445-51.

OGUNDANA,S.K., NAQVI,S.H.Z. & EKUNDAYO,J.A. (1975). Changes in carbohydrate contents of yams (*Dioscorea* spp.) due to storage rot-fungi in Nigeria. *International Biodeterioration Bulletin* 11, 36-40.

OLSON,S.W. (1967). The application of microbiology to cosmetic testing. *Journal of the Society of Cosmetic Chemists* 18,191-8.

O'NEILL,L.A. & SKINNER,C.E. (1966). Laboratory tests for resistance of paint films to mould growth under tropical conditions. In *Microbiological deterioration in the tropics, Society of Chemical Industry Monograph* No.23, 170-8.

ONIONS,A.H.S. (1975). Organisms for biodeterioration testing-moulds and fungi. In *Microbial aspects of the deterioration of materials*, ed. R.J. Gilbert & D.W. Lovelock, *The Society for Applied Bacteriology Technical Series No.9*, pp.1-22. Academic Press: London, New York, San Francisco.

ORLITA,A. (1968). Biodeterioration in leather industry. In *Biodeterioration of materials*, ed.A.H. Walters & J.J. Elphick, pp.297-301. Elsevier Publishing Co.Ltd.: Amsterdam, London, New York.

PANKHURST,E.S., DAVIES,M.J. & BLAKE,H.M. (1972). The ability of polymers or materials containing polymers to provide a source of carbon for selected micro-organisms. In *Biodeterioration of materials*, ed. A.H.Walters & E.H. Hueck-van der Plas, Vol. 2. pp.76-90. Applied Science Publishers Ltd.: London.

PARVATHAPPA,H.C., POORNIMA,P., RAGUNATHAN,A.N. & MAJUMDER, S.K. (1970). Physical and biochemical changes in sorghum (*Sorghum vulgare). International Biodeterioration Bulletin* 6, 95-9.

PITT,J.I. (1975). Xerophilic fungi and spoilage of foods of plant origin. In *Water relations in food*, ed. R.B. Duckworth, pp.273-307. Academic Press: London, New York, San Francisco.

PITT,J.I. & CHRISTIAN,J.H.B. (1968). Water relations of xerophilic fungi isolated from prunes. *Applied Microbiology* 16, 1853-8.

PIXTON,S.W. (1967). Moisture control - its significance and measurement in stored products. *Journal of Stored Products Research* 3, 35-47.

POISSON,J., CAHAGNIER,B. & CHANET,B. (1972). Microbiological and biochemical developments in rapeseed cakes during storage. In *Biodeterioration of materials*, ed. A.H.Walters & E.H.Hueck-van der Plas, Vol.2,pp.507-14. Applied Science Publishers Ltd.:London.

RAPER,K.B. & FENNELL,D.I. (1965). *The genus Aspergillus*. The Williams and Wilkins Co.: Baltimore.

RAYMOND,W.D. (1966). The importance of moulds in the deterioration of tropical foods and feeding stuffs. In *Microbiological deterioration in the tropics, Society of Chemical Industry Monograph* No.23, 21-30.

REESE,E.T., CRAVETZ,H. & MANDELS,G.R. (1955). Activity of fungi on oils. *Farlowia* 4, 409-21.

REESE,E.T. & DOWNING,M.H. (1951). Activity of the *Aspergilli* on cellulose, cellulose derivatives and wool. *Mycologia* 43,16-28.

ROOK,J.J. (1955). Microbiological deterioration of vulcanized rubber. *Applied Microbiology* 3, 302-9.

ROSS, R.T.(1963). Microbiology of paint films. *Advances in Applied Microbiology* 5, 217-33.

ROSS,R.T., SLADEN,J.B. & WIENERT,L.A. (1968). Biodeterioration of paint and paint films. In *Biodeterioration of materials*, ed. A.H.Walters & J.J.Elphick, pp.317-25. Elsevier Publishing Co.Ltd.: Amsterdam, London, New York.

SAVIN,J.A. (1967). The microbiology of topical preparations in pharmaceutical products. I. Chemical aspects. *Pharmaceutical Journal* 199, 285-8.

SAVORY,J.G. (1954). Breakdown of timber by Ascomycetes and Fungi Imperfecti. *Annals of Applied Biology* 41, 336-47.

SCHEFFER,T.C. (1973). Microbial degradation and the causal organisms. In *Wood deterioration and its prevention by preservative treatments*, ed. D.D. Nicholas, pp.31-106. Syracuse University Press.

SCHINDLER,A.F., ABADIE,A.N., GECAN,J.S., MISLEVIC,P.B. & BRICKEY,P.M. (1974). Mycotoxins produced by fungi - isolated from inshell pecans. *Journal of Food Science* 39, 213-4.

SCOTT,D.B. (1965). Toxigenic fungi isolated from cereal and legume products. *Mycopathologia et Mycologia Applicata* 25, 213-22.

SCOTT,P.M. (1973). Mycotoxins in stored grain, feeds and other cereal products. In *Grain storage: part of a system*, ed. R.N. Sinha & W.E. Muir, pp.343-65. Avi Publishing Co.Inc.: Westport, Connecticut.

SEMENIUK, G., HARSHFIELD,G.S.,CARLSON,C.W., HESSELTINE,C.W. & KWOLEK,W.F. (1971). Mycotoxins in *Aspergillus*. *Mycopathologia et Mycologia Applicata* 43, 137-52.

SHAPOSHNIKOV,V.N., KOZLOVA,E.I. & AZOVA,L.G. (1964). Degradation of wool by micro-organisms. *U.S.Clearing House Translation No. AD* 719550 (1971). From *Vestnik Moskovskogo Universiteta No.2*, 58-63.

SHARP,R.F. (1975). The colonisation of perfumed beech stakes by soil micro-organisms. *Mycopathologia et Mycologia Applicata* 55, 185-92.

SHARP,R.F. & EGGINS,H.O.W. (1970). The ecology of soft-rot fungi. 1. Influence of pH. *International Biodeterioration Bulletin* 6, 53-64.

SHERIDAN,J.E. & SOTEROS,J.J. (1974). A survey of fungi in jet aircraft fuel systems in New Zealand. *International Biodeterioration Bulletin* 10, 105-7.

SINHA,R.N. (1973). Interrelations of physical, chemical and biological variables in the deterioration of stored grains. In *Grain storage: part of a system*, ed. R.N.Sinha & W.E. Muir, pp.15-47. Avi Publishing Co. Inc.: Westport, Connecticut. U.S.A.

SIU,R.G.H. (1951). *Microbial decomposition of cellulose*. Reinhold Publishing Corporation: New York.

STANDARD,P.G., MACKEL,D.C. & MALLISON,G.F. (1971). Microbial penetration of muslin and paper wrapped sterile packs stored on open shelves and in closed cabinets. *Applied Microbiology* 22, 432-7.

STANDARD,P.G., MALLISON,G.F. & MACKEL,D.C. (1973). Microbial pentration through three types of double wrappers for sterile packs. *Applied Microbiology* 26, 59-62.

SUTIC,M., AYRES,J.C. & KOEHLER,P.E. (1972). Identification and aflatoxin production of moulds isolated from country cured hams. *Applied Microbiology* 23, 656-8.

TABER,R.A. & SCHROEDER,H.W. (1967). Aflatoxin-producing potential of isolates of the *Aspergillus flavus-oryzae* group from peanuts (*Arachis hypogaea*). *Applied Microbiology* 15, 140-4.

TEITELL,L., BERK,S. & KRAVITZ,A. (1955). The effects of fungi on the direct current surface conductance of electrical insulating materials. *Applied Microbiology* 3, 75-81.

TENENBAUM,S. (1973). Microbial content of cosmetics and nonsterile drugs. *Cosmetics & Perfumery* 88, 49-53.

THOMAS,A.R. (1973). *Fungal growths in petroleum products*. Ph.D. Thesis, University of Wales. 150 pp.

THOMAS,A.R. & HILL,E.C. (1976*a*). *Aspergillus fumigatus* and supersonic aviation. I. Growth of *Aspergillus fumigatus*. In press.

THOMAS,A.R. & HILL,E.C. (1976*b*). *Aspergillus fumigatus* and supersonic aviation. II. Corrosion. In Press.

THOMAS,A.R. & HILL,E.C. (1976*c*). *Aspergillus fumigatus* and supersonic aviation. III. Survival. In press.

THOMAS,A.R. & HILL,E.C. (1976*d*). *Aspergillus fumigatus* and supersonic aviation. IV. Biocidal control. In press.

TILBURY,R.H. (1966). Microbiological deterioration of raw cane sugar in the tropics. In *Microbiological deterioration in the tropics, Society of Chemical Industry Monograph* No.23, 63-79.

TOMLINS,R.I. & TOWNSEND,R.J. (1968). Biodeterioration of groundnut oil by *Aspergilli*. In *Biodeterioration of materials*, ed. A.H.Walters & J.J.Elphick, pp.703-16. Elsevier Publishing Co. Ltd.: Amsterdam,London, New York.

TURNER, J.N. (1967). *The microbiology of fabricated materials.* J. & A. Churchill Ltd.: London.

UPSHER,F.J. (1972). Microfungi at The Joint Tropical Research Unit, Innisfail, Queensland In *Biodeterioration of materials*, ed. A.H.Walters & E.H. Hueck-van der Plas, Vol.2, 27-34. Applied Science Publishers Ltd.: London.

VINCENTS NISSEN,T. (1967). Fungi from green coffee beans. *International Biodeterioration Bulletin* 3, 77.

WÄLCHLI, O. (1968). Biodeterioration test methodology. In *Biodeterioration of materials*, ed. A.H. Walters & J.J. Elphick, pp.242-51. Elsevier Publishing Co.Ltd.: Amsterdam, London, New York.

WALLACE,H.A.H. (1973). Fungi and other organisms associated with stored grain. In *Grain storage: part of a system*, ed. R.N. Sinha, & W.E. Muir, pp.71-98. Avi Publishing Co.Inc.: Westport, Connecticut.

WALSH, J.H. (1968). Ecological considerations of biodeterioration. *International Biodeterioration Bulletin* 4, 1-10.

WELTY,R.E. & LUCAS,G.B. (1968). Fungi isolated from damaged fine-cured tobacco. *Applied Microbiology* 16, 851-4.

WELTY,R.E. & LUCAS,G.B. (1969). Fungi isolated from fine-cured tobacco at time of sale and after storage. *Applied Microbiology* 17, 360-5.

WU, M.T., AYRES,J.C. & KOEHLER,P.E. (1974). Toxigenic *Aspergilli* and *Penicillia* isolated from aged, cured meats. *Applied Microbiology* 28, 1094-6.

ZYSKA,B.J., RYTYCH,B.J., ZANKOWICZ, L.P. & FUDALEJ, D.S. (1972). Microbiological deterioration of rubber cables in deep mines and the evaluation of some fungicides in rubber. In *Biodeterioration of materials*, ed. A.H.Walters & E.H. Hueck-van der Plas, Vol. 2, pp.256-67. Applied Science Publishers Ltd.: London.

SELECTED BIBLIOGRAPHY FOR ADDITIONAL READING

Biodeterioration of Materials, ed. A.H. Walters & J.J. Elphick.(1968) Elsevier Publishing Co.Ltd., Amsterdam, London, New York.

Biodeterioration of Materials, Vol.2, ed. A.H. Walters & E.H. Hueck-van der Plas. (1972). Applied Science Publishers Ltd., London.

Biodeterioration Research Titles, Biodeterioration Information Centre, University of Aston, Birmingham.

Deterioration of Materials - causes and preventive techniques, ed. G.A. Greathouse & C.J. Wessel. (1954). Reinhold Publishing Corporation, New York.

Fungicides - an advanced treatise, Vol.1, ed. D.C. Torgeson. (1967). Academic Press, New York and London.

International Biodeterioration Bulletin, Biodeterioration Information Centre, University of Aston, Birmingham.

Microbial aspects of the deterioration of materials, ed. R.J. Gilbert & D.W. Lovelock. (1975). The Society for Applied Bacteriology Technical Series No. 9. Academic Press, London, New York, San Francisco.

Microbial deterioration in the tropics. (1966). Society of Chemical Industry Monograph No.23.

Oriental Food Uses of *Aspergillus*

B. J. B. WOOD

Department of Applied Microbiology, University of Strathclyde, Glasgow, U.K.

INTRODUCTION

In the Western World, uses of fungi as food are few in number and bounded by a mixture of legend and taboos which often impinge only peripherally upon reality. Even in continental Europe, more macro-fungi are consumed than in the British Isles. It is automatically assumed that filamentous micro-fungi are necessarily harmful - with the sole exception of the blue and green moulds found in certain types of cheese. The modern discovery of aflatoxins offers some justification for this traditional distrust of the filamentous fungi, but the long-term nature of the effects of these toxins makes it unlikely that there is any causal connection.

In the Orient, on the other hand, there has long been a clear understanding that processes closely allied to putrefaction and biodegradation can, by exercising proper control and the selection of appropriate conditions, convert rather unappetising and distinctly indigestible ingredients, into palatable and nutritious food whose rich and varied flavours contribute much to a largely vegetable diet with a tendency to blandness. The origin of these processes is lost in antiquity, presaging by three millenia our modern use of enzymes in food processing, for it is in their ability to produce substantial amounts of extra-cellular hydrolases when growing on comparatively dry substrates that we find the explanation for what must have seemed the near magical power of these moulds to transform foodstuffs.

To the Occidental observer, the range of foods made with the aid of moulds is a further remarkable aspect of this fascinating story. Hesseltine (1965) in his seminal study "A Millenium of Fungi, Food and Fermentation" has listed over a hundred such fermentations, commenting that "this list offers a wealth of subjects for future investigation". Although fungi belonging to several genera are employed in the various fermentations, the genus *Aspergillus* is pre-eminent both in the number of products made through its activities, and in the commercial importance of certain of these products, notably soy sauce, already familiar to most Westerners, and miso, the generic term for a range of soy-based pastes now becoming available in this country.

At Strathclyde University a series of studies have examined the chemical and biochemical changes which occur during soy sauce (shoyu) fermentation, and an account of these changes and of the art of koji production will be given. This will be followed with a comparative examination of some other fermentations, including miso, tamari, saké and tempeh, and with some discussion of the subsequent processing steps and the properties of the finished products.

KOJI

This is a generic term of great antiquity, sometimes used in a rather confusing manner to encompass the mould, the starter culture, and the main fermentation mass. To avoid such confusion, the term 'koji-mould' will be used for the organisms, 'seed-koji' for the starter culture, and 'koji' for the main fermentation mass. The production and use of koji has certain similarities with the production and use of malted cereal grains in the West, and I shall start by briefly summarising the latter process. Malt is produced by soaking sound grain (normally barley, although wheat is occasionally employed) then subjecting it to conditions of warmth, moisture and adequate aeration. Uniform conditions are provided and local overheating and anaerobiosis are prevented by turning the heaps of grain from time to time, a process which also helps to prevent the rootlets of the seeds matting together, and thus simplifies the subsequent handling of the finished malt. When the process of germination has reached the desired stage, it is halted, either by drying the grain, or by wet-grinding it for immediate use. The timing of this step is crucial to the success of the whole operation; if done too soon, both enzyme formation and degradation of the protein and starch reserves of the seeds will be insufficient to meet subsequent requirements; if left too late, then losses of reserves through respiration and through conversion to un-available materials such as cellulose, will be too great with adverse effects on process economics. Regulating and timing the malting process demands a subtle combination of scientific analysis and shrewd observation based on long practical experience. All of these considerations apply in similar fashion to the quite different process of koji production. In both cases the objective is to effect maximum enzyme production and hydrolysis of the raw materials, while minimising undesirable changes. These objectives are to some extent in conflict, and must be met by careful regulation of moisture, temperature and aeration, and by halting the process just short of the initiation of undesirable changes, i.e. shoot (acrospire) emergence in malting, sporulation in koji.

A number of moulds may be employed in koji production, the most commonly employed being strains of *Aspergillus oryzae* or *A. soyae*.

Depending on the process involved, many different substrates may be used, for example, rice, soybeans, beans mixed with rice, wheat or barley, and so forth. In soy sauce production, a mixture of soybeans and wheat flour is commonly used. Beans are first washed, then soaked for 16 to 24h in water, which is changed every few hours to prevent excessive bacterial growth and consequent souring of the mixture. The prime purpose of this soaking is to permit the beans to take up their full quota of water. Little work seems to have been done on the nature and extent of biochemical changes occurring in the bean during this phase of the process, or on the influence of such changes on subsequent stages and on the quality of the final product. However, it would seem reasonable to postulate significant differences between freshly harvested beans and those held in storage for long periods.

The soaked beans are then cooked, either by boiling at ambient pressure until soft, or in a pressure vessel at 10 p.s.i.g. for 1h. There seems to be some conflict of opinion over the desirability of pressure-cooking the beans, but most modern authorities seem to agree that these mild conditions are quite acceptable, and have the additional advantage of sterilising the beans; we have used these conditions throughout our work.

Meanwhile, the wheat is prepared. Soft (i.e. low-protein) wheat is normally preferred. Whole wheat or wheat flour may be taken. Whole wheat is toasted until crisp, then crushed while wheat flour may be steamed or be subjected to dry heat('parched'). In the present studies we have used wheat flour heated to 150° for 12h.

Wheat and beans are then thoroughly mixed, using roughly equal quantities on an original weight basis. The beans should remain separate and have an even coating of flour, giving a rather dry surface. The mould inoculum is then mixed in; this can be either a portion of a previous koji, or a specially prepared seed-koji, or a spore suspension. A spore suspension or seed-koji are preferred since they permit more exact microbiological control. In our work the mould is maintained on rice agar or malt wort agar slopes. Good sporulation occurs in 4 to 5 days at 30° under these conditions, and the spores are washed off with water containing a small amount of Tween detergent, and this suspension used to inoculate the substrate. Inoculated material is then spread into shallow trays, traditionally made of woven bamboo, although we have found expanded aluminium to be satisfactory. An open mesh structure for the trays is essential to permit adequate aeration. Trays are then placed in a humid room at 30°. Spores germinate rapidly and hyphae permeate the substrate. Exactly as in malting, it is necessary to turn the koji at intervals in order to prevent local overheating, and even so a layer one inch deep will attain a temperature in excess of 35°. Turning also prevents the development of excessively dry conditions, conducive to sporulation, on the outside of the mass, helps promote aeration of the entire koji, and keeps the beans reasonably separate. Indeed, if mixing is inadequate the whole trayful soon becomes one solid mass, bound tightly together by the mould mycelium. Mixing is normally effected at 24 and 48h after inoculation, and the koji is terminated after 72h. Unlike malt, the finished koji is never dried but is immediately used in the next stage of the process, in this respect resembling whisky distillery rather than brewery practice. If the koji stage is not terminated after about 72h, then sporulation rapidly sets in. No detailed study of the reasons why sporulation is initiated at this stage seems so far to have been carried out. There is ample food present in the form of the hydrolysis products which the mould enzymes have formed from the starches and proteins in the substrate, most of which are therefore wasted from the point of view of the mould. A possible explanation would seem to be that the koji experiences considerable water loss due to evaporation induced by the metabolic heat produced by the mould, and that it is this reduction in available water which induces sporulation.

If significant sporulation does occur then the soy sauce subsequently produced has a mouldy off-flavour, and an excessive level of ammonia. This latter seems a most remarkable state of affairs. In liquid culture, *Aspergillus* species normally start to sporulate only when the ammonia concentration has reached a very low level, yet, in koji, ammonia is produced in some quantity at the onset of sporulation. It seems unlikely that the organism should need to de-aminate amino acids to obtain organic carbon, when it is surrounded by assimilable carbohydrates. Clearly, the behaviour of *Aspergillus oryzae* in koji has much to interest the microbial physiologist.

Having described the basic process, let us now examine in greater detail the biochemical changes involved. Yong (1971) and

Yong & Wood (1976) have shown that there is a precise sequence of changes in experimental kojis. After about 20h incubation, the temperature begins to rise from the initial value of 30°, reaching a maximum of 38° to 40° some 10h later, thereafter declining slowly to 30°. It appears that spore germination and mycelial growth begins after about 20h incubation and that all other biochemical changes therefore begin at that time. The pH rises fairly steadily from then on, initially being around pH 6.5, reaching pH 7.3 after 72h and pH 7.5 after 96h. As has been previously mentioned, the moisture content declines from an initial value around 50%, to a final (72h) value between 30 to 35% depending on the strain.

The pattern of change in the level of reducing sugars is particularly interesting. From a low initial value (around 3 mg/g dry weight koji as glucose) it rises very rapidly to a value of 20 mg/g in the middle of the fermentation, then a secondary peak around 20 to 25 mg/g toward the end of the fermentation, thereafter holding steady or showing a slight downward drift if the fermentation is prolonged beyond 72h. The sharp initial rise occurs some time before significant amounts of amylase can be detected in the koji, due to the release of a sucrose-hydrolysing enzyme (Yong & Wood, 1975) which later work (Goel, 1974) has shown to be a fructofuranoside transferase. The enzyme produces reducing sugars from the sucrose and related saccharides which constitute up to 20% of the dry weight of the bean.

The latter rise will represent the hydrolysis of starch from the wheat flour by amylases. Both α- and β-amylases are produced by *A. oryzae* and increase in amount steadily during the fermentation (Yong, 1971; Goel, 1974), whereas the sucrose-hydrolysing enzyme reaches a maximum after 30 to 50h, thereafter declining in concentration (Yong, 1971) although if the vegetative mycelium be challenged by transfer to fresh sucrose-containing medium it will release further enzyme (Goel, 1974).

Cellulase is also released into the koji (Goel, 1974) and we believe that this enzyme may play a significant role in aiding the disintegration of the beans during the secondary fermentation in brine (the moromi stage).

The fact that reducing sugar levels decline but little during the late phases of the koji fermentation contributes to the view that the appearance of ammonia during sporulation is not due to a simple lack of assimilable carbon necessitating the de-amination of amino acids.

Total soluble nitrogen increases around three-fold during the course of the fermentation of koji, to a final value around 14 to 16 mg N/g (Yong, 1971). In fermentations prolonged beyond 72h there is no significant further change in this parameter. Simultaneous measurement of amino-nitrogen by the Sörensen Formol method revealed a more complex picture. Initially scarcely detectable, this material increased very rapidly between 20 to 30h of the fermentation to around 3 mg nitrogen/g dry weight of koji (further increasing to 5 mg N/g in one strain of *A. oryzae* being examined), then dropping almost as rapidly to half its peak value, rising to a minor peak at 65 to 70h, a second decrease and thereafter a steady increase until the fermentation was terminated at around 100h. It is clear from the great deal of work that has been published on the proteases and peptidases of the *Aspergilli* that these moulds generate a complex battery of these enzymes, with a range of pH optima and substrate specificities. Unfortunately, it has not been possible to carry out a detailed study of the pattern of protein-hydrolysing enzyme production by these moulds, and we have been compelled as a compromise to employ a general assay measuring casein hydrolysis at pH 6.5. This showed a pattern of changes quite similar to, although not identical with that exhibited by the amino-nitrogen level. There is clearly a relationship here, although at present there is not sufficient data to say to what extent a causal relationship exists, and one would like to know far more about the uptake of amino acids and simple peptides by the mould.

In the production of soy sauce, the 'mature' (72h old) koji is mixed with 18% brine, and this mixture (called moromi) undergoes a prolonged fermentation in which first lactic acid bacteria and then yeasts are involved. The finished product is pressed, clarified and pasteurised, bottled and marketed either alone or in a mixture with caramel, molasses and other additives, depending on the preferences of the market being served. We think that the enzymes produced by the mould during the koji stage remain active in the moromi, slowly hydrolysing the proteins and polysaccharides present in the mash, and so enhancing the quality of the finished product.

Soy-sauce fermentation is clearly a very complex process, and this is reflected in the very diverse studies which are carried out in Japan. Thus a recent review of the fermentation of soy sauce, miso, glutamic acid and nucleic acid related substances by Natsuno & Okada (1974) cited 604 references, while the review by Yong & Wood (1974) of the soy sauce fermentation cited 259. A further complication for the Occidental worker is the amount of material which - not surprisingly - is published in the Japanese language. English-language abstracts, no matter how carefully prepared, cannot possibly answer all the questions which will strike someone active in the field. For example the paper by Yamamoto, Yangida & Suminoe (1972) reports on studies of changes in enzyme levels during the preparation of the koji for soy sauce. Their general conclusions resemble those of Yong (1971) and Goel (1974), as judged from the abstract in "Chemical Abstracts". However, there is an intriguing reference to the enzyme phosphatase. Within the confines of an abstract it would be impossible to explain this fully, but the significance of the presence of this enzyme in koji is not readily apparent. Again, Oba (1974) reports that the enzyme tyrosinase was produced in a rice koji, and reached a maximum value after 40h incubation. In contrast, Yong (1971) could not detect tyrosinase activity at any stage of the soy sauce koji fermentation.

What of future developments in the shoyu fermentation? One obvious line for study is the possibility of using artificial mixtures of industrial enzymes produced in bulk to replace the traditional koji. Work on miso (Kitaoka, 1972) suggests that the entire replacement of koji by enzymes would result in an unsatisfactory produce with an inferior flavour. On the other hand, a combination of koji and enzymes could be very beneficial (Sato, Ito, Kawano & Endo, 1973; Takeuchi & Yokoo, 1974).

A recent patent (Beatrice Foods Co., 1973), the review by Yokotsuka (1973) and the work by Goel (1974) indicate that more elaborate microbiological control of the processes will both accelerate the production of shoyu and miso, and also enhance the quality of the final product (Shibasaki & Hesseltine, 1962).

Finally, to those who live in countries where soybeans cannot be grown commercially, producing products similar to shoyu and miso from indigenous crops is attractive both in the potential saving on imports, and in the chance to upgrade materials of marginal utility in human nutrition. A report from Kansas State University (Kao, 1974) shows success in producing miso and tempeh from chick pea and horsebean, both of which seeds gave satisfactory products. There is evident scope for similar research using seeds produced in the United Kingdom.

The importance of the soy sauce market to the East is clear from the production in Japan alone of about 110 million kilolitres per year (Yokotsuka, 1972).

TAMARI

In the course of discussions with commercial concerns who distribute and market soy sauce, I became aware that the term 'tamari' is sometimes used interchangeably with 'shoyu'. At first I dismissed this as a quirk of the 'macrobiotic' food cult, but more detailed study has revealed a rather interesting situation. Hesseltine (1965) citing Yokotsuka (1960) describes tamari as a "type of soy sauce especially common in China in which a greater amount of soybeans is used than wheat". The latter author states that three types of soy sauce are available in Japan; tamari, a type more common in China, where it accounts for most of the soy sauce, raw material mainly soybeans; koikuchi, accounting for 90% of the Japanese soy sauce, consisting of a blend of fermented soy sauce and a chemical hydrolysate of defatted soybeans; and usukuchi, made from soybeans, wheat and a small amount of rice (Yokotsuka, 1972). Prestcott & Dunn (1959) describe tamari as "a sauce prepared from soybeans, often with the addition of other materials, such as rice". They continue "the flavour of tamari sauce differs from that of soy sauce. This difference in flavour is due to the use of *A. tamarii*, this mould being the dominant microorganism in this process. The fermentation period is shorter than that of the soy sauce fermentation".

Murakami (1971) examining the classification of koji moulds by a computer-generated cluster analysis of 20 characters of each of 659 strains of *Aspergilli* recognised two groups, the *A. flavus* group and the *A. oryzae* group. *A. tamarii* was placed alongside *A. oryzae* and *A. sojae* and in the *A. oryzae* group, which included all the koji moulds examined. He states that *A. tamarii* is now isolated from koji.

This information suggests that it would be interesting to compare the behaviour of *A. oryzae* and *A. tamarii* in controlled koji fermentations, since there is clearly an area of confusion present, at least in the English language literature on the subject.

MISO

This is a preparation of equal importance with shoyu in Japan, although it is at present little known in Europe. Basic raw materials are soybeans, grain and salts and it is marketed as a thick paste, and is used as a source of flavour and salt in cooking and food preparation. Shibasaki & Hesseltine (1962) have described the making of this product, and its characteristics, although it is rather misleading to use the term 'product', since they describe several varieties of miso, and when even the United Kingdom suppliers carry three types of miso. There is evidence to suggest that the miso fermentation predates the production of soy sauce, and it is perhaps not surprising therefore that it encompasses a more diverse range both of substrates and of fermentation conditions, leading to products that are readily differentiated by even the untrained Occidental palate.

Three basic types of miso can be recognised (Hesseltine, 1965; Shibasaki & Hesseltine, 1962, on whose excellent reports the following summary is based). There is kome miso, made from beans and rice; mugi miso, made from beans and barley; and mame miso (sold as hatcho miso in the United Kingdom) made from soybeans alone. For the production of kome and mugi misos, the koji is produced by soaking polished rice overnight in water, steaming it for about an hour, then inoculating with mould spores. Incubation is for 50h at 35°, with thorough stirring at least twice. The

A. oryzae strains used for these fermentations should be higher in amylolytic activity than is necessary for shoyu production.

The koji is then mixed with an appropriate mixture of soaked, cooked soybeans, barley or rice, and salt. Thus the process differs significantly from shoyu production in that the mould does not grow on the beans. Even in the case of mame miso, only a portion of the beans are moulded. The precise proportions of beans and grain used, and also the extent to which the beans are cooked, depends on the type of miso being made. Thus rice-based misos (kome miso) are further sub-divided by Shibasaki & Hesseltine (1962) into 5 types.

The mixture of koji, beans, rice or barley, and salt is then inoculated. Traditionally, this is done with a portion of an old miso, but Shibasaki & Hesseltine (1962) obtained better results by using a pure culture of *Saccharomyces rouxii*. After fermentation at 28°, followed by a period at 35°, the finished miso if further blended and packaged for sale.

As illustration of the range of conditions which can be used, Shibasaki & Hesseltine (1962) give the details summarised in Table 1. Naturally, the lighter, less salty miso types have a shorter shelf-life than do the dark and salty varieties.

Table 1. Condition for producing different types of rice miso

Type	Colour	Nature	Fermentation time	Ratio of beans: rice: salt v/v
White		Sweet	1 week	10: 10 to 20: 2.5 to 3.5
Edo	Pale red-brown	Sweet	2 weeks	10: 8 to 10: 0.25 to 0.4
Sendai	Red-brown	Sweet, salty	1 year or more	10: 4: 0.4 to 0.5
Shinshu	Light yellow-brown	Salty	1 year or more	10: 5: 4.5
Mame	Deep red-brown	Salty	2 years	Beans and salt only

Shibasaki & Hesseltine (1962) report fairly high microbial counts in the early stages of miso fermentation, and a decline in the later stages. They comment on the contribution of lactic acid bacteria to the sour taste of miso, and to the prevention of spoilage during storage, yet made very good miso without these organisms. In a preliminary examination of three miso samples on sale in the United Kingdom, I found rather a low microbial count, with the mame miso having rather more organisms than did the kome or mugi miso samples.

The remarks about future developments in the soy sauce fermentation apply with at least equal force to the miso fermentation. In addition, miso is still a rather rare item on the United Kingdom market, and probably also on the European market. As with yoghurt, its primary incursions seem to be through the 'Health Food' trade, and there is no reason why it should not progress to a wider market in the same successful manner.

TEMPEH

Tempeh is a foodstuff common in Indonesia and some other countries in that part of South East Asia. Most authorities agree that the mould concerned belongs to the genus *Rhizopus*. Most of the strains examined belong to the species *R. oligosporus*, although 5 other species have been demonstrated (Hesseltine, 1965). However, the same author cites an early report (Burkill, 1935) which states that *Aspergillus oryzae* may be involved in the fermentation. In the early part of our own studies on soy sauce, we obtained three strains of *Aspergillus*. Two were soy sauce strains from China supplied to us by Dr. Hesseltine, the third was a strain procured from Singapore and described by Yong (1971) as follows; "*A. oryzae* strain A01 was isolated from tempeh obtained in Singapore. The identity of this fungus is still doubtful but Tan (Miss S.K. Tan of the Singapore Institute of Standards and Industrial Research) (personal communication, 1970), has managed to produce soy sauce of good quality using this organism".

We subjected this strain to the same tests on a typical sauce koji substrate, as were used for the two soy sauce strains of *Aspergillus*, and described above. The results of Yong (1971) show significant differences between strain A01 and the two Chinese strains. Whereas the soy sauce strains gave a fairly steady increase in pH over the whole duration of the fermentation, A01 gave a very rapid increase from pH 6.55 at 22h incubation, to 7.5 at 29.5h, a pH only reached by the soy sauce strains after 90h of fermentation. From the 20h to 65h, the pH remained fairly steady, then actually decreased to 7.2, a behaviour which was never observed in soy sauce mould kojis. Strain A01 produced nearly three times as much α-amylase as did the other strains, but its overall protease level was rather lower, while lipase level was much lower.

The A01 koji showed a rather similar pattern of changes in reducing sugar level to that shown by the other two moulds, except that the second maximum was both earlier and greater in the A01 kohi, as would be expected from the higher amylase levels which were observed. Invertase production by strain A01, on the other hand, was very much lower than that by the other strains. Total soluble nitrogen levels followed those in the other two kojis, and amino nitrogen levels were close to those of the less productive soy sauce strain. Final ammonia levels were however slightly higher in the koji made with strain A01.

If strain A01 is typical of tempeh moulds, it is not particularly obvious what advantages the observed differences in biochemical activities would confer. It is however desirable that further comparative work be carried out, using conditions appropriate to tempeh production, before any firm conclusions be drawn.

SAKÉ

Saké is the traditional alcoholic beverage made from rice, known as 'rice wine' in the West because its alcoholic strength approximates that of a wine. Amerine & Cruess (1960) quote comparative Japanese production figures for 1952 as Grape Wine, 25 million U.S. gallons; Saké, 100 million U.S. gallons. Murakami (1972) reports that in 1970 about 1.6 million kilolitres of saké were made.

Saké production depends upon the hydrolysis of rice by a koji produced by growing *Aspergillus oryzae* on a portion of the rice. We may see analogies with the production of Grain Whisky in Scotland, where malted barley is used to hydrolyse much larger quantities of unmalted grains, or indeed with the sort of modern beer production where adjuncts such as maize are hydrolysed by malted barley assisted by mould enzymes. Saké production does however have many points of interest, and it is clear from the numerous references to Japanese language papers, that it is in a phase of rapid technical development, for example into continuous culture fermentation at the alcohol production stage.

Arima (1957) has given a detailed account of saké production. He shows that it is but one member of a family of such 'wines' (and spiritous distillates made therefrom) to be found throughout the East. Rice is the cereal most commonly used, although he reports that distilled drinks are made from sorghum and sweet potatoes. Variety may be introduced by employing adjuncts; for example a red wine is obtained by adding Ang-kak, which is rice moulded with *Monascus purpureus* which produces a very attractive red pigment.

When producing saké, conditions are so arranged that, as with the fermentation stage of Scotch whisky production, enzymic hydrolysis of starch continues during fermentation. An important difference from whisky production is that rather low temperatures (around 10°) are favoured for the initial stages of saké. The presence of lactic acid bacteria in controlled amounts is essential for a satisfactory saké fermentation, since the pH of the basic mash is too high for good yeast growth, or for a satisfactory flavour in the finished product. It is interesting to note that there is a school of thought in the whisky industry which regards a controlled lactic fermentation as desirable for optimum flavour development in their finished product. It is of course essential that the lactic fermentation should not be excessive, since a very acid mash will inhibit yeast growth.

According to Arima (1957), saké production is conducted as follows. First the rice is highly polished, to remove some 12 to 15% of the original weight (even more for the finest grades of saké). This is because rice bran has adverse effects on the colour and flavour of saké. The polished rice is then steeped in water for several hours, steamed at atmospheric pressure for one hour, then cooled and the moisture content adjusted to 40 to 50% w/w.

For koji production, steamed rice is cooled to 40°; inoculated with *Aspergillus oryzae* spores; left in piles in a room at 28° until the spores germinate (12 to 15h); spread in shallow boxes and incubated for a further 30 to 36h, with stirring every 5 to 7h. As with some other kojis, mould growth is enhanced if a small proportion of wood ash made from selected timbers, is added to the rice; 0.5% is a suitable rate of addition. This procedure presumably represents supplementation with potash and possibly phosphates.

The short incubation period means that little, if any, sporulation will have taken place and that extracellular enzyme levels will be maximal. For best results in the subsequent fermentation, it is desirable that the koji be made with mould strains which give high amylase titres, a difference from soy sauce strains which are selected for proteinase production.

The koji is now used in the alcoholic fermentation. Steamed rice (90 1) koji (36 1) and water of medium hardness and chloride content of 30-70 mg per litre (110 1) are mixed and held at a temperature of 5° to 6° for 6 to 7 days with occasional stirring. Starch hydrolysis and a lactic fermentation occur. The mash is next inoculated with *Saccharomyces saké*, and a vigorous fermentation ensues, with the temperature rising to 18° to 20°. Arima (1957) shows that the interactions between bacteria and yeasts are considerably more complex than this simplified account might indicate, although in more modern processes the lactic fermentation may be replaced by addition of commercial lactic acid.

The mash is then 'fed' with three successive additions of

increasing amounts of steamed rice, koji and water, and after about 20 days the fermentation comes to an end, with a final alcoholic content between 20 to 22%. After filtration and a period of storage to permit maturation and clarification, the saké is diluted to an alcohol content of 15 to 16%, pasteurised and dispensed for sale.

A noteworthy aspect of the fermentation is the very high alcohol level which the yeast tolerates, which may be explained in part by the way in which the process operates to keep the sugar concentration low at all times.

As Hayashida, Kamachi & Hongo (1972) observe, saké's unusual feature of involving a fermentation of a thick mash high in solids, prepared without a sterilisation step, generally fermented by natural mixtures of bacteria and yeasts in an open vessel, create special problems in any attempt at a continuous fermentation process. However they have succeeded in operating a semi-continuous process, involving passage of the mash through a series of vessels arranged as a casdade.

SOME MINOR FERMENTATIONS

Throughout the East there is a wealth of fermentations, many of which are admittedly but small variations on the important fermentations already discussed. The information on them is frequently contained in unusual, obscure or very old sources, and it is difficult adequately to describe the importance of Hesseltine's contribution in securing and examining these references, and in collating his results in such a convenient form (Hesseltine, 1965). For the sake of brevity, therefore, an annotated list of these fermentations are set out; unless otherwise specified, the information given is drawn from Hesseltine (1965), where the original references may be found. It is worth noting that many authors claim that these mould fermentations represent a significant addition of B-group vitamins to the diet in addition to their more obvious roles in increasing the palatability and digestibility of their substrates.

Chee-fan. A type of soy-curd cheese (sufu) in which *A. glaucus* may participate, as well as the more usual *Mucor* and *Actinomucor*.

Cheese. Nakanishi (1973) has published a review of his patents on Japanese cheese production utilising *A. oryzae*. Unfortunately for the present author, this interesting-sounding paper is written in Japanese. He is presumably not concerned with the ordinary Chinese soy-bean curd cheese, called 'sufu' and fermented with *Actinomucor elegans*.

Fermented Minchin. A fermentation of wheat gluten, practiced in Northern China. The gluten is allowed to mould in a closed container, then mixed with 10% of salt and further aged. At least seven fungal genera, including *Aspergillus* spp. are thought to participate.

Hamanatto. Intact soybeans are soaked, cooked, and moulded with *A. oryzae* in the usual manner. After mycelium development is complete, the beans are dried in the sun to a moisture content of 12%, then steeped in 18% brine, along with strips of ginger, for up to a year. The beans retain their integrity, in marked contrast to their behaviour in the shoyu or miso fermentations; this is presumably due to their retaining the seed-coat, and to effects of the drying process. After ageing in brine, the beans are again dried. They are said to be soft and sweet, with a flavour otherwise like that of soy sauce or miso. Hesseltine (1965) states, and conversations with Japanese students confirm, that this interesting fermentation is of extremely local occurrence in Japan. Commercial concerns in the United Kingdom have expressed interest in it as a possible premium product. Some preliminary work in our laboratories suggests that air-drying may be a suitable alternative to sun-drying on completion of mould growth. For a fuller discussion of the published information on this fermentation see Wood & Yong (1975).

Kaffir Beer. Although an African product, this is of interest here because the grains used (sorghum, maize, etc.) are saccharified through the action of *Aspergillus flavus* and *Mucor rouxii*. After the starch has been hydrolysed, the mixture undergoes a mixed lactic acid bacteria plus yeast fermentation. Hesseltine (1965) stresses the importance of this beverage as a source of dietary niacin and riboflavin.

Katsoubushi. This is an extraordinary fermentation of fish by *Aspergillus ochraceus* and possibly *A. glaucus*.

Ketjap. This is a form of soy sauce made in Indonesia, using *A. oryzae*.

Mirin. A mixture of distilled ethanolic liquor, rice koji and steamed glutinous rice, is allowed to age. The alcohol content is around 20%. The resulting produce is described as an extremely fragrant condiment.

Mizaume. Prescott & Dunn (1959) describe this as a sugar syrup made in Japan through the action of *A. oryzae* on rice.

Ragi. An Indonesian product made from rice fermented by yeasts, bacteria and moulds, and used to hydrolyse the starch in rice or cassava. Hesseltine (1965) states that the moulds present are *Rhizopus oryzae* and *Chlamydomucor oryzae*, but Stanton & Wallbridge (1969) list *Aspergillus oryzae* among the organisms present in this product.

Tao-Cho. Yet another *Aspergillus* fermentation of soybeans and rice flour reported from the Malay Peninsula.

Tao-Si. A Phillipine version of soy sauce made with *A. oryzae*.

FERMENTED FOODS AND MYCOTOXINS

The modern scientist, confronted with accounts of these traditional fermentations is at once driven to think of aflatoxins and similar products of mould fermentation. That secondary metabolite formation does occur is strongly suggested by the many references to the beneficial effects of moulded foodstuffs on digestive disorders and other diseases. Indeed it has been suggested that the health of the inhabitants of the Japanese town of Noda (which seems to be soy sauce production as Burton-on-Trent is to beer brewing) benefits from the high concentration of mould spores in the atmosphere.

Japanese workers have naturally been very concerned with the problem of aflatoxins in these traditional foodstuffs which are so important to their diet. The general conclusion of their many studies is that there is no evidence at all of the presence of toxic materials in properly produced foods fermented with *Aspergilli*. This is a most remarkable situation when we consider the great antiquity of these fermentations, far pre-dating the development of modern microbiology, with its range of pure culture techniques, isolation and identification procedures, etc. The real danger from many mycotoxins lies in their carcinogenicity, which may take many years to manifest itself. Acute poisoning is easy to detect and even primitive people rapidly learn the appropriate lessons. But how did the soy sauce and miso makers select out the beneficial strains of *A. oryzae* from all the superficially similar but ultimately deadly moulds?

It is not simply that the conditions found in Koji do not permit harmful organisms to grow, or that they prevent toxin formation. Shotwell *et al.* (1966) have obtained high yields of aflatoxin by growing *Aspergillus flavus* on solid rice. Manabe & Matsuura (1972) have shown that aflatoxins are stable under conditions found in miso, yet Manabe, Onuma & Matsura (1972) were unable to find any aflatoxins in the commercial and home-made miso samples which they examined. A few of the samples showed a fluorescent substance when extracts were examined by thin layer chromatography, but this material was not toxic to chick embryos. Similarly, in their examination of 214 Japanese industrial strains of *Aspergillus*, Murakami, Takase & Ishii (1967) and Murakami, Takase & Kuwabara (1968) found that while some organisms produced chloroform-extractable, fluorescent compounds, none produced detectable amounts of material possessing the spectral properties of aflatoxins.

Yokotsuka *et al.* (1968) have examined shoyu, miso and mirin for pyrazine compounds. On prolonged cultivation, 26 of 68 industrial *Aspergillus* strains produced aspergillic acid, but if incubation was restricted to only 2 or 3 days, then contamination with this material could be avoided. They point out that some strains produce it earlier than do others, and suggest that the latter be used for food production. They also adduse evidence that ether extracts of shoyu at dose rates corresponding to a year's consumption of shoyu by a Japanese adult (12 1 of shoyu) were not lethal to mice upon intraperitoneal injection.

It is therefore clear that the available evidence all tends to the conclusion that properly prepared foodstuffs made by controlled fermentations with species of *Aspergillus* are as safe and wholesome as they are appetising and enjoyable.

ACKNOWLEDGEMENTS

I should like to record my thanks to Associated British Malsters Ltd. for their help with information on analytical methods, to the British Council for support to S.K. Goel and Yong Fook-Min, and the Singapore Institute of Standards and Industrial Research for their support of Yong Fook-Min, students who have done much to educate me in the subjects discussed in this paper.

REFERENCES

AMERINE, M.A. & CRUESS, W.V. (1960). *The technology of wine making*. Avi Publishing Co.: Westport, Conn.

ARIMA, K. (1957). Saké and similar yeasts. *Yeasts* (Roman, W., Editor), pp. 143-154. Dr. W. Junk, Publishers: Den Haag.

BEATRICE FOODS CO. INC. (1973). Soy Sauce. French Patent No. 2,172,812. Cited in *Chemical Abstracts 80* (1974) 94395.

BURKILL, I.M. (1935). A dictionary of the economic products of the Malay Peninsula. Crown Agents for the Colonies, London, Vol. 1, pp. 1080-1086. Cited by Hesseltine (1965), *q.v.*

GOEL, S.K. (1974). *Studies on microbiological and biochemical aspects of soy sauce fermentation*, Ph.D. Thesis, University of Strathclyde.

HAYASHIDA, A., KAMACHI, T. & HONGO, M. (1972). Semi-continuous saké fermentation. *Proceedings of the IVth International Fermentation Symposium, Fermentation Technology To-day*, pp. 645-649.

HESSELTINE, C.W. (1965). A millennium of fungi, food and fermentation. *Mycologia 57*, 1-148.

KAO, C. (1974). Fermented foods from chickpea and horsebean. *Dissertation Abstracts International B 35* (5), 2250.

KITAOKA, K. (1972). Soybean miso manufacturing with enzyme preparations. *Miso No Kagaku To Gigutsu*, No. 218, pp. 19-24. Cited in *Chemical Abstracts 80* (1974) 81,090.

MANABE, M. & MATSUURA, S. (1972). Fluorescent compounds in fermented foods. IV. Degradation of added aflatoxin during miso fermentation. *Nippon Shokuhin Kago Gakkai-Shi 19* (6), 275-279. Cited in *Chemical Abstracts 81* (1974), 167,979.

MANABE, M., ONUMA, S. & MATSUURA, S. (1972). Fluorescent compounds in fermented foods. II. Tests on aflatoxin contamination of miso and miso-koji in Japan. *Nippon Shokuhin Kogo Gakkai-Shi 19* (2), 76-80. Cited in *Chemical Abstracts 81* (1974), 167,973.

MURAKAMI, H. (1971). Classification of the koji mould. *Journal of General and Applied Microbiology 17*, 281-300.

MURAKAMI, H. (1972). Some problems in saké brewing. *Proceedings of the IVth International Fermentation Symposium: Fermentation Technology To-day*, pp.639-643.

MURAKAMI, H., TAKASE, S. & ISHII, T. (1967). Non-productivity of aflatoxin by Japanese industrial strains of *Aspergillus*. I. Production of fluorescent substances in agar slant and shaking cultures. *Journal of General and Applied Microbiology 13*, 323-34.

MURAKAMI, H., TAKASE, S. & KUWABARA, K. (1968). Non-productivity of aflatoxin by Japanese industrial strains of *Aspergillus*. II. production of fluorescent substances in rice koji and their identification by absorption spectrum. *Journal of General and Applied Microbiology 14* 97-110.

NAKANISHI, T. (1973). New technique in cheese making using micro-organisms. *Shokukin Koggo 16* , (18), 34-40. Cited in *Chemical Abstracts 80* (1974), 2341.

NATSUNO, S. & OKADA, H. (1974). Fermentation of soy sauce, miso, glutamic acid and nucleic acid related substances. *Nippon Jazo Kyokai Zasshi 69* (5), 283-301. Cited in *Chemical Abstracts 81*, (1974) 103166.

OBA, T. (1974). Simple determination of tyrosinase in rice-koji and its production. *Nippon Jazo Kyokai Zasshi 69* (1), 56-8. Cited in *Chemical Abstracts 81* (1974), 36395.

PRESCOTT, S.C. & DUNN, C.G. (1959). *Industrial Microbiology*. McGraw Hill Book Company Inc.: New York.

SATO, Y., ITO, K., KAWANO, K. & ENDO, H. (1973). Miso manufacturing using enzyme preparations. *Miso No Kagaku to Gijutsu* No. 236 pp. 20-30. Cited in *Chemical Abstracts 81* (1974), 69030.

SHIBASAKI, K. & HESSELTINE, C.W. (1962). Miso fermentation. *Economic Botany 16* (3), 180-95.

SHOTWELL, O.L., HESSELTINE, C.W., STUBBLEFIELD, R.D. & SORENSON, W.G. (1966). Production of aflatoxin on rice. *Applied Microbiology 14* (3), 425-8.

STANTON, W.R. & WALLBRIDGE, A. (1969). Fermented food processes. *Process Biochemistry 4* (4), 45-51.

TAKEUCHI, T. & YOKOO, Y. (1974). Peptides in miso and soy sauce. XI. Effect of koji addition on proteolytic patterns of soybean miso manufactured with an enzyme preparation. *Hakko Kogaku Zasshi 52* (1), 58-61. Cited in *Chemical Abstracts 80* (1974), 81093.

WOOD, B.J.B. & YONG, F.M. (1975). Oriental food fermentations. *The Filamentous Fungi, Vol. I,* pp. 265-280. Smith, J.E. & BERRY, D.R., Editors). Edward Arnold (Publishers) Ltd.: London.

YAMAMOTO, Y., YANGIDA, F. & SUMINOE, K. (1972). Changes of enzyme activities during preparation of the koji for raw soy sauce. *Nippon Shokuhin Kogyo Gakkai-Shi 19* (1), 29-30. Cited in *Chemical Abstracts 81* (1974), 150,264.

YOKOTSUKA, T. (1960). Aroma and flavour of Japanese soy sauce. *Advances in Food Research 10*, 75-134.

YOKOTSUKA, T. (1972). Some recent technological problems related to the quality of Japanese shoyu. *Proceedings of the IVth International Fermentation Symposium; Fermentation Technology To-day* pp. 659-62.

YOKOTSUKA, T. (1973). Development of techniques for food utilising microorganisms. Quality improvement of soy and microorganisms. *Shokukin Kogyo 16* (18), 20-25. Cited in *Chemical Abstracts 80* (1974), 2339.

YOKOTSUKA, T., ASAO, Y., SASAKI, M. & OSHITA, K. (1968). Pyrazine compounds produced by moulds. *Proceedings of the First U.S.-Japan Conference on Toxic Microorganisms, Honolulu, Hawaii* (Herzberg, M., Editor), pp. 133-42.

YONG, F.M. (1971). *Studies on soy sauce fermentation,* M.Sc. Thesis, University of Strathclyde.

YONG, F.M. & WOOD, B.J.B. (1974). Microbiology and biochemistry of soy sauce fermentation. *Advances in Applied Microbiology 17*, 157-94.

YONG, F.M. & WOOD, B.J.B. (1975). Sucrase from soy sauce moulds. *Transactions of the British Mycological Society 64* (1), 1-3.

YONG, F.M. & WOOD, B.J.B. (1976). Microbial succession in experimental soy sauce fermentations. *Journal of Food Technology 11*, 1-12.

Aspergillus Mycotoxins

M. O. MOSS

Department of Microbiology, University of Surrey, Guildford, U.K.

INTRODUCTION

It is relatively easy to give the definition of a mycotoxin as a metabolite produced by a mould which is able to elicit a toxic response when administered to an animal. The term is usually used in the context of higher animals such as mammals, birds and fish.

It is more difficult to be precise about the group of illnesses of man and his domesticated animals referred to as mycotoxicoses. One would like to use the term to describe those diseases caused by the consumption of food or feeds on which moulds have grown, producing specific toxic compounds, the essential symptoms of the disease being reproduced by consumption of the toxins alone. Perhaps such a situation could be described as primary mycotoxicosis (Fig. 1). It is possible, however, for domestic animals to consume mycotoxins with their feed without showing overt signs of illness. The toxins may then be, either absorbed and bound unchanged in these tissues, or they may be metabolised and secreted in the milk. In either case toxic metabolites of fungal origin may enter the human food chain with foods which the consumer need have no reason to suspect as having been associated with moulds. Thus ochratoxin is sufficiently stable that it may remain in the tissue of pigs long after slaughter and processing (Krogh, 1974), whereas aflatoxin B_1 is readily metabolised to a toxic derivative (aflatoxin M_1) by the cow and secreted in that form in milk

(Allcroft *et al.*,1966).

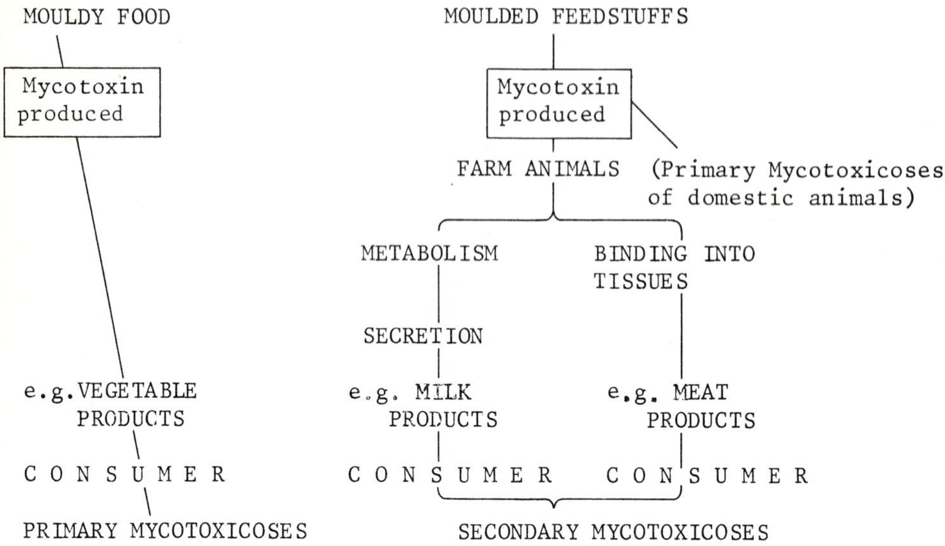

Fig. 1. Pathways by which mycotoxins may reach the consumer through food chains.

A further situation which might be covered by the term mycotoxicosis, if one allows the definition to become more diffuse, is that in which a toxin is absorbed by inhalation either with fungal spores or because it is volatile. Some mould metabolites may also cause irritation, and even more severe symptoms, by contact with the skin.

It must be recognised that the growth of moulds on food materials will alter their nature, perhaps removing vitamins already at a limiting concentration or, so change the organoleptic quality as to make the food unpalatable. Such changes may lead to chronic ill health without the production of an overtly toxic metabolite.

To emphasise how wide a range of fungi may be implicated as aetiological agents of non-infective diseases it is worth recalling two historic examples - ergotism caused by *Claviceps purpurea*, a plant parasite capable of synthesising a number of complex alkaloids in the inert looking sclerotial stage (Fig. 2) - and facial eczema of sheep caused by the saprophytic mould *Pithomyces chartarum* (Fig. 3). Three genera stand out as being particularly associated with a large number of mycotoxicoses - *Fusarium*, *Penicillium* and *Aspergillus*.

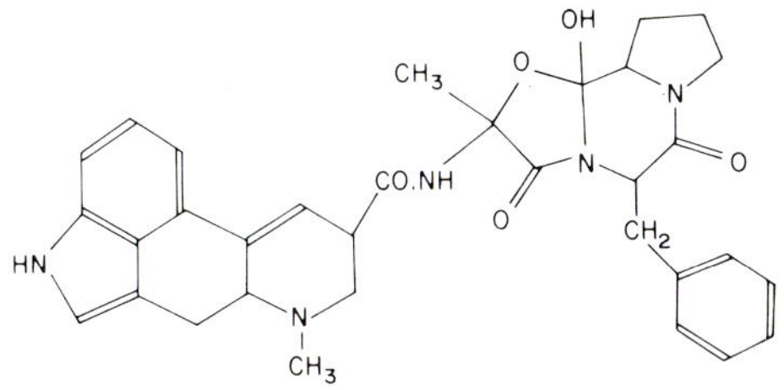

Fig. 2. Ergotamine, one of the toxic alkaloids of *Claviceps purpurea* associated with ergotism.

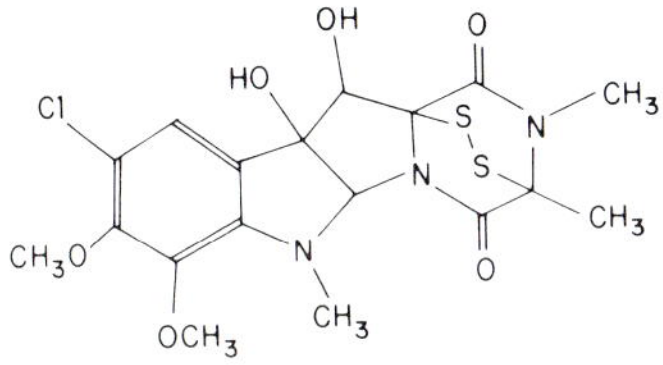

Fig. 3. Sporidesmin A, one of the toxic metabolites of *Pithomyces chartarum* associated with facial eczema of sheep.

Members of the genus *Aspergillus* are world wide in their distribution and, although a few species have adapted to the specialised niches of plant (Wood, Ballio & Graniti, 1972) and animal parasitism (Austwick, 1965), the majority are saprophytes growing on an extensive range of organic substrates. Indeed, aspergilli form a major component of those fungi associated with the moulding of materials especially post harvest, vegetable products in storage.

The biochemical activity of several species of aspergilli has been harnessed to the conversion of readily available carbohydrates into such products as citric, itaconic and gluconic acids, as well as for the industrial production of enzymes such as amylases, proteinases and pectinases. For many centuries strains of the flavus-oryzae group have been used in the manufacture of foods in the Far East and, more recently, such fungi as *Aspergillus niger* are being used for the conversion of vegetable and fruit wastes into potentially useful microbial biomass.

When moulds such as aspergilli are grown on relatively rich media in laboratory culture it is frequently observed that they produce, not only biomass, but also a range of metabolites, some of which are retained in the mycelium or fruiting structures, but the majority of which are secreted into the medium. The fact that some of these metabolites have antimicrobial activity led to a vigorous study of their production, structure and biological properties. As part of the evaluation of each new antibiotic it needs to be extensively tested for toxicity with the result that such promising antibiotics as patulin and citrinin had to be abandoned by the pharmaceutical industry because of their high mammalian toxicity.

Despite a detailed knowledge of the toxicity of many mould metabolites amongst those working in the pharmaceutical industry mycotoxins had, as it were, to be discovered again as the causative agents of diseases of previously unknown aetiology. Mycotoxicoses cannot be recognised by any circumscribed set of symptoms because the only factor they have in common is the consumption of food contaminated by toxic mould metabolites. Even within the single genus *Aspergillus* there is a wide diversity of metabolites producing an equally wide diversity of toxic effects (Fig. 4).

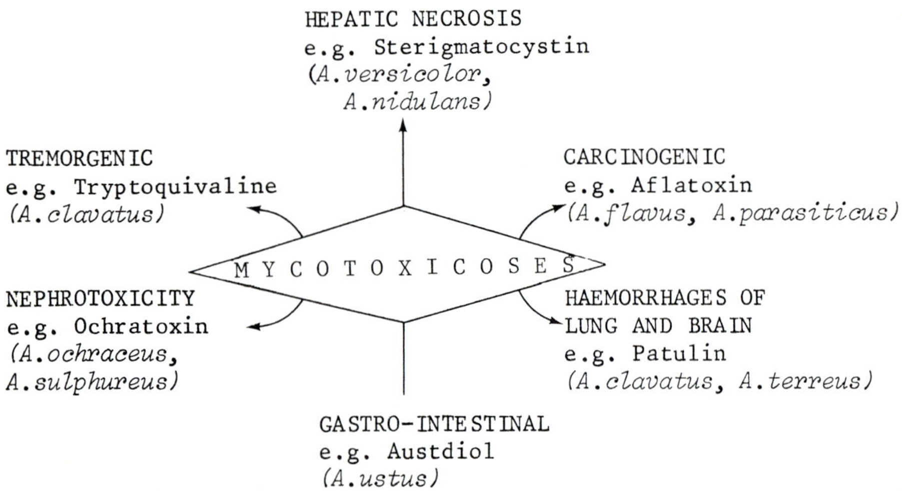

Fig. 4. The diversity of symptoms of mycotoxicoses associated with the genus *Aspergillus*.

During a study of the known metabolites of species of *Aspergillus* it has been fascinating to note biochemical relationships between species within the groups recognised by Raper & Fennell (1965) as well as between the groups themselves.

One of the smallest, though not necessarily the simplest, molecule associated with mycotoxicoses is patulin (I) (Fig. 5), a metabolite produced by a large number of species of *Penicillium*, by *Byssochlamys nivea* and by three species of *Aspergillus* - *A.clavatus*, *A.giganteus* and *A.terreus*. The many aspects of this versatile compound have been reviewed by Scott (1974).

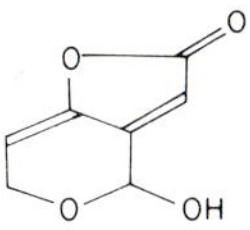

(I) Patulin

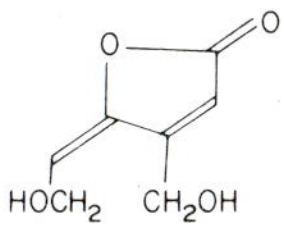

(II) Ascladiol

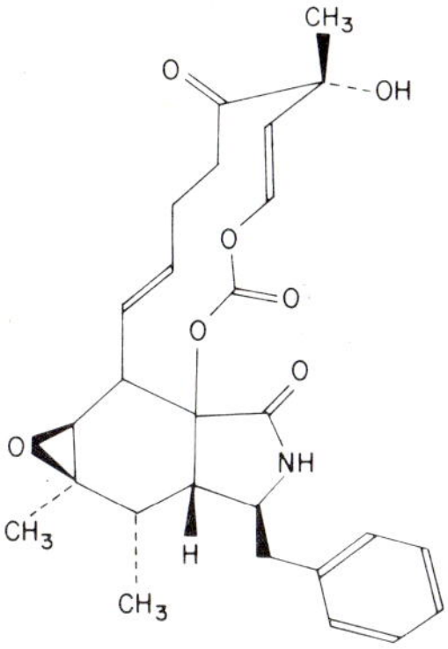

(III) Cytochalasin E

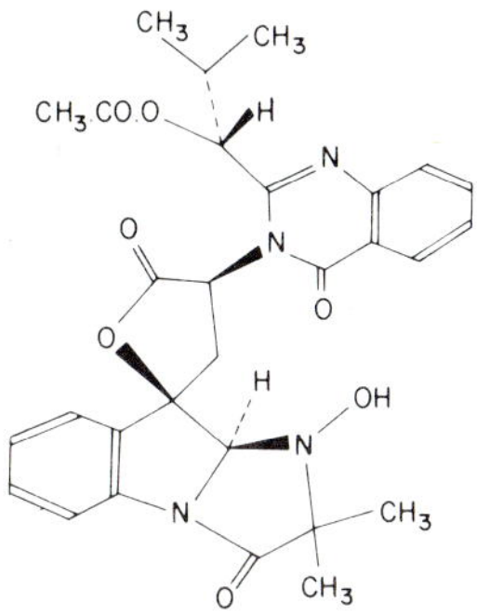

(IV) Tryptoquivaline

Fig. 5. Toxic metabolites isolated from strains of *Aspergillus clavatus*.

Patulin showed remarkable promise as a wide spectrum antibiotic before its toxicity was appreciated. Although a great deal is known about its properties it is still often circumstantial evidence which links the metabolite with field outbreaks of mycotoxicosis. It is not a very stable compound but its stability is increased at the lower pH of many plant juices and it is particularly stable in apple juice, and concentrations as high as 1 mg/l have been demonstrated in commercial apple juice (Scott, Miles, Toft & Dube, 1972). However, that particular substrate is usually associated with *Penicillium expansum*, a particularly good patulin producer, and need not concern us here.

Grossbard (1952) has hown that *A.clavatus* can produce an antibiotic, presumed to be patulin, when growing on vegetable material such as wheat straw and an outbreak of poisoning of cattle in Germany, following consumption of germinated malt contaminated with a toxigenic strain of *A.clavatus*, might well have been due to patulin (Schultz, Motz, Schafer & Baumgart, 1969). Even when a patulin producing mould has been isolated from such toxic feeds it is not always possible to detect patulin itself in the feed.

Ascladiol (II) represents a related, more stable, but less toxic, metabolite which has been isolated from *A.clavatus* obtained from a sample of wheat flour (Tanabe & Suzuki, 1970). It was shown to have only a quarter of the toxicity of patulin itself.

The isolation of cytochalasin E (III) from *A.clavatus* by Glinsukon, Shank, Wogan & Newberne (1975) is an interesting example of the demonstration that quite widely different fungi may produce the same highly complex secondary metabolite, despite ones instinctive feeling that the very complexity may ensure a degree of specificity. The cytochalasins are a group of metabolites with bizarre effects on the process of cell division, frequently resulting in the production of multinucleate or even anucleate cells (Carter, 1967). The cytochalasins have been isolated from such genera as *Helminthosporium, Phoma, Metarrhizium, Zygosporium* and cytochalasin E itself was first isolated from *Rosellinia necatrix* (Aldridge, Burrows & Turner, 1972). The structure first assigned to this compound by these authors has since been modified by Buchi *et al.*(1973). Cytochalasin E may be exceptional amongst this group of compounds in not inducing the formation of anucleate cells.

Since the report by Wilson, Wilson & Hayes (1968) that *Penicillium cyclopium* and related moulds could synthesise compounds with neurotoxic properties causing tremors in experimental animals, tremorgenic metabolites have been extensively studied. Such activity has been found to be associated with a number of metabolites of *Aspergillus* one example of which is tryptoquivaline (IV) produced by a toxigenic strain of *A.clavatus* (Clardy *et al.*, 1975). Although the molecule looks complex it is interesting to note the presence of an intact tryptophan residue in its structure.

We find tryptophan, even less disguised, forming the nucleus of the echinulins, metabolites produced by a number of species classified together as the *Aspergillus glaucus* group. Echinulin (V) (Fig. 6) and neoechinulin (VI) are examples demonstrating a relatively straight forward molecular architecture based on the formation of a cyclic anhydride of tryptophan and alanine with the addition of a number of isopentenyl residues presumably derived from mevalonic acid.

Aspergillus chevalieri, a species of the *glaucus* group, can produce a toxic metabolite shown to be identical with the diisonitrile xanthocillin X (VII) (Feuell, 1966). This compound was first isolated from *Penicillium notatum* and studied intensively because of its antibiotic properties (Rothe, 1950; Rothe, Sous, Keller & Muckter, 1956). More recently Ishida, Hamasaki & Hatsuda (1972) have reported the isolation of an analogous dinitrile compound from *Aspergillus nidulans*. This compound has been called emerin (IX) (Fig. 7) and, unlike xanthocillin, it is without any demonstrable toxicity, at least to mice. *A.chevalieri* also produces at least two diterpene compounds with antibacterial and antiviral activity (Ellestad, Kunstmann, Mirando & Morton, 1972). These compounds (see formula VIII) also inhibit the growth and motility of *Tetrahymena pyriformis*, an organism which has been studied as a possible candidate for the bioassay of mycotoxins, for example the rubratoxins (Wyatt & Townsend, 1974).

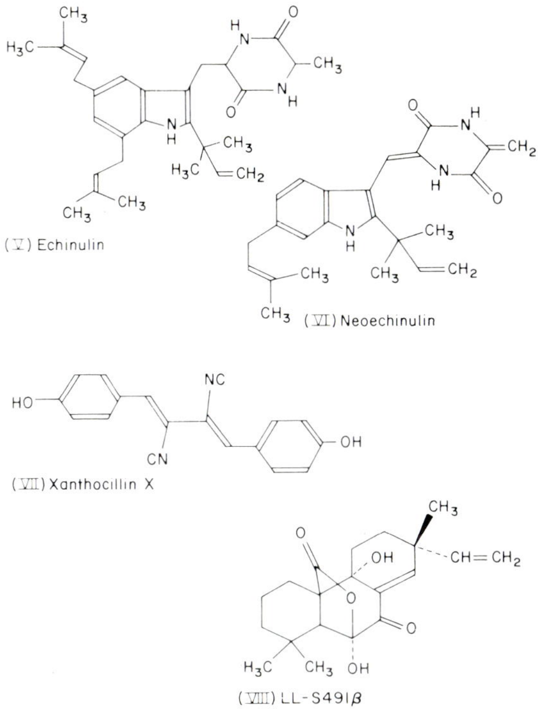

Fig. 6. Metabolites produced by members of the *Aspergillus glaucus* group.

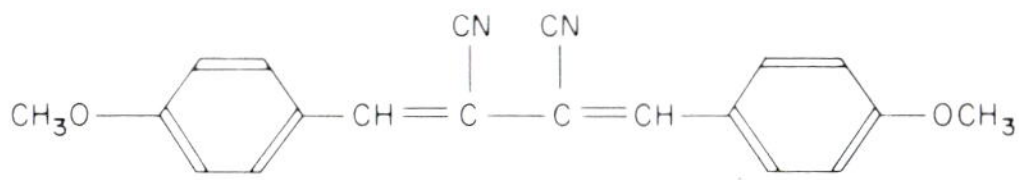

(IX)

Fig. 7. Emerin, a metabolite of *Aspergillus nidulans*.

Although the echinulins are not recognised as mycotoxins they bear a striking structural relationship with the fumitremorgins (X and XI) which are tremorgenic toxins produced by *Aspergillus fumigatus* studied by both Japanese (Yamazaki, Fujimoto & Kawasaki, 1975) and American (Eickman,Clardy, Cole & Kirksey, 1975) scientists. First isolated from Japan from miso, the fumitremorgins became of interest in America because of their similarity to verrucologen (XIV) (Fig. 9) a tremorgenic compound produced by (*Penicillium verraculosum* (Fayos *et al.*, 1974). A particularly interesting feature of fumitremorgin A (X) (Fig. 8) is the presence of a peroxide group in the structure of the molecule. *A.fumigatus* is also noteworthy in the number of epoxide containing metabolites which it produces of which the toxic antibiotic fumagillin (XIII) is a particularly complex example (Tarbell *et al.*, 1960).

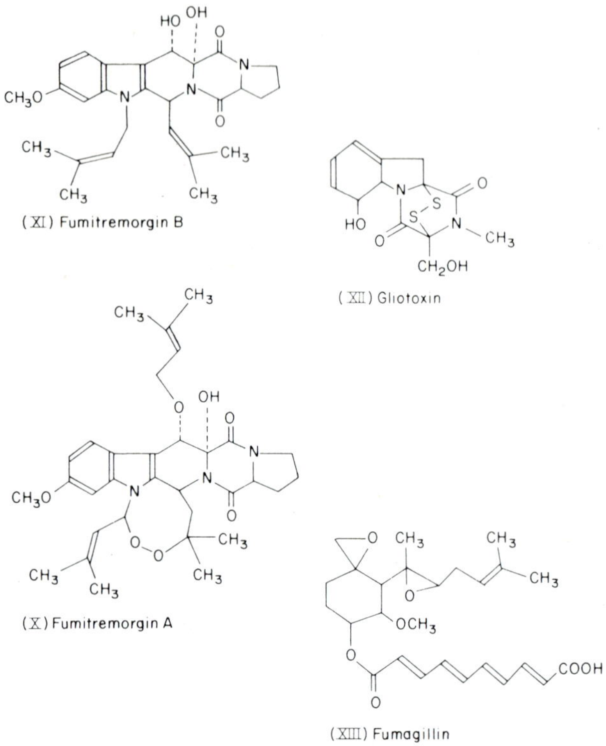

Fig. 8. Toxic metabolites produced by *Aspergillus fumigatus*.

Gliotoxin (XII) occasionally creeps into accounts describing the isolation of toxigenic fungi from food materials although it has never positively been proven to be implicated. It is produced by a wide range of moulds and frequently accounts for their observed antagonism to other fungi. In the genus *Aspergillus* gliotoxin is reported to be produced by *A.fumigatus* (Menzel, Wintersteiner & Hoogerheide, 1944), *A.chevalieri* (Wilkinson & Salisbury, 1965) and *A.terreus* (Miller *et al.*, 1968). It is interesting to note in passing that the presence of a disulphide bridge across the diketopiperazine ring is a feature shared with sporidesmin (Fig. 3) and aranotin (XV) (Fig. 10). The latter is an interesting antiviral compound produced by *A.terreus* and *Arachniotus aureus* (see Nagarajan *et al.*, 1968).

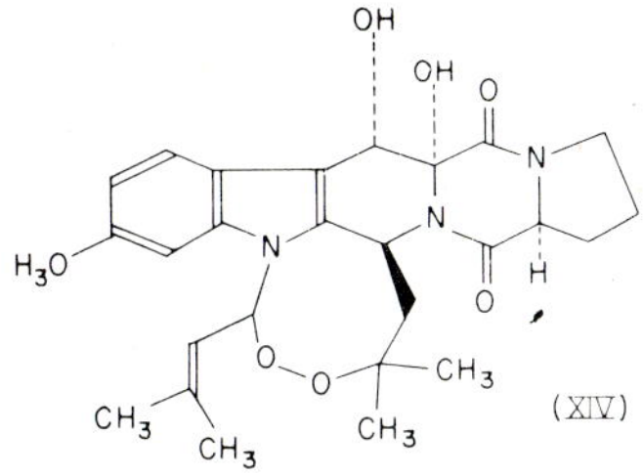

(XIV)

Fig. 9. Verruculogen, a tremorgenic toxin produced by *Penicillium verruculosum*.

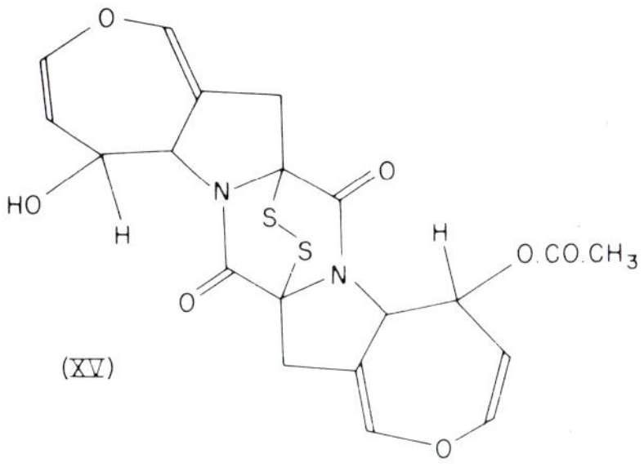

(XV)

Fig. 10. Aranotin, an antiviral metabolite of *Aspergillus terreus* and *Arachniotus aureus*.

Further examples of toxins, involving the incorporation of amino acids in their biosynthesis, are to be found amongst the metabolites of *Aspergillus ochraceus*. The destruxins are best known as insecticides produced by such insect pathogens as *Metarrhizium anisopliae* and *Oospora destructor*, but one of these cyclic peptidolactones, destruxin B (XVI) (Fig. 11) is produced by a strain of *Aspergillus ochraceus* (Tamura, Kuyama, Kadaira & Higashikawa, 1964). Two model amino acid anhydrides have also been isolated from this organism involving proline and valine (XVII) and proline and leucine (XVIII) (Chen, 1960; Kodaira, 1961).

The best known and most widely studied toxin of *Aspergillus ochraceus* is undoubtedly ochratoxin A (XIX), a phenylalanine derivative of a biogenically interesting dihydroisocoumarin. As Harwig (1974) points out, ochratoxin was discovered as a result of a laboratory screen of fungal isolates for the production of toxic metabolites. It was isolated and characterised by members of the group working in South Africa who have done a great deal to further our understanding of mycotoxins. Not all strains of *A.ochraceus* are toxigenic and it was a strain isolated from sorghum grain from which ochratoxin was first isolated and characterised (Van der Merwe *et al.*, 1965). Although ochratoxin was first isolated from laboratory cultures it has subsequently been isolated directly from a sample of low quality maize by Shotwell, Hesseltine & Goulden (1969) in the United States and from mouldy Canadian wheat by Scott, van Walbeek, Harwig & Fennell (1970).

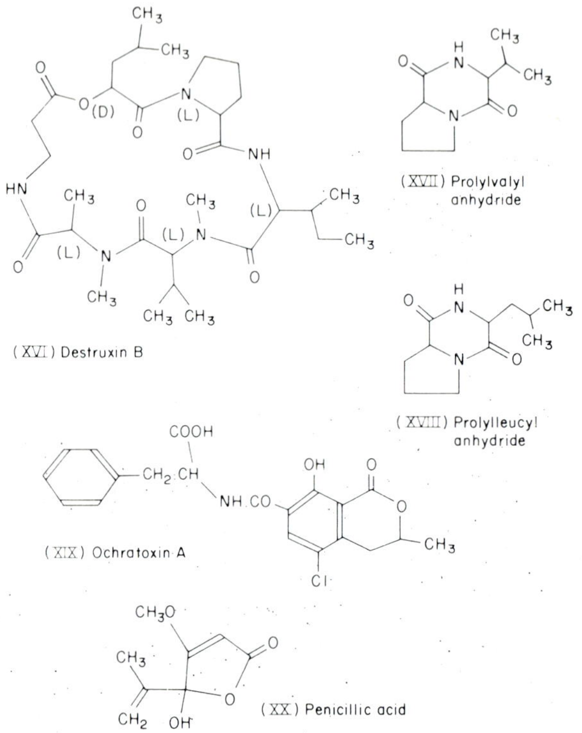

Fig. 11. Toxic, and related, metabolites of *Aspergillus ochraceus*.

As so frequently happens the ochratoxins turned out to be a closely interrelated family of compounds of which ochratoxin A is the major component and the most toxic of the group (the oral LD 50 being of the order of 150 μg/duckling). Continuous studies on the biological activity of ochratoxin have led to the general impression that, although it will give rise to pathological effects in several organs of the body, it is as a nephrotoxin that it is particularly important. Indeed ochratoxin has taken its place side by side with the now classical nephrotoxin, citrinin, in considerations of porcine nephropathy, a disorder of some economic importance in major bacon producing countries such as Denmark (Krogh, 1974). Although citrinin was first isolated from *Penicillium citrinum* by Hetherington & Raistrick (1931) it has since been isolated from many other species from all the major groups of *Penicillium* and from at least three species of *Aspergillus* - *A.candidus*, *A.niveus* and *A.terreus*. Ochratoxin is not only an amino acid derivative but the isocoumarin moiety is a clear example of the polyketide route of biosynthesis which gives rise to such a wide diversity of secondary metabolites, including citrinin (Fig. 12). However, before returning to this theme, it is worth examining some further examples of toxic amino acid derived metabolites.

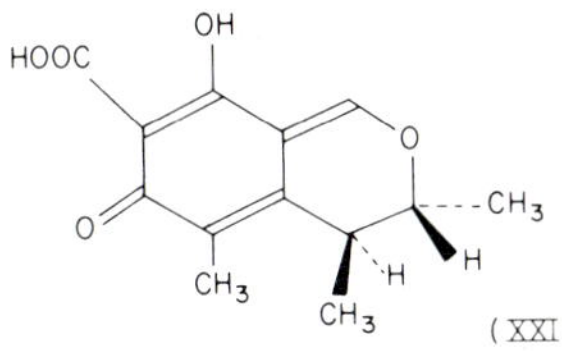

(XXI)

Fig. 12. Citrinin, toxic metabolite of *Penicillium citrinum* and several species of *Aspergillus*.

Aspergillus niger and related moulds are well known as biodeteriogens, as useful sources of enzymes and primary metabolites and as possible plant pathogens. The malformins (Fig.13) are a group of cyclic peptides, produced by *A.niger*, which have been extensively studied because of their effects on the growth and morphology of the actively growing stems and petioles of plants (Curtis, 1958). Malformin A_1 (XXII) has recently been shown to be toxic to mammals as well as causing malformations of plant tissues. Yoshizawa, Tsuchita, Morooka & Sawada (1975) have shown that malformin A_1 has an intraperitoneal LD_{50} in male mice of 3.1 mg/kg body weight, although oral doses of up to 50 mg/kg failed to show acute toxicity. Even more recently, Anderegg, Biemann, Büchi & Cushman (1976) reported the isolation and characterisation of malformin C (XXIII) from *A.niger* with a significantly higher mammalian toxicity (intraperitoneal LD_{50} 0.90 mg/kg body weight of newborn rats and 0.87 mg/kg in the case of 28 day old animals.

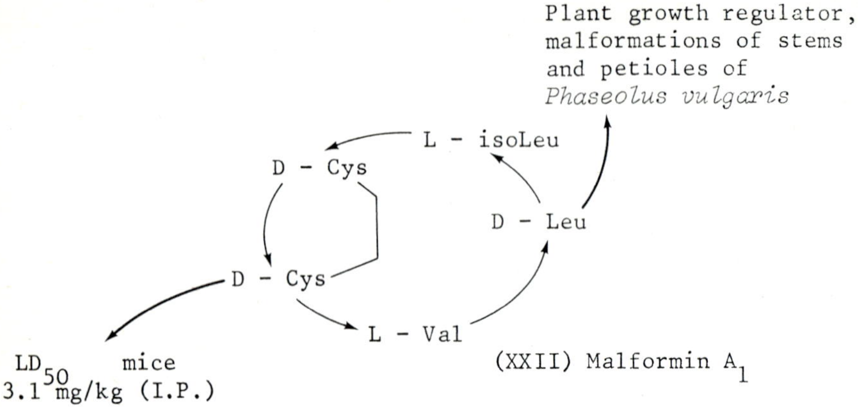

(XXII) Malformin A_1

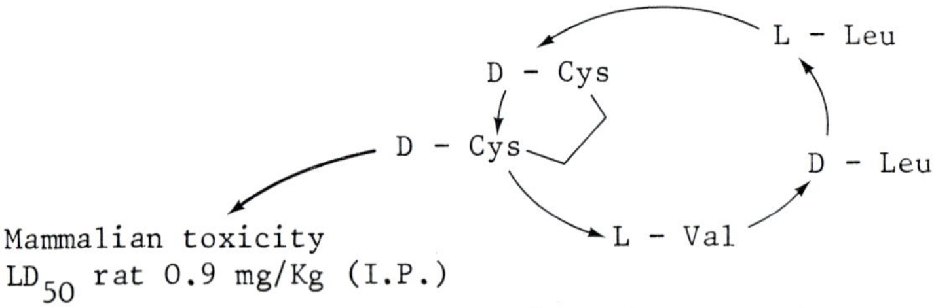

(XXIII) Malformin C

Fig. 13. The malformins, toxic metabolites isolated from strains of *Aspergillus niger*, *A.awamori*, *A.ficuum* and *A. phoenicis*.

The one organism which was responsible for mycotoxin studies becoming truly international is *Aspergillus flavus* and so much has been written about its toxic and carcinogenic metabolites - the aflatoxins (Fig. 14) that I would like to dwell on them only to make three points:-

i) They make a useful vehicle for discussing an important aspect of toxicology, namely the enormous variability of reaction from one species to another of higher animals to the consumption of a single toxic compound. This is shown, not only in the acute toxicity of aflatoxin B_1 (Table 1), but also in the response to the undoubted carcinogenic properties of the metabolite. Thus, for the rat, aflatoxin B_1 must be considered to be one of the most carcinogenic natural products known, as little as 40 μg/day for a period of 10 days being sufficient to induce hepatic carcinoma (Wogan & Newberne, 1968), whereas it has not been possible to induce tumours in some other animals. As for man it is necessary to make use of extensive epidemiological studies to establish the role of these compounds in disease. Some of the evidence indicating that they do is reviewed by Moss (1975) and their occurrence and significance in food is examined by Jarvis (1975).

Species	LD_{50} (mg/kg)
Chick embryo	0.025 ug/embryo
Duckling	0.35 - 0.46
1 day old rat	1.0
21 day old rat	5.5
Adult rat (male)	7.2
Adult rat (female)	17.9
Rabbit	0.3
30 day old hamster	10.2
Guinea pig	1.4
Sheep	1.0
Pig	0.6
Dog	0.5 - 1.0
Cat	0.6
Mouse	9.0

Table 1. Acute toxicity of aflatoxin B_1

ii) This variability in response becomes clearer if it is realised that animal tissues have a considerable potential for further altering the toxic molecule in a number of different ways. Patterson & Allcroft (1970) have reported that the liver of different animal species metabolises aflatoxin B_1 at quite different rates. Some of the pathways by which this metabolite may be metabolised further are indicated in Fig. 15.

iii) Although this is not the place to discuss in detail the biosynthesis of individual mould metabolites, the aflatoxins do offer an opportunity to demonstrate that apparently complex molecules may be derived by beautifully elegant pathways. The polyketide origin of the aflatoxins is now firmly established and, although it is still necessary to clarify the details of the origin of the fused difuran ring system, the kernal of the molecule is almost certainly derived by a series of rearrangements from anthraquinone to xanthone to aflatoxin B and hence to aflatoxin G. (Fig. 16). These considerations are relevant to the question of how widespread is aflatoxin production. Previous examples have indicated that quite complex molecules may be produced by a range, not only of species within a genus, but of taxonomically diverse genera of moulds. Despite intensive studies (for example those of Frank, 1972) it has not been possible to demonstrate conclusively that aflatoxins are produced by moulds other than those of the *Aspergillus flavus* group. One might expect that where a group of secondary metabolites are produced by a series of fairly dramatic changes in molecular architecture, the ability to carry out the changes could be restricted to sequentially fewer groups of organisms. Within the genus *Aspergillus*, and using the Groups described by Raper & Fennell (1965), it is possible to find records of anthraquinone production by members of 6 groups, xanthone production by members of 4 groups and aflatoxin production by members of only one group.

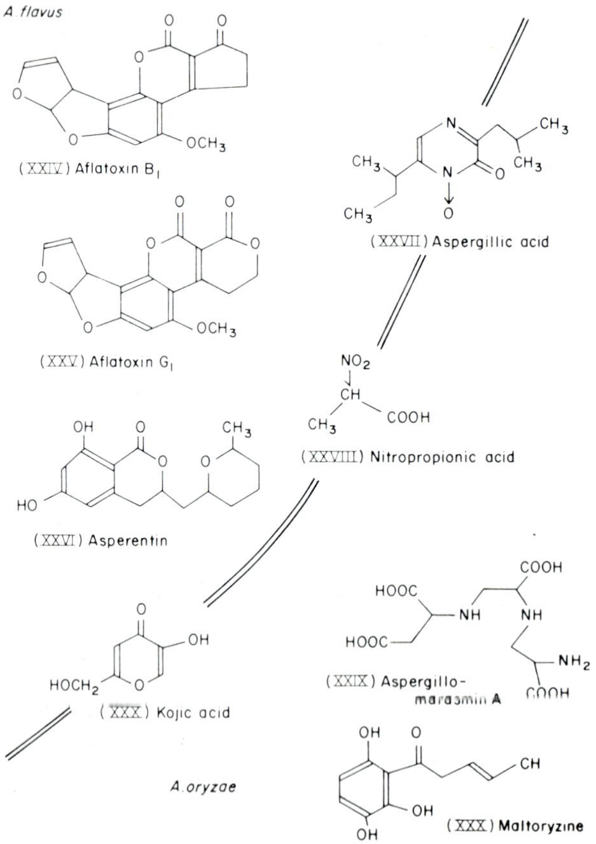

Fig. 14. Toxic metabolites of the *Aspergillus flavus/oryzae* group.

The aflatoxins are dealt with in considerable detail by

Goldblatt (1969), Detroy, Lillehoj & Ciegler (1971), Moss (1972) and Butler (1974) as well as in books edited by Herzberg (1970) and Purchase (1971).

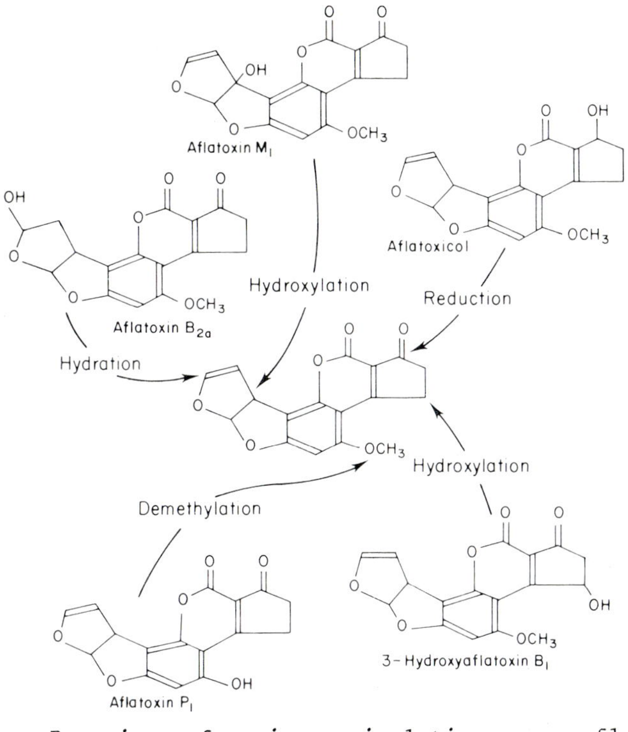

Fig. 15. Reactions of various animal tissues on aflatoxin B_1

The aflatoxins are not the only toxic metabolites of *A.flavus*, and its close relative *A.parasiticus*, toxins such as aspergillic acid (XXVII) (Fig.14), beta nitropropionic acid (XXVIII) and kojic acid (XXXI) being also produced by *A.oryzae*. Grove (1972) has isolated and characterised a number of derivatives of asperentin (XXVI) from entomogenous strains of *A.flavus* which do not produce aflatoxin.

Although *A.oryzae*, which is so extensively used in food fermentations of the Far East, does not produce aflatoxins it may produce other toxic metabolites. Many variants of the amino acid anhydride, aspergillic acid (XXVII), are known (Yokotsuka, Asao, Sasaki & Oshita, 1970), and the phytotoxic aspergillomarasmin A (XXIX) represents yet another type of amino acid derivative

(Robert *et al.*, 1962). A strain referred to as *A.oryzae* var. *microsporus*, isolated from poisonous malt sprouts, has been shown to owe its toxicity to the production of the metabolite maltoryzine (XXX) (Iizuka & Iida, 1962).

In many respects *Aspergillus versicolor* seems to occupy a central key position in the pattern of metabolite diversity amongst the aspergilli. The toxic metabolite, sterigmatocystin (XXXII) (Fig.17), has also been isolated from *A.unguis* (Mislivec, Dieter & Bruce, 1975), *A.nidulans* (Ishida *et al.*, 1972) and *A.variecolor* (Chexal, Holker, Simpson & Young, 1975) as well as from *A.versicolor* (Bullock, Roberts & Underwood, 1962). The methyl ether of sterigmatocystin has also been isolated from *A.flavus* (Burkhardt & Forgacs, 1968). Nidurufin (XXXIII), first isolated from *A. nidulans*, is one of a large number of anthraquinone derivatives isolated from *A.versicolor* (Aucamp & Holzapfel, 1970) and sterigmatin (XXXIV), described by Hamasaki, Matsui, Isono & Hatsuda (1973), offers a link with *A.ustus*, being structurally related to the austocystins.

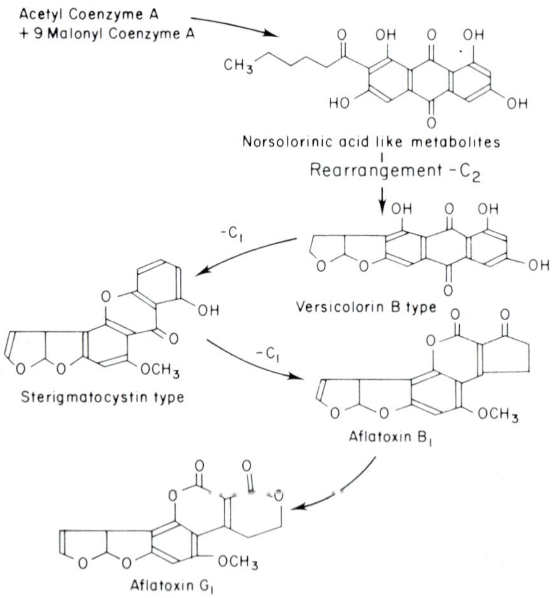

Fig. 16. Types of intermediate which possibly arise during the biosynthesis of aflatoxin.

From amongst the many known metabolites of *A.versicolor* two deserve further mention in the context of mycotoxins. Versimide (XXXV), described by Brown (1970), is an interesting derivative of a dicarboxylic acid, possibly derived from the citric acid cycle, and a lightly disguised alanine residue. It has been shown by Cole & Rolinson (1972) to be a novel contact insecticide with a high and specific activity against *Drosophila melanogaster*. Cyclopiazonic acid (XXXVI), best known as a toxic metabolite of *Penicillium cyclopium* (Holzapfel, 1968) has, rather surprisingly, been isolated from *A.versicolor* by Ohmono, Sugita & Abe (1973) bringing us back to the tryptophan derived metabolites. As an indication of the possible relevance of cyclopiazonic acid to mycotoxicoses it is worth recalling an outbreak of disease of calves reported in England in 1970 (Harrison, 1971). The animals displayed muscular tremors of the hind quarters followed by convulsions and eventually died. The feedstuff was crushed barley and readily yielded *Penicillium cyclopium*. Cyclopiazonic acid itself was isolated from the barley and identified by thin layer chromatography.

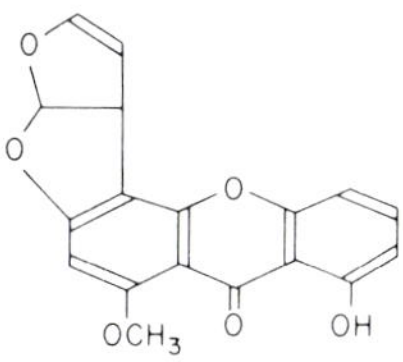

(XXXII) Sterigmatocystin

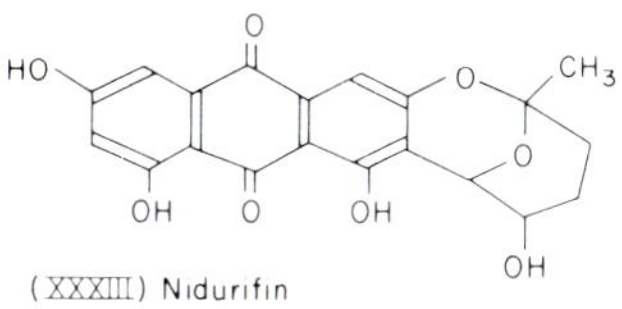

(XXXIII) Nidurifin

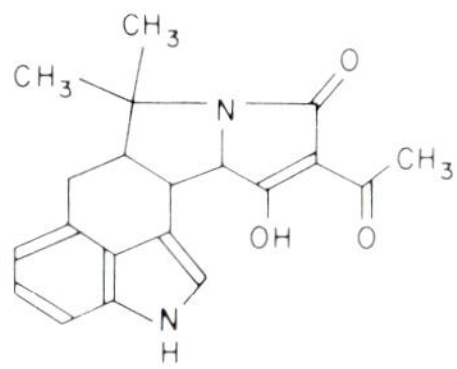

(XXXVI) Cyclopiazonic acid

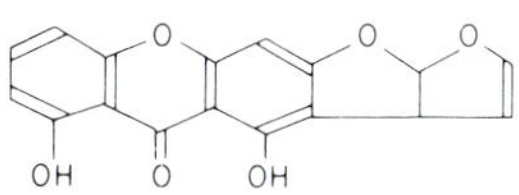

(XXXIV) Sterigmatin

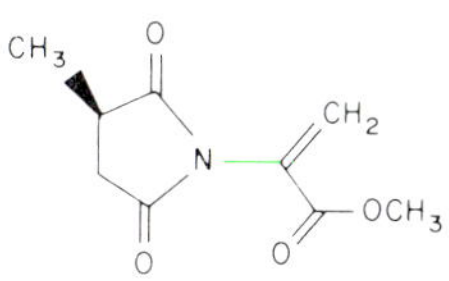

(XXXV) Versimide

Fig. 17. Toxic metabolites of *Aspergillus versicolor*.

Toxigenic strains of *Aspergillus ustus* have recently been isolated from stored foodstuffs. From cultures of this organism a number of toxic metabolites have already been characterised representing several different groups. Six closely related compounds referred to as austocystins have been characterised by Steyn & Vleggaar (1974). . Austocystin A (XXXVII) (Fig.18) is the only member of the group to contain a chlorine substituent but several, such as austocystin B - E, contain a hydroxy isopentyl derivative (XXXVIII - XLI) while austocystin F (XLII) is simply the hydroxy derivative of sterigmatin (XXXIV) already mentioned as a metabolite of *A.versicolor*. Austamide (XLIII), described by Steyn (1973), and deoxybrevianamide, described by Ritchie & Saxton (1975), are two further examples of tryptophan derived metabolites containing the diketopiperazine ring system characteristic of amino acid anhydrides.

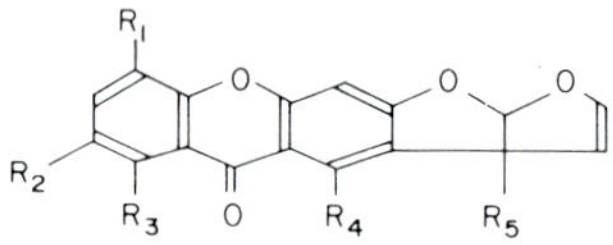

The Austocystins

		R_1	R_2	R_3	R_4	R_5
(XXXVII)	A	H	Cl	OCH_3	OCH_3	H
(XXXVIII)	B	P	H	OH	OH	H
(XXIX)	C	P	H	OH	OCH_3	H
(XL)	D	P	H	OH	OH	OH
(XLI)	E	P	H	OH	OCH_3	OH
(XLII)	F	H	H	OH	OH	OH

P = $CH_2.CH_2.C(OH)(CH_3)CH_3$

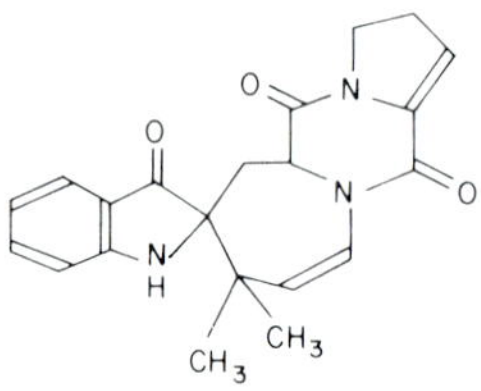

(XLIII) Austamide

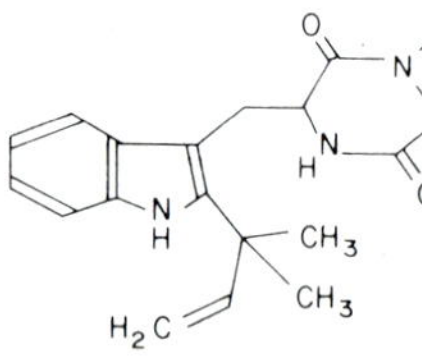

(XLIV) Deoxybrevianamide

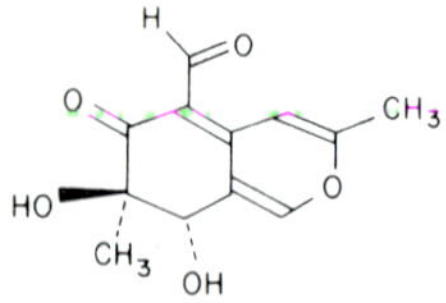

(XLV) Austdiol

Fig. 18. Toxic metabolites of *Aspergillus ustus*.

The major toxin of *Aspergillus ustus* is reported to be austdiol (XLV), a pentaketide comparable in its biosynthesis to citrinin and to the isocoumarin nucleus of ochratoxin (Fig.19).

The constitution and absolute configuration of austdiol have been described by Vleggaar, Steyn & Nagel (1974) who suggest that the compound acts as a gastro-intestinal toxin. The three toxins are not only all pentaketides but the carbon skeleton is folded in the same way. They differ in the manner in which further carbon atoms have been added from the biochemically active one-carbon pool (Fig.20).

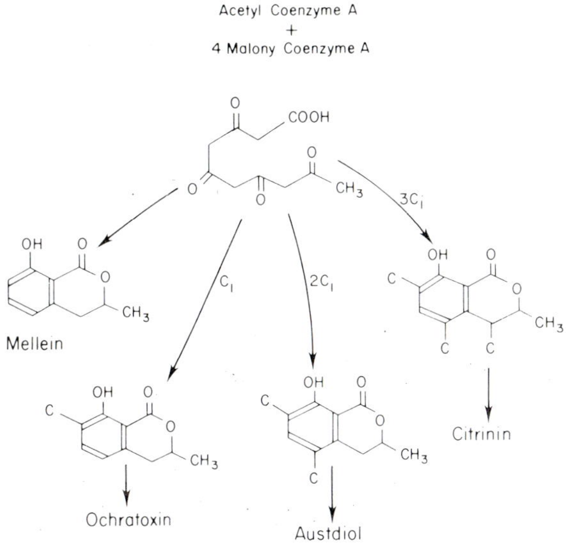

Fig. 19. Biogenetic relationship between pentaketide metabolites produced by members of the aspergilli

Although the majority of polyketide metabolites, as exemplified by austdiol, citrinin and ochratoxin, are formed by the addition of two carbon units from malonyl coenzyme A to a single primer molecule of acetyl coenzyme A there are a number of examples in which a different primer molecule is involved. Viriditoxin (XLVI), the green toxin of *Aspergillus viride-nutens*, a member of the *niger* group, has been described by Lillehoj & Ciegler (1972). An examination of its structure would suggest a polyketide origin in which the methyl ester of malonyl coenzyme A is the priming molecule (Fig.20).

This brief review can only have indicated a small part of the wealth of study which has been made on the natural history of the toxic metabolites of the aspergilli. Some will continue to be discovered through the screening programme of the pharmaceutical industry for new antimicrobial agents, others by the deliberate search for toxic compounds. Chlorflavonin (XLVII) was isolated from a strain of *A.candidus* because of its potentially useful antifungal activity (Munden, Butterworth, Hanscomb & Verrall, 1970). The same species has also been studied as a result of its isolation during a survey for toxic mould metabolites (Takahashi *et al.*, 1974) with the result that the terphenyl compound (XLVIII), previously described by Marchelli & Vining (1973), was shown to be responsible for some of the morphological changes of HeLa cells used as the screening agent.

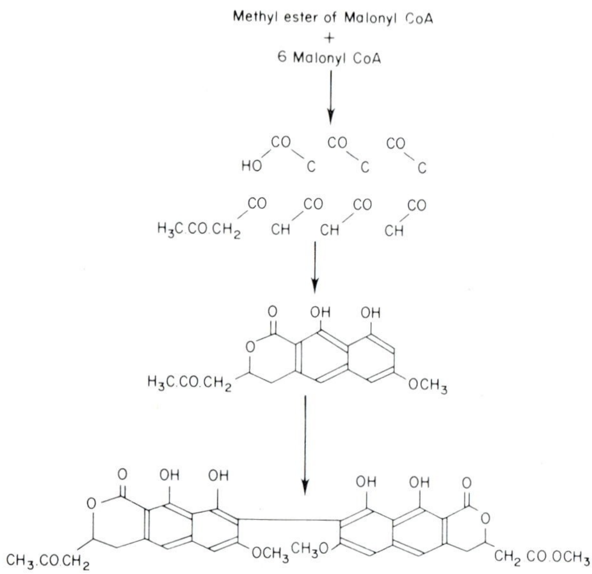

(XLVI) Viriditoxin

Fig. 20. Viriditoxin, a toxic metabolite of *Aspergillus viride-nutans* and *A.brevipes*

Inevitably there are a number of cases in which toxicity has been ascribed to isolates of aspergilli without the isolation and characterisation of a toxic metabolite. *Aspergillus amstelodami* has been demonstrated to show toxicity to poultry and rabbits (Rabie, De Klerk & Terblance, 1964) and an interesting example of the influence of temperature on toxicity came to light during studies on *A.wentii* (Rabie & Terblance, 1967). One isolate produced much more toxin at 20 - 25°C than at 30°C, although it is not clear whether this simply reflects the amount of growth or not. Probably the best example of the effect of temperature on toxin production, as distinct from growth, is shown in the involvement of species of *Fusarium* in alimentary toxic aleukia. The possiblity of complex interactions between environmental parameters and expression of toxicity has been further demonstrated by Bacon, Sweeney, Robbins & Burdick (1973) in which they found that the optimum conditions for the production of the two toxins, penicillic acid and ochratoxin, by a single strain of *A.ochraceus* were widely different.

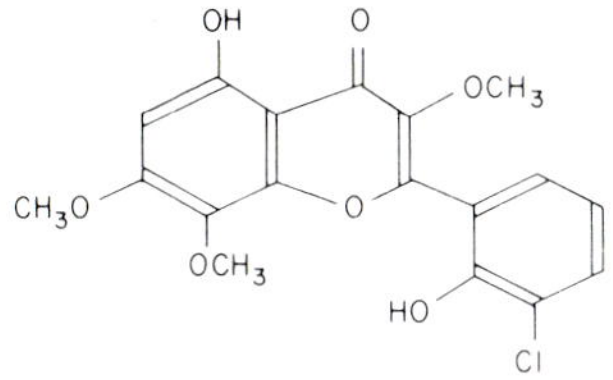

(XLVII) Chlorflavonin

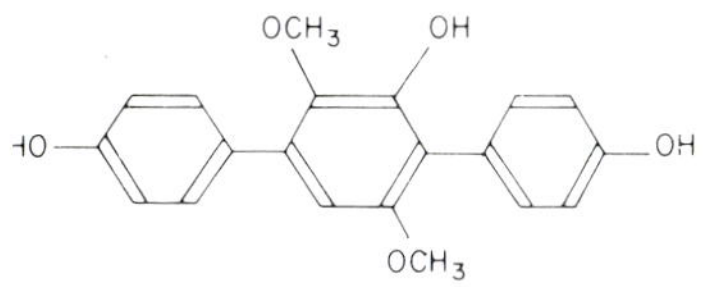

(XLVIII) Terphenyllin

Fig. 21. Metabolites of *Aspergillus candidus*.

With the reservation that I would like to draw the distinction between mycotoxins as toxic mould metabolites, some of which may only be produced in the special environment of the laboratory culture, and mycotoxicoses as diseases of man and other higher animals following consumption of, or contact with, those mycotoxins actually produced in our environment, there is no denying the importance of the aspergilli in the study of mycotoxins.

REFERENCES

ALDRIDGE,D.C., BURROWS,B.F. & TURNER,W.B. (1972). The structure of the fungal metabolites cytochalasins E & F. *Journal of Chemical Society, Chemical communication* 148-9.

ALLCROFT,R., ROGERS,H., LEWIS,G., NABNEY,J. & BEST,P.E. (1966). Metabolism of aflatoxin in sheep: excretion of the "milk toxin". *Nature,London* 209, 154-5.

ANDEREGG,R.J., BIEMANN,K., BUCHI,G. & CUSHMAN,M. (1976). Malformin C, a new metabolite of *Aspergillus niger*. *Journal of American Chemical Society* 98, 3365-70.

AUCAMP,P.J. & HOLZAPFEL,C.W. (1970). Polyhydroxyanthraquinones from *Aspergillus versicolor, Aspergillus nidulans* and *Bipolaris* species. Their significance in relation to biogenetic theories on aflatoxin B_1. *Journal of South African Chemical Institute* 23, 40-56.

AUSTWICK,P.K.C. (1965). Pathogenicity. In *The Genus Aspergillus*, pp.82-126. K.B. Raper & D.I.Fennell, Williams & Wilkins Co.

BACON,C.W., SWEENEY,J.G., ROBBINS,J.D. & BURDICK,D. (1973). Production of penicillic acid and ochratoxin A on poultry feed by *Aspergillus ochraceus:* temperature and moisture requirements. *Applied Microbiology* 26, 155-60.

BROWN,A.G. (1970). Versimide, a metabolite of *Aspergillus versicolor*. *Journal of Chemical Society (C)* 2572-3.

BUCHI,G., KITAURA,Y., YUAN,S.S., WRIGHT,H.E., CLARDY,J., DEMAIN,A.L., GLINSUKON,T., HUNT,W.& WOGAN,G.N. (1973). Structure of cytochalasin E, a toxic metabolite of *Aspergillus clavatus*. *Journal of American Chemical Society* 95, 5423-5.

BULLOCK,E., ROBERTS,J.C. & UNDERWOOD,J.G. (1962). Studies in mycological chemistry. XI. The structure of isosterigmatocystin and an amended structure for sterigmatocystin. *Journal of Chemical Society* 4179-83.

BURKHARDT,H.J. & FORGACS,J. (1968). O-Methlsterigmatocystin, a new metabolite from *Aspergillus flavus* Link ex Fries. *Tetrahedron* 24, 717-20.

BUTLER,W.H. (1974). Aflatoxin. In *Mycotoxins* pp.1-28. ed. I.F.H. Purchase, Elsevier Scientific Publishers Co. Amsterdam.

CARTER,S.B. (1967). Effects of cytochalasins on mammalian cells. *Nature, London* 213, 261-4.

CHEN,Y.S. (1960). Studies on the metabolic products of *Rossellinia necatrix*. I. Isolation and characterisation of several physiologically active neutral substances. *Bulletin of the Agricultural Chemical Society (Japan)*24, 372-81.

CHEXAL,K.K., HOLKER,J.S.E., SIMPSON,T.J. & YOUNG,K. (1975). The biosynthesis of fungal metabolites. Part V. Structure of variecoxanthones A,B and C, metabolites of *Aspergillus variecolor;* conversion of variecoxanthone A into (±) - De-C-prenylepishamixanthone. *Journal of Chemical Society, Perkin I* 543-8.

CLARDY,J., SPRINGER,J.P., BUCHI,G., MASUO,K. & WIGHTMAN,R. (1975). Tryptoquivaline and tryptoquivalone, two tremorgenic metabolites of *Aspergillus clavatus*. *Journal of American Chemical Society*, 97, 663-5.

COLE,M. & ROLINSON,G.N. (1972). Microbial metabolites with insecticidal properties. *Applied Microbiology* 24, 660-2.

CURTIS,R.W. (1958). Root curvatures induced by culture filtrates of *Aspergillus niger*. *Science* 128, 661-2.
DETROY,R.W., LILLEHOJ,E.B. & CIEGLER,A. (1971). Aflatoxin and related compounds. In *Microbial toxins, Vol.6*, pp3-178, ed. E.Ciegler, S.Kadis & S.J. Ajl, Academic Press: New York,London.
EICKMANN,N., CLARDY,J., COLE,R.J. & KIRKSEY,J.W. (1975). The structure of fumitremorgin A. *Tetrahedron Letters* 1051-4.
ELLESTAD,G.A., KUNSTMANN,M.P., MIRANDO,P. & MORTON,G.O. (1972). Structure of fungal diterpene antibiotics LL - S491β and γ *Journal of American Chemical Society* 94, 6206-8.
FAYOS,J., LOKENSGARD,D., CLARDY,J., COLE,R.J. & KIRKSEY,J.W. (1974). Structure of verruculogen, a tremor producing peroxide from *Penicillium verruculosum*. *Journal of American Chemical Society* 96, 6785-7.
FEUELL,A.J. (1966). Toxic factors of mould origin. *Canadian Medical Association Journal* 94, 574-81.
FRANK,H.K. (1972). Doubts about aflatoxin formation by *Penicillium*. *Zeitshrift für Lebensmittel Untersuchung und Forschung* 150, 151-3.
GLINSUKON,T., SHANK,R.C., WOGAN,G.N. & NEWBERNE,P.M. (1975). Acute and subacute toxicity of cytochalasin E in the rat. *Toxicology and Applied Pharmacology* 32, 135-46.
GOLDBLATT,L.A. (ed)(1969). *Aflatoxin, scientific background control and implications*, Academic Press: New York,London.
GROSSBARD,E. (1952). Antibiotic production by fungi on organic manures and in soil. *Journal of General Microbiology* 6, 295-310.
GROVE,J.F.(1972). New metabolic products of *Aspergillus flavus*. Part I. Asperentin, its methyl ethers, and 5' hydroxy asperentin. *Journal of Chemical Society, Perkin I*. 2400-5.
HAMASAKI,T., MATSUI,K., ISONO,K. & HATSUDA,Y.(1973). A new metabolite from *Aspergillus versicolor*. *Agricultural and Biological Chemistry* 37, 1769-70.
HARRISON,J. (1971). Food moulds and their toxicity. *Tropical Science* 13, 57-63.
HARWIG,J. (1974). Ochratoxin A and related metabolites. In *Mycotoxins* pp345-67, ed.I.F.H. Purchase,Elsevier Scientific Publishing Co.: Amsterdam.
HERZBERG,M.(ed)(1970). *Proceedings of the 1st U.S.-Japan Conference on Toxic Microorganisms*. UJNR Joint Panels on Toxic Micro-organisms & U.S. Dept.of the Interior.
HETHERINGTON,A.C. & RAISTRICK,H. (1931). Studies in the biochemistry of microorganisms. Part XIV. On the production and chemical constitution of a new yellow colouring mater, citrinin, produced from glucose by *Penicillium citrinum* Thom. *Philosophical Transactions of the Royal Society, London* B220, 269-95.
HOLSAPFEL,C.W.(1968). The isolation and structure of cyclopiazonic acid, a toxic metabolite of *Penicillium cyclopium* Westling.

Tetrahedron 24, 2101-9.
IIZUKA,H. & IIDA,M. (1962). Maltoryzine, a new toxic metabolite produced by a strain of *Aspergillus oryzae* var.*microsporus* isolated from the poisonous malt. *Nature,London* 196, 681-2.
ISHIDA,M., HAMASAKI,T. & HATSUDA,Y. (1972). A new metabolite from *Aspergillus nidulans*. *Agricultural and Biological Chemistry* 36, 1847-8.
JARVIS,B.(1975). Mycotoxins in food - their occurrence and significance. *International Journal of Environmental Studies* 8, 187-94.
KODAIRA,Y. (1961). Toxic substances to insects produced by *Aspergillus ochraceus* and *Oospora destructor*. *Agricultural and Biological Chemistry* 25, 261-2.
KROGH,P. (1974). Mycotoxic nephropathy. In *Mycotoxins*.pp419-28. ed. I.F.H. Purchase, Elsevier Scientific Publishing Co.: Amsterdam.
LILLEHOJ, E.B. & CIEGLER, A.(1972). A toxic substance from *Aspergillus viridi-nutans*. *Canadian Journal of Microbiology* 18, 193-7.
MARCHELLI,R. & VINING,L.C. (1973). Biosynthesis of flavanoid and terphenyl metabolites by the fungus *Aspergillus candidus*. *Journal of Chemical Society, Chemical Communications* 555-6.
MENZEL,A.E.O., WINTERSTEINER,O. & HOOGERHEIDE,J.C.(1944). The isolation of gliotoxin and fumigacin from culture filtrates of *Aspergillus fumigatus*. *Journal of Biological Chemistry*, 152, 419-29.
MILLER,P.A., TROWN,P.W., FULMAR,W., MORTON, G.O. & KARLIMER, J.(1968) An epidithiapiperazinedione antiviral agent from *Aspergillus terreus*. *Biochemical and Biophysical Research Communications* 35, 219-21.
MISLIVEC,P.B., DIETER,C.T. & BRUCE,V.R. (1975). Mycotoxin producing potential of mold flora of dried beans. *Applied Microbiology* 29, 522-6.
MOSS,M.O. (1972). Aflatoxin and related mycotoxins. In *Phytochemical ecology*,pp125-144. ed. J.B. Harborne, Phytochemical Society Symposium Series No.8, Academic Press: London.
MOSS,M.O. (1975). Role of mycotoxins in disease of man. *International Journal of Environmental Studies* 8, 165-70.
MUNDEN,J.E., BUTTERWORTH,D., HANSCOMB,G. & VERRALL,M.S.(1970). Production of chlorflavonin, an antifungal metabolite of *Aspergillus candidus*. *Applied Microbiology* 19, 718-20.
NAGARAJAN,R., HUCKSTEP,L.L., LIVELY,D.H., DELONG, D.C., MARSH,M.M. & NEUSS,N. (1968). Aranotin and related metabolites from *Arachniotus aureus*. I. Determination of structure. *Journal of American Chemical Society* 90, 2980-2.
OHMONO,S., SUGITA,M. & ABE,M. (1973). Production of alkaloids and related substances by fungi. XI. Isolation of cyclopiazonic imine

and bis(secodehydrocyclopiazonic acid) from the cultures of *Aspergillus versicolor*. *Nippon Nogei Kagaku Kaishi* 47, 57-63, (*Chemical Abstracts* 79, 103625t, 1973).

PATTERSON,D.S.P. & ALLCROFT,R. (1970). Metabolism of aflatoxin in susceptible and resistant animal species. *Food and Cosmetics Toxiocology* 8, 43-53.

PURCHASE,I.F.H. (ed) (1971). *Mycotoxins in human health*, Macmillan: London.

RABIE,C.J. & TERBLANCHE,M. (1967). Influence of temperature on the toxicity of different isolates of *Aspergillus wentii* Wehn. *South African Journal of Agricultural Science* 10, 263-6.

RABIE,C.J., DEKLERK,W.A. & TERBLANCHE,M. (1964). Toxicity of *Aspergillus amstellodami* to poultry and rabbits. *South African Journal of Agricultural Science* 7, 341-6.

RAPER,K.B. & FENNELL,D.I. (1965). *The Genus Aspergillus*, Williams & Wilkins Co: New York.

RITCHIE,R. & SAXTON,J.E. (1975). Synthesis of deoxybrevianamide E. *Journal of the Chemical Society, Chemical Communications* 611-2.

ROBERT,M., BARBIER,M., LEDERER,E., ROUX,L., BIEMANN,K. & VETTER,W. (1962). Two new natural phytotoxins: aspergillomarasmine A and B and their relation to lycomarasmine and its derivatives. *Bulletin Societe Chimica Francais* 187-8.

ROTHE,W. (1950). New antibiotic. *Pharmazie* 5, 190.

ROTHE,W., SOUS,H., KELLER,H. & MUCKTER,H. (1956-57). Xanthocillin: experimental and clinical studies. *Antibiotics Annual* 140-3.

SCHULTZ,J., MOTZ,R., SCHAFFER,M. & BAUMGART,W. (1969). Experimental investigations into *Aspergillus clavatus* poisoning in cattle. *Monatsh. Veterinary Medicine* 24, 14-17.

SCOTT,P.M. (1974). Patulin. In *Mycotoxins*, pp383-403 ed. I.F.H. Purchase, Elsevier Scientific Publishing Co.: Amsterdam.

SCOTT,P.M., MILES,W.F., TOFT,P. & DUBE,J.G.(1972). Occurrence of patulin in apple juice. *Journal of Agricultural and Food Chemistry* 20, 450-1.

SCOTT,P.M., VAN WALBEEK,W., HARWIG,J. & FENNELL,D.I. (1970). Occurrence of a mycotoxin, ochratoxin A, in wheat and isolation of ochratoxin A and citrinin producing strains of *Penicillium viridicatum*. *Canadian Journal of Plant Science* 50, 583-5.

SHOTWELL,O.L., HESSELTINE,C.W. & GOULDEN,M.L.(1969). Ochratoxin A: occurrence as natural contaminant of a corn sample. *Applied Microbiology* 17, 765-6.

STEYN,P.S. (1973). The structure of five diketopiperazines from *Aspergillus ustus*. *Tetrahedron* 29, 107-20.

STEYN,P.S. & VLEGGAAR,R. (1974). Austocystins. Six novel dihydrofurd (3', 2' : 4,5)furo(3, 2-6)xanthenones from *Aspergillus ustus*. *Journal of the Chemical Society, Perkin Transactions* I 2250-6.

TAKAHASHI,C., YOSHIKIRA,K., NATORI,S., UMEDA,M., OHTSUBO,K. & SAITO,M. (1974). Toxic metabolites of *Aspergillus candidus*. *Experientia* 30, 529.

TAMURA,S., KUYAMA,S., KODAIRA,Y. & HIGASHIKAWA,S. (1964). The structure of destruxin B, a toxic metabolite of *Oospora destructor*. *Agricultural & Biological Chemistry* 28, 137-8.

TANABE,H. & SUZUKI,T.(1970). A new mycotoxin produced by *Aspergillus clavatus*. *Proceedings of 1st U.S. - Japan Conference on Toxic Microorganisms* pp127-8. ed. M. Herzberg, UJNR Joint Panels on Toxic Microorganisms and the U.S. Dept. of the Interior.

TARBELL,D.S., CARMAN, R.M., CHAPMAN,D.D., HUFFMAN,K.R. & McCORKINDALE,N.T. (1960). The structure of fumagillin. *Journal of the American Chemical Society* 82, 1005-7.

VAN DER MERWE,K.J., STEYN,P.S., FOURIE,P.S., SCOTT, DE B. & THERON, J.J. (1965). Ochratoxin A, a toxic metabolite produced by *Aspergillus ochraceus* Wilk. *Nature, London* 205, 112-3.

VLEGGAAR,R., STEYN,P.S. & NAGEL,D.W. (1974). Constitution and absolute configuration of austdiol, the main toxic metabolite from *Aspergillus ustus*. *Journal of the Chemical Society, Perkin Transactions I.* 45-9.

WILKINSON,S. & SALISBURY,J.F. (1965). Gliotoxin from *Aspergillus chevalieri* (Mangin) Thom & Church. *Nature, London* 206, 619.

WILSON,B.J., WILSON,C.H. & HAYES,A.W. (1968). Tremorgenic toxin from *Penicillium cyclopium* grown on food materials. *Nature,London* 220, 77-78.

WOGAN,G.N. & NEWBERNE,P.M. (1968). Dose-response characteristics of aflatoxin B carcinogenesis in the rat. *Cancer Research* 27, 2370-6.

WOOD,R.K.S., BALLIO,A. & GRANITI,A. (ed)(1972). *Phytotoxins in plant diseases*, Academic Press: New York,London.

WYATT,T.D. & TOWNSEND,R.J.(1974). A bioassay of rubratoxins A & B using *Tetrahymena pyriformis* Strain W. *Journal of General Microbiology* 80, 85-92.

YAMAZAKI,M., FUJIMOTO,H. & KAWASAKI,T. (1975). The structure of a tremorgenic metabolite from *Aspergillus fumigatus* Fres.: fumitremorgin A. *Tetrahedron Letters* -241-4.

YOKOTSUKA,T., ASAO,Y., SASAKI,M. & OSHITA,K.(1970). Pyrazine compounds produced by moulds. In *Proceedings of the 1st U.S.-Japan Conference on Toxic Microorganisms*, pp133-142, ed. M.Herzberg, UJNR Joint Panels on Toxic Microorganisms and the U.S.Dept. of the Interior.

YOSHIZAWA,T., TSUCHITA,Y., MOROOKA,N. & SAWADA,Y. (1975). Malformin A, as a mammalian toxicant from *Aspergillus niger*. *Agriculture and Biological Chemistry* 39, 1325-6.

Medical Aspects of *Aspergillus*

J. H. EDWARDS and T. S. Al-ZUBAIDY

M.R.C. Pneumoconiosis Unit, Llandough Hospital, Penarth, Glamorgan, U.K.

Department of Biology, Science College, Mosul University, Iraq

INTRODUCTION

The *Aspergillus* genus is a group of free living saprophytic moulds with a widespread distribution. Its distribution, on a world-wide basis, varies from 0.1 to 22% of the total airspora samples (Mullins, 1974), being the most common organism in some countries to 15th in order of abundance in others. Figures from the Cardiff area (Mullins, Harvey & Seaton, 1976) reveal an average concentration of 15 per cu.m. for *A.fumigatus* with higher counts in the winter months (Fig. 1). In comparison, the counts of *Cladosporium* averaged 2133 per cu.m. *Sporobolomyces*. 262, ascospores, 1156, and basidiospores, 1343 per cu.m. It was considered that the growth of *A.fumigatus* in compost during the winter gave rise to the higher counts. *A.fumigatus* is also found in indoor air, sometimes in high numbers. Noble & Clayton (1963) found 2400 per cu.m. in a hospital ward, whereas Mullins *et al.* (1976) found less than 15 per cu.m. average in the hospital ward they studied.

The versatility of the genus in metabolizing substrates that other organisms utilize poorly or not at all, has been put to good use and Aspergilli have found themselves providing key roles in several industrial processes. However, some species are capable of causing disease either by straight-forward tissue invasion or colonization, or by means of reactions involving the

the immune system.

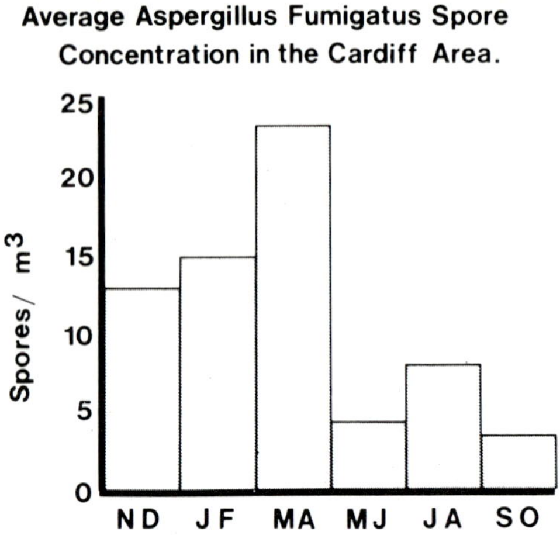

Fig. 1. The average bimonthly *Aspergillus fumigatus* spore concentration in the Cardiff area taken over a period of 3 years. Note the higher winter month figures compared with summer figures. By courtesy of Dr. J. Mullins, Asthma Research Unit, Sully Hospital, S. Glamorgan.

Since the spores of the organism (approx. 5µm in diameter for *A.fumigatus*) can become airborne, the most frequently affected organ is the lung and respiratory tract and as early as 1856 Virchow described human pulmonary disease due to *Aspergillus*. Later *A.niger* was reported by Cramer (1859) from ear infections. *A.nidulans* and *A.flavus* were also recovered from otomycosis (Siebenmann, 1889) and *A.terreus* by Langeron (1922). *A.fumigatus* had already been implicated as an important human pathogen by Gaucher & Sergent (1894) in people employed to chew grain for feeding squabs. *Aspergillus* also infects animals; it can produce mycotic placentitis in cattle (Austwick & Venn, 1962) and is an important avian pathogen.

There are several ways in which *Aspergillus* can cause disease in man depending not only on the pathogenicity of the organism but often on the state of the individual, i.e. healthy or on drugs affecting the immune response, again the genetic predisposition of the individual can play an important role, e.g. atopicity.

In general disease may be produced by one (or more) of 4 roles of the organism. These are:

Aspergillus as an infective agent
Aspergillus growing saprophytically
Aspergillus producing hypersensitivity reactions
Aspergillus producing mycotoxins

ASPERGILLUS AS AN INFECTIVE AGENT

Very rarely does Aspergillus infect normal humans; the capacity of the lung to deal with Aspergillus spores is exceptionally good and exposure even to vast numbers may be without resulting disease. Presumably the macrophage system and ciliary clearance are sufficient to eliminate the spores since the incidence of an immune response to Aspergillus in normal individuals is relatively low.

Occasionally, however, *A.fumigatus* may cause a widespread invasion of tissues (Burke, Storring & Parry, 1970) that is rapidly fatal even in normal individuals, or the invasion may be localized in the lungs producing a necrotizing pneumonia. *A. niger* may infect the outer ear, sometimes affecting the tympanic membrane and rarely, Aspergillus invades the mucosa of the nasal passages spreading to the meninges and brain with fatal results. More often than not the pulmonary or systemic invasion by *A.fumigatus* is enhanced by contributory factors such as an impaired immune response and the organism may be considered as an opportunist pathogen under these circumstances. Patients with leukaemia, or Hodgkin's disease, or those with drug (corticosteroids, immunosuppressives) or radiation induced immune suppression may succumb to this form of invasion. Indeed, one U.S.A. survey showed several patients dying following transplantation surgery, had evidence of significant Aspergillus infection. Possibly multiple antibiotic therapy may contribute to the progression of invasion (Kennedy, Malone & Blyth, 1970). In general, experimental work on the capacity of Aspergilli to invade tissues has suggested that the virulence could not be attributed to such characters as size of conidia or thermotolerance of different strains of the same species (Scholer, 1974). The opportunistic nature of Aspergillus infection is again borne out by Aspergillus infection following trauma, e.g. burns or *A.terreus* infection of the eyes following corneal trauma.

Aspergillus can cause endocarditis, particularly where a heart valve has been replaced by a prosthetic valve or homograft (Kammer & Utz, 1974); this is usually attributable to airborne Aspergillus spores settling and developing on the valve. Petheram & Seal (1976) describe 7 fatal cases of Aspergillus endocarditis occurring in 206 open heart operations involving 180 prosthetic valve inserts. The patients were given broad spectrum antibiotics pre- and post-operatively but no anti-fungal agents. At post mortem all 7 were found to have vegetative growths of *A.fumigatus* either narrowing or occluding the lumen of the valve or interfering with the seating of the valve (Fig. 2). Also in 3 cases there was good evidence to suspect emboli in spleen, brain and either kidney or spleen, similar to the findings of Andriole,

Kravetz, Roberts & Utz (1962) and Kammer & Utz (1974). One case of Aspergillus endocarditis caused by *A.ustus* was cured by surgery, amphotericin B and long term 5 fluorocystosine therapy (Carrizosa, Levison, Lawrence & Kaye, 1974), other Aspergilli involved as infective agents in endocarditis have been *A.flavus, A.niger, A.glaucus, A.sydowi and A.terreus* (Carrizosa *et al.*, 1974).

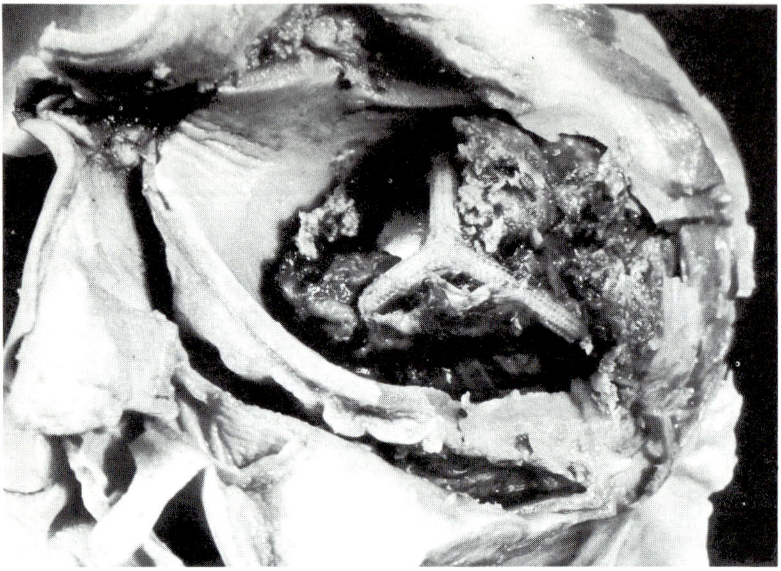

Fig. 2. Heart prosthetic valve occluded by *A.fumigatus* mat, removed at autopsy. By courtesy of Dr. I.S.Petheram, Cardiology Dept., Sully Hospital, S. Glamorgan.

Aspergillus infections may be concommitant with other infections, e.g. bacterial, viral, fungal, and this may again be related to a reduced immune status induced by the other infections. A diagnosis of infection may be extremely difficult due to the fulminating nature of the infections in poorly competent individuals. In the 7 cases with prosthetic valve replacements, Petheram & Seal (1976) were unable to culture fungi from the blood or demonstrate precipitins to *A.fumigatus* extracts.

In pulmonary infections by Aspergilli the situation is marginally less difficult since mycelium may be observed or Aspergilli cultivated from sputum. Also a lung biopsy may reveal mycelium to be present and if the patient survives long enough and precipitins appear in the serum against Aspergillus a more certain diagnosis may be made. In these cases treatment with nystatin, amphotericin B or clotrimazole may control the infection.

ASPERGILLUS GROWING SAPROPHYTICALLY

Aspergillus (usually *A.fumigatus* but occasionally *flavus, niger* or *nidulans*) can colonize old tuberculous or abscess cavities (Voisin *et al.*, 1964) and other pulmonary cavities e.g. caused by carcinoma, infarction, bronchiectasis (Krakowka *et al.*, 1970) producing a fungal ball. The aspergilloma occasionally termed mycetoma is rarely found other than in the lungs (occasionally the nasal sinuses) and need not necessarily merit treatment but may spontaneously regress. Patients may be asymptomatic but 60% show haemoptysis and a persistent cough may be a feature. Sometimes fever and malaise are the symptoms. Mostly (nearly 50%) aspergilloma occur in old tuberculous cavities (Suter, 1974) and there may be multiple foci. The prognosis in aspergilloma of old tuberculous cavities does not appear to be any worse than if the cavity is not colonized, although such people have a high mortality rate (18%) it is no higher if aspergilloma complicates the picture (British Thoracic and Tuberculosis Association, 1970). However, patients with aspergilloma are more prone to develop chronic haemoptysis.

The apparent saprophytic nature of the colonization may be as a result of a delicate balance between the Aspergillus and the immune system and becomes an increasing hazard if resistance to infection is reduced by drugs. The fungal ball seems to act as a large foreign body with the fungus surrounded by inflammatory cells within a cavity which has a complete epithelial lining (Fig. 3). There is nearly always a strong immune response to the fungal ball in the form of precipitating antibody, 98% cases of aspergilloma were precipitin positive with high titre antibodies i.e. 77% formed more than 6 precipitin lines against Aspergillus extracts (Pepys, 1969). Drouhet (1974) found 215 of 224 cases of aspergilloma to have precipitins and commented that some of the 9 serologically negative cases had other diseases, e.g. diabetes, systemic lupus etc. or that the local lesion was not in intimate contact with the pulmonary tissue i.e. presence of fibrotic tissue surrounding the fungal ball. He found immunodiffusion to be a better discriminating test in terms of identifying the causative species than immunofluorescence, presumably there are interspecies common antigens on the fungal hyphae whereas metabolites of the organism, most of which have enzyme function (Tran Van Ky, Uriel & Rose, 1966) are antigenically distinct. In the presence of such high titre antibody the skin reaction is surprisingly poor - an immediate and/or late reaction may be present. Pepys (1969) considers that the late (type III) reaction may be dependent on antigen-antibody ratios in that an excess of antibody is present in most aspergilloma cases. It was found that in an aspergilloma patient with type I reaction and weak type III skin response, injecting the patient's serum at 1/10 and 1/100 into test sites prior to the test injection, evoked a strong type III reaction on testing that was greater in the 1/100 prepared site, thus it may require an antigen excess at the test site for a strong type III reaction.

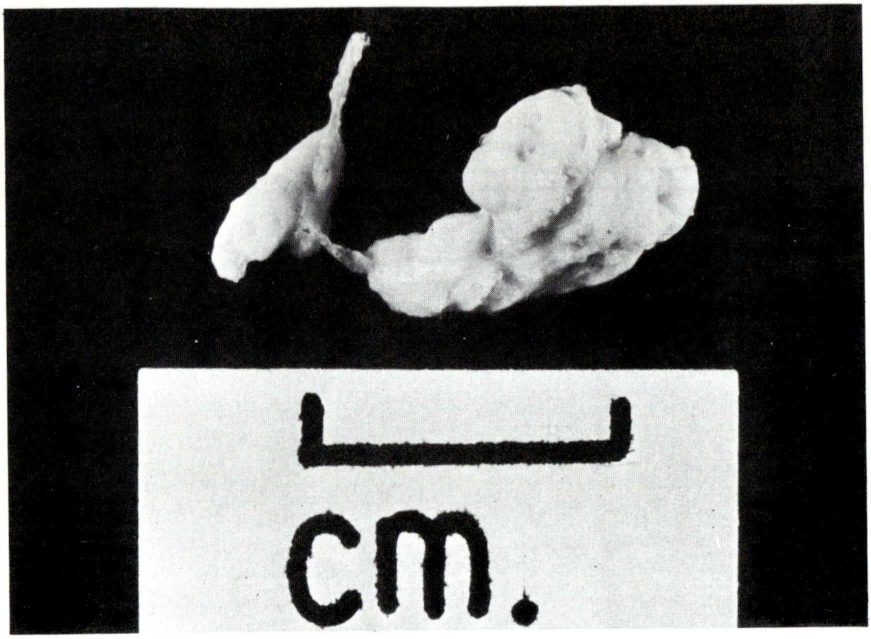

Fig. 3. Sputum plug from a case of bronchopulmonary aspergillosis. By courtesy of Mr. D. Llewellyn, Llandough Hospital, S. Glamorgan.

Diagnosis in the case of aspergilloma is on radiographic grounds aided by serology, whereas skin testing and sputum growth of Aspergillus are of lesser value. Treatment is usually not necessary but is dependent on the clinical picture. Amphotericin B is probably the best treatment but is often badly tolerated intravenously and is nephrotoxic. The reports of treatment with local instillation of amphotericin (as a paste) or natamycin show that they are only occasionally effective. Local drainage of the cavity has been tried, followed by drug instillation, however, the classic treatment of aspergilloma is its surgical removal by lobectomy, segmental or wedge resection depending on the ability of the patient to withstand such an operation.

Corticosteroids in moderate doses have also been tried

(Hilvering, Stevens & Orie, 1970), with encouraging results, certainly the fever that may be associated with the aspergilloma is reduced.

ASPERGILLUS PRODUCING HYPERSENSITIVITY REACTIONS

The hypersensitivity respons(s) to Aspergillus spores or products is governed in many ways by pre-existing genetic immune status.

Division is made between atopics (approx. 10% population) and non-atopics. In general atopics are predisposed to develop infantile eczema, rhinitis and asthma; they develop greater amounts of IgE antibody (reagin) and may become sensitized by every day exposure to commonplace allergens, e.g. house dust mites, pollens, fungal spores. Non-atopics tend to produce IgG antibody but this may only be evident after intensive or prolonged exposure; sensitization to commonplace allergens may not be apparent. There are certain individuals who overlap, e.g. a non-atopic being reagin sensitized to a particular allergen and also some non-atopics who produce high titre IgG antibody after exposure to small amounts of organic material.

Asthmatic responses

Similarly to grass pollen, house dust mite, etc., Aspergilli can cause an immediate attack is mediated by the interaction of specific IgE antibody and inhaled *A.fumigatus* (or other Aspergillus species) spores, etc. The spores (or organism metabolites) react with adjacent IgE molecules on mast cells (or basophils) causing an allosteric conformational change that triggers a sequence of energy dependent enzyme steps culminating in the release of pharmacologically active amines - histamine, 5 hydroxytryptamine (5HT), also slow reacting substances (SRS). These affect smooth muscle in the bronchioles causing constriction of the lumen thereby decreasing airflow. The reaction is termed a type I or immediate response and the concommitant effect of the vasoactive amines i.e. increase in capillary permeability, may be important in aiding the development of immune complexes (type III) mediated tissue damage (McCarthy & Pepys, 1971*a*) in those with the appropriate IgG antibody against Aspergillus, although such antibody is not normally a feature of this type of straightforward allergy to Aspergillus (Longbottom & Pepys, 1964). 10-20% of all asthmatics show evidence of sensitization to *A.fumigatus* although they may also respond to other species, *flavus*, *nidulans*, *niger*, *terreus*, etc.

Diagnosis is made by skin testing, clinical history

and examination. A degree of bronchial relief may be obtained with antihistamines or prophylactically with disodium cromoglycate; corticosteroids may have to be used in severe conditions.

Allergic bronchopulmonary aspergillosis - asthma and pulmonary eosinophilia (APE)

A complication of the asthmatic response to *A.fumigatus* may be a transient pulmonary infiltration and consolidation with eosinophilia in blood (>1000 per cu.mm) and sputum. In these cases the asthmatic response, the production of mucoid plugs in the bronchioles and the presence of eosinophils probably serve to retain the organism within the lung where a limited degree of development occurs, this can be seen when the plugs (0.5-1.5cm long) are expectorated (Fig. 4). Mycelia may be observed in up to 90% of these plugs, also eosinophils and their production is highly characteristic of this form of allergy. The infiltrations may last days or weeks and a characteristic bronchiectasis appears (Scadding, 1967) presumably due to the presence of the plugs and immunological reaction to the Aspergillus and its products in the plugs. These are usually located in medium sized bronchi and may be seen on X-ray. Also the organism may be cultivated from the sputum, 21/22 cases were positive when 3 specimens of sputum were provided (Henderson, 1968). McCarthy (1970) found 30/36 sputa positive in cases with active pulmonary infiltration.

Skin tests are highly informative, over 90% cases had a positive immediate (type I) skin reaction on prick testing (Longbottom & Pepys, 1964) generally followed by an Arthus (type III) response peaking at 7-8 h.

Gel diffusion testing indicates precipitins in over 60% cases of APE increasing to 90% when serum is concentrated. Often only one line of precipitation is seen even with concentrated sera.

That there is an immunological basis to the disease can be shown by the effect of corticosteroids on the pulmonary infiltrations. These drugs do not inhibit type I responses but exert considerable influence on type III reactions, possibly by raising cyclic AMP levels (Lavin, Rachelefsky & Kaplan, 1976) or by stabilizing lysosomes (Weissmann & Thomas, 1962). They can control the pyrexial episode sometimes seen with APE and reduce or abolish the purulent sputum production supporting the role of an allergic response retaining spores in the respiratory tract. The potential effect of Aspergillus on pulmonary function can be seen when cases of APE are challenged with an aerosol of Aspergillus extract. Apart from the initial, i.e. immediate, type I reaction, manifest by a fall in the forced expiratory volume in one second ($FEV_{1.0}$), a later (starting at 6 h) and often more severe fall can take place. The initial fall in FEV can be prevented by disodium cromoglycate inhalation prior to the test, the late response may be reduced, sometimes abolished (Pepys, 1969). Disodium cromoglycate, appears to have an action of cyclic AMP phosphodiesterase thereby causing a rise in intracellular cyclic AMP (Lavin *et al.*, 1976). The liberation of histamine, 5HT, SRS, is then inhibited. Disodium cromoglycate only acts at this level and does not block the effects of histamine, 5HT, SRS, once released. *In vitro* studies have shown the drug to inhibit the anaphylactic release of histamine from mast cells and this is reflected *in vivo* by the prevention of the bronchospasm.

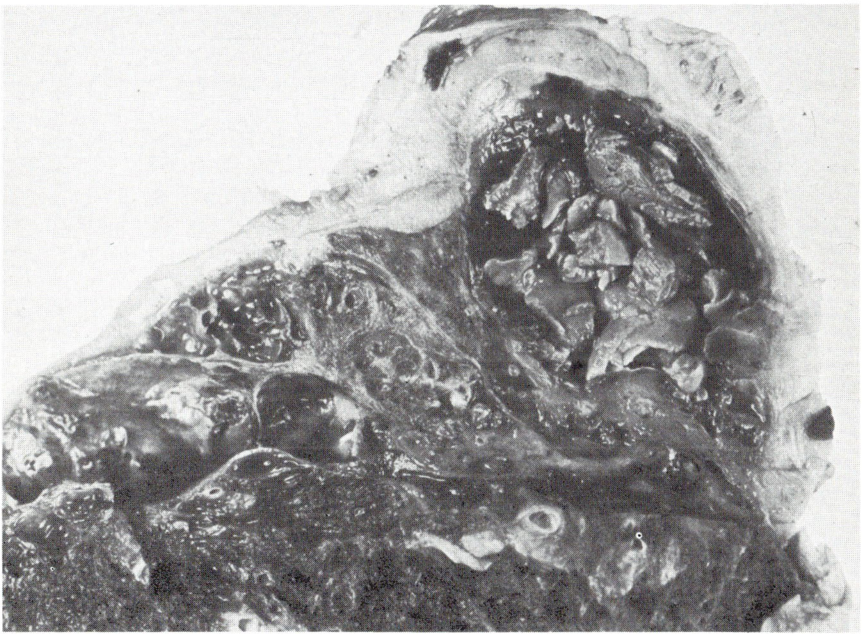

Fig. 4. Lung section from case of aspergilloma (with some fibrosis). Note fungus ball lying in an old tubercolous cavity. By courtesy of Mr. R. Boothby, Llandough Hospital, S. Glamorgan.

The onset of pulmonary eosinophilia after the start of asthma bears an inverse relationship to the atopicity of the individual, highly atopic patients, i.e. those with asthmatic symptoms before 10 years of age, took about 24 years for pulmonary eosinophilia to appear. Those with asthmatic symptoms starting between 11 and 30 years of age (less atopic) had pulmonary eosinophilia occurring after about 11 years and the low atopics (asthma starting after 31 years) had pulmonary eosinophilia starting after about 3½ years (McCarthy & Pepys, 1971*b*).

Aspergillus producing extrinsic allergic alveolitis (EAA)

Exposure to high levels of Aspergilli spores, usually associated with an industrial process, can sensitize non-atopics and produce characteristic episodes of EAA some hours after exposure. The features of the disease do not include an asthmatic component nor eosinophilia but are largely akin to other diseases of this type caused by organic dusts, i.e. dyspnoea, fever, malaise. *A.fumigatus* may be involved either as a contaminant of cereal or as part of an industrial process. Vallory-Radot & Giroud (1928) described symptoms typical of EAA in maltworkers exposed to high levels of *A.fumigatus* spores that had contaminated the grain. Horejsi *et al.*(1960) reported on a syndrome resembling farmer's lung (the most frequently encountered disease in the EAA group) in workers inhaling spores of Aspergillus and Penicillia that had been produced after the use of these organisms for the fermentation of molasses in the industrial production of citric acid. In one group of 38 subjects exposed to mould spores every third day, 35 experienced symptoms after working with the moulds, another group of 23 workers had discontinued this work due to the symptoms. It was noted that symptoms were more pronounced after a holiday and when *A.fumigatus* (or Penicillia) spores were inhaled. With *A.niger* the complaints were fewer. Horejsi *et al.*(1960) considered that 55% of the workers and 87% of the past workers had experienced acute episodes similar to farmer's lung. Yet there was no evidence of fungal infection in the 61 people employed at the plant even though 90% of the workers consistently produced positive cultures of moulds. Riddle & Grant (1967) reported on EAA in a maltworker where *A.clavatus* was isolated from germinating barley in the maltings. The patient had precipitins against *A.clavatus* and the inhalation of spores produced a pyrexial episode after 6 h. Following this study a further case was found again with precipitins against *A.clavatus* also a type III response was observed on skin testing, the authors considered that the pulmonary response was probably a precipitin mediated type III reaction (Riddle *et al.*, 1968). In an excellent study on maltworker's lung Grant, Blackadder, Greenberg & Blyth (1976) found 5.2% of employees in the Scottish malting industry to have symptoms of EAA. The disease was more prevalent in the older type maltings, e.g. open floor (6.8%) and Saladin (6.2%) than the modern mechanical method (1.1%) and with the emphasis on increasing mechanization it is likely that the prevalence of the disease will decrease. 82% of the cases of EAA were found to have precipitins against *A. clavatus* compared with 11% asymptomatic workers. The proportion of maltworkers with EAA was the same in atopics as non-atopics and respiratory function testing indicated normal $FEV_{1.0}$ and FVC (forced vital capacity) in 23/28 cases with a mild restrictive ventilatory defect in 6 cases and an obstructive defect in 8 cases.

EAA to other Aspergilli, e.g. *A.flavus* has been recorded by Patterson, Sommers & Fink (1974) and Rhudy, Burrell & Morgan (1971) described a case where *A.versicolor* growing on a wet dog blanket provoked an attack in the dog's owner, again precipitins were found against the offending organism. The mechanism producing EAA is generally accepted to be initially the formation of immune complexes (inhaled antigen and antibody) following by activation of the complement system. The C567 trimolecular component of the complement system is chemotactic for polymorphonuclear leucocytes and recruitment of these cells is an integral part of the Arthus (type III) mechanism. The polymorphonuclears phagocytose the immune complexes and the primary lysosomes within the cell migrate and fuse with the vacuole forming a secondary lysosome. Powerful hydrolases are secreted onto the phagocytosed complexes but there is often incomplete fusion of the external cell membrane and by a process termed 'regurgitation during feeding' the hydrolases leak out damaging the architecture of the lung (or other organ) causing haemorrhage and thrombus formation etc. The pyrexia observed in EAA may be due to pyrogen granule release from the polymorphonuclears. Lately we have observed cases identical with EAA that are precipitin negative and have determined that certain substances can activate the complement system in the absence of antibody by the so-called 'alternative pathway of complement' mechanism (Edwards, Baker & Davies, 1974). It is possible that heavy exposure to fungi may do this (Edwards, Seaton & Seal, 1976) and explains in many ways the absence of precipitins in some cases.

Other work on EAA due to pigeon antigen inhalation suggests cellular hypersensitivity (Hanson & Penny, 1974) to be the causative mechanism where inhaled antigen reacts with specifically sensitized lymphocytes and the tissue response is caused by lymphokine release from the lymphocytes.

However, it is possible that cells and serum in the peripheral circulation do not reflect totally the cells and fluids within the lung, hence the problem of precipitin or cell mediated response may have to rely on bronchial lavage for resolution.

The best treatment for the diseases is to avoid exposure or if this is not possible a suitable respirator or mask may eliminate sufficient material so that the acute response does not occur. Otherwise corticosteroids are effective in an attack of EAA and may help prevent the fibrotic reaction that may follow acute attacks. Lately aerosols of corticosteroids have been used and help provide a local concentration of the steroids in the lung.

ASPERGILLI PRODUCING MYCOTOXINS

Aspergilli can produce mycotoxins and mediate disease thereby. (see next chapter).

The capacity of man to prevent colonization by Aspergilli is usually effective unless the immune system is impaired, but exposure or partial colonization of the respiratory tract may result in sensitization. The subsequent allergic manifestations are governed largely by the atopicity of the individual and the variety of response seen to this genus has afforded a great deal of insight into the mechanisms of sensitization and allergic response.

REFERENCES

ANDRIOLE,V.T., KRAVETZ,H.M.,ROBERTS,W.C. & UTZ,J.D.(1962). Candida endorcaditis, clinical and pathological studies. *American Journal of Medicine* 32, 251-85.

AUSTWICK,P.K.C. & VENN,J.A.J. (1962). Mycotic abortion in England and Wales 1954-1960. *Proceedings of the 4th International Congress on Animal Reproduction* 3, 562-8.

BRITISH THORACIC AND TUBERCULOSIS ASSOCIATION (1970). Aspergilloma and residual tuberculous cavities - the results of a resurvey. *Tubercle* 51, 227-45.

BURKE,B.J., STORRING,F.K. & PARRY,T.E. (1970). Disseminated Aspergillosis. *Thorax* 25, 702-7.

CARRIZOSA,J., LEVISON,M.E., LAWRENCE,T. & KAYE,D. (1974). Cure of *Aspergillus ustus* endocarditis on a prosthetic valve. *Archives of Internal Medicine* 133, 486-90.

CRAMER,C. (1859). Ueber eine neue fadenpilzgattung, Sterigmatocystis. *Vierteljahresschrift Naturforschende Gesellschaft in Zurich* 4, 325-37.

DROUHET,E. (1974). Precipitating and fluorescent antibodies in human aspergillosis. In *Proceedings of the 4th International Symposium on Aspergillosis and Farmer's Lung in Man and Animal. Davos 1971. pp.*117-123. Ed. R.de Haller & F. Suter, Hans Huber: Bern.

EDWARDS,J.H., BAKER,J.T. & DAVIES,B.H. (1974). Precipitin test negative farmer's lung - activation of the alternative pathway of complement by mouldy hay dusts. *Clinical Allergy* 4, 379-88.

EDWARDS,J.H., SEATON,A. and SEAL,R.M.E. (1976). Mycotoxicosis? *Chest* 69, 569 (Letter).

GAUCHER,E. & SERGENT,E. (1894). Un cas de pseudotuberculose aspergillaire simple chez un gaveur de pigeons. *Bulletins et memoirs de la Societe medicale de Hopitaux de Paris* 11, 512-21.

GRANT,I.W.B., BLACKADDER, E.S., GREENBERG,M. & BLYTH,W. (1976). Extrinsic allergic alveolitis in Scottish maltworkers.

British Medical Journal 1, 490-3.
HANSON,P.J. & PENNY,R.(1974). Pigeon-breeder's disease. Study of the cell-mediated immune response to pigeon antigens by the lymphocyte culture technique. *International Archives of Allergy* 47, 498-507.
HENDERSON,A.H. (1968). Allergic Aspergillosis. Review of 32 cases. *Thorax* 23, 501-12.
HILVERING,C., STEVENS,E.A.M . & ORIE,N.G.M. (1970). Fever in Aspergillus mycetoma. *Thorax* 25, 19-24.
HOREJSI, M., SACH,J., TOMSIKOVA,A., MECL,A., BLAHNIKOVA,D., TUMOVA,M. & VALISOVA,A. (1960). A syndrome resembling farmer's lung in workers inhaling spores of Aspergillus and Penicillia moulds. *THORAX* 15, 212-7.
KAMMER,R.B. & UTZ,J.P. (1974). Aspergillus species endocarditis. *The American Journal of Medicine* 56, 506-21.
KENNEDY,W.P.U., MALONE,D.N. & BLYTH,W. (1970). Necrotizing pulmonary Aspergillosis. *Thorax* 25, 691-701.
KRAKOWKA,P., TRACZYK,K., WALCZAK,J., HALWEG, H., ELSNER,Z. & PAWLICKA,L. (1970). Local treatment of aspergilloma of the lung with paste containing nystatin or amphotericin B. *Tubercle* 51, 184-91.
LANGERON,M. (1922). Sur un champignon d'une otomycose bresilienne Sterigmatocystis Hortai. *Bulletin de la Societe de pathologie exotique et de ses filiales* 17, 345-7.
LAVIN, N., RACHELEFSKY,G.S. & KAPLAN,S.A. (1976). An action of disodium cromoglycate. Inhibition of cyclic 3',5'-AMP phophodiesterase. *Journal of Allergy and Clinical Immunology* 57, 80-8.
LONGBOTTOM, J.L. & PEPYS, J.(1964). Pulmonary aspergillosis: diagnostic and immunological significance of antigens and C-substance in *Aspergillus fumigatus*. *Journal of Pathology and Bacteriology* 88 141-51.
McCARTHY,D.S. (1970).Clinical immunology of allergic bronchopulmonary Aspergillosis. M.D. Thesis,London.
McCARTHY,D.S. & PEPYS,J. (1971*a*). Allergic bronchopulmonary aspergillosis. Clinical immunology: skin, nasal and bronchial tests. *Clinical Allergy* 1, 415-32.
McCARTHY,D.S. & PEPYS,J. (1971*b*). Allergy bronchopulmonary aspergillosis. Clinical immunology: clinical features. *Clinical Allergy* 1, 261-86.
MULLINS,J. (1974). The ecology of *Aspergillus fumigatus* (Fres). Ph.D. Thesis, University of Wales.
MULLINS,J., HARVEY,R. & SEATON,A. (1976). Sources and incidence of airborne *Aspergillus fumigatus* (Fres). *Clinical Allergy* 6, 209-17.
NOBLE,W.C. & CLAYTON,Y.M. (1963). Fungi in the air of hospital wards. *Journal of General Microbiology* 32, 397-402.
PATTERSON,R., SOMMERS,H. & FINK,J.N. (1974). Farmer's lung following inhalation of *Aspergillus flavus* growing in mouldy corn.

Clinical Allergy 4, 79-86.
PEPYS,J. (1969). Monographs in Allergy, 4, Karger, Basel.
PETHERAM, I.S. & SEAL,R.M.E. (1976). Aspergillus prosthetic valve endocarditis. *Thorax* 31, 380-90.
RIDDLE,H.F.V. & GRANT,I.W.B. (1967). Allergic alveolitis in a malt worker. *Thorax* 22, 478.
RIDDLE,H.F.V., CHANNELL,S., BLYTH,W., WEIR,D.M., LLOYD,M., AMOS,W.M.G. & GRANT,I.W.B. (1968). Allergic alveolitis in a maltworker. *Thorax* 23, 271-80.
RHUDY,J., BURRELL,R.G. & MORGAN,W.K.C. (1971). Yet another cause of allergic alveolitis. *Scandinavian Journal of Respiratory Diseases* 51, 177-80.
SCADDING,J.G. (1967). The bronchi in allergic aspergillosis. *Scandinavian Journal of Respiratory Diseases* 48, 372-7.
SCHOLER,H.J. (1974). Thermophilia (thermotolerance) of the aspergilli in relation of their pathogenicity. In *Proceedings of the 4th International Symposium on Aspergillosis and Farmer's Lung in Man and Animal. Davos 1971.* pp.35-40. Ed. R. de Haller & F. Suter, Hans Huber: Bern.
SIEBENMANN,F.(1889). Die Schimmelmykosen des menschichen Ohres. pp.118. Bergmann Wiesbaden.
SUTER,F. (1974). Aspergilloma - surgical aspects. In *Proceedings of the 4th International Symposium on Aspergillosis and Farmer's Lung in Man and Animal. Davois 1971.* pp.148-150. Ed. R. de Haller & F.Suter, Hans Huber: Bern.
TRAN VAN KY, P., URIEL,J. & ROSE,F. (1966). Caracterisation de types d'activities enzymatiques dans des extraits antigeniques d'*Aspergillus fumigatus* apres electrophorese et immunoelectrophorese en agarose. *Annals Institute Pasteur* III, 162-70.
VALLORY-RADOT,P. & GIROUD,P. (1928). Sporomycose des pelleteurs de grains. *Bulletins et memoirs de la Societe medicale de Hopitaux de Paris* 52, 1632-45.
VIRCHOW,R. (1856). Beitrage zur Lehre von den beim Menschen vork ommenden pflanzlichen Parasiten. *Archiv Pathologische Anatomie und Physiologie Virchow's* 9, 557-93.
VOISIN,C., BIGUET,J., TRAN VAN KY,P., SCHOULLER, A., SERGEANT,Y.H. & GERNEZ-RIEUX,C. (1964). Aspergilloses latentes et tuberculose pulmonaire chronique. *Revue de la Tuberculose et de Pneumologie (Paris)* 28, 1311-6.
WEISSMANN,G. & THOMAS,L. (1962). Studies on lysosomes. I. The effects of endotoxin, endotoxin tolerance and cortisone on the release of acid hydrolases from a granular fraction of rabbit liver. *Journal of Experimental Medicine* 116, 433-50.

Species Index

Subject Index